HP ASE FlexNetwork Solutions Integrator
OFFICIAL CERTIFICATION STUDY GUIDE
(Exam HP0-Y47)

First Edition

Miriam Allred

HP Press
660 4th Street, #802
San Francisco, CA 94107

HP ASE FlexNetwork Solutions Integrator
Official Certification Study Guide (Exam HP0-Y47)
Miriam Allred

Published by:

HP Press
660 4th Street, #802
San Francisco, CA 94107

ISBN: 978-1-937826-75-8

Printed in United States of America

WARNING AND DISCLAIMER

This book provides information about the topics covered in the Deploying HP FlexNetwork Core Technologies (HP0-Y47) certification exam. Every effort has been made to make this book as complete and as accurate as possible, but no warranty or fitness is implied.

The information is provided on an "as is" basis. The author, HP Press, and Hewlett-Packard Development Company, L.P., shall have neither liability nor responsibility to any person or entity with respect to any loss or damages arising from the information contained in this book or from the use of the discs or programs that may accompany it.

The opinions expressed in this book belong to the author and are not necessarily those of Hewlett-Packard Development Company, L.P.

TRADEMARK ACKNOWLEDGEMENTS

All terms mentioned in this book that are known to be trademarks or service marks have been appropriately capitalized. HP Press or Hewlett-Packard Inc. cannot attest to the accuracy of this information. Use of a term in this book should not be regarded as affecting the validity of any trademark or service mark.

GOVERNMENT AND EDUCATION SALES

This publisher offers discounts on this book when ordered in quantity for bulk purchases, which may include electronic versions. For more information, please contact U.S. Government and Education Sales 1-855-4HPBOOK (1-855-447-2665) or email sales@hppressbooks.com.

Feedback Information

At HP Press, our goal is to create in-depth reference books of the best quality and value. Each book is crafted with care and precision, undergoing rigorous development that involves the expertise of members from the professional technical community.

Readers' feedback is a continuation of the process. If you have any comments regarding how we could improve the quality of this book, or otherwise alter it to better suit your needs, you can contact us through email at feedback@hppressbooks.com. Please make sure to include the book title and ISBN in your message.

We appreciate your feedback.

Publisher: HP Press

Contributors and Reviewers: Wim Groeneveld, Don McCracken, Chris Peacock, Jens Egger

HP Press Program Manager: Michael Bishop

HP Headquarters

Hewlett-Packard Company
3000 Hanover Street
Palo Alto, CA
94304–1185
USA

Phone: (+1) 650-857-1501
Fax: (+1) 650-857-5518

About the Author

Miriam Allred has spent the last nine years configuring, testing, and troubleshooting HP wired and wireless networks. Miriam combines this wide range of technical expertise with pedagogy and instructional design training, allowing her to create technical training courses for both advanced and entry-level networking professionals. Miriam Allred has a master's degree from Cleveland State University and a bachelor's degree from Brigham Young University.

Introduction

This study guide is based on the Deploying HP FlexNetwork Core Technologies course. Although this guide is designed to help you study for the related exam (HP0-Y47), you can also use it to learn about fundamental networking technologies such as VRRP, OSPF, BGP, ACLs, port authentication, and multicast and HP technologies such as HP Intelligent Resilient Framework (IRF) and distributed trunking. If you have a background in using other vendors' equipment, you can use this guide to learn how these technologies are implemented on HP switches.

HP ExpertOne Certification

HP ExpertOne is the first end-to-end learning and expertise program that combines comprehensive knowledge and hands-on real-world experience to help you attain the critical skills needed to architect, design, and integrate multivendor and multiservice converged infrastructure and cloud solutions. HP, the largest IT company in the world and the market leader in IT training, is committed to help you stay relevant and keep pace with the demands of a dynamic, fast-moving industry.

The ExpertOne program takes into account your current certifications and experience, providing the relevant courses and study materials you need to pass the certification exams. As an ExpertOne certified member, your skills, knowledge, and real-world experience are recognized and valued in the marketplace. To continue your professional and career growth, you have access to a large ExpertOne community of IT professionals and decision-makers, including the world's largest community of cloud experts. Share ideas, best practices, business insights, and challenges as you gain professional connections globally.

To learn more about HP ExpertOne certifications, including storage, servers, networking, converged infrastructure, cloud, and more, please visit hp.com/go/ExpertOne.

Audience

This study guide is designed for networking professionals who want to demonstrate their expertise in managing HP FlexNetwork solutions by passing the HP0-Y47 certification exam. It is also helpful for networking professionals who want to extend their understanding of fundamental networking technologies and protocols and learn how to implement them on HP switches.

Assumed Knowledge

To understand the technologies and protocols covered in this study guide, networking professionals should have a basic understanding of how Ethernet networks function. They should understand IP addressing, virtual local area networks (VLANs), link aggregation, Spanning Tree Protocol (STP), and static IP routing, basic Open Shortest Path First (OSPF) operations, and basic HP IRF operations. General familiarity with the HP Comware and ProVision CLI is also helpful.

Minimum Qualifications

Although anyone can take the HP0-Y47 exam, it is recommended that candidates have some "on the-job" experience. This study guide provides foundational knowledge, but candidates are also expected to have real-world experience implementing the technologies covered in this study guide.

Relevant Certifications

After you pass these exams, your achievement may be applicable toward more than one certification. To determine which certifications can be credited with this achievement, log in to The Learning Center and view the certifications listed on the exam's More Details tab. You might be on your way to achieving additional HP certifications.

Preparing for Exam HP0-47

This self-study guide does not guarantee that you will have all the knowledge you need to pass the exam. It is expected that you will also draw on real-world experience. You may also benefit from completing the hands-on lab activities provided in the instructor-led training.

The HP0-47 exam is designed to test your mastery of the technologies and protocols outlined in this study guide. To pass the certification exam, you must be able to answer multiple-choice questions about how these technologies and protocols function on HP switches. You must also be able to go beyond simply identifying how the technologies and protocols function. You must be able to *apply* what you know to specific network environments. For some questions, you will be given a scenario and asked to answer questions about it. For example, you may need to determine the correct way to implement a technology for a particular network environment, or you may need to evaluate the configuration settings for a network and be able to identify the expected behavior. (However, you are not expected to recall command syntax.) The practice test at the end of this study guide will help you determine if you are prepared to take the actual exam.

Recommended HP Training

Recommended training to prepare for each exam is accessible from the exam's page in The Learning Center. See the exam attachment, "Supporting courses," to view and register for the courses.

Obtain Hands-on Experience

You are not required to take the recommended supported courses, and completion of training does not guarantee that you will pass the exams. HP strongly recommends a combination of training, thorough review of courseware and additional study references, and sufficient on-the-job experience prior to taking an exam.

Exam Registration

To register for an exam, go to hp.com/certification/learn_more_about_exams.html.

CONTENTS

1 HP FlexNetwork Architecture

EXAM OBJECTIVES

In this chapter, you learn to:

✓ Describe two-tier and three-tier HP FlexCampus and HP FlexFabric solutions.

✓ Explain the basic hardware architecture of HP switches, in particular, distinguishing between crossbar and CLOS fabrics.

ASSUMED KNOWLEDGE

Before studying this chapter, you should be familiar with the FlexNetwork architecture and its components:

- FlexCampus
- FlexFabric
- FlexBranch

INTRODUCTION

This chapter outlines reference topologies for each component of the HP FlexNetwork architecture. It focuses in particular on two-tier (core and access layer) topologies but also presents three-tier topologies, which have been optimized for today's networks.

HP FlexNetwork

The HP FlexNetwork consists of three separate but interlocking components, each designed to meet the needs of a particular enterprise environment (see Figure 1-1):

- FlexCampus solutions provide flexible, unified access for users in an enterprise campus LAN that must support a variety of applications from data to voice to rich media.

- FlexFabric solutions establish a powerful, highly available, resilient, and energy-efficient foundation for enterprise data centers.
- FlexBranch solutions extend a consistent user experience across a company's remote sites with easy-to-deploy solutions.

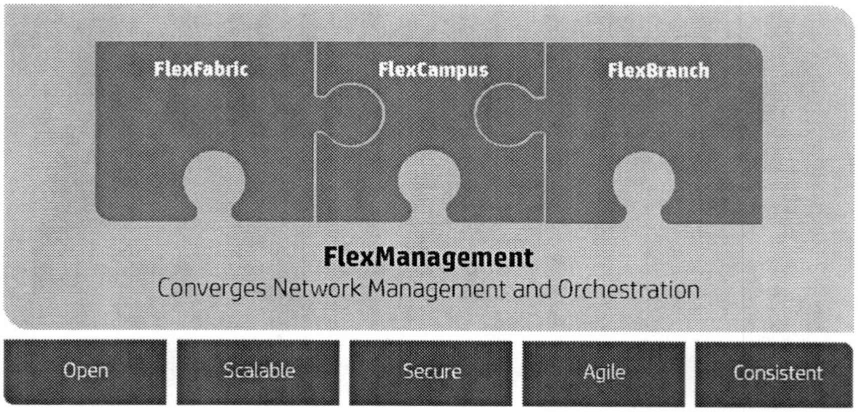

Figure 1-1: FlexNetwork

With HP FlexManagement tools, IT staff members manage all components of the network from a centralized location. FlexManagement helps to simplify and automate tasks, moving the focus of network management away from day-to-day tasks toward meeting business-level needs.

HP FlexCampus two- and three-tier reference architectures

The HP FlexCampus two-tier reference architecture, shown in Figure 1-2, includes a core of two to four HP 10500 modular switches, operating as a single HP Intelligent Resilient Framework (IRF) virtual switch. At the access layer, you can choose from several options. Three options are typically used for an enterprise campus LAN, although you might choose just one of those options:

- HP fixed-form switches in an IRF virtual switch (HP 5500 EI/HI Switch Series)
- HP fixed-form switches in a backplane stack (HP 3800 or HP 2920 Switch Series)
- HP modular switches (HP 5400 zl or HP 8200 zl Switch Series for high availability)

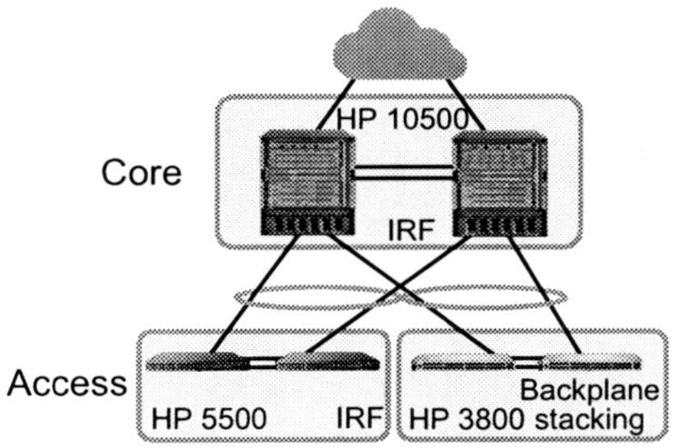

Figure 1-2: FlexCampus two-tier reference architecture

Note

The switch models listed in the "HP FlexCampus two- and three-tier reference architecture" section are typically recommended models for the campus access layer as of the publication of this study guide. For up-to-date information, visit the HP website.
Also note that the FlexCampus solutions feature security and wireless components which are not shown, being beyond the scope of this study guide.

HP recommends the two-tier topology whenever the environment supports it. In this study guide, you will learn how technologies such as backplane stacking and IRF work within such a topology.

Sometimes, however, legacy solutions or cabling restraints force you to use a three-tier topology, and HP offers optimized solutions for these cases as well. Figure 1-3 shows a recommended FlexCampus three-tier topology. The added distribution layer consists of several pairs of HP 7500 modular switches. Figure 1-3 shows one pair as an example.

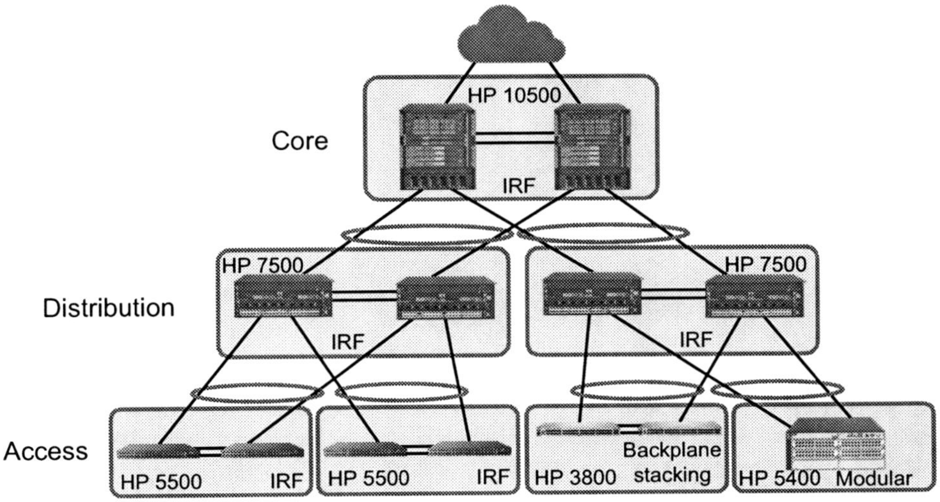

Figure 1-3: FlexCampus three-tier reference architecture

In either the two-tier or three-tier architecture, the switches at each layer can connect to the next layer using aggregated links, due to the use of IRF at the core layer and also at the distribution layer in the three-tier topology. The access layer might use IRF or backplane stacking, but even if the access layer switches do not support these technologies, they can use aggregated links to connect to switches in the upstream IRF virtual switch. These aggregated links are a key feature of the FlexCampus architectures because they eliminate complicated and issue-prone spanning tree designs.

IRF is generally recommended for the enterprise campus core and the distribution layer. However, for some small to medium businesses, you can create a redundant core using two HP ProVision 8200 zl switches, as shown in Figure 1-4. This study guide explains how to implement core technologies on HP ProVision switches as well as on HP Comware switches.

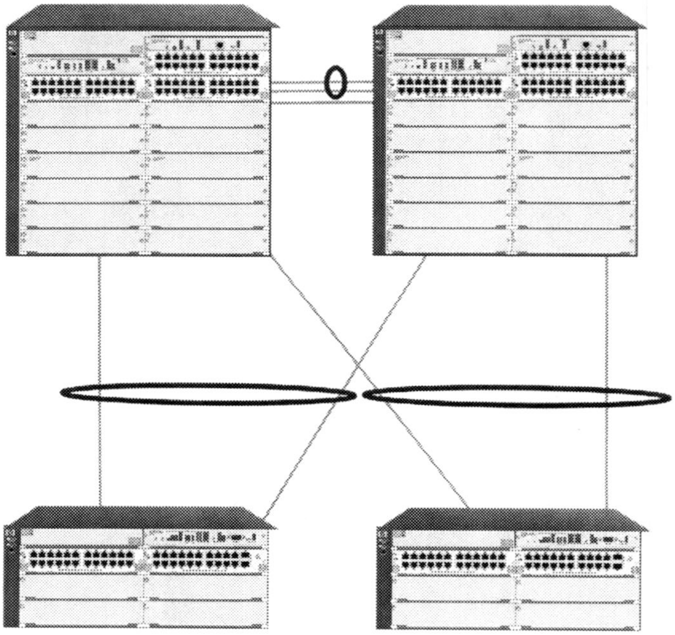

Figure 1-4: FlexCampus architecture for a smaller network

HP FlexFabric reference architectures

You will now examine HP FlexFabric architectures, designed specifically to meet the needs of contemporary data centers, which must host rich media services, anything as a server (XaaS), cloud-based services, big data, high-performance computing, and more.

As Figure 1-5 illustrates, HP 12900 Series switches can be combined in an IRF virtual switch at the core. These switches, with a high density of 10G/40G ports now and 100G ports in the future, provide the high-speed, low-latency, highly available core that forms a solid foundation for an enterprise's services.

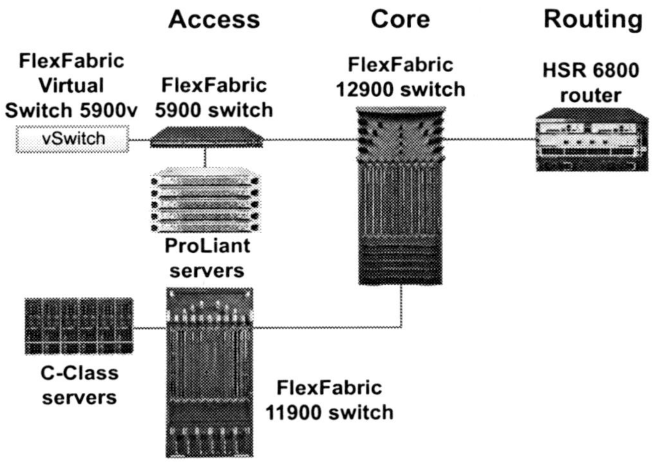

Figure 1-5: FlexFabric reference architecture

To support rack servers, HP provides several choices, including HP 5900 AF Series switches, which operate in redundant IRF virtual switches and support the customer's choice of Gigabit or 10G ports. (You can even install a virtual version of this switch within VMWare hosts.) The HP 11900 Series switches can provide the access layer for blade servers installed in HP C-Series blade enclosures.

Note

The FlexFabric reference architecture includes these switch models as of the publication of this study guide. You should check the HP website for additional options.

Although HP provides three-tier data center solutions for enterprises that must meet legacy requirements or have very large data centers, HP recommends flattening the architecture when possible.

A two-tier topology built on IRF presents several benefits:

- Device-level redundancy all the way to the server access layer, which provides Gigabit or 10G connections, as required by the enterprise
- Fewer hops and lower latency for the server-to-server (east–west) traffic that composes the majority of traffic in contemporary data centers

- A redundant technology that is built on aggregated links rather than links blocked by spanning tree
- Aggregated links provide a more efficient balancing of traffic loads and faster failover than Spanning Tree Protocols. The architecture also avoids the potential issues of spanning tree.
- A simple flexible architecture that minimizes issues as administrators extend VLANs wherever they are required

  Enterprises need to be able to add or scale services by deploying new servers, physical or virtual, wherever power, space, and other facility requirements dictate. To scale and to optimize virtualized services, companies rely on technologies such as vMotion and Microsoft live migration. These technologies, in turn, rely on Layer 2 connectivity between the servers that host migrating virtual servers, so a simpler topology that reduces potential issues is crucial.
- Better energy efficiency through devices that not only are energy efficient themselves but also reside in an architecture that supports the service and application load with fewer devices (no distribution layer)

Note

FlexFabric solutions also include storage solutions and HP TippingPoint Intrusion Prevention Systems (IPSs); however, Figure 1-5 does not show these components because they are beyond the scope of this study guide.

Supplemental information about HP BladeSystem solutions

HP BladeSystem solutions can connect to a two-tier FlexFabric network. The server access layer typically consists of a pair of high-speed, high-performance switches such as HP 5900 Series or HP 11900 Series switches. The pair of switches operates as a highly resilient IRF virtual device, enabling server administrators to connect the blade enclosure to two separate physical devices while still allowing the blade servers within to use any form of NIC teaming that is desired.

With its BladeSystem solutions, HP can even flatten the network to a single tier, enhancing all the benefits of a two-tier architecture. In either case, the BladeSystem itself fosters a flexible, quickly deployed, and easily managed system:

- The HP C7000 blade enclosures provide blade servers with everything that they need from power to cooling to network connectivity.
- The C7000's Virtual Connect (VC) modules simplify provisioning by managing the blade server's identity, both internal (serial number) and external (MAC address and World Wide Name [WWN], an address used for storage technology Fibre Channel [FC]), transforming that identity into a virtual resource that persists through hardware refreshes and moves.

- The VC modules also help administrators to provision and manage blade servers' network connections more easily.
- The VC Flex-10 and FlexFabric modules enable administrators to transform a single 10G NIC into as many as four FlexNICs, each of which shares the bandwidth but operates like an independent NIC. This technology reduces the equipment required to provide blade servers with network connectivity by as much as 75 percent—drastically reducing cost and power consumption.

HP FlexBranch solutions

For branch offices, companies seek seamless, consistent access to centralized services and assurance that any problems can be resolved quickly despite limited onsite IT staff. The HP FlexBranch solutions meet both goals with components that can be provisioned, managed, and supported remotely. FlexBranch solutions consist of standardized components that scale from very small remote offices of several users to large offices with about a hundred employees.

Figure 1-6 suggests HP Multi-Service Routers (MSRs) and switches for branch offices of various sizes. You can eliminate components as required for your solution. For example, you might eliminate the MSR if the branch office connects to the central data center over an Ethernet WAN, and the connection does not require advanced VPN services.

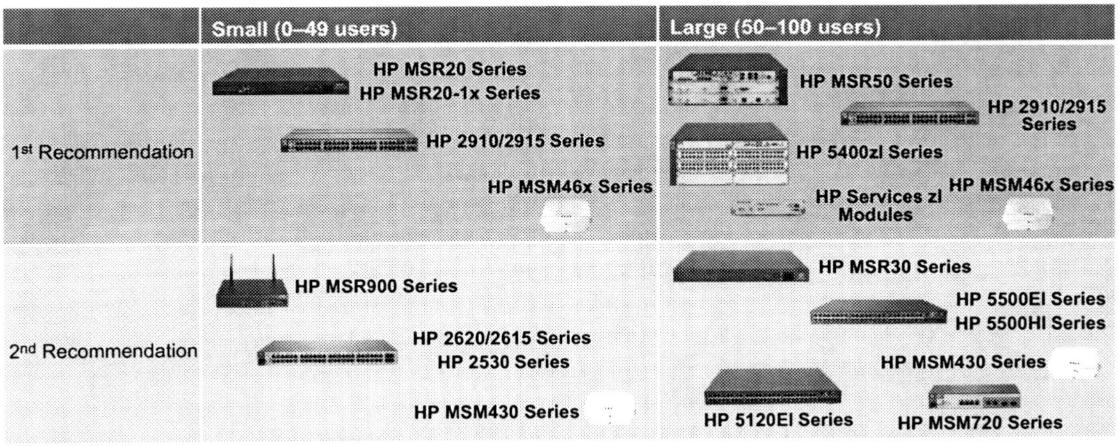

Figure 1-6: FlexBranch solutions

Whatever the infrastructure, the branch office supports a consistent set of endpoints and devices, including VoIP phones that participate in a centralized—but survivable—Unified Communications and Collaboration (UC&C) solution; desktops; HP ProLiant servers, which can be maintained remotely; and printers.

The modular 5400 zl switches or MSR routers can also host modules, which run a variety of services, creating a one- or two-box solution for survivable, high-quality voice, security, WAN acceleration, and so on.

Note

Figure 1-6 shows the models recommended for FlexBranch solutions as of the publication of this study guide. Check the HP website for updates.

Note

The FlexBranch solutions also include security, storage, and wireless components, which are not shown, being beyond the scope of this study guide.

HP FlexManagement

HP FlexManagement provides a consistent set of tools that help companies move from ad hoc management of the network infrastructure to full-scale business orchestration, which helps the enterprise to obtain the business benefits they desire from the network resources in which they have invested. (See Figure 1-7.)

This study guide focuses on HP Intelligent Management Center (IMC), which forms the foundation of the FlexManagement solutions. IMC discovers and inventories network devices, including network infrastructure devices from a variety of vendors, desktops, servers, and even virtual servers. Network administrators can deploy configurations and software to switches, easing upgrades and expansions. IMC also alerts administrators to various problems with its extensive alarm system. As you will learn in Chapter 2, an extensive set of add-on components help companies to meet specific business needs from supporting voice, video, and other sensitive traffic to managing wireless networks to controlling and customizing user access.

Switch planes

You will now learn more about the physical hardware of HP switches. Understanding this architecture will help you to implement the protocols and features outlined in this study guide.

As Figure 1-8 shows, switches divide functions across the following planes:

- **Management plane**—Each switch has a centralized management plane. When you access the switch to manage or configure it, you interact with this plane, whether you access the switch through a console, Telnet, Secure Shell, FTP, or Simple Network Management Protocol (SNMP) session.

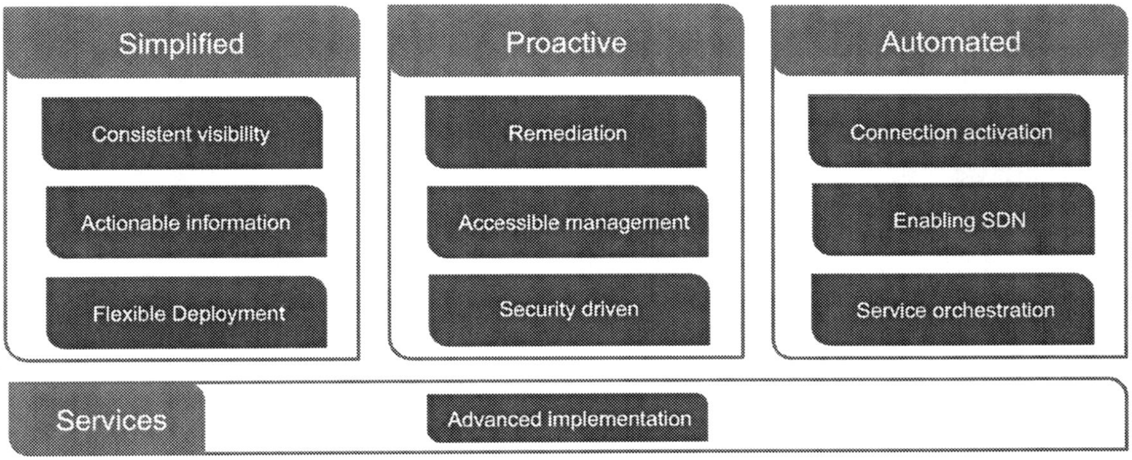

Figure 1-7: FlexManagement

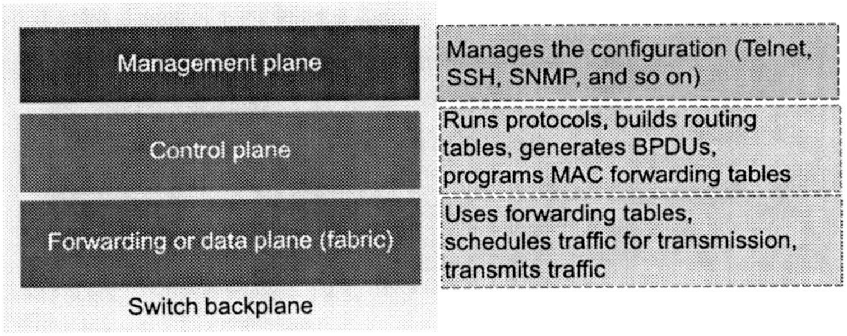

Figure 1-8: Switch planes

- **Centralized control plane**—This plane runs routing protocols and builds the routing table. It also runs Layer 2 protocols such as Spanning Tree Protocol (STP) and its variants. As part of its duties, this plane generates traffic source from the switch such as routing protocol updates and Bridge Protocol Data Units (BPDUs). The centralized control plane also handles the file system, including configuration files. Finally, it manages the hardware, monitoring its temperature, power, fan, and modules.
- **Forwarding plane or data plane**—The forwarding plane, sometimes called the data plane, contains the data channels that receive and transmit traffic. This plane also includes control elements such as packet processors and schedulers, which you will examine in a moment.

A fixed-port switch typically contains these three elements in the backplane; each port connects into the backplane's control and forwarding plane. The management plane is logically or physically separated from the other planes.

Modular switch planes

A modular switch has a management module, or main processing unit (MPU), which contains the management plane, the centralized control plane, and a forwarding plane. As Figure 1-9 shows, a modular switch might have two management modules, each of which contains these three elements. One of the management modules is active, while the other is in standby mode. However, Comware switches can load balance traffic through a standby module's forwarding plane.

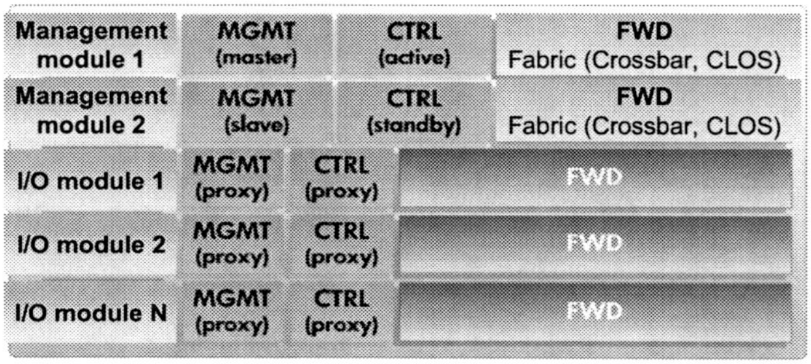

Figure 1-9: Modular switch planes

Each of a modular switch's input/output (I/O) modules includes its own forwarding plane with portions of the MAC forwarding table programmed by the centralized control plane. The I/O module might also include a routing table proxied from the centralized control plane so that the I/O module can select the correct egress port for routed traffic without sending the traffic through the management module.

The management modules and the I/O modules connect together through fabric modules in the backplane.

General forwarding plane architecture

First, you will examine the switch forwarding plane in more detail. This fabric consists of data channels for transmitting traffic as well as control elements called schedulers and packet processors.

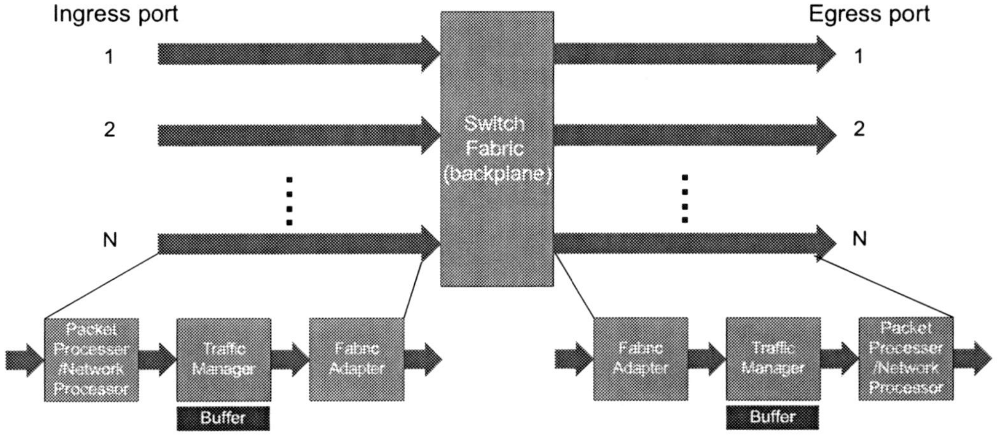

Figure 1-10: General forwarding switch plane

Figure 1-10 illustrates the basic architecture of any fabric:

1. A packet arrives on the physical port.
2. A packet processor, sometimes called a network processor, reads the packet header and makes filtering and forwarding decisions.

   The packet processor switches traffic based on the MAC forwarding table programmed by the control plane, which has built this table using STP/RSTP/MSTP and MAC learning functions. A routing switch's forwarding plane also has a forwarding information base (FIB), which holds IP routes.

   After processing, the packet has an egress port, or it is sometimes destined to the control plane for further processing.

Note

You might need to analyze a port or I/O module's ability to operate at line rate, which refers to the component's ability to process packet headers. If a port receives a great deal of small packets, it must process more headers when operating at the same bandwidth as a port that receives larger packets. A port is considered to operate at line rate if it can process as many 64-byte packets per second as its bandwidth implies.

---

3. An ingress traffic manager might handle queuing and buffering the packets at the ingress port before transmission across the fabric. You will learn more about queuing in Chapter 8.
4. The ingress fabric adapter (FA) readies the traffic to flow through the fabric to the correct egress port.
5. At this point, a scheduler, or arbiter, might signal when the FA can transmit the traffic.

6. At the egress port, the packet flows through the same elements in reverse. It leaves the fabric through the egress FA. The egress traffic manager queues and buffers the traffic for transmission out the correct port, and the packet processer copies or adds the correct headers.

In a simple bus fabric, each port has an ingress and egress connection into the forwarding plane. Traffic passes from the ingress port across the plane until it reaches the egress port. You will look at other types of fabric in a moment. First, however, you should understand how the more complex modular switch backplane functions.

Crossbar fabric

The backplane fabric could consist of a simple bus through which all traffic passes. In this scheme, the egress ports typically handle all priority queuing and scheduling. This approach requires deep egress buffers in case multiple ingress ports need to send traffic out the same egress port. To handle all the packets, the fabric must also be oversubscribed (provide more forwarding capacity than the ports), which limits the scalability of the system.

Crossbar fabrics were designed to provide this scalability. They have become the traditional fabric for switches with a high density of high-speed ports and the need to apply more granular scheduling to meet QoS requirements.

In this architecture, each line card has an ingress FA port, which connects to a fabric port on a crossbar (see Figure 1-11). Other crossbar fabric ports connect to egress FA ports, which each line card has as well. For simplicity, these connections are referred to as ingress ports and egress ports in this study guide.

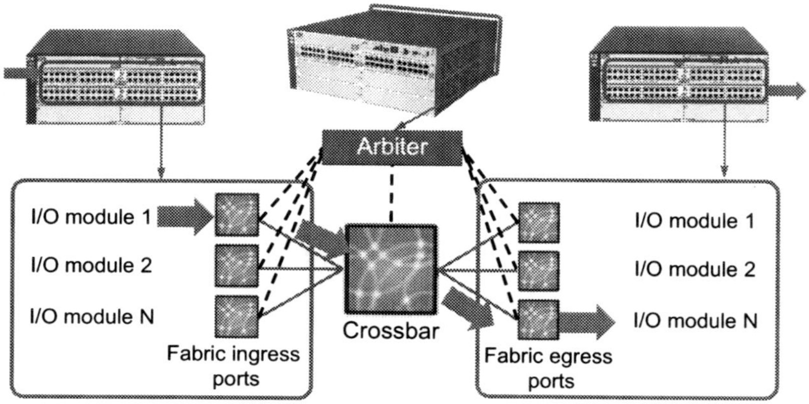

Figure 1-11: Crossbar fabric

The bandwidth for the fabric ports determines how much traffic the I/O module can support. For example, a fabric module might provide two 40G ingress ports and two 40G egress ports to each I/O module, providing up to 80G transmission—as long as the I/O module's ingress and egress FA ports support that bandwidth as well.

Each ingress port, egress port, and crossbar also connects to an arbiter, which schedules the traffic flow using an explicit request system. Scheduling is required because congestion can occur at the egress port. For example, assume that three ingress ports need to forward a packet through the same egress port. The ports have their own connections to the crossbar, so no congestion occurs there, but the egress port can only receive one packet at a time.

Therefore, the ingress port asks the arbiter for permission to send traffic to a specific egress port. The arbiter considers the congestion, which the egress port has reported. Only if the egress port can handle the traffic, the arbiter sends a message to the ingress port granting permission to use the crossbar. The ingress port then transmits the traffic across the crossbar to the egress port.

HP crossbar-based switches use combined input output queuing (CIOQ) to buffer the traffic at *both* the ingress port and the egress port:

- At the ingress port, the FA implements virtual output queuing (VOQ). Consider what might happen without VoQ. The ingress port needs to transmit a packet at the front of the ingress queue, but the arbiter does not send permission because the packet's egress port is congested. The FA cannot forward other packets even if their egress ports are open. This situation, called head-of-line blocking, spreads congestion from one fabric port across the entire switch. VoQ prevents head-of-line blocking by dividing the ingress queue in different queues based on the traffic's egress port (and by priority, as you will learn in Chapter 8). The ingress port can then forward packets in ingress queues for which the egress port is open while waiting to forward a packet with a congested egress port.
- The egress fabric port also buffers the traffic, dividing the traffic into the proper queues and scheduling it for transmission based on the particular QoS mechanism applied on the port.

A CIOQ crossbar design meets the needs of many environments. The fabric applies fine-grained scheduling at the ingress port and across the fabric usually using complex, software-based algorithms. However, the complexity of these algorithms can prevent scaling up to the highest traffic volumes. A crossbar fabric's need for explicit scheduling can create a bottleneck, in particular when the switch needs to schedule packets in priority queues to implement QoS.

Some crossbar fabrics feature multiple crossbars, as part of an effort to increase the fabric's scalability. The I/O module then has multiple ingress and egress fabric ports, each pair of which connects to one crossbar. However, these fabrics still experience some problems with scalability. First, particular traffic flows are statically assigned to particular crossbars. One crossbar might become congested while another remains unused. In addition, the arbiter remains a potential bottleneck.

Therefore, while the crossbar fabric can work well, it cannot handle the highest volume of 10G traffic or scale to a 40G/100G environment.

CLOS fabric

The HP 12900, HP 12500, HP 11900, and HP 10500 Switch Series leverage a CLOS fabric, which the industry agrees is necessary for scaling to 40G/100G environments. A CLOS fabric consists of a three-level architecture that includes many crossbar switching elements. Stage 1 elements receive traffic from the ingress ports, and Stage 3 elements transmit traffic on the egress ports. Several Stage 2 elements connect these stages, providing multiple paths between each Level 1 and Level 3 element.

Note

CLOS is not an acronym. Rather, the name refers to Charles Clos, who developed the switching architecture on which this design is based.

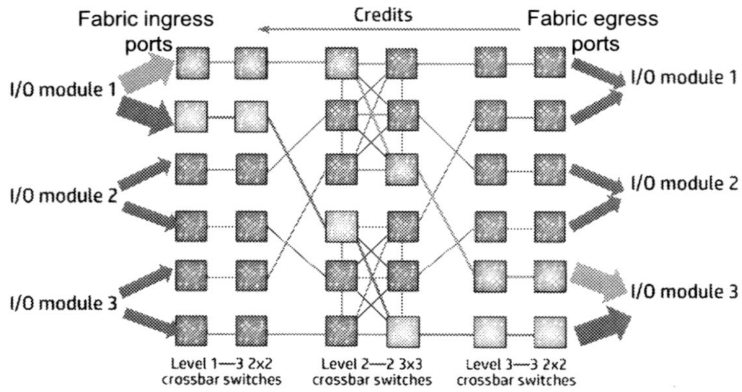

Figure 1-12: CLOS fabric

As you see in Figure 1-12, each Stage 1 or Stage 3 element connects to each Stage 2 element. Each Stage 1 and Stage 3 element is a kxk switch with k indicating the number of Stage 2 elements. Conversely, each Stage 2 element is an nxn switch with n indicating the number of ingress fabric ports (or egress fabric ports). In this way, the CLOS fabric offers multiple paths through the fabric from each ingress port and egress port. The CLOS fabric exceeds the switching capacity of a multiple-crossbar fabric, which also offers multiple paths, because the CLOS uses different scheduling mechanisms.

First, the ingress port can divide packets and send pieces of the packets, called cells, to different Stage 2 switches. The egress port receives the cells and reassembles them into a packet. By dynamically load balancing traffic streams across the multiple data channels, the switch can use the fabric's full capacity. The fabric can also support higher-bandwidth ports such as 40G and 100G ports, which require multiple data channels.

Second, in HP switches with CLOS fabrics, the ingress port is allowed to send traffic without explicit consent from an arbiter. Instead, each egress port assesses its own congestion and, based on that assessment, assigns a credit to each ingress port queue, which it periodically updates. The ingress port sends traffic for as long as it has credits available for a queue. (QoS adds another layer to this scheme; that is, every priority queue has a credit for forwarding traffic.) This scheme removes the arbiter as the sole element for scheduling all traffic through the fabric, making the system much more scalable.

Note that another three-level CLOS fabric can replace the Stage 2 element, making the architecture indefinitely scalable.

Enhancing resiliency in networks with HP ProVision cores

This study guide explains how to implement a variety of technologies that enhance availability, resiliency, security, and QoS for HP FlexNetwork solutions.

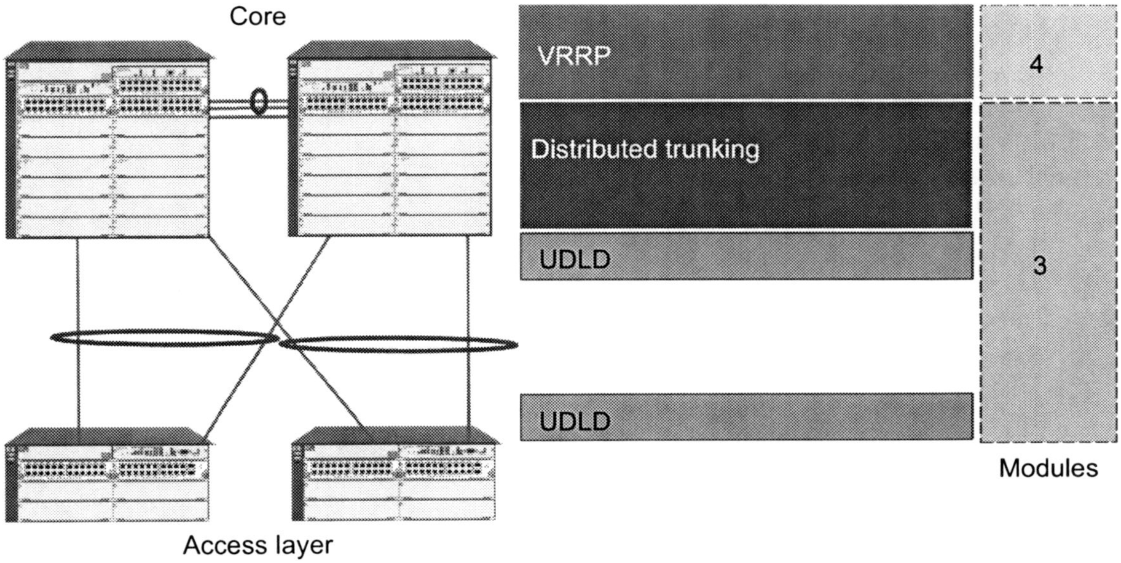

Figure 1-13: Enhancing resiliency in networks with ProVision cores

For example, Figure 1-13 shows the reference architecture for a smaller campus LAN with HP 8200 zl Series switches at the core. Traditionally, technologies such as Multiple Spanning Tree Protocol (MSTP) and Rapid Per-VLAN Spanning Tree Plus (RPVST+) have eliminated dangerous loops in redundant Layer 2 technologies. Chapter 3 briefly reviews these technologies and then shows how you can replace these them with distributed trunking on HP ProVision core (or distribution layer) switches, promoting greater stability, better utilization of bandwidth, and faster failover.

Chapter 4 explains how to extend redundancy to Layer 3 using Virtual Router Redundancy Protocol (VRRP). Multiple devices at the two-tier core (or the three-tier distribution layer) participate in a virtual router, with one device actively routing and the others ready to take over when required.

Note

Chapter 3 also explains a bit about other mechanisms that promote high availability at the data link layer, including technologies for detecting and mitigating unidirectional links.

Enhancing resiliency in networks with HP Comware cores

When you have HP Comware switches at the core, as recommended in FlexFabric and larger FlexCampus solutions, IRF eliminates the need for spanning tree-based topologies, distributed trunking, and VRRP by transforming multiple physical switches into a single virtual one.

As Figure 1-14 shows, Chapter 5 builds on your knowledge of IRF, explaining how IRF functions within a network architecture and how you can use this innovative technology to promote highly available services. You will learn how to implement Multi-Active Detection (MAD) to protect against split brain situations, how to minimize service maintenance time using In-Service Software Upgrade (ISSU), and generally how to manage the IRF virtual switch.

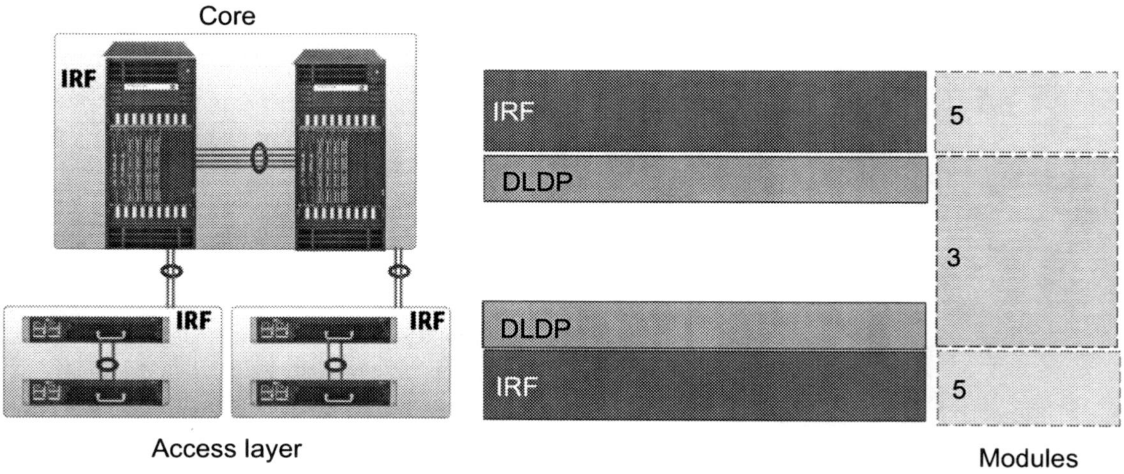

Figure 1-14: Enhancing resiliency in networks with Comware cores

Enhancing security and QoS

The next chapters guide you through enhancing security from the core to the access layer. In Chapter 6, you learn how to configure access control lists (ACLs) and apply those lists to filter traffic. Your knowledge of ACLs will continue to pay off as you use ACLs to select traffic for other features. Chapter 7 provides solutions for protecting and customizing the point of connection.

With the network secured, you ensure quality of service (QoS) for applications, in particular time-sensitive ones such as voice and video, in Chapter 8.

Figure 1-15 shows reference architectures for campus LAN solutions. However, these technologies also apply in the branch, as does ACLs and QoS in the data center.

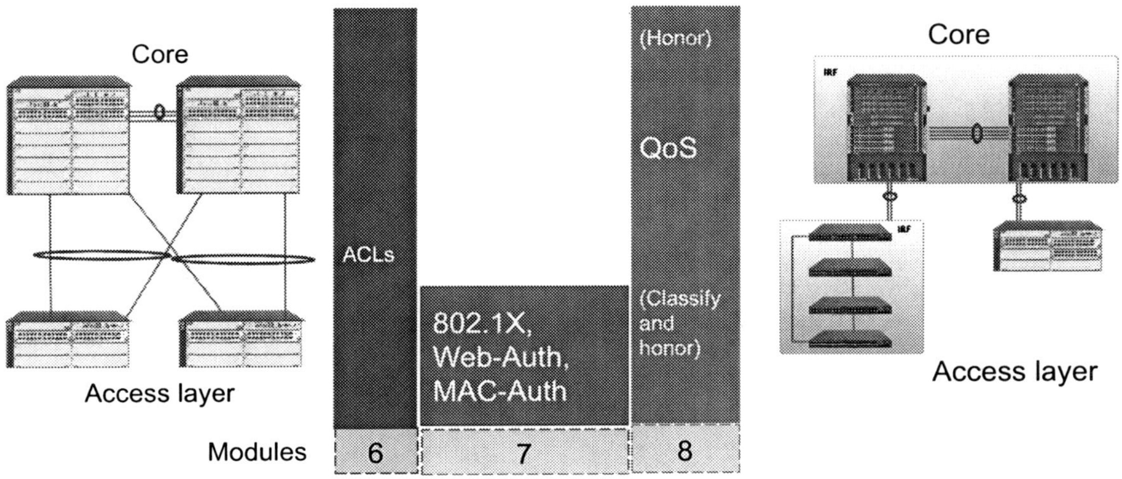

Figure 1-15: Enhancing security and QoS

Implementing dynamic routing

You then turn your attention to more complicated network topologies that feature multiple routing core (or core and distribution) devices and perhaps multiple branches, each with its own routing device. In architectures such as these, a well-designed dynamic routing protocol simplifies configuration, streamlines the addition of new sites and subnets, and optimizes routing paths in both normal operation and failover situations.

As Figure 1-16 shows, Chapter 9 explains how to implement Open Shortest Path First (OSPF) with multiple areas, helping you to optimize the solution and enhance its availability. Chapter 10 focuses on companies that require BGP specifically to manage redundant connections to an ISP.

Whenever a company needs to run a multicast application across routed links, you must implement multicast routing. Chapter 11 explains how to set up Internet Group Management Protocol (IGMP) and Protocol-Independent Multicast (PIM) to meet these needs.

Managing, monitoring, and troubleshooting

You can monitor, manage, and troubleshoot all these components of your FlexNetwork solution, whatever the architecture, with IMC. As you can see in Figure 1-17, Chapter 2 describes the open-standard management and monitoring technologies that IMC uses—technologies such as Simple Network Management Protocol (SNMP), traffic mirroring, and traffic sampling. It also explains how IMC manages and monitors discovered network resources, helping you to understand how you can use IMC to monitor and manage the features that this study guide teaches you how to implement.

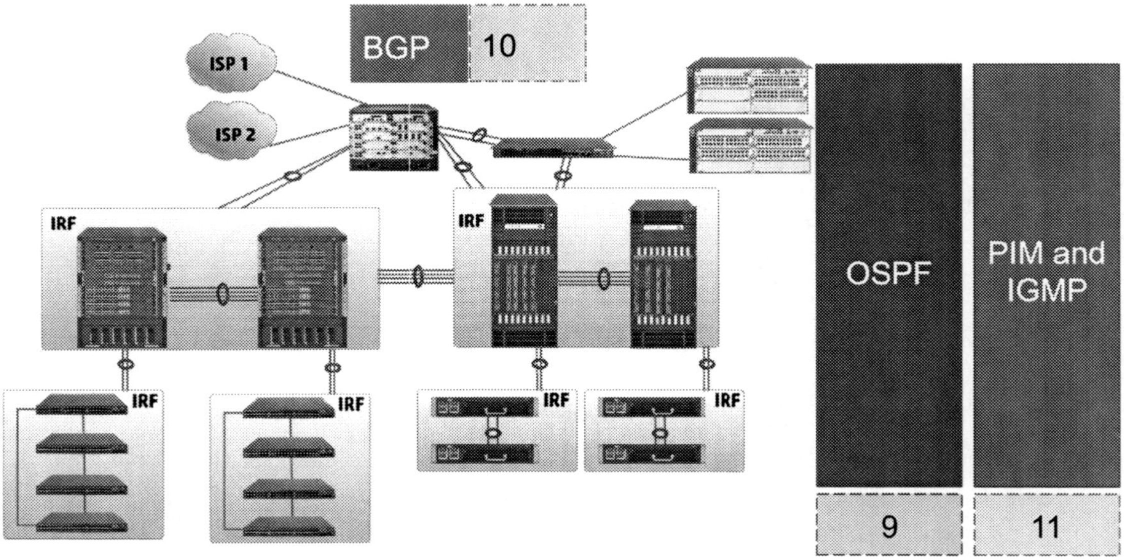

Figure 1-16: Implementing dynamic routing

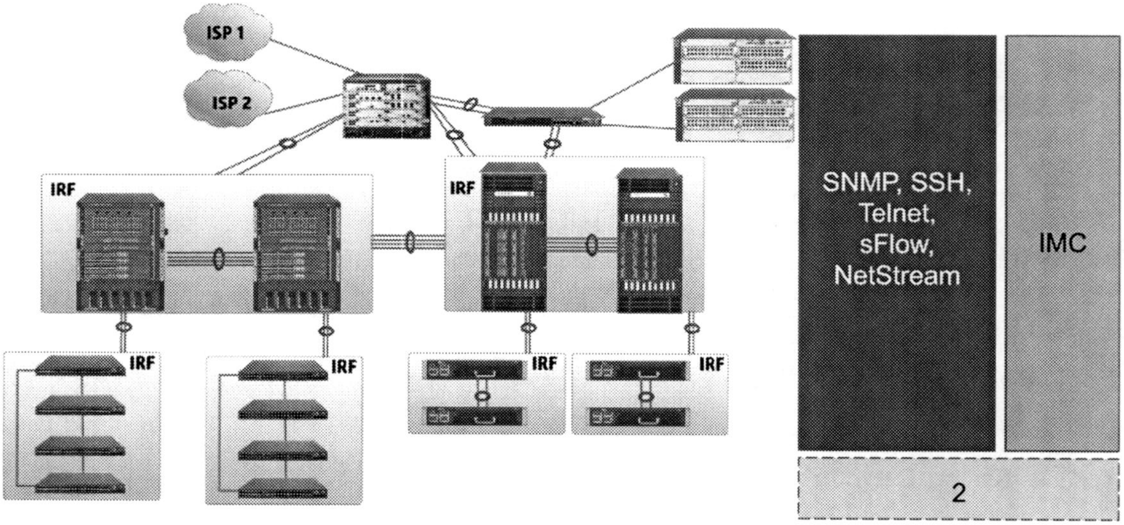

Figure 1-17: Managing, monitoring, and troubleshooting

Meeting other business needs

You can further enhance the FlexNetwork solutions to meet specific business needs, including:

- Supporting Bring Your Own Device (BYOD) environments using the IMC add-on component, User Access Manager (UAM) BYOD server
- Virtualizing the network infrastructure using software-defined networking (SDN) and HP OpenFlow
- Promoting higher stability and availability in data center environments using data center bridging (DCB) technologies such as Transparent Interconnection of Lots of Links (TRILL) and Shortest Path Bridging (SPB)
- HP switches that run Comware 7 support these technologies.
- Designing networks that meet the needs of today's enterprises

HP offers a wide variety of training to teach you how to implement the solutions that interest you. For more information, visit http://www.hp.com/expertone.

Summary

You have examined reference architectures for the FlexCampus, FlexFabric, and FlexBranch solutions, each specifically designed to meet the needs of contemporary campus LANs, data centers, or branch offices.

You then focused on the most foundational unit of the network: an HP switch chassis. With a bit of knowledge about switch planes and fabrics, you can explain how traffic flows through the switch fabric, whether that fabric uses a crossbar or CLOS design. You also understand that, before an ingress fabric port even transmits traffic through the fabric, packet processors have analyzed the packet and made a number of filtering, forwarding, and routing decisions—all based on technologies such as those covered in this study guide.

You understand at a high level when, where, and why you would implement the technologies covered in this study guide, and you are ready to focus on each one in turn—beginning with the open-standard technologies and HP solutions that help you to manage the network as a whole.

Learning check

Answer the following questions:

1. You are creating a FlexCampus two-tier architecture for a larger campus LAN. You want a redundant core. Which switches are recommended for the core, and how should you deploy them?

2. Discuss options for the campus access layer for a larger two-tier FlexCampus LAN.

3. Which statements are true of the FlexFabric architecture? Select two.

   a. It is built on aggregated links that provide a more efficient balancing of traffic loads than Spanning Tree Protocols.

   b. It supports a three-tier network topology, but not a two-tier topology.

   c. It can be flattened to a single tier with BladeSystem solutions.

   d. It provides separate solutions for wireless and wired access control, each supporting the unique needs of that type of connection.

   e. It provides high-speed network connectivity for users at a large campus as well as users in remote offices of various sizes.

4. What are the main functions of the three switch planes?

   - Management plane
   - Control plane
   - Forwarding plane or data plane

5. A switch has a crossbar fabric. How does an I/O module transmit traffic across the crossbar for transmission out another module? (Remember to describe the role of arbiter.)

Answers

1. You are creating an FlexCampus two-tier architecture for a larger campus LAN. You want a redundant core. Which switches are recommended for the core, and how should you deploy them?

   Deploy two to four HP 10500 modular switches, operating as a single Intelligent Resilient Framework (IRF) virtual switch. (The number depends on the size of the network.)

2. Discuss options for the campus access layer for a larger two-tier FlexCampus LAN.

   For the access layer switches, you can choose from a variety of options, including:

   - HP fixed-form switches in an IRF virtual switch (HP 5500 EI/HI)
   - HP fixed-form switches in a meshed stack (HP 3800 or HP 2920)
   - HP modular switches (HP 5400 zl or HP 8200 zl for high availability)

   Note that the recommendations might have changed since the publication of this study guide. If necessary, update the information.

3. Which statements are true of the FlexFabric architecture? Select two.

   a. It is built on link aggregations that provide a more efficient balancing of traffic loads than Spanning Tree Protocols.

   b. It supports a three-tier network topology, but not a two-tier topology.

   c. It can be flattened to a single tier with BladeSystem solutions.

   d. It provides separate solutions for wireless and wired access control, each supporting the unique needs of that type of connection.

   e. It provides high-speed network connectivity for users at a large campus as well as users in remote offices of various sizes.

4. What are the main functions of the three switch planes?

   - Management plane

     You interact with this plane when you access the switch to manage or configure it.

   - Control plane

     The control plane:

     - Runs routing protocols and builds the routing table
     - Runs Layer 2 protocols such as Spanning Tree Protocol (STP)

    - Generates traffic source from the switch such as routing protocol updates and Bridge Protocol Data Units (BPDUs)
    - Handles the file system, including configuration files
    - Manages and monitors the hardware

- Forwarding plane or data plane

  The forwarding plane, sometimes called the data plane, contains the data channels that receive and transmit traffic. This plane also includes control elements such as packet processors and schedulers.

5. A switch has a crossbar fabric. How does an I/O module transmit traffic across the crossbar for transmission out another module? (Remember to describe the role of arbiter.)

   The I/O module has ingress fabric port that connects to the crossbar. The egress I/O module has an egress port that also connects to the crossbar. Both of these ports and the crossbar also connect to an arbiter, which schedules the traffic flow using an explicit request system. Scheduling is required because congestion can occur at the egress port.

2 Network Management

EXAM OBJECTIVES

In this chapter, you learn to:

- ✓ Explain the differences between local and remote port mirroring on HP Comware and HP ProVision switches.
- ✓ Explain how technologies such as sFlow, NetFlow, NetStream, and Remote Monitoring (RMON) allow you to monitor network traffic.
- ✓ Describe the differences between SNMPv2c and v3 and configure SNMP settings on Comware and ProVision switches.
- ✓ Explain the advantages of using HP Intelligent Management Center (IMC) to manage a network.
- ✓ Use IMC to discover and manage network devices.
- ✓ Explain how HP IMC Network Traffic Analyzer (NTA) allows you to view information about network traffic.

ASSUMED KNOWLEDGE

Before studying this chapter, you should have a working knowledge of virtual LANs (VLANs) and be able to configure them on Comware and ProVision switches. You should also have a basic understanding of SNMPv1 and v2c and know the types of capabilities that IMC provides.

INTRODUCTION

This chapter explains the technologies that allow you to troubleshoot problems or examine network traffic patterns or network data more closely. It also covers Simple Network Management Protocol (SNMP) and IMC, which allows you to manage HP and third-party infrastructure devices from a single management console.

Port mirroring

Port mirroring allows a switch to copy frames that arrive on or are sent from a particular port or virtual LAN (VLAN) and forward them to another port. You can use port mirroring to monitor network traffic to detect threats, troubleshoot problems, or manage the network. For example, you might want to send certain traffic to a security appliance such as an Intrusion Detection System (IDS) device, which can examine the traffic and detect possible threats. Or you may want to send traffic to a network protocol analyzer, which you can use to examine the traffic and troubleshoot a network problem.

HP switches support two types of port mirroring:

- **Local port mirroring**—The mirror ports or ports in the mirror VLANs and the monitor port are located on the same switch.
- **Remote port mirroring**—The mirror ports or the ports in the mirror VLANs are located on a different switch than the monitor port (see Figure 2-1).

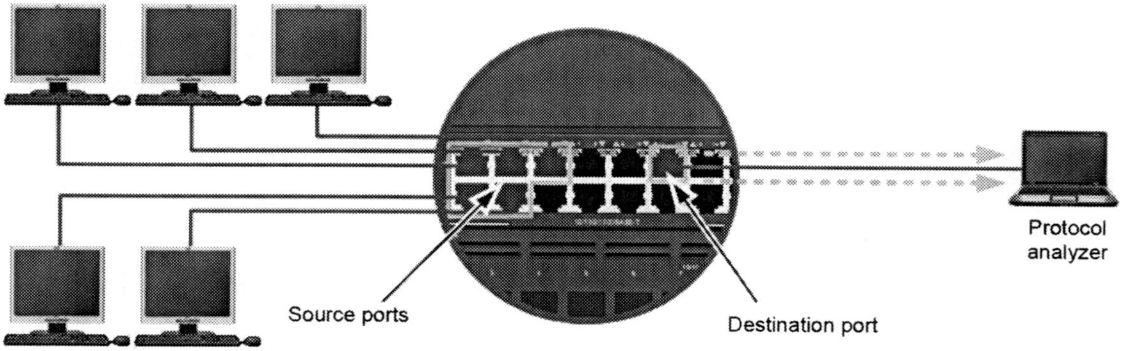

Figure 2-1: Local port mirroring

Note

Because there is no industry standard for port mirroring, each vendor determines how to implement port mirroring on its own switches. You cannot implement remote mirroring between switches that run different software.

Port mirroring on HP Comware switches

HP Comware switches support both local and remote port mirroring. Comware switches also support mirroring traffic based on a quality of service (QoS) policy and allow you to mirror traffic to an interface or a switch CPU. This study guide, however, focuses only on local and remote port mirroring. (For information about more advanced mirroring options, check your switch configuration.)

Configuring local port mirroring on Comware switches

Configuring local mirroring is fairly straightforward. You first configure a local mirroring group:

```
[Switch] mirroring-group <group-id> local
```

You then assign mirror ports or mirror VLANs to that group and specify whether the switch will mirror inbound traffic, outbound traffic, or both inbound and outbound traffic:

```
[Switch] mirroring-group <group-id> mirroring-port <port-list> [both |
inbound | outbound]
```

```
[Switch] mirroring-group <group-id> mirroring-vlan <vlan-list> [both |
inbound | outbound]
```

You then assign a monitor port to the mirroring group:

```
[Switch] mirroring-group <group-id> monitor-port <port-id>
```

Note

If you are implementing local mirroring and want to send the mirrored traffic to more than one monitor port, you can use the remote probe VLAN (a feature of remote mirroring). With the remote probe VLAN, the switch broadcasts the mirrored traffic to all ports that are members of the VLAN. Check your switch documentation for more details about this configuration.

Configuring remote port mirroring on Comware switches

All Comware switches support remote mirroring at Layer 2. Some Comware switches can also implement Layer 3 remote port mirroring through the addition of a Generic Routing Encapsulation (GRE) tunnel. (This study guide does not cover Layer 3 remote mirroring. Check your switch documentation to see if your switch supports this feature.)

To enable the source switch to forward the mirrored packets to a destination switch at Layer 2, you must configure a remote probe VLAN on the source and destination switches. The ports that connect the source switch and the destination switch must support this VLAN. The connection can be direct or indirect; that is, the two switches can be directly connected, or there might be intervening switches. However, if there are any intermediate switches in the path between the source and destination switches, you must configure the remote probe VLAN on those switches and make sure the ports that connect to other switches can forward the traffic in this VLAN.

As Figure 2-2 illustrates, you configure a reflector port on the source switch. (On some switches such as the 5800 Switch Series, the reflector port is called the *monitor egress port*. Check your switch documentation.) The source switch copies the frames from the mirror ports or mirror VLANs and forwards them to the reflector port, which then broadcasts the frames in the remote probe VLAN. As a result, every port that you assign to the remote probe VLAN carries the mirrored traffic.

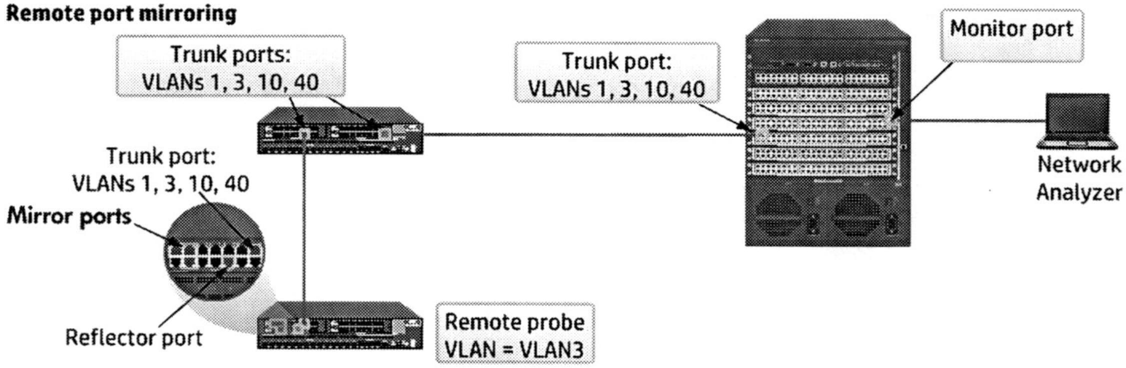

Figure 2-2: Remote port mirroring on Comware switches

You should not connect a device to the reflector port, which must be an access port in the default VLAN (1). You should also disable functions on the reflector port, including all forms of Spanning Tree Protocol (STP), Internet Group Management Protocol (IGMP) snooping, 802.1X, static ARP, and MAC address learning.

Upon receiving a mirrored frame, the destination switch checks the VLAN ID in the 802.1Q field of the Ethernet frame. If the 802.1Q field contains the ID number for the remote probe VLAN, the destination switch forwards the packet to the monitor port. This port must be an access port and should not support STP (it is disabled on that port).

Note

If a switch is part of an HP Intelligent Resilient Framework (IRF) stack, there are a couple of limitations for port mirroring: For example, the switch does not support port mirroring for VLANs or cross-chassis mirroring with IRF. That is, the mirroring port and the monitor port cannot be on different chassis. However, you can mirror the traffic to the cross-chassis aggregate interface.

Configuring the source switch for remote mirroring

Configuring remote mirroring on Comware switches is similar to configuring local mirroring. For remote mirroring, however, a few additional steps are required. And as you would expect, you configure some steps on the source switch and some on the destination switch. The example commands below set up remote mirroring with a reflector port.

On the source switch, you first configure a remote mirroring group:

```
[Switch] mirroring-group <group-id> remote-source
```

You then assign mirror ports or mirror VLANs to the remote mirroring group:

```
[Sourceswitch] mirroring-group <group-id> mirroring-port <port-list>
[both | inbound |outbound]
```

```
[Sourceswitch] mirroring-group <group-id> mirroring-vlan <vlan-list>
[both | inbound |outbound]
```

For remote mirroring, you must also configure a reflector port on the source switch—the switch that contains the mirror ports or mirror VLANs:

```
[Sourceswitch] mirroring-group <group-id> reflector-port <port-id>
```

(If your switch uses the term monitor egress, the command will be `mirroring-group <group-id> monitor-egress <port-id>`.)

You must also create a remote probe VLAN on both the source switch and the remote switch and ensure that any intervening switches support this remote probe VLAN, allowing the traffic to be forwarded to the destination switch.

To create the remote probe VLAN on the source switch and assign it to the remote mirroring group, use the following commands:

```
[Sourceswitch] vlan <vlan-id>

[Sourceswitch-vlanx] quit

[Sourceswitch] mirroring-group <group-id> remote-probe vlan <vlan-id>
```

Configuring the destination switch for remote mirroring

On the destination switch, you configure a remote destination mirroring group:

```
[Destinationswitch] mirroring-group <group-id> remote-destination
```

You then assign the monitor port to this group:

```
[Destinationswitch] mirroring-group <group-id> monitor-port <port-id>
```

Just as you did on the source switch, you must create the remote probe VLAN and assign this VLAN to the remote mirroring group:

```
[Destinationswitch] mirroring-group <group-id> remote-probe-vlan <vlan-id>
```

When the destination switch receives a frame with the remote probe VLAN ID in the 802.1Q field, it forwards the frame to the monitor port.

The last step is to configure the monitor port to support the remote probe VLAN. Use the port access command if the monitor port is an access port; use the port trunk command if the monitor port is a trunk port:

```
[Destinationswitch] interface <interface-type/number>

[Destinationswitch-<interface>] port access vlan <vlan-id>

[Destinationswitch-<interface>] port trunk permit vlan <vlan-id>
```

Port mirroring on HP ProVision switches

As Figure 2-3 shows, when you configure port mirroring on HP ProVision switches, you can select the traffic that will be mirrored based on:

- Port or multiple local ports
- Trunk (link aggregation)
- VLAN or multiple VLANs

You can also select the traffic to be mirrored based on MAC source and destination address or classifier-based service policy. (The configuration for these options is not covered in this study guide. For more information, check your switch documentation.)

Note

With switch software version K.14.00 and below (which runs on the HP 8200 zl, 6200 yl, 6600, 5400 zl, and 3500 Switch Series), you could use ACLs to select the inbound traffic that you wanted to mirror. In software K.14.01, using ACLs was replaced with classifier-based mirroring policies.

When you mirror traffic based on port, VLAN, or MAC address, you can specify the direction of the traffic:

- Inbound
- Outbound
- Both inbound and outbound

The destination for these mirror sessions can be a port on the local switch or a remote switch. In the ProVision documentation, the destination port is called the *exit port.*

A ProVision switch can be the originator of up to four mirror sessions. Each remote switch can also be the destination point for up to 32 total mirror sessions. Four of these 32 mirror sessions can be local, originating from the switch itself.

The destination switch sends these data streams to the port you specify such as the port to which the IDS/IPS device or network analyzer is attached.

Configuring local mirroring on ProVision switches

If you are configuring local traffic mirroring, you enter the **mirror** command, followed by the session number—1 to 4—and then specify the exit port. You can configure one or more mirror sessions to use the same exit port, or you can load balance mirrored traffic across multiple exit ports.

You must complete this step before you define the originating interface. If you try to configure the originating interface first, the CLI displays a message, explaining that it cannot apply the command until you configure the destination for the specified session number:

```
LocalSwitch(config)# mirror session [1-4] <session-name> port <port_ID>
```

After you define the mirror session (including the destination) on the source switch, you can define the source interface for the mirror session. You can specify ports, trunks, or VLANs as the originating interface. The mirrored packets can be those inbound to the interface, those outbound from the interface, or both:

```
LocalSwitch(config)# interface <port or trunk | vlan vid>] monitor all
[in | out | both] mirror session <session_ID>
```

Specify **in**, **out**, or **both** to select traffic based on the direction in which it is being transmitted. If you want to select traffic that is entering the source switch on the selected interface, include the **in** option. If you want to select traffic that is leaving the interface, select **out**, and if you want all traffic being transmitted on that interface, use the **both** option.

The session number tells the switch the destination for the mirrored data. Depending on the **mirror** command you entered earlier, the destination might be another switch or an exit port on the source switch.

Configuring remote mirroring on ProVision switches

If you want to configure remote mirroring, the first step is to configure the destination switch. If the destination switch is not ready to handle the mirrored traffic forwarded by another switch, performance may be adversely affected. You then configure the source switch.

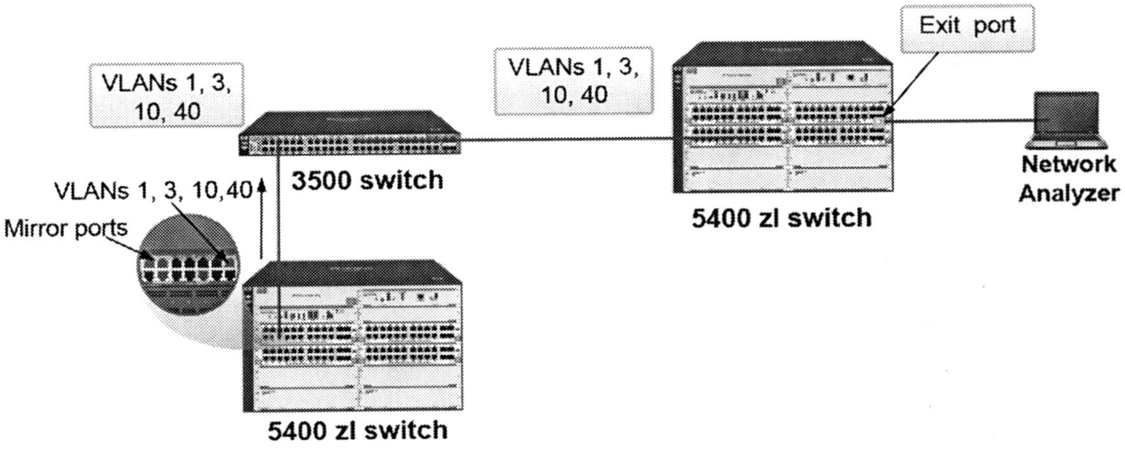

Figure 2-3: Remote port mirroring on ProVision switches

On the destination switch, you define the session number, the settings for the source and destination switch, the UDP port for the mirror session, and the port for the mirror destination:

```
RemoteSwitch(config)# mirror endpoint ip <src-ip> <src-udp-port> <dst-ip>
<exit-port> [truncation]
```

When you configure the *<src-ip-add>*, *<src-udp-port>*, and *<dst-ip-add>* options, you must match exactly the settings that you will configure for the mirror session on the source switch. If the settings do not match, the destination switch will not recognize the mirror session and will not know how to handle the mirrored data stream.

You replace *<src-ip>* with the IP address of the VLAN or subnet on which the mirrored traffic enters or leaves the source switch.

You replace *<src-udp-port>* with a unique UDP port number to use for the mirror session. This unique port number helps the destination switch identify the session from the source switch.

You replace *<dst-ip>* with the IP address of the VLAN or subnet for the exit port on this—the destination—switch.

You replace *<exit-port>* with the exit port on the destination switch.

On the source switch, you then define the source interface for the mirror session and the direction of the traffic that you want to monitor, just as you would for local mirroring:

```
LocalSwitch(config)# mirror <session> remote ip <src-ip> <src-udp-port>
<dst-ip>
```

Include **remote ip** to identify the destination as a remote switch:

- Replace *<src-ip>* with the IP address of the VLAN or subnet on which the mirrored traffic enters or leaves this—the source—switch.
- Replace *<src-udp-port>* with the unique UDP port number to use for the session.
- Replace *<dst-ip>* with the IP address of the VLAN or subnet for the exit port on the destination switch.

When you enter this **mirror** command, the CLI displays a reminder message, telling you that you must first configure the destination switch for the remote mirror session. You must confirm that you have already configured the destination switch to apply the command.

Enabling jumbo frames or using the truncation option

When the source switch sends the mirrored data stream to the destination switch, it adds a 54-byte proprietary header to the information field, increasing the total size for the frame. On a typical network using the default maximum transmission unit (MTU) of 1500 bytes, the switch can mirror untagged frames with information fields that are smaller than or equal to 1446 bytes. This is because the 54-byte proprietary header, inserted into the Ethernet frame of the data field, when coupled with the header and footer of the MAC frame, may exceed the limits of the hardware, either 1518 bytes (normal untagged Ethernet frame) or 1522 bytes (802.1Q tagged).

Consequently, a switch could receive a frame larger than what it is configured to handle. If the switch receives frames that have data fields larger than 1446 bytes, in normal operation, it will drop the frames because the new header makes these packets exceed the maximum frame size for legacy Ethernet, either 1518 or 1522. (If this happens, the switch may log that it has received "giants.") The switch will forward the frames to their regular destination—as long as the received frames do not exceed the default MTU, which is typically set by sending devices at 1500 bytes for legacy Ethernet.

You can handle large frames in one of two ways:

- **Enable jumbo frames**—If you want to capture all interesting traffic, you should enable jumbo frames. The switch can then transmit a frame that is a total of 9220 bytes, which must include the 54-byte header and the 4-byte 802.1Q tag (if applicable). Remember that jumbo frames are supported only on ports operating at 1 Gbps or greater.

  To allow these larger frames to be transmitted across your network, you must enable all the switches that might carry the mirrored traffic to support jumbo frames. These switches include both the source and destination switches and any switches in between. If you have redundant paths between the switches (if spanning tree is running in your environment, for example), you must enable jumbo frames on those switches as well.

  On each switch, enable jumbo frames on the VLAN that carries the mirrored traffic with this global configuration mode context command:

  ```
  Switch(config)# vlan <vlan_id> jumbo
  ```

- **Enable truncation of oversized frames**—You can include the **truncation** option in your **mirror** command. In this case, the switch will truncate the part of the frame that is larger than the MTU size you specify. When the switch truncates the frame, it will make the frame size a multiple of 18 bytes. For example, suppose you specify that the MTU is 1000 bytes and the switch receives a frame that is larger than 1000 bytes. The switch would truncate the frame to 990 bytes (55 x 18 bytes).

Packet-sampling protocols

You will now consider packet-sampling technologies, which can provide insight into traffic patterns and bandwidth consumption.

Both ProVision and Comware switches support sFlow and RMON. Some Comware switches also support NetFlow and NetStream. (As you will learn, NetFlow is a proprietary technology developed by Cisco.) Check your switch documentation to determine which packet-sampling technologies it supports.

HP IMC supports all these packet-sampling technologies.

sFlow

Comware and ProVision switches both support sFlow, an industry-standard technology that uses statistical sampling to enable network monitoring. sFlow relies on two components:

- sFlow agents
- sFlow collector (which is also sometimes called a receiver)

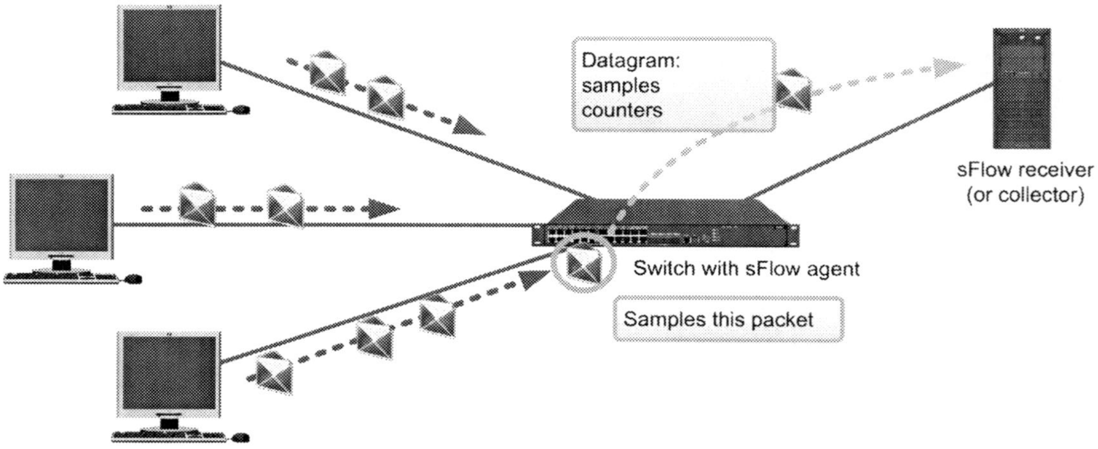

Figure 2-4: sFlow

As shown in Figure 2-4, sFlow agents run on network devices—such as switches, routers, and access points—gathering information about the packets transmitted on these devices and sending that information to the sFlow collector for analysis. Specifically, agents can send traffic samples and polling counters.

For traffic samples, the sFlow standard defines all the packets that a network device receives on one interface and are forwarded to another interface as a "flow." Using a statistically accurate algorithm, the sFlow agent examines on average one of every *n*th packets, where n is the number of packets. The samples are taken "on average" because sFlow employs some randomness to avoid sampling packets at precise intervals that might coincide with certain traffic patterns.

An sFlow agent packages sampled information into small datagrams, which include Layer 2 through 7 information, such as:

- Source and destination interface
- Sampled packet header
- Length of sampled packet
- Forwarding decision the switch made for this packet

An sFlow agent can also use counter polling to gather traffic statistics. It polls data sources at specified intervals, adds this information to the sFlow datagrams, and sends them to a specified sFlow collector. On the Comware and ProVision switches, the data sources are ports (Ethernet interfaces).

The agents can include information from several packets into a single datagram. Because traffic samples are compact and agents sample only a small percentage of traffic, sFlow does not require a large amount of network bandwidth or processing power on network devices.

To forward the datagram to the sFlow collector, the sFlow agent appends a UDP header and an IP header to the datagram and encapsulates the packet in an Ethernet frame.

Comware and ProVision switches implement sFlow in hardware—which is typically how sFlow is implemented. This allows the switch to deliver the information quickly and efficiently.

The sFlow collector analyzes the packaged information from each sFlow agent on a wireless and wired network. From this information, the sFlow collector can create a statistical model of network traffic that can be used for a variety of purposes, including network traffic management, security auditing, or troubleshooting.

HP IMC, which you will learn more about in this chapter, is an sFlow collector. Network Traffic Analyzer (NTA), an add-on module for IMC, uses sFlow information to provide detailed information about network traffic.

Note

For more information about sFlow, visit http://sflow.org.

Configuring sFlow on Comware switches

When you configure sFlow on an Comware switch, HP recommends that you assign an IP address to the sFlow agent. The switch checks for this address periodically, and if an IP address has not been configured for the agent, the switch selects one of the interface IP addresses to use as the sFlow agent address. However, HP recommends that you configure the IP address that the sFlow agent uses. When the switch selects this IP address, it does not save it, and it is not displayed as the sFlow agent's IP address in the startup-config.

To specify an IP address for the sFlow agent, enter:

```
[Switch] sflow agent ip <ip_address>
```

Next, you specify a collector:

```
[Switch] sflow collector <collector_id> ip <ip_address>
```

You can then configure the sampling rate on an interface:

```
[Switch] interface <interface_type/interface_number>
[Switch-<interface>] sflow sampling-rate <rate>
```

You can also specify the sFlow collector for traffic sampling on this interface:

```
[Switch-<interface>] sflow flow collector <collector_id>
```

Finally, you can configure counter sampling on the interface and specify the sFlow collect for counter polling:

```
[Switch-<interface>] sflow counter interval <interval time>
[Switch-<interface>] sflow counter collector <collector_id>
```

Configuring sFlow on ProVision switches

To configure sFlow on ProVision switches, you configure the IP address of the sFlow receiver:

```
Switch(config)# sflow <receiver-instance> destination <ipv4 | ipv6> <udp-port-num>
```

Replace `<receiver-instance>` with the number 1, 2, or 3. The ProVision switches support three sFlow instances.

Specify if sFlow should use IPv4 or IPv6.

Replace `<udp-port-num>` with a UDP port. If you do not include this option in the command, the ProVision switch will use the default UDP destination port, 6343.

You can also specify which ports should be sampled and the sampling rate, using the following command:

```
Switch(config)# sflow <receiver-instance> sampling <port-list> <sampling rate>
```

Again, replace `<receiver-instance>` with the number 1, 2, or 3.

Replace `<port-list>` with the ports that you want to have sFlow sample.

Replace `<sampling rate>` with the rate you want to use for sampling. The lower the value, the higher the percentage of packets that are sampled.

Finally, you can specify the polling interval:

```
Switch(config)# sflow <receiver-instance> polling <port-list> <polling interval>
```

Replace `<receiver-instance>` with the number 1, 2, or 3.

Replace `<port-list>` with the ports that you want to have sFlow poll.

Replace `<polling interval>` with a value that determines how frequently (in seconds) sFlow polls the interfaces.

NetFlow

Like sFlow, NetFlow collects information about network traffic, which can be used to gain insight into network traffic. However, NetFlow uses a different approach to gather information about network traffic.

Figure 2-5 illustrates the NetFlow architecture, which includes:

- **NetFlow exporter**—The NetFlow exporter identifies flows, gathers data about those flows, converts this information to NetFlow records, and sends these records to a NetFlow collector.
- **NetFlow collector**—The NetFlow collector receives records from the exporters, saves these records to its local database, and then forwards them to the NetFlow analyzer.
- **NetFlow analyzer**—The NetFlow analyzer compiles records and sorts through the collected data, presenting that data in a meaningful way. For example, an analyzer could display bandwidth usage and highlight the devices using the most traffic on the network.

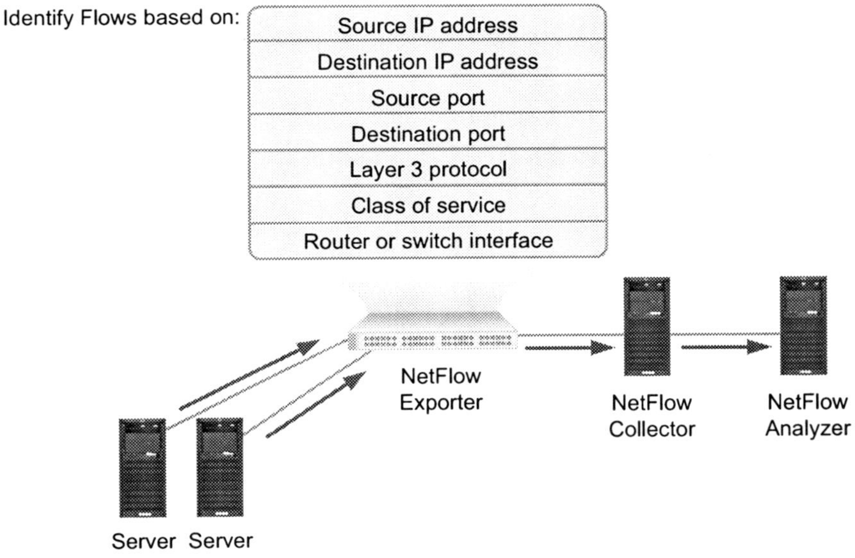

Figure 2-5: NetFlow

A NetFlow exporter examines network traffic and identifies flows. Each flow is a group of packets that share the same values for the following attributes:

- IP source address
- IP destination address
- Source port
- Destination port
- Layer 3 protocol (IP)
- Class of service
- Router or switch interface

Once the exporter identifies a flow, it saves information about the attributes that identify the flow and begins to tally the packets and bytes, thereby determining the amount of traffic in that flow. The exporter also timestamps the flow, records the next hop IP address, and examines TCP flags (indicating TCP handshakes).

Initially, the NetFlow data is held in cache. The NetFlow exporter sends the data to the collector when a flow is inactive for a specified period of time. For long-lived flows, the exporter waits until the active timer has expired before sending data about that flow.

The NetFlow exporter can convert the information it gathers about traffic flows into different formats, which are referred to as export versions. Many NetFlow exporters use export version 5, but when this study guide was published, version 9 was the most recent version. The version 9 format incorporates templates, which define the format of the data that is collected. These templates are designed to be extensible, enabling them to support enhancements in new versions of NetFlow without affecting previous versions.

NetFlow is a proprietary protocol that was developed by Cisco. Subsequently, the IETF developed the Internet Protocol Flow Information Export (IPFIX) standard, which is based on NetFlow version 9.

IMC supports NetFlow, in addition to sFlow and NetStream, which you will learn about next.

NetStream

HP IMC and some Comware switches and routers support NetStream. As Figure 2-6 shows, NetStream uses the same type of architecture as NetFlow:

- **NetStream data exporter (NDE)**—An NDE analyzes and processes network flows, extracts flows that meet conditions for statistics, and sends the data to the NetStream Collector (NSC).
- **NetStream collector (NSC)**—The NSC parses packets from the NDE and saves statistics to the database. The NSC can collect data exported from multiple NDEs and filter and aggregate the data.

- **NetStream data analyzer (NDA)**—As a traffic analysis tool, NDA extracts statistics from the NSC, processes the statistics, and generates a report, which can be used for traffic accounting, network planning, and attack monitoring. The NDA allows users to easily obtain, check, and analyze the collected data.

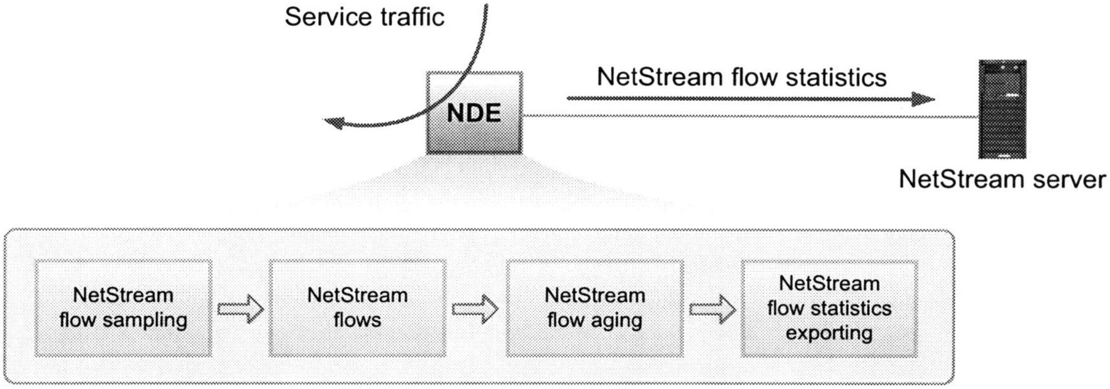

Figure 2-6: NetStream

Some Comware switches, such as the HP 7500 Switch Series, support a NetStream module, which can be installed into the switch. You can then configure the switch, which functions as the NDE, to mirror NetStream flow information to the NetStream module, which functions as the NSC.

NetStream provides statistics about flows, which are groups of packets that have the same value. For IPv4 traffic, flows have the same value for destination address, source IP address, destination port number, source port number, protocol number, ToS, and inbound or outbound interface.

You can have the NDE collect data about all flows, or you can apply filters to NetStream and gather data only on specific traffic. To enable filtering, you apply an access control list or a quality of service policy.

You can also configure the NDE to collect statistic on fewer packets by enabling sampling. This option reduces the impact collecting statistics may have on the NDE, but may not provide as complete a picture of the network.

Once the NDE identifies a flow and gathers data about it, it can export that data in one of the following ways:

- **Traditional data export**—The NDE uses flow aging to determine when to export data to the NSC. When the NDE identifies a traffic flow, it creates an entry for the flow in cache and starts an entry timer. When that timer expires, the NDE exports the data to the NSC.
- **Aggregation data export**—The NDE merges the statistics for several flows before exporting the data to the NSC. You can control how the NDE merges these statistics by configuring the

aggregation mode. For example, you can configure the NDE to use protocol-port aggregation, which bases aggregation on the protocol number, source port, and destination port. Check your switch documentation for a complete list of aggregation modes.

NetStream also supports several export data formats, including version 5, version 8 (which supports aggregation data exports), and version 9 (which provides the most flexible format, allowing you to define templates that have different statistics fields).

For configuration information, consult the documentation for the Comware switch or NetStream module.

RMON

Like sFlow, NetFlow, and NetStream, Remote Monitoring (RMON) can run on a network infrastructure device and monitor traffic sent and received on that device's ports. However, as implemented on HP devices, RMON's purpose differs a bit from that of NetFlow, NetStream, and sFlow. Those sampling protocols profile the sessions and applications in use in a network and also provide information about port statistics such as link utilization. RMON focuses on collecting port statistics but, as Figure 2-7 shows, also alerts network management solutions when particular thresholds are crossed. Thus, RMON and the traffic sampling protocols fulfill somewhat overlapping but also distinct functions.

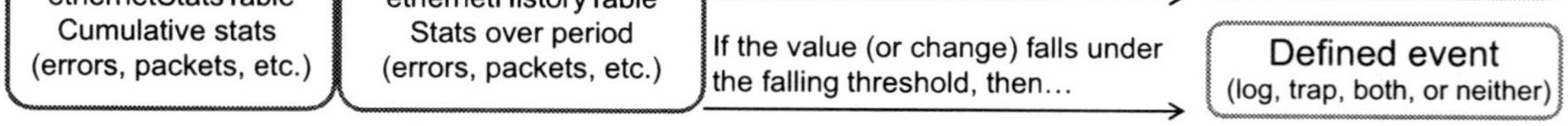

Figure 2-7: RMON

RMON works in conjunction with Simple Network Management Protocol (which is described in more depth later in this chapter). RMON was designed to inform SNMP solutions about traffic-related information in a more efficient manner. For example, an SNMP solution might need to monitor when ports' utilization exceeds a specific percentage. Because SNMP traps do not provide this information, the SNMP solution would need to send many periodic get messages querying network devices. The solution might also need to compare and otherwise analyze information collected at different times, adding to its processing burdens.

RMON removes these burdens. Each local device monitors its own ports' statistics, consolidating and storing the information to be queried as required. The local device also automatically sends alarms when various thresholds are exceeded.

RMON support on Comware switches

Many Comware switches can act as RMON agents, which support four RMON groups. (An RMON group is a defined object that implements a specific RMON function.) To configure RMON on the switch, you set up the information that RMON collects. You also define alarms as well as the actions taken when alarms are triggered (events).

Collect information

You enable an RMON monitoring function on an interface, which creates a table to store various statistics for that interface.

Comware devices support two types of monitoring functions (defined in RMON groups):

- **Ethernet statistics group**—When you define this group on an interface, an ethernetStatsTable stores statistics such as network collisions, CRC alignment errors, undersize/oversize packets, broadcasts, multicasts, bytes received, packets received, and so on. Every time that RMON checks the interface, it overwrites its table with a cumulative sum of statistics. An interface can support only one statistics group entry.
- **Ethernet history group**—When an interface is configured with this RMON group, an ethernetHistoryTable stores statistics such as bandwidth utilization, number of errors, and number of packets. For history group entries, every time that RMON checks the interface, it creates a new bucket with statistics for just the latest period. (You specify the period when you define the history group entry on the interface. You also specify the number of buckets; RMON writes over the oldest bucket when it has filled the specified number.) An interface can support multiple history entries, but they must use different periods. For example, you might want to track statistics over both 5-minute and 1-hour periods.

When you define an entry on an interface, you define the entry's number. The number must be unique so that the switch can maintain a unique table for each interface and set of statistics.

The command syntax for creating an Ethernet group entry is:

```
[Comware-<interface ID>] rmon statistics <entry number> [owner <name>]
```

The command syntax for creating a history group entry is:

```
[Comware-<interface ID>] rmon history <entry-number> buckets <number>
interval <sampling-interval> [owner <name>]
```

Note

The optional owner keyword defines a user allowed to retrieve the statistics.

Network management solutions can now use SNMP get messages to collect the information in these tables as required. You can also view the tables using these commands:

```
<Comware> display rmon {statistics | history} <interface ID>
```

You might also want to set up alarms that automatically inform the network management solution of relevant information. The next section explains how.

Define alarms and events

You can define an alarm that triggers when the statistics collected for an entry pass a specific threshold:

1. Create the Ethernet statistics or history entry on the desired interface, as explained above.
2. Next create one or more events. Events are another defined RMON group; an event group specifies the actions that RMON will take when an alarm is triggered. RMON can log the alarm, enabling the SNMP solution to read the alarm when it next collects the log, or RMON can send an immediate SNMP trap to a specific SNMP community. The event can also specify both actions (log and immediate trap). Finally, the event can call for no action at all; as you will see, an alarm indicates an event for exceeding or falling below specific values, and you might not want to associate an action with one of those thresholds.

   Every event must have a unique entry index.

   The command syntax for creating an event is:

   ```
   [Comware] rmon event entry-number [ description <string> ] { log | log-trap <log-trapcommunity> | none | trap <trap-community> } [owner <name>]
   ```
3. Create the RMON alarm, another RMON group, which specifies:
   - The MIB OID associated with the specific statistic in the table that you want to monitor
   - A sampling interval
   - The rising threshold value
   - The event that defines the action that RMON should take when the rising threshold value is exceeded
   - The falling threshold value
   - The event that defines the action that RMON should take when the statistic falls below the falling threshold value (can be the same or different as the rising threshold event)
   - Whether the threshold values refer to the actual value in the table (absolute) or to a change from the value in the previous sample (delta)

Note

The alarm triggers when the value first crosses a threshold but not again—unless it has subsequently crossed the other threshold. Therefore, a rising alarm is always succeeded by a falling alarm and vice versa. In other words, when the value first exceeds the rising threshold value, the alarm triggers, but for as long as the value continues to exceed the rising threshold value, no more alarms occur. A new alarm only triggers when the value falls below the falling threshold value.

The command syntax for creating an alarm is:

```
[Comware] rmon alarm <entry-number> <MIB OID> <sampling-interval> { absolute
| delta } risingthreshold <rising threshold value> <rising threshold
event entry> fallingthreshold <falling threshold value> <falling thresh-
old event entry> [ owner <name>]
```

Note

Comware switches also support private alarms, which offer a meta-monitoring capability. They monitor alarms and trigger additional alarms based on a formula that assesses whether an excessive number of alarms seem to be triggered.

You can also create alarms associated with other MIBs, not related to the Ethernet and history statistics collected on an interface. Simply follow steps 2 and 3.

Note

Refer to your Comware device's *Network Management and Monitoring Configuration Guide* for information about how many RMON entries, events, and alarms it supports.

RMON on ProVision switches

ProVision switches also provide RMON agents that support the same four RMON groups as the Comware switches. In other words, their RMON monitoring functions resemble those of the Comware switches, including the collection of cumulative Ethernet statistics and the collection of historical Ethernet statistics (a series of periodic snapshots). As on Comware switches, RMON can trigger alarms when specific values cross a specific threshold, and those alarms cause events, such as SNMP traps, to occur.

However, on ProVision switches, the network management solution always configures these settings. You follow the same basic procedure:

1. Enable the collection of Ethernet statistics, Ethernet history statistics, or both on a port, specifying the sampling period for history statistics.
2. If you want to use alarms, set up events that define actions.
3. Set up alarms, specifying the monitored value (MIB), the sampling interval, the rising threshold values, the falling threshold values, whether values are absolute or relative (delta), and the events that the alarm triggers if values cross the rising or the falling threshold.

Note
In addition to absolute and delta sampling values, IMC, the HP network management solution, supports speed sampling values, which divide the delta value by the sampling interval. This choice allows you to monitor how quickly a value is changing.

SNMP

SNMP is an industry-standard protocol that allows you to manage and monitor a variety of network devices from a central location. Using an SNMP application, you can configure network devices and apply consistent security and management policies to all the devices on your network.

You can also monitor SNMP-compliant devices for conditions requiring administrative attention—for example, a server has gone down, a LAN or WAN link is no longer available, or a link has become congested. You can track device uptime, link states, and many other device information variables. When a problem occurs, SNMP-compliant devices send an alert message to one or more designated SNMP servers, and you can then take steps to resolve the problem.

As Figure 2-8 shows, SNMP works in a client–server relationship: SNMP agents, which function as clients, run on managed devices, and the SNMP server is a management application that requests, handles, and analyzes the information from the managed devices. Typically, you access the SNMP server from a centralized management console.

SNMPv1 and v2c

You should first understand SNMPv1 and v2c operations and communities.

SNMP operations

SNMPv1 includes four protocol operations:

- Get—The SNMP server uses Get operations to request information about one or more object instances from an agent. (Get operations are sometimes referred to as GetRequest operations.)

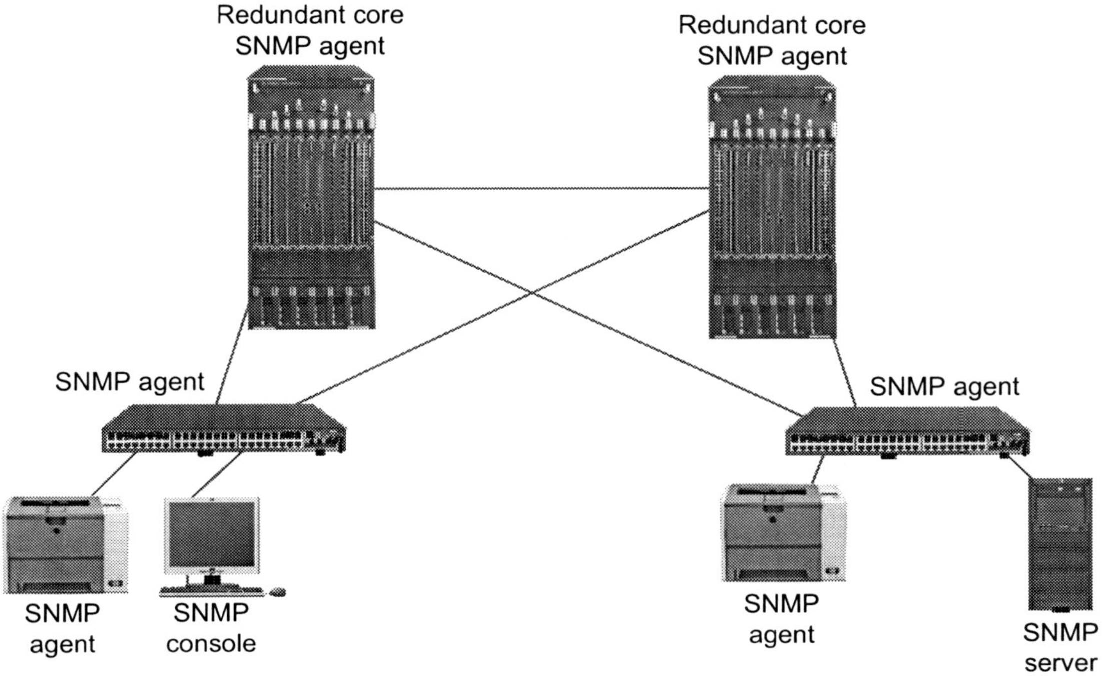

Figure 2-8: SNMP

- GetNext—The SNMP server uses GetNext operations to request information about the next object instance in the table.
- Set—The SNMP server uses this operation to "set" the value of object instances within a trap. (Set operations are sometimes referred to as SetRequest operations.)
- Trap—Agents use trap operations to inform, or notify, the SNMP server of a significant event.

SNMPv2c supports two additional protocol operations, which are designed to make the protocol more efficient:

- GetBulk—The SNMP server uses GetBulk to request large blocks of data. This operation was added to enable the SNMP server to send one request to gather multiple pieces of information with a single request.
- Inform—The SNMP server uses the Inform operation to send trap information to servers. Inform was created to add a capability that the trap operation does not provide: with Inform, the SNMP server acknowledges it received the information.

SNMP communities

SNMP versions 1 and 2c use community strings to restrict SNMP access (see Figure 2-9). To enable different access levels, you can configure one of the following communities:

- Read only—This community limits the user to reading SNMP information only.
- Read write—This community provides full access to SNMP functions, including the ability to make changes on managed devices.
- Trap—A trap allows the managed device to spontaneously send an update to a trap receiver, usually in response to an alarm.

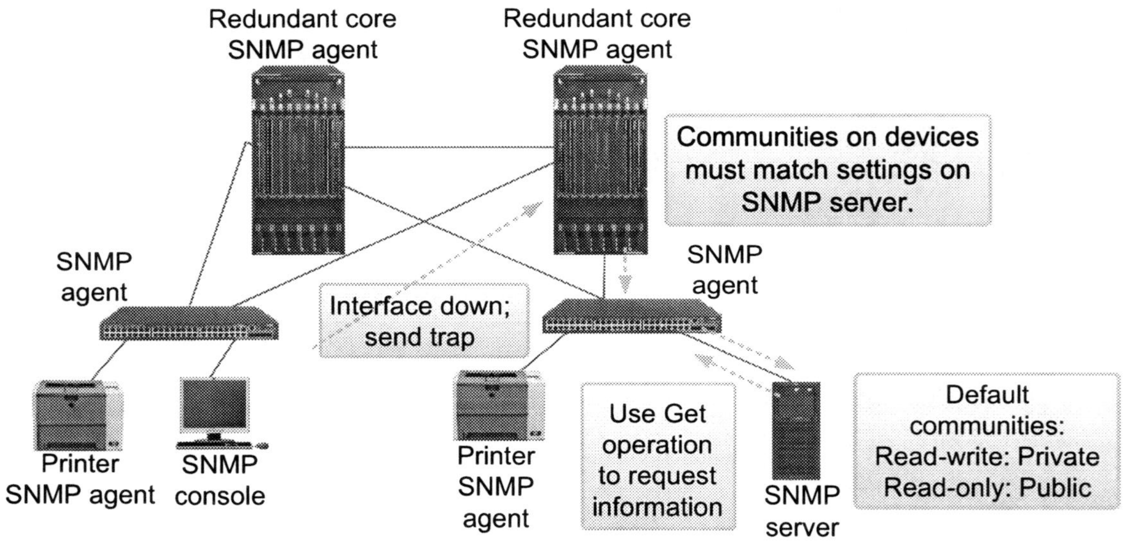

Figure 2-9: SNMPv2c operations and communities

SNMP-compliant devices typically use public as the default read-only community and private as the default read-write community. When you enable SNMP on devices, you should always change these default settings. Because many organizations do not change these default settings, however, their managed devices and SNMP servers are vulnerable to hackers.

SNMP communities do not provide strong access control for SNMP agents and servers. In addition, SNMP versions 1 and 2c do not include any security features to protect the data exchanged between the devices and the SNMP server. Because the packets and the community strings are not encrypted, a hacker could intercept and read them. And because these versions do not provide message integrity measures, a hacker could intercept messages, tamper with the data, and then send it to the SNMP server.

Because SNMP packets contain information about the network, it is important to secure these packets. The early versions of SNMP are vulnerable to attacks such as:

- Man-in-the-middle—An attacker can alter SNMP messages generated on behalf of an authorized user in such a way as to affect management operations. An attacker with read-write access can infiltrate any SNMP-managed device.
- Impersonation—By assuming the identity of an user who has the appropriate authorizations, an attacker can gain read-write access to management operations.
- Reconnaissance—Because early implementations do not encrypt the community string or SNMP packet information, an attacker can eavesdrop on the exchanges between SNMP agents and an SNMP server. The attacker can then collect information about the network or discover the read-write community string.
- Unauthorized access—Early versions of SNMP are vulnerable to replay attacks. During natural operation, most networks will experience packets that are reordered, delayed, or replayed. Knowing this, an attacker can maliciously reorder, delay, or resend packets to gain unauthorized access to management operations.
- Brute force—An attacker can use a brute-force attack to discover network community strings.

SNMPv3

SNMPv3 addresses the major security flaws in SNMP versions 1 and 2c. It defines a User-Based Security Model, allowing you to enforce both privacy and integrity for communications between SNMP agents and the SNMP server. You can, however, enforce only authentication or not impose any security, depending on which of the following security levels you decide to implement:

- AuthPriv—This security level provides authentication and privacy, enforcing the strongest security for SNMP.
- AuthNoPriv—This level provides authentication but no privacy.
- noAuthNoPriv—This level does not provide authentication or privacy.

To enforce authentication and privacy, SNMPv3 uses usernames and passwords, as shown in Figure 2-10. When a user provides an authentication password, the password is converted into a localized key. This key, along with the SNMP engine timestamp and the actual message, are compressed into a message digest, which is forwarded with the packet.

Each user is also given a privacy password, which is used to encrypt the message packet. SNMPv3 uses one of two encryption algorithms, AES and 3DES, to encrypt the localized key and the SNMP packet. This ensures that intercepted SNMP messages cannot be read or tampered with.

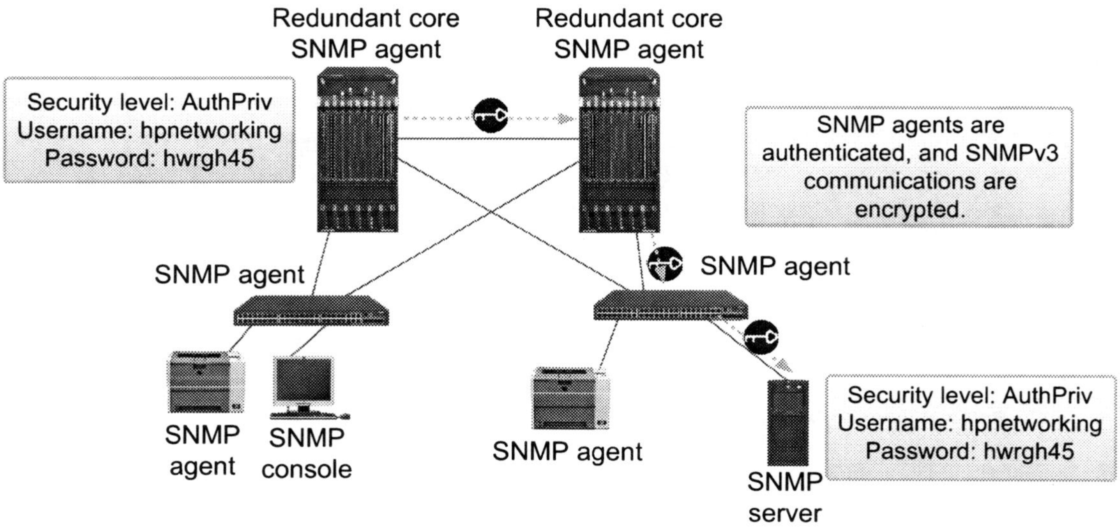

Figure 2-10: SNMPv3 security features

Note

For more information about SNMPv3 and the USM, see RFC 3414 at http://www.ietf.org/rfc/rfc3414.txt.

Configuring SNMPv3 on Comware switches

The SNMP agent on Comware switches is disabled by default. To enable the SNMP agent on an Comware switch, enter:

```
[Switch] snmp-agent
```

By default version 3 is enabled.

To configure system information and location, you enter the following command. Because you are configuring SNMPv3, you do not have to configure a version number:

```
[Switch] snmp-agent sys-info [contact <sys-contact> | location <sys-location> | version [all | v1 | v2c | v3]
```

For SNMPv3, you configure an SNMP agent group and specify if authentication and privacy will be enforced:

```
[Switch] snmp-agent group v3 <group name> [authentication | privacy] [read-view <read-view>] [write-view <write-view] [notify-view <notify-view>] [acl <acl-number>]
```

Use ACLs if you want to apply these settings to certain traffic. (You will learn how to configure ACLs later in Chapter 6.)

Finally, you add a user to the SNMP agent group:

```
[Switch] snmp-agent usm-user v3 <username> <groupname> [cipher] authen-
tication-mode [md5 | sha] <auth-password> [privacy-mode [3des | aes128}
<priv-password> acl <acl-number>
```

Note that the **md5** and **sha** options determine the algorithm used for authentication. The SHA-1 algorithm is more secure than MD5. Likewise, the **3des** and **aes128** options determine the algorithm used to protect the data integrity. The AES 128-bit algorithm is more secure than 3DES.

Configuring SNMPv3 on ProVision switches

To support SNMPv3, ProVision switches must be configured with an IP address and subnet mask. You must then:

1. Enable SNMPv3:

   ```
   Switch(config)# snmpv3 enable
   ```

   You can limit the switch to using only SNMPv3 so that the public community is inaccessible to network applications such as autodiscovery, traffic monitoring, and SNMP trap generation. To do so, use the following command:

   ```
   Switch(config)# snmpv3 only
   ```

2. Configure the SNMP users:

   ```
   Switch(config)# snmpv3 user <user_name> [auth <md5 | sha> <auth_pass>]
   [priv <des | aes> <priv_pass>]
   ```

   Configuring authorization and privacy settings is optional, but provides tighter security. For authorization, the sha option provides the better security; for privacy, the aes option provides the better security.

3. Assign users to groups:

   ```
   Switch(config)# snmpv3 group <group_name> user <user_name> sec-model
   <ver1 | ver2c | ver 3]
   ```

   Use Table 2-1 to select the group in which you want to place the user.

The views allow users to look at or modify certain MIBs. For example, with ManagerReadView, a user has access to all managed objects. With ManagerWriteView, a user has access to all managed objects except the following: vacmContextTable, vacmAccessTable, and vacmViewTreeFamilyTable. With OperatorReadView, a user does not have access to icfSecurityMIB, hpSwitchIpTftp-Mode, vacmContextTable, vacmAccessTable, vacmViewTreeFamily-Table, usmUserTable, and snmpCommunityTable. With Discovery View, a user has access limited to the samplingProbe MIB.

Table 2-1: Group level access

Group name	Group access type	Group read view	Group write view
Managerpriv	Ver3 must have authentication and privacy	ManagerReadView	ManagerWriteView
managerauth	Ver3 must have authentication	ManagerReadView	ManagerWriteView
operatorauth	Ver3 must have authentication	OperatorReadView	DiscoveryView
operatornoauth	Ver3 no authentication	OperatorReadView	DiscoveryView
commanagerrw	Ver2c or ver1	ManagerReadView	ManagerWriteView
commanagerr	Ver2c or ver1	ManagerReadView	DiscoveryView
comoperatorrw	Ver2c or ver1	OperatorReadView	OperatorReadView
comoperatorr	Ver2c or ver1	OperatorReadView	DiscoveryView

NETCONF

Although SNMP was developed to provide centralized management and configuration for network devices, as well as network monitoring–in short, write as well as read functions—in practice, most management solutions use SNMP primarily for monitoring. Network Configuration Protocol (NETCONF), defined in RFC 6241 (which obsoletes RFC 4741), has emerged as another method for centralizing the editing of configurations with customized scripts or a network management solution. Like SNMP, NETCONF uses operations that read (get) and write (edit) configurations. However, it uses XML encoding for its operations in an attempt to more closely approximate the encoding that the device itself uses.

Note

NETCONF is not intended as a replacement for SNMP but rather as an alternative for just one SNMP function. NETCONF deals only with device configurations and does not attempt to monitor devices' status or state information.

NETCONF uses a Remote Procedure Call (RPC) message scheme to manage communications between an application such as a network management solution (the client) and the network infrastructure device (the server). The basic structure for RPC is simple. The application sends RPC messages with operations that affect the network device's configuration. The device sends a reply, acknowledging that it has applied the operation, or an error, indicating that it cannot do so.

Figure 2-11 illustrates the NETCONF structure in more detail:

- **Transport layer**—The client and server must exchange establish a session for transporting the RPC messages. NETCONF requires devices to support Secure Shell (SSH) as a transport protocol

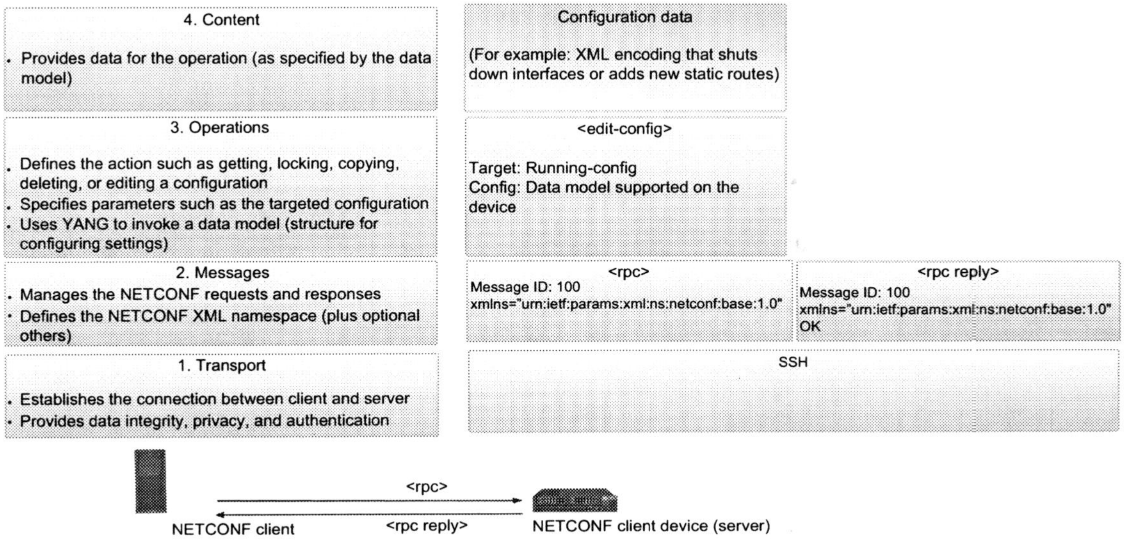

Figure 2-11: NETCONF

but allows other transport protocols as well, as long as the protocol provides data integrity, privacy, and authentication. Transport Layer Security (TLS) provides an acceptable alternative. In the past, some NETCONF solutions also used Simple Object Access Protocol (SOAP) and Blocks Extensible Exchange Protocol (BEEP), but these protocols are mostly obsolete for this function.

- **Message layer**—RPC manages the NETCONF exchanges. The client, or network management solution, always sends RPC messages, while the network device sends RPC replies. The RPC messages define the NETCONF XML namespace (xmlns). You can think of an XML namespace as a dictionary that lists and defines "words" for the XML script—in this case, the NETCONF operations at the next layer.

  The RPC message can define additional XML namespaces and invoke particular elements and attributes in those namespaces; in this way, vendors can extend the basic functionality with customized operations for their network management solutions.

  NETCONF also defines a notification message layer, which lets the network device initiate an exchange, sending information to the application. (These notifications do not use NETCONF operations, but simply include notification data.)

- **Operation layer**—NETCONF defines a variety of operations for managing network device configurations. For example, the "get" operation asks the device to send all or part of its configuration to the application. The "lock" operation locks the device's configuration so that only the client can edit it, while the "partial_lock" operation (a new operation in the latest NETCONF version) locks just part of the configuration, allowing multiple applications to manage a device at the same time. The "config_edit" operation applies changes to the configuration, the "unlock" operation releases the configuration, the "copy" operation copies a configuration form one location to another, and so on.

Particular operations have particular attributes and parameters associated with them. For example, you can define a device's running-config as the target of various operation.

When an application is editing a configuration, of course, it must use XML encoding that the device understands and accepts. Therefore, the "config_edit" operation also includes a "config" attribute that defines the data model for this encoding. NETCONF itself does not define data models, but it refers to the YANG standard (RFC 6020) for this purpose.

- **Content layer**—This layer includes the configuration data for the operation. For example, the application is shutting down the device's interfaces with the "config_edit" operation. The configuration data consists of XML script that specifies the interfaces in question and applies the appropriate "shutdown" attribute, as defined by the invoked data model.

  If the NETCONF message is a notification message from the device to the application, the actual notification would reside in the content layer.

This introduction to NETCONF gives you a basic idea of this protocol's flexibility; with it, centralized applications can script a wide array of configuration changes, as required on the fly.

HP IMC overview

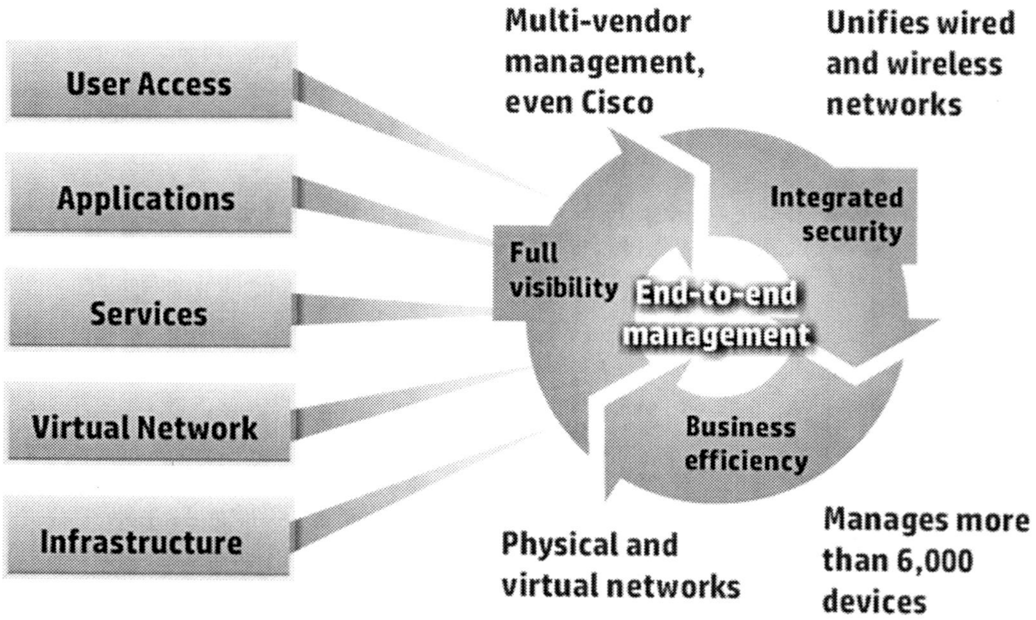

Figure 2-12: IMC capabilities

As Figure 2-12 illustrates, HP IMC consolidates the management functions needed to manage today's networks while providing the tools needed to address emerging networking trends and applications. IMC allows you to manage both your physical and virtual networks. Because IMC automates the configuration of virtual and physical network infrastructures, it eliminates service interruptions caused by virtual/physical network configuration errors. It also accelerates the delivery of new applications.

Specifically, IMC provides the following support for virtual environments:

- Virtual and physical topology views and status indicators for network resources
- VM and virtual switch resource management, including the creation of virtual switches and port groups
- Automatic reconfiguration of network policies that "move" with virtual applications and workloads
- Discovery of VMs and virtual switches and their relationship to the physical network

HP IMC also provides single pane-of-glass network management for multivendor environments. From a single interface, you can monitor and manage network traffic and devices, including HP and third-party network infrastructure devices. HP IMC can manage more than 6,000 devices from more than 200 manufacturers, including Cisco.

HP IMC is designed to provide greater visibility of the entire network, allowing you to control and monitor user access, manage applications and services, and the network infrastructure end to end.

IMC fits any organization's needs

HP IMC meets the needs of any organization—no matter what its size or network requirements. It is available in a variety of packages, as shown in Table 2-2.

Designed for large enterprises, the IMC Enterprise Software Platform provides the scalability and high availability large enterprises need. Using a flexible distributed deployment model; IMC Enterprise Software can be deployed across multiple servers in a hierarchical architecture to provide increased scalability and resilience. This means that all IMC components and add-on modules are installed on the master IMC server, but some components and modules can be deployed on slave servers. The master server is the management center for IMC; slave servers are responsible for specific management tasks. You access the master server to complete all management tasks, and it interacts with slave servers as needed to manage the network.

In addition, the IMC Enterprise Software Platform includes the Network Traffic Analyzer, which you will learn more about in the next section, and a license for extended APIs, which enable organizations to integrate IMC with other applications.

The IMC Enterprise Software Platform is also extensible. You can purchase add-on modules to extend its already impressive capabilities.

Table 2-2: Group level access

Version	Number of devices
IMC Enterprise Software Platform • Network Traffic Analyzer • License for extended APIs • Hierarchical architecture	100 with base product; additive licenses (unlimited) available
IMC Standard Software Platform	100 with base product; additive licenses (unlimited) available
IMC Basic Software Platform	50 (fixed)
IMC Basic WLAN Manager Software Platform	50 with base product; additive licenses available for wireless
IMC Smart Connect Virtual Appliance Edition (BYOD solution)	100 devices and 200 users with base product; additive licenses available
IMC Smart Connect with WLAN Manager Virtual Appliance Edition	100 devices, 200 users, & 50 APs with base product; additive licenses

The IMC Standard Software Platform is designed to meet the needs of medium to large organizations. It provides the same capabilities as the IMC Enterprise Software Platform with one exception: the Standard Software Platform does not support the distributed deployment model.

Created specifically for small- to medium-sized businesses, the IMC Basic Software Platform provides a simple-to-use interface. Limited to 50 nodes, the IMC Basic Software Platform provides sFlow traffic analysis, centralized reports, performance monitoring, network monitoring, and device management (including discovery, baseline configurations, software management, version tracking, and alerts). Designed for companies that need basic management functionality, it does not support the IMC add-on modules.

For small- to medium-sized businesses that centralized control over their wireless networks, HP offers the IMC Basic WLAN Manager Software Platform, which reduces the time needed to deploy configuration changes and provides uniformity throughout the wired and WLAN infrastructure. The IMC Basic WLAN Manager Software Platform provides:

- Wireless LAN (WLAN) device configuration
- Topology maps
- Performance monitoring
- RF heat mapping
- WLAN service reports

The IMC Basic WLAN Manager Software Platform ships with a 50-device license; it supports additive licenses for wireless devices.

HP IMC Smart Connect Virtual Appliance Edition offers a comprehensive management solution for organizations that need to support Bring Your Own Devices. Preconfigured and available as a virtual machine (VM), IMC Smart Connect Virtual Appliance Edition provides onboarding, provisioning, and monitoring BYOD clients and users. With this BYOD solution, organizations can enforce a single policy and manage access across both wired and wireless networks. It also includes tools to help organizations plan capacity for BYOD users and meet and prove regulatory compliance.

HP provides HP IMC Smart Connect with WLAN Manager Virtual Appliance Software for small- to medium-sized companies that need to manage their wireless network and offer BYOD access to their users.

IMC add-on modules

Figure 2-13 shows how IMC and its add-on modules fit within the Fault-management, Configuration, Accounting, Performance, and Security (FCAPS) framework. Created by the International Organization for Standardization (ISO), this framework focuses on five network management levels.

As you can see from the figure, the IMC platform provides capabilities at each level.

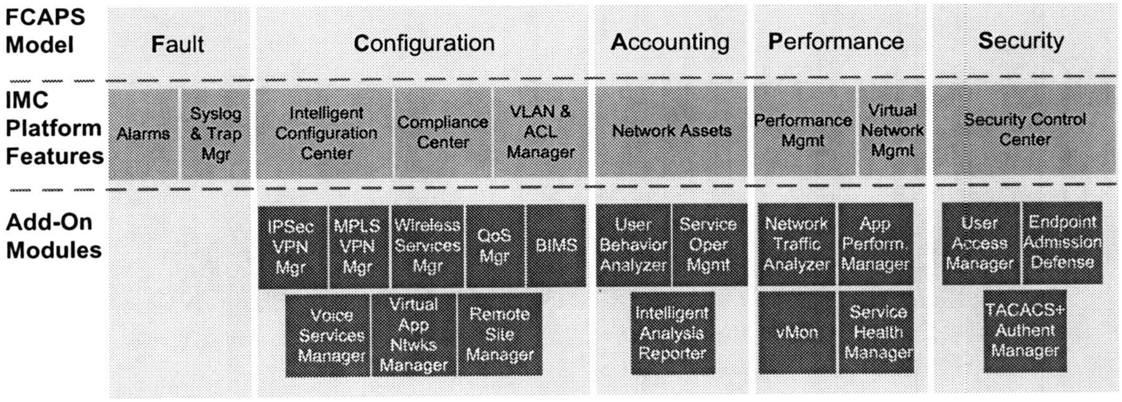

Figure 2-13: IMC add-on modules

As mentioned earlier, the add-on modules extend the capabilities that IMC provides. You can see how these add-on modules provide capabilities at the FCAPS levels. Whether you need to solve your BYOD challenges, manage remote sites with greater efficiency, or gain more insight into how your network is performing, IMC has network management modules to increase the depth and breadth of coverage to meet your business requirements.

In this study guide, you will learn more about three IMC add-on modules:

- Quality of Service (QoS) Manager—You will learn about the IMC QoS Manager in "Chapter 8: Quality of Service."
- User Access Manager (UAM)—You will learn about UAM in "Chapter 7: Port-Based Authentication."
- Network Traffic Analyzer (NTA)—You will learn more about NTA in this module.

IMC Deployment Monitoring Agent

To help you monitor IMC and deploy the components and add-on modules, IMC includes a Deployment Monitoring Agent (see Figure 2-14).

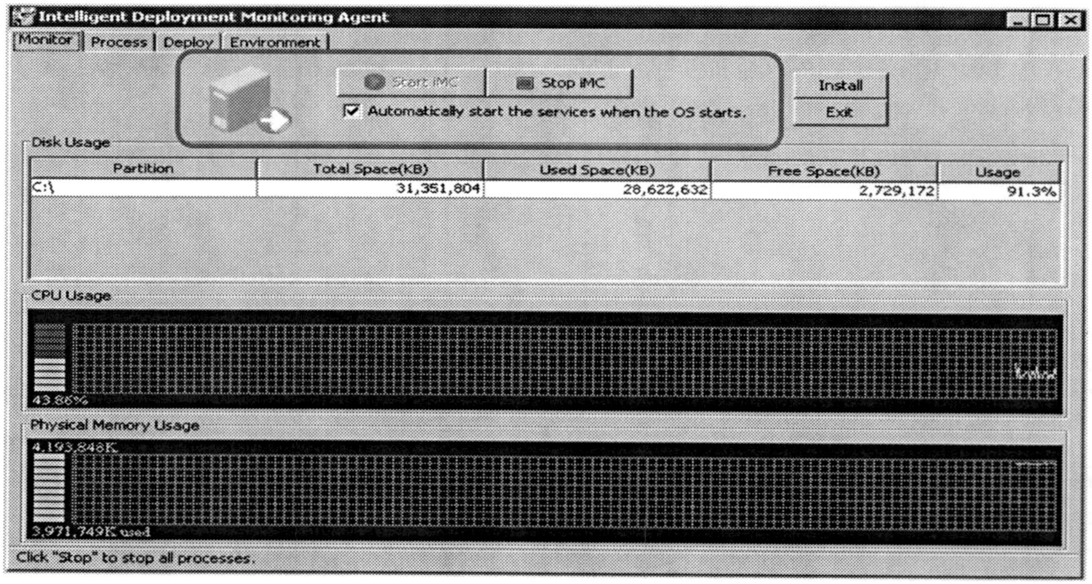

Figure 2-14: IMC Deployment Moritoring Agent

The agent allows you to perform tasks such as:

- Installing the IMC server and IMC modules (see Figure 2-15)
- Starting and stopping the IMC server
- Viewing the CPU usage and memory usage
- Determining the status (such as starting, started, or stopped) of IMC processes

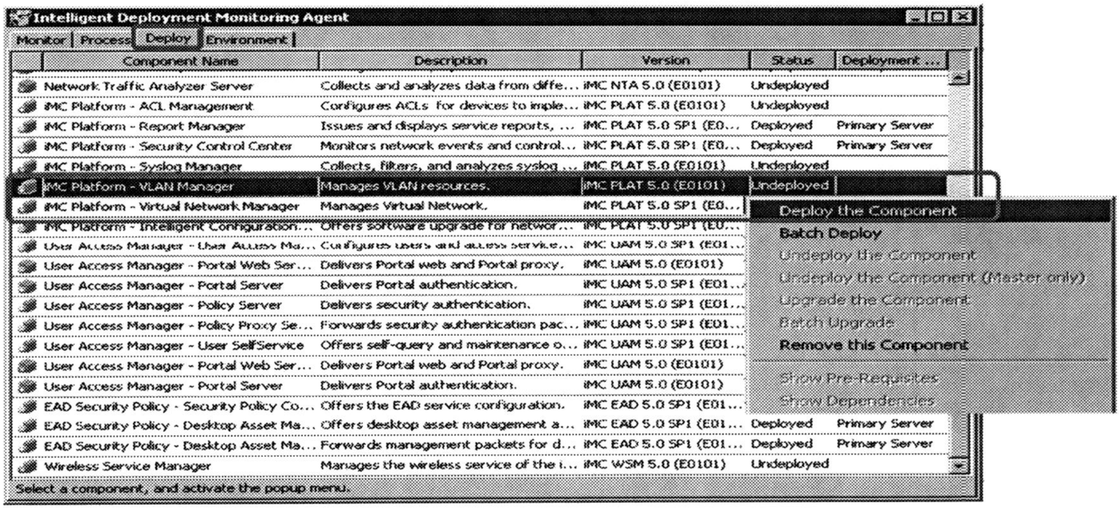

Figure 2-15: Deploying an IMC module

In addition, the Deployment Monitoring Agent provides information about the server on which it is running. For example, it lists the amount of:

- Database space available and used
- Log file space available and used

The agent also provides details about the platform operating system and database used. For example, it might list the Windows Server 2008 R2 operating system and the Microsoft SQL Server 10.50.1600.

SNMP, SSH, and Telnet templates in IMC

To manage your network devices through IMC, you must ensure that the management settings on the network devices match the management settings configured in IMC. The IMC SNMP, SSH, and Telnet templates allow you to configure these settings once, ensure that the settings are correct, and then apply them to your devices. If you later need to change a setting, you can simply modify the template.

You can even configure multiple SNMP, SSH, or Telnet templates and apply the templates to different groups of devices (see Figure 2-16).

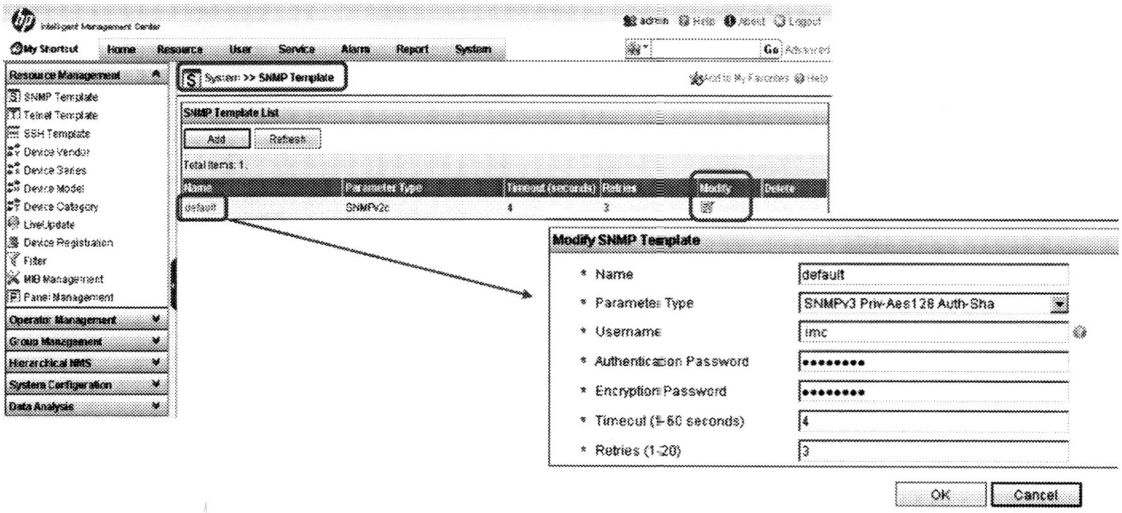

Figure 2-16: SNMP, SSH, and Telnet templates in IMC

IMC Discovery

IMC provides a Basic Auto Discovery option, which requires you to specify a Start IP and End IP as well as the segment address. IMC also supports a number of advanced discovery options, including routing-based, ARP-based, and network segment-based discovery.

Once devices are discovered in IMC, you can easily check their status, as shown in Figure 2-17.

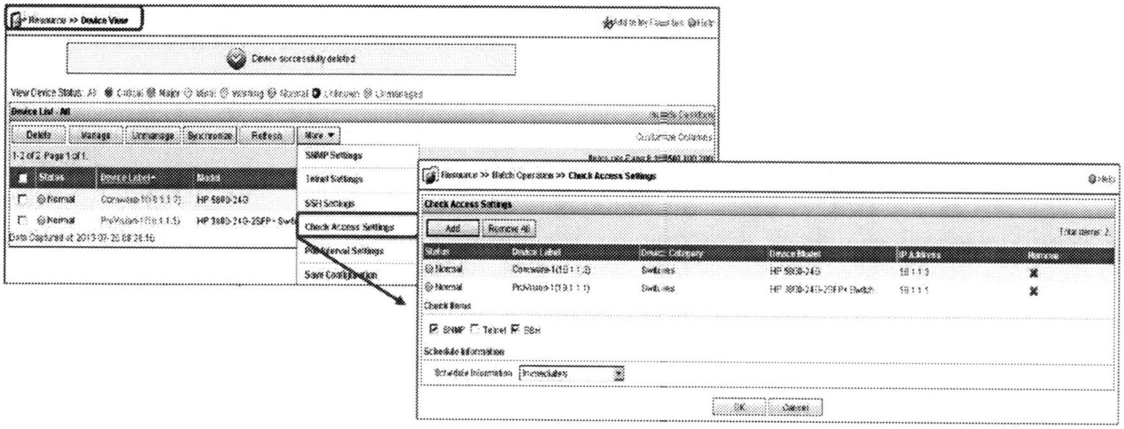

Figure 2-17: IMC Discovery

IMC Network Traffic Analyzer

The NTA is a graphical network-monitoring tool that provides real-time information about users and applications consuming network bandwidth. You can use NTA to plan, monitor, enhance, and troubleshoot networks, as well as identify bottlenecks and apply corrective measures for enhanced throughput. NTA also allows you to monitor Internet egress traffic, analyzing the bandwidth usage of specific applications and monitoring the impact of nonbusiness applications (such as network games) on user productivity (see Figure 2-18).

In addition, NTA can help you protect the network against virus attacks. It provides granular, network-wide surveillance of complex, multilayer switched and routed environments and helps you rapidly identify and resolve network threats.

To gather information about your network traffic, NTA supports sFlow, NetFlow, and NetStream. That is, NTA functions as an sFlow collector, a NetFlow collector/analyzer, and a NetStream collector/analyzer.

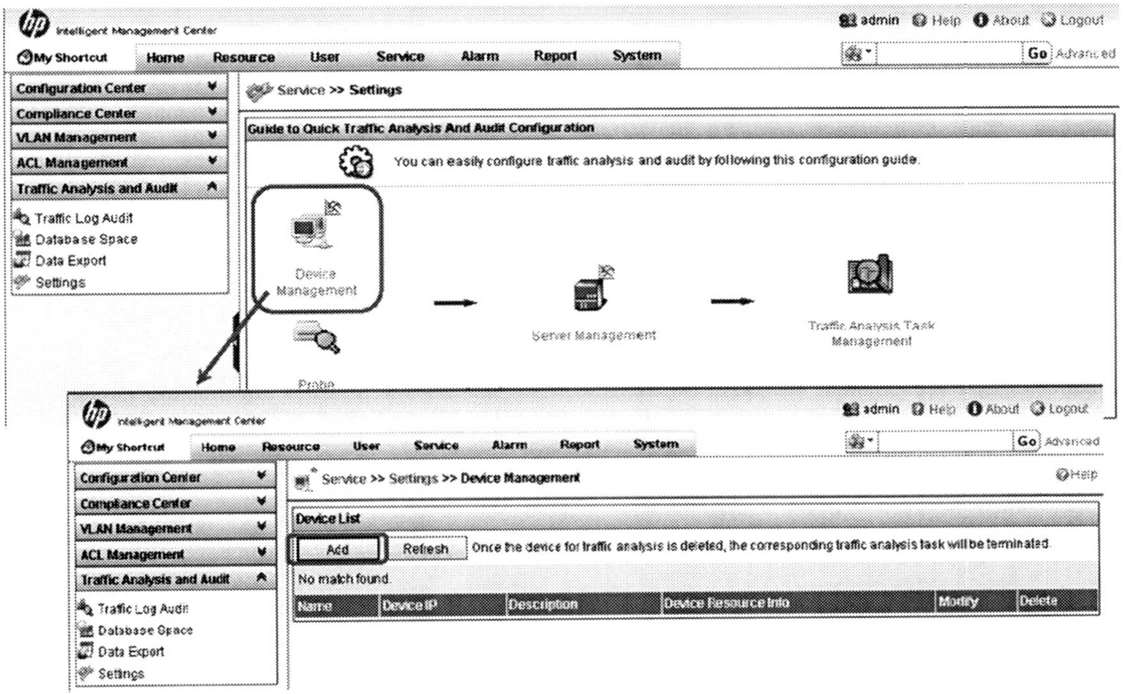

Figure 2-18: IMC Network Traffic Analyzer

Summary

In this chapter, you learned about key technologies that can be used in conjunction with applications and tools to manage your network. You learned about port mirroring, sFlow, NetFlow, RMON, and SNMP.

In addition, you learned how IMC can help you manage the network, and you performed some basic tasks using this management tool. You also learned to use NTA to analyze network traffic.

Learning check

Discuss these questions:

1. What is the process for configuring local mirroring on a ProVision switch?

2. Which statement best describes Remote Monitoring (RMON)?

   a. It collects port statistics and alerts network management solutions, such as Simple Network Management Protocol (SNMP), when particular thresholds are crossed.

   b. It relies on agents that run on network devices to gather information about transmitted packets and send the data to the collector component for analysis.

   c. It includes an exporter that identifies flows, a collector that receives and saves records, and an analyzer that compiles the records and presents them in a meaningful way.

   d. It can identify and gather data about a flow and then export the information either through traditional data export methods or through aggregation data export.

3. What security flaws in SNMP versions 1 and 2c does SNMPv3 address?

4. What are some of the functions the Network Traffic Analyzer (NTA) IMC add-on module provides?

Answers

1. What is the process for configuring local mirroring on a ProVision switch?

   Before you define the originating interface, you enter the **mirror** command, followed by the session number—1 to 4—and then specify the exit port. You then define the source interface for the mirror session and select whether the originating interface is a port, trunk, or VLAN and if you want the mirrored packets to be those inbound to the interface, outbound, or both. You can also specify multiple sessions to send mirrored traffic on two or more exit ports.

2. Which statement best describes Remote Monitoring (RMON)?

   a. It collects port statistics and alerts network management solutions, such as Simple Network Management Protocol (SNMP), when particular thresholds are crossed.

   b. It relies on agents that run on network devices to gather information about transmitted packets and send the data to the collector component for analysis.

   c. It includes an exporter that identifies flows, a collector that receives and saves records, and an analyzer that compiles the records and presents them in a meaningful way.

   d. It can identify and gather data about a flow and then export the information either through traditional data export methods or through aggregation data export.

3. What security flaws in SNMP versions 1 and 2c does SNMPv3 address?

   SNMP communities do not provide strong access control for SNMP agents and servers. In addition, SNMP versions 1 and 2c do not include any security features to protect the data exchanged between the devices and the SNMP server. Because the packets and the community strings are not encrypted, a hacker could intercept and read them. And because these versions do not provide message integrity measures, a hacker could intercept messages, tamper with the data, and then send it to the SNMP server.

4. What are some of the functions the Network Traffic Analyzer (NTA) IMC add-on module provides?

   It provides real-time information about users and applications consuming network bandwidth. It helps users to plan, monitor, enhance, and troubleshoot networks, as well as identify bottlenecks and apply corrective measures.

   In addition, NTA can help protect the network against virus attacks. It provides granular, network-wide surveillance of complex, multilayer switched and routed environments and helps you rapidly identify and resolve network threats.

3 Data Link Layer Redundancy

EXAM OBJECTIVES

In this chapter, you learn to:

- ✓ Explain how Per-VLAN Spanning Tree Plus (PVST+) and Rapid PVST+ (RPVST+) enhance convergence times and allow more efficient use of bandwidth, as compared to STP.
- ✓ Compare PVST+ and RPVST+ with MSTP.
- ✓ Describe how Unidirectional Link Detection (UDLD) and Device Link Detection Protocol (DLDP) detect and handle unidirectional links.
- ✓ List each component in an HP distributed trunk and describe its role in establishing and maintaining the distributed trunk.
- ✓ Configure distributed trunking on an HP ProVision switch that supports this feature.

ASSUMED KNOWLEDGE

You should know that if a network does not provide redundancy, a link failure could prevent users from accessing essential network services. However, simply adding redundant links can create loops, which result in broadcast storms that make the network inaccessible. You should understand the role of various Spanning Tree Protocols in creating a redundant and loop-free network environment. You should be familiar with Spanning Tree Protocol (STP), Rapid Spanning Tree Protocol (RSTP), and Multiple Spanning Tree Protocol (MSTP).

INTRODUCTION

This chapter describes the role of Per-VLAN Spanning Tree Plus (PVST+), introduces you to Rapid PVST+, and explains how to configure these technologies on ProVision and HP Comware switches.

Link redundancy that relies on spanning tree can sometimes introduce instability and inefficiency. This chapter introduces distributed trunking, which allows two ProVision switches to establish a single link aggregation, eliminating the need for spanning tree on that link aggregation while also potentially increasing throughput and providing redundancy.

Note

Emerging data link layer redundancy and loop-protection mechanisms such as Transparent Interconnection of Lots of Links (TRILL) and Shortest Path Bridging (SPB) are beyond the scope of this study guide.

To remain resilient, a network needs a way to detect problems that can occur on links. This chapter also introduces you to Unidirectional Link Detection (UDLD) and Device Link Detection Protocol (DLDP), which help to detect and mitigate issues that can occur on fiber links. It also describes how to configure these technologies on your HP switches.

STP review

This chapter provides a brief review of the Spanning Tree Protocols: STP, RSTP, and MSTP.

The 802.1D-1990 standard, STP, enables Ethernet bridges, or switches, to:

- Establish a loop-free path that connects every bridge in the network
- Block redundant links
- Automatically bring up the appropriate blocked links if a failure occurs

The first step in the convergence process is for the bridges to elect a root bridge (by transmitting Bridge Protocol Data Units [BPDUs] that contain their priority). The root bridge serves as the central point (or root) of the spanning tree. It becomes solely responsible for generating BPDUs, which help other bridges determine their best path to the root. As downstream bridges receive a BPDU, they propagate the BPDU down the tree, adding the cost associated with the link as they do. The root bridge is also responsible for receiving notifications of topology changes and notifying other switches of any STP status and topology changes. The other devices forward these BPDUs as needed.

If a change occurs in the Layer 2 network—for example, if a link goes down, a new link or switch is added, or a switch fails—the STP algorithm is recalculated and a new loop-free topology is defined.

Review questions

Answer the questions based on what you know about STP.

1. What is the state of an STP port when it first comes up? Can frames received on the port affect the switch's MAC forwarding table? Will the switch flood broadcasts, multicasts, or unknown multicasts out the port? Why?

2. Figure 3-1 shows the states for each of the ports and links on switch C after STP converges. Explain the states in terms of their role in the spanning tree, the BPDUs that they send and receive, and whether they can transition to learning and finally forwarding frames.

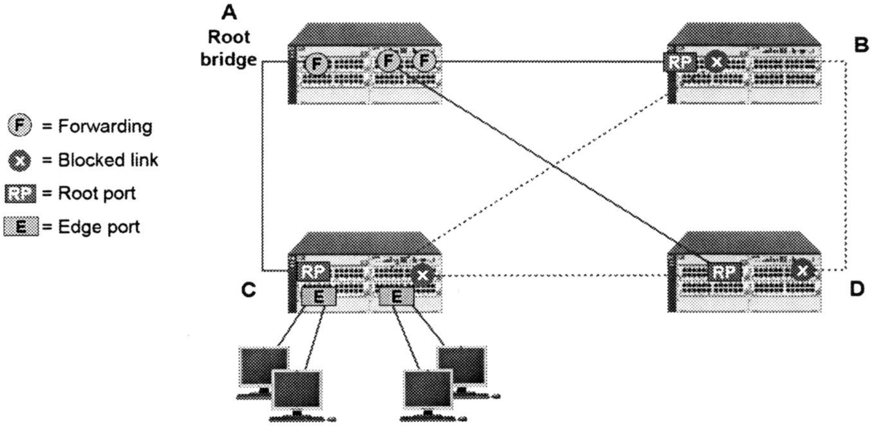

Figure 3-1: STP review

Answers

1. What is a port's state when it first comes up? Can frames received on the port affect the switch's MAC forwarding table? Will the switch flood broadcasts, multicasts, or unknown multicasts out the port? Why?

   When a port on a switch that implements spanning tree first comes up, it listens for BPDUs, but it does not learn MAC addresses or forward traffic. During this state, the switch listens for BPDUs to establish the port's proper role. This ensures that the switch knows whether that port will create a loop *before* it learns the port as the proper forwarding port for a particular MAC address and *before* it starts flooding broadcasts, multicasts, or unknown unicasts out of it.

2. Figure 3-1 shows the states for each of the ports and links on switch C after STP converges. Explain the states in terms of their role in the spanning tree, the BPDUs that they send and receive, and whether they can transition to learning and finally forwarding frames.

   After STP converges, switch C has assigned its ports these roles:

- **Root port**—The port that connects to switch A is a root port. This port offers this switch its lowest-cost path to the elected root. In other words, this port receives the best BPDU as compared to the BPDUs received on other ports. A switch has one and only one root port, and root ports are in the forwarding state (they can learn MAC addresses and forward frames).

- **Designated port**—The port that connects to switch D is a designated port. It offers a better path to the root for switch D than switch D offers to switch C. (This port sends a better BPDU than switch D's port would.) The ports that connect to the computers are also designated ports. Like root ports, designated ports are in the forwarding state (they can learn MAC addresses and forward frames).
- **Blocking port**—All other ports are blocking. Specifically, a blocking port does not offer the lowest-cost path to the root. However, the connected port (switch B) offers a better path to the root for this switch than this switch offers for the connected port. (The port receives a superior BPDU to the one that it would otherwise send.) Note that the blocking port continues to receive this BPDU from the upstream switch, which has a designated port on its side. In this way, switch C knows to continue blocking the port.

  Ports that the switch has selected to become root and designated ports transition to the listening state (intended to prevent the switches from learning MAC addresses while the topology was unstable), the learning state (learning MAC addresses but not forwarding frames), and finally the forwarding state (learning MAC addresses and forwarding frames). Other ports remain in the blocking state in which they neither learn MAC addresses nor forward frames.

RSTP review

Take a moment to review what you know about RSTP (which is outlined in the 802.1w standard):

- RSTP is an extension to STP that uses essentially the same convergence process to elect a root bridge and identify the root path, but enables faster convergence in the event of a change in the topology. Convergence can occur in 1 second or less, but typically occurs within 6 seconds. STP convergence, on the other hand, can take up to 50 seconds.
- RSTP improves convergence in several ways.
- RSTP also adjusted a variety of other behaviors beyond the scope of this review. (For example, designated ports generate their own BPDUs and send them more often. Only more important topology changes are communicated and so on.)
- The RSTP standard also updated path costs to reflect the availability of faster speeds. The various path costs are:
  - STP uses the cost values defined by the 802.1D-1998 standard.
    - Fast Ethernet—19
    - Gigabit—4
    - 10 Gigabit—2

- RSTP and MSTP—which you will review next—use the 802.1t-2001 standard:
  - Fast Ethernet—200,000
  - Gigabit—20,000
  - 10 Gigabit—2,000

ProVision switches use the RSTP/MSTP cost structure by default. Comware switches use their own cost structure by default. In legacy mode, the port calculation for Comware switches is:

- Fast Ethernet—200
- Gigabit—20
- 10 Gig—2

To set the path costs to the 802.1t-2001 standard on Comware switches, you must enter the system command **stp pathcost-standard dot1t**.

MSTP review

Finally, you will review how MSTP (802.1Q-2005) differs from both STP and RSTP and why.

STP and RSTP operate without regard to a network's VLAN configuration: They maintain one Common Spanning Tree throughout a bridged network, and switches configured to use these protocols must forward all traffic in that spanning tree. If a link is blocked, therefore, it is blocked for all VLANs. Depending on how much traffic is generated, active links can become congested while blocked links remain idle. In addition, with STP and RSTP, more traffic is forwarded over links closer to the root bridge, which increases the demand on root bridge resources.

With users putting more demands on network bandwidth, companies cannot afford to have idle links. They want to use redundant links to transmit traffic, while still being able to rely on the links for redundancy.

MSTP creates multiple spanning trees in a network, which significantly improves network resource utilization while maintaining a loop-free environment. Each MSTP spanning tree instance can have a different root bridge and block different redundant links, and MSTP applies a per-instance forwarding topology. Thus, a port may be dynamically blocked in one spanning tree instance but in a forwarding state in another instance.

Each spanning tree instance can include one or more VLANs. Figure 3-2 shows two different views of the same network:

- Instance 1 supports VLANs 1, 8, 10, and 16.
- Instance 2 supports VLANs 4, 5, and 6.

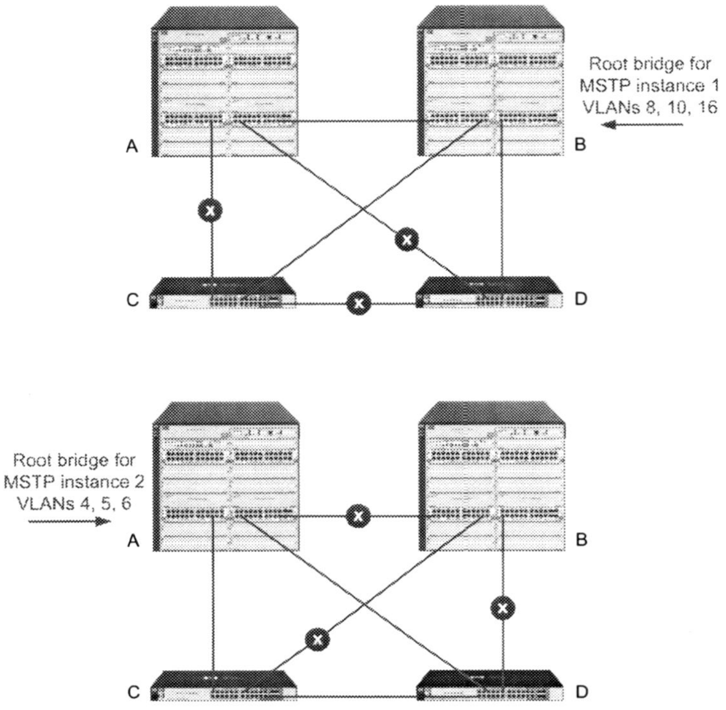

Figure 3-2: MSTP review

When the switch needs to flood a frame assigned to a particular VLAN, it determines which ports are forwarding by referring to the port's state in the instance to which the VLAN is associated.

Switches use MSTP BPDUs, which are backward compatible with RSTP and STP BPDUs to build and maintain the multiple spanning trees. A single untagged BPDU carries information for every instance. This enables load sharing across the network while keeping the switch's CPU load at a moderate level. You should be aware that although the instance specifies the associated VLANs, the instance spanning tree is built without regard to the VLANs that are actually carried on the ports. To ensure a true loop-free topology, every switch must agree on the tree that carries traffic for a specific VLAN. Therefore, MSTP only operates within an MSTP region, in which every switch uses the same region name, configuration revision number, and VLAN-to-instance mappings.

In the example network, switch B is the root bridge for instance 1, and switch A is the root bridge for instance 2. Each instance has different active links. For example, the link between switch A and switch C is active for instance 2, and the ports are in a forwarding state; the link is blocked for instance 1, and the ports are in a discarding state.

MSTP uses RSTP mechanisms for fast convergence. It is also completely interoperable and compatible with RSTP and STP and will emulate the behaviors of these protocols when encountering devices that run them. Similarly, it uses RSTP behaviors to interoperate with switches in a different MSTP region.

Not all Cisco switches and IOS versions support MSTP. Most companies with Cisco networks use the default mode on Cisco, the proprietary protocol, PVST+ or Rapid PVST+ (RPVST+), a version of PVST+ that is fully compatible with PVST+. Many vendors' devices support these proprietary protocols. Therefore, when administering a heterogeneous network, network administrators who are very familiar with PVST+ may not wish to switch to MSTP even if the latter is an option.

PVST+ spanning tree instances

Rather than implementing MSTP, some companies use RPVST+. You can implement RPVST+ on HP switches so that you can deploy them in heterogeneous environments that implement this protocol or PVST+. The HP switches' support for RPVST+ depends on the model and software version:

- ProVision switches running K.15.09 software or higher support RPVST+.
- RPVST+ support was also added to HP Comware v5.

Note

This information is accurate as of the publication of this study guide; however, supported features can change. You should refer to the HP networking website to find and check the features supported on particular switch models in the latest software update.

Cisco based PVST+ on STP (the standard at the time PVST+ was developed) but enhanced its functionality to address two needs:

- Allowing the use of redundant links during normal operation, easing overburdened networks
- Speeding STP's slow convergence

This chapter first examines the way in which PVST+ handles overburdened networks.

As you just reviewed, STP and RSTP run one spanning tree instance: Redundant links cannot be blocked based on VLAN, and frames belonging to all VLANs are forwarded along the same spanning tree. MSTP can run one or more instances, with one or more VLANs in each spanning tree.

Like MSTP, PVST+ can establish multiple spanning trees (see Figure 3-3). With PVST+, each VLAN has a root bridge and a spanning tree rooted at that bridge. By establishing different root bridges on different VLANs, you can ensure that a downstream switch can forward traffic in some VLANs on one uplink and traffic other VLANs on another link. Although multiple links forward traffic, each VLAN (broadcast domain) remains free of loops.

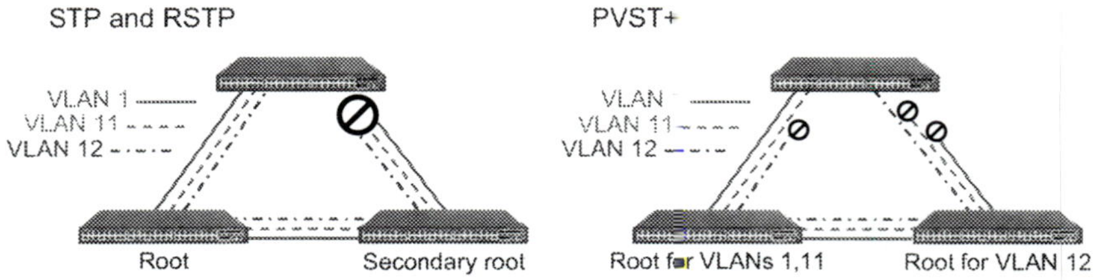

Figure 3-3: Comparing STP and RSTP with PVST+

Thus, PVST+ offers some of the same benefits that MSTP does in terms of load balancing. However, PVST+ must always create one spanning tree for each VLAN, even if many VLANs use the same paths, as is often the case. Therefore, the load balancing comes with some trade-offs; the switches need to process more BPDUs and maintain more spanning trees, creating a higher CPU utilization that can cause issues.

Background information about PVST

Cisco actually developed PVST before PVST+. PVST switches used Cisco Inter-Switch Link (ISL) encapsulation to carry information about frames' VLAN assignment rather than 802.1Q VLAN tags. Therefore, switches using PVST could not interoperate with open-standard-based switches using 802.1Q and STP.

Cisco developed PVST+ as an extension to PVST that enabled such interoperability. Rather than running on ISLs, PVST+ runs on 802.1Q trunks (tagged links), defined as Ethernet links that use 802.1Q tags to identity frames' VLANs.

PVST+ became the default protocol for Cisco switches when Cisco discontinued ISL.

Spanning tree and PVST+ BPDUs

You can better understand how PVST+ establishes the spanning trees by examining its BPDUs and comparing them with the standard BPDUs.

STP (802.1D-1990), RSTP (802.1D-2004), and MSTP (802.1Q-2004) BPDUs use a standard bridge multicast MAC address—01:80:c2:00:00:00. All switches that run STP, RSTP, or MSTP examine BPDUs destined to this MAC address.

RSTP BPDUs are backward compatible with STP BPDUs, and MSTP BPDUs are backward compatible with both. When a device that supports only RSTP or STP receives an MSTP BPDU, it can interpret all of the BPDU except the MSTP-specific data.

MSTP BPDUs (as well as STP and RSTP BPDUs) are sent untagged and are not attached to any VLAN. An HP switch that implements one of the standard Spanning Tree Protocols always sends the untagged BPDUs, even if the port does not have an untagged VLAN assignment (unless spanning tree is specifically disabled or filtered on that port).

Note

The fact that MSTP BPDUs contain information about all instances reduces overhead for BPDU management. There is no need to restrict the number of VLANs created per switch.

PVST+, on the other hand, is not part of the STP IEEE 802.1D standard, and PVST+ BPDUs are different from the standard BPDUs. First, a port on a switch that implements PVST+ sends and receives *multiple* BPDUs, one for each VLAN and spanning tree.

The PVST+ BPDUs use a Cisco destination MAC address, multicast address 01:00:0c:cc:cc:cd, as shown on the right side of Figure 3-4. The BPDU is tagged or untagged based on the configuration of the port. Because a port supports at most one untagged VLAN, most BPDUs are tagged. Often the untagged VLAN is VLAN 1, so the PVST+ port sends the VLAN 1 BPDU as untagged. However, this is not required.

However, VLAN 1 does have a special role: PVST+ switches use this VLAN to transmit STP BPDUs for compatibility with STP. The PVST+ port sends the STP BPDUs to the standard multicast MAC address 01:80:c2:00:00:00. To send these BPDUs, the port must support VLAN 1 (whether tagged or untagged).

You will look at more detailed information about the precise BPDUs transmitted by an HP switch port when you examine RPVST+ BPDUs, as HP switches support this version of PVST+.

What is most important for you to understand is that:

- As Figure 3-4 shows, PVST+ uses *different* BPDUs to construct the spanning tree for each VLAN.
- The BPDUs carried on each port depend on the VLANs supported on those ports.
- Therefore, the topologies constructed by PVST+ are VLAN aware. In other words, the spanning tree for a particular VLAN only extends as far as that VLAN extends. This provision can make it easier for you to design the network.

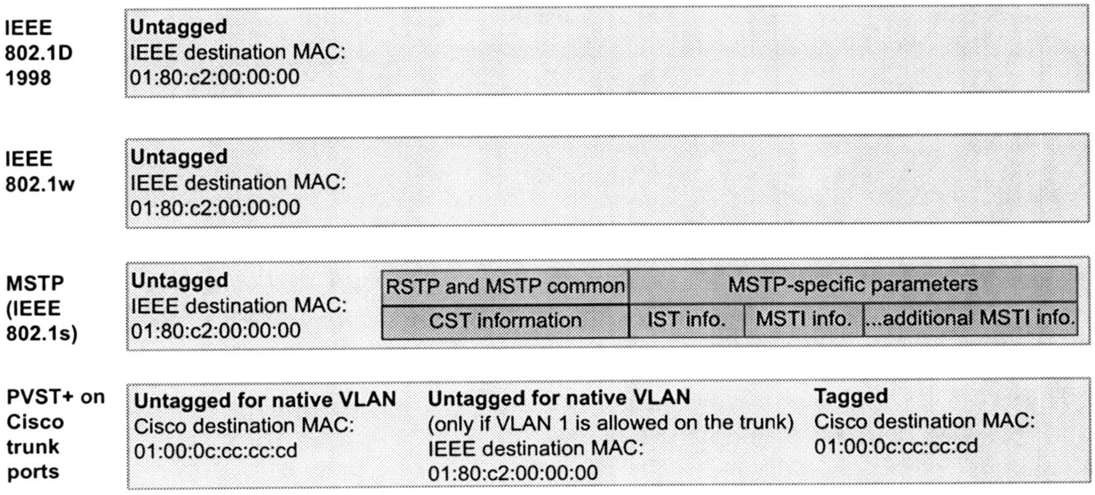

Figure 3-4: Spanning tree and PVST+ BPDUs

Important

Because PVST+ BPDUs are sent in each VLAN, they create overhead. It is often recommended that you reduce the number of VLANs on the switch to reduce overhead. The forwarding and reception of PVST+ BPDUs depends on whether a VLAN exists on a particular link.

Other PVST+ enhancements

You now understand how PVST+ provides load balancing on redundant links to relieve overburdened networks (in much the same manner as MSTP, the open-standard protocol, is designed to meet many of the same needs). The second goal of PVST+ is to speed up convergence, and it achieves this using the following Cisco proprietary technologies:

- PortFast
- UplinkFast
- BackboneFast

Although the HP switches do not support these technologies, RSTP incorporated similar features, so standard Spanning Tree Protocols now have much the same functionality without additional configuration.

PortFast

PortFast is suitable only for switch ports connected directly to a single workstation or server and is roughly equivalent to the edge port definition in RSTP and MSTP. This technology causes a Layer 2 LAN interface configured as an access or trunk port to enter the forwarding state immediately, bypassing the listening and learning states and enabling the device to immediately connect to the network rather than waiting for the spanning tree to converge.

UplinkFast

UplinkFast, implemented on ports that connect to upstream switches in the spanning tree, provides faster convergence for those ports.

As you recall, if a switch's root port fails and the switch is using 802.1D, the switch cannot determine by itself which port it should open and which it should leave blocked to avoid loops. The convergence process must be initiated, and convergence can take up to 50 seconds to complete.

RSTP (and MSTP), as you reviewed earlier, helped to speed convergence in this situation by replacing the "blocking" port role with the "backup" and "alternate" port roles. The alternate port rapidly provides an alternate path in the direction of the root bridge in the event that the root port fails.

UplinkFast speeds up convergence for PVST+ in a similar fashion. UplinkFast improves convergence time by determining the possible paths to the root bridge before there is an uplink failure. It lets a switch understand that it has a set of uplinks that potentially provide access: the root port and an alternate port (or alternate ports). When the root port fails, the alternate port with the next best path to the root immediately comes up rather than transitioning through all the STP states. The new path can begin to transmit traffic in about 1 to 5 seconds.

BackboneFast

Earlier you read about how RSTP takes advantage of point-to-point links to help new designated ports to rapidly transition to the forwarding state. Such a port completes a proposal agreement handshake to ensure that it does not come up until the downstream switch has guaranteed a loop-free topology. There is no longer a need to wait for protocol timers to expire.

Cisco based BackboneFast on the same logic.

As you recall, a switch can be signaled to an indirect failure (a failure of the root port on their designated bridge) by unexpectedly receiving an inferior BPDU on blocking port.

Because inferior BPDUs signal a failure in the network, these BPDUs could help the network recover from the failure more quickly. Ignoring inferior BPDUs is an important part of STP stability, however, because it is impossible to validate if the inferior BPDU information is based simply on locally available information.

The BackboneFast feature introduces a mechanism to explicitly verify inferior BPDU information. Essentially, a switch that receives an inferior BPDU sends queries on all other ports on which it has been receiving BPDUs. The queries propagate back to the root, and switches wait for responses confirming that the root is indeed still up. While slightly different from RSTP's agreement proposal messages, the effect is somewhat similar. What is important to understand is that because the switch can determine whether or not it has a link back to its current known root switch, it can ignore the max age timer and start the listening and learning process on a port immediately upon receiving an inferior BPDU.

Introduction to RPVST+

RPVST+ offers faster convergence for PVST+ without the need for the proprietary PortFast, UplinkFast, or BackboneFast features. Basically, RPVST+ is RSTP operating per-VLAN in a single Layer 2 domain.

As with PVST+, each RSTP spanning tree can have a different root switch and span the network through shared or different links. A redundant link might be active for one topology but blocked for another. In the simple network shown in Figure 3-5, for example, the link between switch B and switch C is blocked for VLAN 10 but in a forwarding state for VLAN 20. Likewise, the link between switch A and switch C is blocked for VLAN 20 but active for VLAN 10.

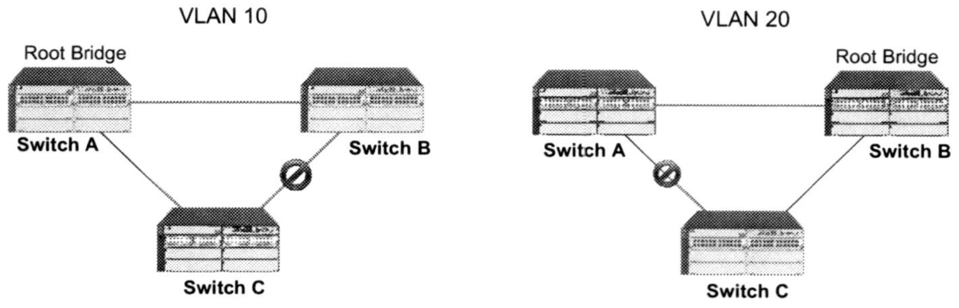

Figure 3-5: PVST+: Allowing different root switch for each VLAN

Because there is a separate spanning tree for each VLAN, topology changes affecting a particular VLAN are propagated only inside that VLAN. However, in a network with a large number of per-VLAN spanning tree instances, RPVST+ can increase the load on the switch's CPU. As a result, in networks with 100 or more VLANs, MSTP is the recommended spanning tree choice.

If you want to implement a feature like PortFast option in RPVST+, you can define a port as an edge port, manually or automatically. With RPVST+, you do not need to take any steps to obtain the basic functionality of UplinkFast and BackboneFast. Remember, similar behaviors are built into RSTP, and RPVST+ uses RSTP to regulate the behavior within each VLAN's spanning tree.

When you configure an HP switch in PVST+ mode, it always uses RPVST+, ensuring fast convergence. In contemporary networks, many PVST+ switches use the RPVST+ version. However, if the HP switch does connect to a switch that uses only basic PVST+, the HP switch interoperates in much the same way that an RSTP switch interoperates with an STP switch.

RPVST+ BPDUs

An HP switch that runs RPVST+ on a VLAN sends RPVST+ BPDUs in that VLAN. It sends them as tagged or untagged frames, depending on the VLAN setting for the transmitting port. These BPDUs use the same MAC address as the PVST+ BPDUs. However, their protocol version indicates Rapid Spanning Tree.

To provide compatibility with RSTP, RPVST+ switches also send RSTP BPDUs in VLAN 1. To send these BPDUs, a ProVision port or a Comware trunk port *must* support VLAN 1. VLAN 1 can be either tagged or untagged on the port, but the switches always send the RSTP BPDUs as untagged frames.

Consider the example port on the root bridge for the RPVST+ network shown in Figure 3-6. This port is configured to support untagged traffic for VLAN 1 and tagged traffic for VLAN 15. An equivalent Comware port would be a trunk port with PVID 1 that permits VLANs 1 and 15. In either case, the port sends the following BPDUs:

- RSTP BPDUs, which are untagged frames in VLAN 1 (of course, untagged frames do not include VLAN information but the BPDUs use VLAN 1's costs and priorities)

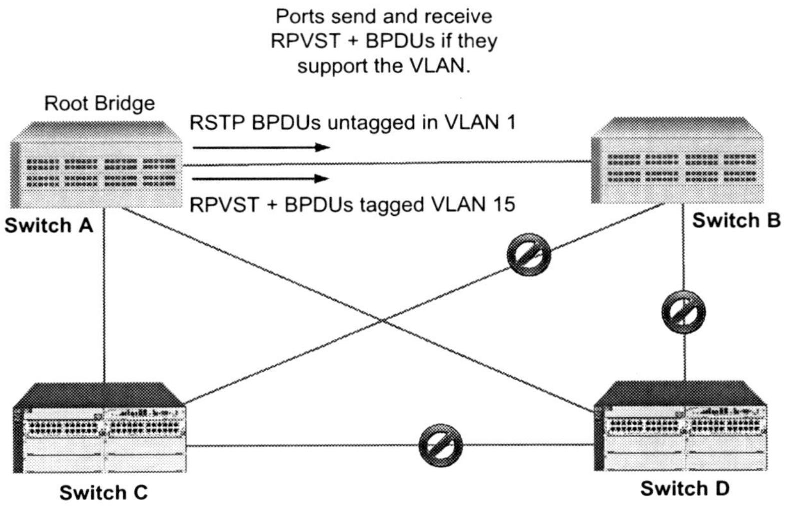

Figure 3-6: RVST+ BPDUs

- RPVST+ BPDUs, which are untagged frames in VLAN 1
- RPVST+ BPDUs, which are tagged frames in VLAN 15

If the port supports tagged traffic for VLAN 1 and untagged traffic for VLAN 15, on the other hand, it would send the following BPDUs:

- RSTP BPDUs, which are untagged frames in VLAN 1
- RPVST+ BPDUs, which are untagged frames in VLAN 15
- RPVST+ BPDUs, which are tagged frames in VLAN 1

A Comware access port behaves a bit differently. It sends untagged RPVST+ BPDUs in the port's VLAN, but it also sends untagged RSTP BPDUs in that VLAN *even if that VLAN is not VLAN 1.* (However, switch-to-switch ports are usually trunk ports.)

Interoperation between protocols

Until now, you have considered HP switches that are all running RPVST+. Sometimes, however, you might need to deploy some HP switches that run RSTP or MSTP as well as some HP switches that implement RPVST+. These protocols can interoperate, but you need to plan carefully.

Two primary factors govern interoperability:

- First, the chosen Spanning Tree Protocols must use a standard IEEE untagged BPDU format.
- Second, the untagged BPDU in the correct format must have the IEEE standard destination MAC address.

As you learned, an HP switch port that implements RPVST+ can transmit an untagged RSTP BPDU, which fits these requirements. You just need to ensure that the ProVision port supports VLAN 1, that the Comware port is an access port, or that the Comware port is a trunk port that permits VLAN 1.

In any of these cases, the HP switches use these BPDUs to establish the Common Spanning Tree (CST) defined in STP, RSTP, and MSTP. The CST extends across the RPVST+ network and the STP/RSTP/MSTP network.

Because RPVST+ BPDUs use a different MAC address, they are only understood by switches running PVST+ or RPVST+. If an STP/RSTP/MSTP switch receives a tagged PVST+/RPVST+ BPDU on a port that supports the VLAN ID, it will not examine the content, but instead forward it, flooding it as a regular multicast. Therefore, the PVST+/RPVST+ per-VLAN spanning trees treat the STP/RSTP/MSTP segment like a shared Ethernet link with one designated port and multiple alternate and backup ports.

Note

If the network includes any Cisco switches, you should note that Cisco switches require the switch ports to support the native VLAN in order to transmit the standard untagged BPDU.

Path cost differences

When implementing any version of STP, you should consider the port cost differences for the STP protocols, as shown in Figure 3-7. ProVision switches that run RPVST+ use the 802.1t-2001 standard (200,000/20,000/2,000) by default. No matter what the STP mode, on the Comware switches, the default type for path cost is "legacy," but you can set the path costs to the 802.1t-2001 standard with the global command **stp pathcost-standard dot1t**.

You can also set both ProVision and Comware switches to use the 802.1D costs although the newer costs are recommended.

Path Cost	STP/PVST+	RSTP/MSTP
Fast Ethernet	19	200,000
Gigabit	4	20,000
10 Gigabit	2	2,000
Enabling	802.1D-1998	802.1t-2001 (must use **stp pathcost-standard dot1t** command on Comwareswitches)

Figure 3-7: Path costs

Configuring support for RPVST+

The following sections provide an overview for configuring RPVST+ on Comware and ProVision switches.

Configuring RPVST+ support on HP Comware switches

On HP Comware switches, STP is not enabled by default. To enable support for RPVST+, you should follow these steps:

1. Configure all the VLANs required on the switch. Set up the ports that will connect to other switches as trunk ports with the appropriate VLANs permitted.

2. Enable STP:

   ```
   [Comware]stp enable
   ```

3. Change the STP mode to RPVST+ (the default mode on Comware switches is MSTP):

   ```
   [Comware]stp mode pvst
   ```

   At this point, the switch automatically attempts to establish a spanning tree in every VLAN. You can manually disable (and re-enable) RPVST+ on a VLAN, however:

   ```
   [Comware][no] stp vlan 2 enable
   ```

4. On most switches, you can accept the default bridge priority for all VLANs. If the switch is the root or secondary root for any VLANs, set the priority with this command. You can set the priority for multiple VLANs at a time:

   ```
   [Comware]stp vlan <vlan-list> priority <priority value>
   ```

   As in STP, valid priority values range from 0 to 61440 and must be set in steps of 4096. The lower the value, the higher the switch's priority in being elected root bridge for that VLAN.

5. It is recommended that you change the path cost to the RSTP standard:

   ```
   [Comware]stp pathcost-standard dot1t
   ```

   You can also complete optional configurations such as adjusting port costs manually, adjusting timers, and so on. Refer to the configuration guide for more details.

6. You can view settings with commands such as these:

   ```
   [Comware]display stp brief
   VLAN   Port                             Role   STP State   Protection
     1    GigabitEthernet1/0/1             DESI   LEARNING    NONE
     1    Ten-GigabitEthernet1/0/25        DESI   LEARNING    NONE
     2    GigabitEthernet1/0/1             DESI   LEARNING    NONE
     2    Ten-GigabitEthernet1/0/25        DESI   LEARNING    NONE
   ```

```
[Comware]display stp [vlan <ID>]
-------[VLAN 1 Global Info]-------
Protocol Status          :enabled
Bridge ID                :4096.0023-89d9-c398
Bridge Times             :Hello 2s MaxAge 20s FwDly 15s
Root ID / RPC            :4096.0023-89d9-c398 / 0
RootPortId               :0.0
BPDU-Protection          :disabled
TC or TCN received       :0
Time since last TC       :0 days 0h:0m:57s
```

Note

The maximum number of VLANs for which RPVST+ can maintain instances varies by switch model. Consult the configuration documentation for your switch for more information.

When configuring PVST+ on devices of different models in a network, to avoid network failures, make sure that the number of VLANs for which PVST maintains instances does not exceed the lowest number supported.

Configuring RPVST+ support on HP ProVision switches

To configure RPVST+ on HP ProVision switches, follow the succeeding steps. Consult the configuration documentation for your particular switch for more information:

1. Configure all the VLANs required on the switch. Set the appropriate untagged and tagged VLAN memberships on ports that will connect to other switches.
2. Verify that spanning tree is enabled on the switch:

```
ProVision(config)# spanning-tree
```

3. Specify RPVST+ as the spanning tree mode:

```
ProVision(config)# spanning-tree mode rapid-pvst
```

4. On most switches, you can accept the default bridge priority for all VLANs. If the switch is the root or secondary root for any VLANs, set the priority with this command. You can set the priority for multiple VLANs at a time:

```
ProVision(config)# spanning-tree vlan <vlan list> priority <prior-
ity_value | root <primary | secondary>]
```

On the ProVision switches, you set the priority value between 0 and 15, and the priority is multiplied by 4096. You can also set the switch as the primary root (priority 0) or the secondary root (priority 1 or 4096).

You can also complete optional configurations:

- Change the path cost mechanism.
- Customize the RPVST+ options for each VLAN.
- Configure per-port options.

Refer to the configuration guide for more details.

5. You can view STP settings and port states with commands such as these:

```
ProVision-1(config)# display stp brief

 VLAN     Port                             Role    STP State     Protection
    1      2                               ROOT    FORWARDING    None
    1      3                               DESI    FORWARDING    None
    2      2                               DESI    FORWARDING    None
    2      3                               ROOT    FORWARDING    None

ProVision-1(config)# show spanning-tree [vlan <ID>]

 Spanning Tree Information

  STP Enabled                    [No] :Yes
  Mode                                :RPVST
  Extended System ID                  :Enabled
  Ignore PVID Inconsistency           :Disabled
  Switch MAC Address                  :0024a8-e5b9c0

  VLAN ID                             :1
  RPVST Enabled                       :Enabled

  Root MAC Address                    :002389-d9c4dc
  Root Priority                       :0
  Root Path Cost                      :20,000
  Root Port                           :2
  Operational Hello Time (secs)       :2
```

```
Topology Change Count          :1
Time Since Last Change         :4 mins
```

Note

The maximum number of VLANs for which RPVST+ can maintain instances varies by switch model. Consult the configuration documentation for your switch for more information.

When configuring RPVST+ on devices of different models in a network, to avoid network failures, make sure that the number of VLANs for which RPVST maintains instances does not exceed the lowest number supported.

Detecting unidirectional links

Unidirectional Link Detection (UDLD) and Device Link Detection Protocol (DLDP) are Layer 2 protocols that allow switches to monitor the physical status of a link. Typically, these protocols are used to monitor fiber-optic links although they can be used to monitor copper-based links as well.

UDLD can also be used to monitor links between two switches that are not directly connected. For example, two switches, perhaps in different buildings, might be separated by a hub or a repeater, making it more difficult for the switches to verify the connection end to end. In fact, the switches might even be separated by other switches that do not run UDLD.

In the first example shown in Figure 3-8, two switches are connected through a fiber-optic link. Because the fibers are connected correctly, the link is bidirectional: One fiber is used for sending frames from the 10500 switch to the 5800 switch; the other fiber is used to send frames from the 5800 switch to the 10500 switch.

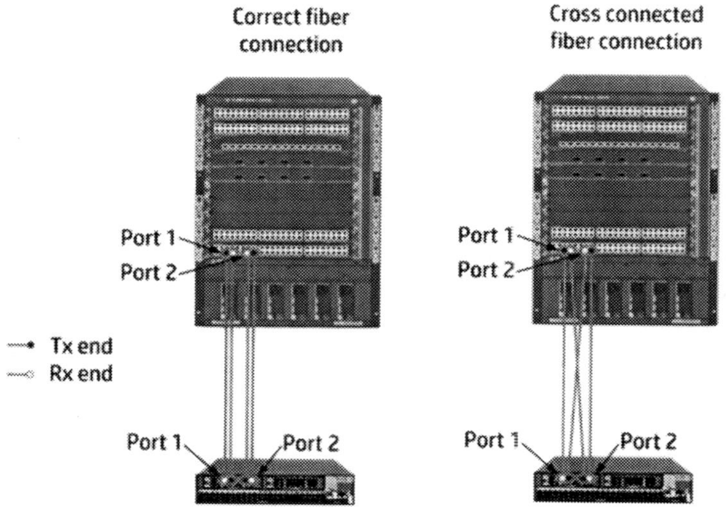

Figure 3-8: Detecting unidirectional links

In the second example in Figure 3-8, however, the fibers are cross-connected, creating a unidirectional link. That is, only one of the two directly connected switches can receive frames from the other directly connected switch. Because the physical connection is still up, however, the switches' normal detection feature may not indicate that the link is down. Although cross-connected fibers are less common of a problem with current cables and transceivers, they still might occur, and you should protect against them. Indeed, the same problem can occur on fiber-optic links if one fiber is disconnected.

Such unidirectional links prevent switches from forwarding traffic as intended and can disrupt the operation of other network protocols. For example, they could cause a loop in a Spanning Tree Protocol (STP) network. If an STP-enabled switch does not receive BPDUs on a link, it determines that it does not have a connection to a switch. With a unidirectional link, however, the switch might be able to forward traffic to the other switch even if it cannot receive the switch's BPDUs. This may cause the switch to open a blocked port, thereby creating a network loop.

UDLD, which runs on ProVision switches, and DLDP, which runs on Comware switches, can detect unidirectional links and shut down the affected ports, thereby safeguarding the network. Because there is no industry standard, vendors have developed several protocols, providing similar capabilities. Unfortunately, the different implementations are not compatible. You can enable UDLD between two ProVision switches and DLDP between two Comware switches, but UDLD and DLDP are not compatible. Also, although the ProVision switches run UDLD, this implementation is not compatible with Cisco's implementation of UDLD.

ProVision switches: UDLD

When you enable UDLD on ProVision switches, each one uses the protocol to identify the switch and the port on the other end of the protected link. You must, therefore, configure the ports on switches to support UDLD. Once the link is established, each switch port sends UDLD frames, as shown in Figure 3-9. These UDLD frames contain the following information:

- Port's own device ID and port ID
- Neighbor's device ID and port ID, which the switch receives through UDLD

When a port receives a UDLD frame, it checks it to ensure that the frame contains its own device ID and port ID. This is often referred to as the "echo" information.

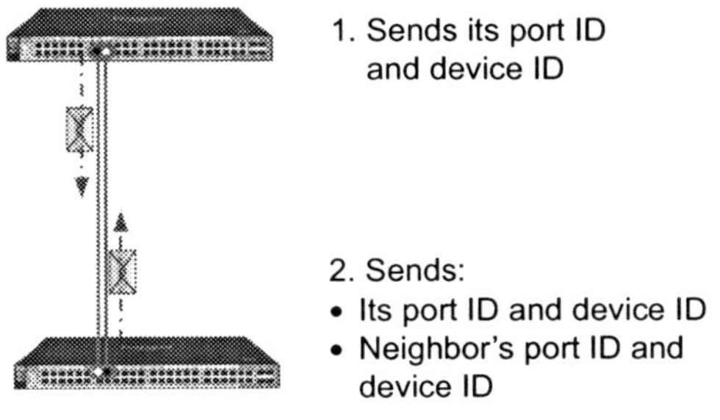

Figure 3-9: UDLD on ProVision switches

If the UDLD frame does not contain this "echo" information, the switch waits for the period of time determined by two settings on the ProVision switches:

- **link-keepalive interval**—This setting determines how frequently the switch sends a UDLD frame. The default setting is every 5 seconds, but you can specify a time from 10 to 100 ms or 1 to 10 seconds. To change the default setting, you use the following command:

```
Switch (config)# link-keepalive interval <interval>
```

- **link-keepalive retries**—This setting determines how many times the switch sends a UDLD frame before determining the link is down and blocking the port. The default is 5 retries. At the default settings, therefore, the switch will wait a total of 25 seconds if it does not receive a valid UDLD frame (5 seconds link-keepalive interval x 5 retries = 25 seconds).

If a port is blocked by UDLD, the switch records the event in its log or sends a Simple Network Management Protocol (SNMP) trap, if you have configured SNMP traps on the switch. The port will remain blocked until:

- The link is unplugged, disabled, or fixed.
- You disable UDLD on the port.

ProVision switches: Guidelines for configuring UDLD

When configuring UDLD, keep the following in mind:

- You must configure UDLD on a per-port basis.
- You must configure and enable UDLD on both ends of the link.
- To configure UDLD on an aggregated link, you must configure the feature on each port individually.
- Dynamic trunking (link aggregation) is not supported.

Note

You can enable UDLD on dynamic link aggregations, and the feature might work. However, this mode of link aggregation is not recommended or supported for UDLD because UDLD requires every member of the link aggregation to support UDLD. In a dynamic link aggregation, a port might be selected that does not support UDLD.

Configuring UDLD

Follow these steps to configure UDLD on a ProVision switch:

1. Enable UDLD on the desired interface or interfaces:

```
ProVision(config)# [no] interface <port-list> link-keepalive
```

2. If you want, adjust the keepalive interval and retry time settings:

   ```
   ProVision(config)# link-keepalive interval <interval>
   ProVision(config)# link-keepalive retries <number>
   ```

   You specify the keepalive interval in 100 ms. For example, **link-keepalive interval 10** sets the interval to 1 second.

Tagged or untagged UDLD frames

Sometimes you might want to implement UDLD on a port that is not an untagged member of any VLANs; it only has tagged assignments. By default, ProVision switches send untagged UDLD frames on such ports. ProVision switches will accept the untagged UDLD frames, but if you are using UDLD to monitor a link between two ProVision switches that are separated by a non-HP device, the non-HP device port may reject these frames.

To prevent this from happening, you can configure the port to send tagged UDLD frames:

```
Switch(eth-<ID>)# link-keepalive interval vlan <vlan_id>
```

You must configure the VLAN that will carry the UDLD frames to extend across all intervening devices. Otherwise, the switches cannot maintain the UDLD-protected link.

When you are setting up UDLD on ports in a link aggregation group (ProVision trunk) and you need to extend UDLD across intervening switches, you should assign a different VLAN for each port. For example, you need to configure UDLD on ports 4 and 5, which are in a link aggregation group (trunk) that supports tagged VLANs 4 and 5. You enter these commands:

```
Switch(eth-4)# int 4
Switch(eth-4)# link-keepalive vlan 4
Switch(eth-4)# int 5
Switch(eth-5)# link-keepalive vlan 5
```

View information about UDLD-enabled ports

To view UDLD information, use these commands:

```
Switch# show link-keepalive
Total link-keepalive enabled ports: 4
Keepalive Retries: 3 Keepalive Interval: 1 sec
```

```
Port     Enabled    Physical    Keepalive    Adjacent        UDLD
                    Status      Status       Switch          VLAN
--------------------------------------------------------------------
1        Yes        up          up           00d9d-f9b700    200
2        Yes        up          up           01560-7b1600    untagged
3        Yes        down        off-line
4        Yes        up          failure
5        No         down        off-line

Switch# show link-keepalive statistics
Port:                     1
Current State:            up          Neighbor MAC Addr: 0000a1-b1-c1d1
Udld Packets Sent:        1000        Neighbor Port:     5
Udld Packets Received:    1000        State Transitions: 2
Port Blocking:            no          Link-vlan:         1
```

Comware switches: DLDP

As Figure 3-10 shows, DLDP provides the same basic functionality as UDLD. Switches use DLDP to identify a neighbor device and monitor a link. If the link becomes unidirectional, DLDP shuts down the port.

DLDP works with the physical layer protocols to determine if the ports at both ends of the link are operating properly at that layer. If they are, DLDP verifies that the link is operating properly at the data link layer.

You can configure DLDP to operate in one of two modes on a link: normal mode or enhanced mode.

In normal mode, DLDP can detect fiber links that are cross-connected, resulting in unidirectional traffic. In addition to cross-connected fibers, DLDP in enhanced mode can detect fiber-optic connections that have one broken fiber or one disconnected fiber.

You can see the difference between normal and enhanced mode by reviewing what happens if a switch detects a new neighbor connected to the DLDP-enabled port. When that happens, the switch—

whether in normal mode or enhanced mode—starts the DLDP entry timer. The next steps the switch takes vary, depending on whether DLDP is operating in normal mode or enhanced mode:

- **Normal mode**—If DLDP is in normal mode and the switch does not receive a DLDP frame with the neighbor's information on the DLDP-enabled port, it begins sending DLDP advertisement frames that include the Re-synchronization (RSY) tag. The switch also removes information about the neighbor from its internal tables. The switch moves the active state.
- **Enhanced mode**—If DLDP is operating in enhanced mode and the switch does not receive a DLDP frame before the entry timer expires, it starts the enhanced timer. The switch then sends up to eight DLDP probe frames to the neighbor, sending one probe frame per second. If the switch does not receive a DLDP echo frame from the neighbor switch, it moves to the disable state.

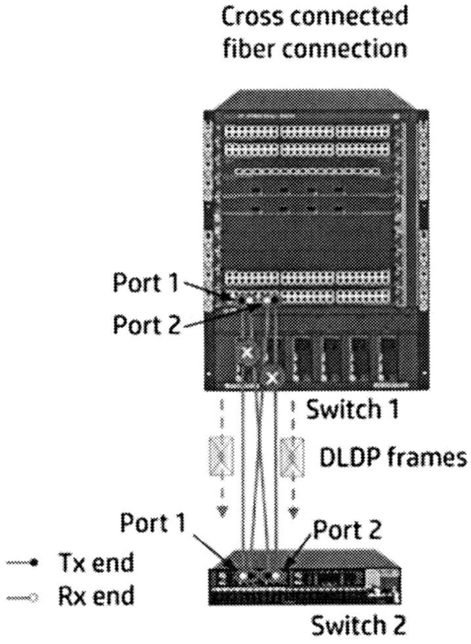

Figure 3-10: DLDP on Comware switches

Comware switches: DLDP states and timers

When a port is enabled to support DLDP, it is in one of the DLDP link states outlined in Table 3-1. In general, the switch advertises information about itself or requests information about its neighbor. In most DLDP states, the switch sends a DLDP frame and sets a DLDP timer, which determines how long it waits for its neighbor to respond. The type of frame the switch sends and the timer it sets depend on its DLDP state.

Table 3-1: DLDP link states and associated timers

DLDP states	Description of DLDP state	DLDP frames sent	DLDP timers set	Description of DLDP timer
Initial state	DLDP is disabled	Flush frames (see note after table)		
Active state	DLDP is enabled; the link is up. A switch may also be in an active state if its neighbor entries have been cleared	Advertisement frame with RSY tag	Active timer	Establishes interval for sending advertisement frames with RSY tags. Default interval: 1 second Max. # of advertisement frames with RSY tags: 5
Advertisement state	Link is bidirectional, or DLDP has been in active state for more than 5 seconds	Normal advertisement frame	Advertisement timer	Determines the interval for sending common advertisement frames. Default interval: 5 seconds
Probe state	Switch received a frame from an unknown neighbor. Switch sends DLDP frame to check if the link is unidirectional	Probe frame	• Probe timer • Echo timeout timer	• Determines interval of probe frames. Default setting: 2 probe frames every second; maximum # of probe frames: 10 • Triggered when switch is in probe state of enhanced mode is detected. If switches does not receive echo frame before echo timer expires, line is unidirectional and state is set to disable
Disable state	Port enters this state when: • A unidirectional link is detected • The contact with the neighbor in enhanced mode gets lost Port does not receive or send packets other than DLDPDUs	Disable frame and the RecoverProbe frame	RecoverProbe timer	Determines interval of RecoverProbe frames to determine if unidirectional link is restored. Default: 1 frame every 2 seconds

Continued

Table 3-1: Continued

DLDP states	Description of DLDP state	DLDP frames sent	DLDP timers set	Description of DLDP timer
DelayDown state	Port in active, advertisement, or probe state moves to this state if it detects a port-down event		DelayDown timer	Triggered by a port-down event. If the switch detects a port-up event before the DelayDown timer expires, the port resumes its previous DLDP state. If not, the switch removes the DLDP neighbor information and moves to the inactive state.
Inactive state	DLDP is enabled, but the link is down			

Note

A switch sends DLDP flush frames when it transitions from the active, advertisement, probe, or DelayDown state to the initial state. A switch does not send DLDP flush frames when it transitions from inactive or disable state to the initial state.

The switch must also process the DLDP frames it receives from its neighbor switch. Table 3-2 outlines how the switch handles different types of DLDP frames.

Table 3-2: Processing DLDP frames received from neighbor

DLDP frame received from neighbor	Processing action	Next step
Advertisement frame with RSY tag	Obtains information about neighbor	• If the switch does not have an entry for the neighbor, it creates it, starts the entry timer, and moves to a probe state • If the switch has an entry for the neighbor, it resets the entry timer and moves to the probe state
Normal advertisement frame	Obtains information about neighbor	• If the switch does not have an entry for the neighbor, it creates it, starts the entry timer, and moves to a probe state • If the switch has an entry for the neighbor, it resets the entry time
Flush frame	Determines if the local port is in disable state	• If yes, no further action is taken • If no, the switch removes the neighbor's entry (if it exists)
Probe frame	Obtains information about neighbor	• If the switch does not have an entry for the neighbor, it creates the entry, moves to the probe state, and sends an echo frame • If the switch has an entry for the neighbor, it resets the entry timer and sends an echo frame

Continued

Table 3-2: Continued

DLDP frame received from neighbor	Processing action	Next step
Echo frame	Obtains information about neighbor	• If the switch does not have an entry for the neighbor, it creates one, starts the entry timer, and moves to a probe state. If the switch has an entry for the neighbor, it will do one of the following: If the neighbor information conflicts with the existing information about the neighbor, the switch drops the frame If the neighbor information does not conflict, the switch sets the flag of the neighbor as bidirectional If its flag is bidirectional, it moves to the advertisement state and disables the echo timer
Disable frame	Determines if the local port is in disable state	• If yes, no further action is taken • If no, the switch moves the port to the disable state
RecoverProbe frame	Determines if the local port is in disable or advertisement state	• If yes, the switch sends RecoverEcho frames to the neighbor • If no, no further action is taken
RecoverEcho frames	Determines if the local port is in disable state	• If yes, the switch checks that the neighbor information is consistent with local port information. If it is, the port moves to the active state • If no, no further action is taken
LinkDown frame	Determines if the local port operates in enhanced mode	• If yes and local port is not in the disable state, the switch moves the local port to the disable state • If no, no further action is taken

Based on the information the switch receives or does not receive from its neighbor, the switch will determine that its neighbor switch is in one of three states:

- **Unknown**—The switch determines its neighbor is in the unknown state if the switch has just detected the neighbor and has sent DLDP probe frames to learn more about it.
- **Two-way**—The switch determines its neighbor is in the two-way state if the switch receives a response from the neighbor.
- **Unidirectional**—The switch determines its neighbor is in the unidirectional state if it determines the link is unidirectional.

To get an idea of which types of frames switches send and when, answer the following questions using Tables 3-1 and 3-2:

1. If a switch port is in the active state, what type of DLDP frame will it send and what timer will it set?

2. If a switch port sends a normal advertisement frame, what state is it in? What type of timer will it set?
3. What happens if a switch port enters the disable state?
4. What action will the switch port take if it receives the following DLDP frames?
   - Advertisement with RSY tag
   - Normal advertisement frame
   - Flush frame
   - Disable frame
   - Probe frame
   - Echo frame

Answers

1. If a switch port is in the active state, what type of DLDP frame will it send and what timer will it set?

   The port sends advertisement frames with the RSY tag. It sends them at the active interval timer (default, 1 second).

2. If a switch port sends a normal advertisement frame, what state is it in? What type of timer will it set?

   The port is in the advertisement state, which is the normal operation DLDP state after the session has been established for as long as no errors are detected. The port uses the advertisement timer as the interval for normal advertisement frames.

3. What happens if a switch port enters the disable state?

   It starts to send disable and RecoverProbe frames in an attempt to receive a response from the other end of the link and to bring the DLDP session back up.

4. What action will the switch take if it receives the following DLDP frames?

   a. Advertisement with RSY tag

      The switch checks for the neighbor's information. It creates an entry for a new neighbor, sets the probe timer, and the port moves to the probe state to attempt to establish the bidirectional session. If the neighbor already exists, the port simply resets the probe timer and moves to the probe state.

   b. Normal advertisement frame

      A normal advertisement frame from a known neighbor prompts the port to update its neighbor entry. Otherwise, the port creates an entry for the neighbor, sets the probe timer, and moves to the probe state.

c. Flush frame

The switch removes the neighbor's entry—unless the port is in the disable state, in which case, the port does nothing.

d. Disable frame

The switch port moves to the disable state unless it is already in that state (in which case it remains in it).

e. Probe frame

The port resets a known neighbor's entry time. It also moves to the probe state and sends an echo frame.

f. Echo frame

If the port has an entry for the neighbor and the information matches, the port declares the link bidirectional and moves to the advertisement state. If the information does not match, the port drops the frame. On the other hand, if the port does not know the neighbor at all, it adds an entry, sets the probe time, and moves to the probe state.

Comware switches: Port shutdown modes

If a DLDP-enabled switch detects a unidirectional link, it can shut down the related ports. By default, UDLD uses auto mode. If a unidirectional link is detected, the port will move to the disable state, generate logs and traps, and set the port state to DLDP Down.

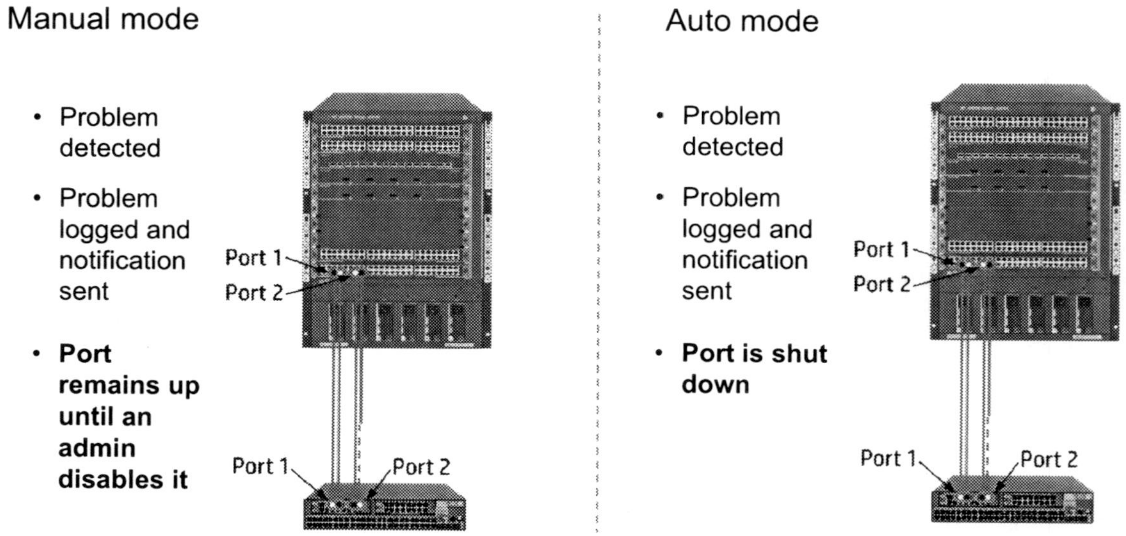

Figure 3-11: UDLD port shutdown modes on Comware switches

If you want to be more cautious about disabling a port, however, you may want to manually shut down a port, rather than having the switch automatically shut down the port. For example, misconfiguring or removing DLDP on one side of the link could cause the other side to detect a unidirectional link falsely. If you are concerned about false unidirectional links, you may want to set the shutdown mode to manual mode. In this mode, DLDP detects unidirectional links and records the event in logs or sends traps to prompt you to shut down the related ports (see Figure 3-11). The switch then moves to the disable state (in which UDLD is not running but the port is still up).

Comware switches: DLDP authentication

If you want to ensure that a switch sends DLDP frames only to authorized DLDP neighbors, you can configure DLDP authentication. This will prevent attackers from using DLDP to gather information about your network that they can use to launch an attack.

You can configure one of the three DLDP authentication modes:

- **Plain text authentication**—When sending DLDP frames, the switch sets the Authentication field to the password configured in plain text and sets the Authentication type field to 1. The switch that receives the DLDP frame checks the values of the two fields and drops any frames in which the two fields conflict with its local authentication settings.
- **MD5 authentication**—When sending DLDP frames, the switch encrypts the user-configured password using MD5, assigns the digest to the Authentication field, and sets the Authentication type field to 2. The switch that receives the UDLD frame checks the values of the two fields and drops any frames in which the two fields conflict with its local authentication settings.
- **Nonauthentication**—When sending DLDP frames, the switch sets the Authentication field and the Authentication type field of DLDP frames to 0. When a switch receives the DLDP frames, it checks the values of the two fields and drops any frames in which the two fields conflict with its local authentication settings.

Both DLDP-enabled switches should be configured to use the same authentication settings.

Comware switches: DLDP configuration steps

To configure DLDP on Comware switches, complete these basic steps:

1. Enable DLDP globally:

   ```
   [Comware] dldp enable
   ```

2. Access the physical interface configuration view for the port on which you want to enable DLDP. (Alternatively, you can configure DLDP on multiple ports in a manual port group.) Enable DLDP on the port or ports:

```
[Comware] interface <interface ID>
[Comware-<interface ID>] dldp enable
[Comware-<interface ID>] quit
```

3. Optionally, adjust the timers. Timers operate on a global basis:

```
[Comware] dldp interval <1-100 seconds>
[Comware] dldp delaydown-timer <1-5 seconds>
```

4. Optionally, configure DLDP authentication, which requires both sides of the link to authenticate to each other:

   - Enable MD5 authentication:

   ```
   [Comware] dldp authentication-mode md5 <password>
   ```

   - Enable simple (plaintext) authentication:

   ```
   [Comware] dldp authentication-mode simple<password>
   ```

   - Disable authentication (default):

   ```
   [Comware] dldp authentication-mode none
   ```

5. Optionally, configure manual mode shutdown so that DLDP simply sends logs rather than shutting down the port on its own:

```
[Comware] dldp unidirectional-shutdown { auto | manual }
```

6. Follow the same steps to configure the other side of the link. You must match the authentication mode.

Distributed trunking on ProVision switches

You will now learn about distributed trunking, a feature that is available on some ProVision switches.

With traditional link aggregation—including the industry-standard Link Aggregation Control Protocol (LACP)—aggregated links are established between two devices, and the physical connections terminate on the same two devices. The HP distributed trunking feature, however, allows you to establish an aggregated link between one downstream device, which is called the distributed trunking device (DTD), and two upstream switches, which are called distributed trunking, or DT, switches. As you can see in Figure 3-12, the DTD forms one side of the aggregated link and the two DT switches form the other side of the aggregated link. To the DTD, the two DT switches appear to be a single device.

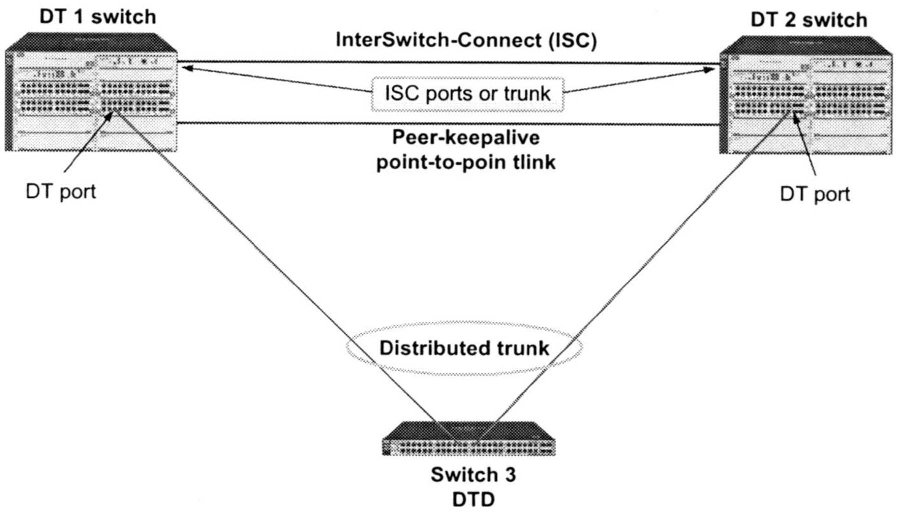

Figure 3-12: Distributed trunking on ProVision switches

Two types of distributed trunking are available: server-to-switch trunking and switch-to-switch trunking. In this chapter, you will examine both types of distributed trunking in more depth.

Distributed trunking is supported on ProVision switches that support the KXX.X and KAXX.X software. For switches that run KXX.XX software, HP networking introduced server-to-switch distributed trunking in software version K.14.09, allowing you to connect a server to two switches to form a distributed trunk. Initially, the feature supported only Link Aggregation Control Protocol (LACP), the industry standard for aggregated links. With K.15.03 or above, however, you can create server-to-switch distributed trunks using HP port trunking as well. The K.15.05.001 software release adds support for switch-to-switch distributed trunking.

The HP 3800 Switch Series runs the KA.XX software and supported server-to-switch distributed trunking in the initial release. The KA.15.09 software release provided support for switch-to-switch distributed trunking.

Server-to-switch distributed trunking

With server-to-switch distributed trunking, you can connect the two DT switches to a server as shown in Figure 3-13. To be part of a distributed trunk, the server must support LACP or be able to form a trunk with a switch that is using HP port trunking. Depending on the server environment, NIC teaming might also be required to provide failover and load balancing on the server. (Check your server documentation.)

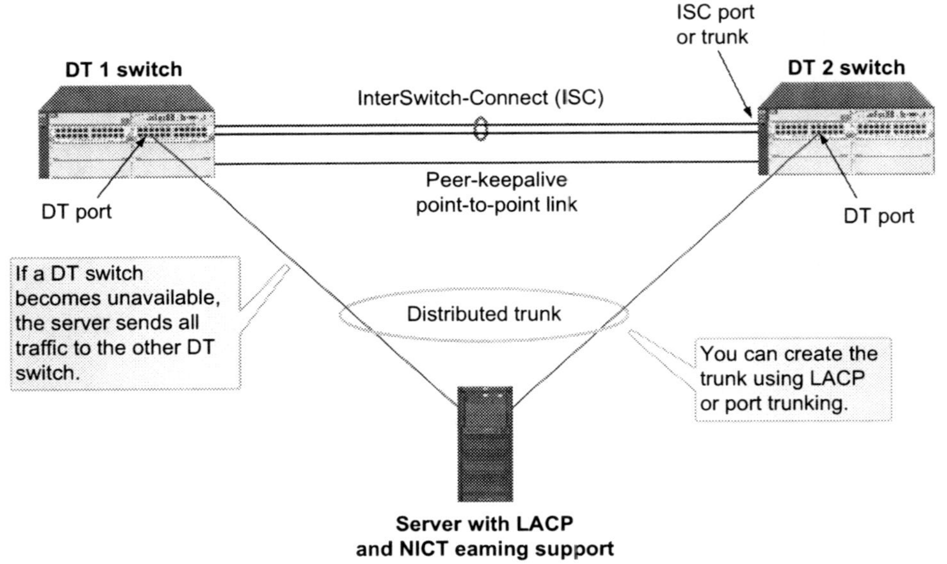

Figure 3-13: Server-to-switch distributed trunking on ProVision switches

Like a traditional aggregated link, the distributed trunk increases bandwidth, thereby improving performance. It further improves performance because the DT switches load share traffic across the links in the distributed trunk and traffic can be processed by one of the two DT switches.

In addition to increasing bandwidth, distributed trunking provides device-level redundancy and link-failure protection. If one of the DT switches becomes unavailable, the DTD continues to send and receive traffic through its connection to the other DT switch. Likewise, if one of the links in the distributed trunk becomes unavailable, the server simply sends and receives traffic over the other links in the distributed trunk.

Switch-to-switch distributed trunking

You can also connect two DT switches to another switch as shown in Figure 3-14. The DTD can be another HP switch (running Comware or ProVision) or a non-HP switch. The switch functioning as the DTD must support LACP or be able to form a trunk with a ProVision switch that uses port trunking.

Later in this chapter, you will examine two examples of distributed trunking: one with a Cisco device and one with an HP Intelligent Resilient Framework (IRF) virtual device.

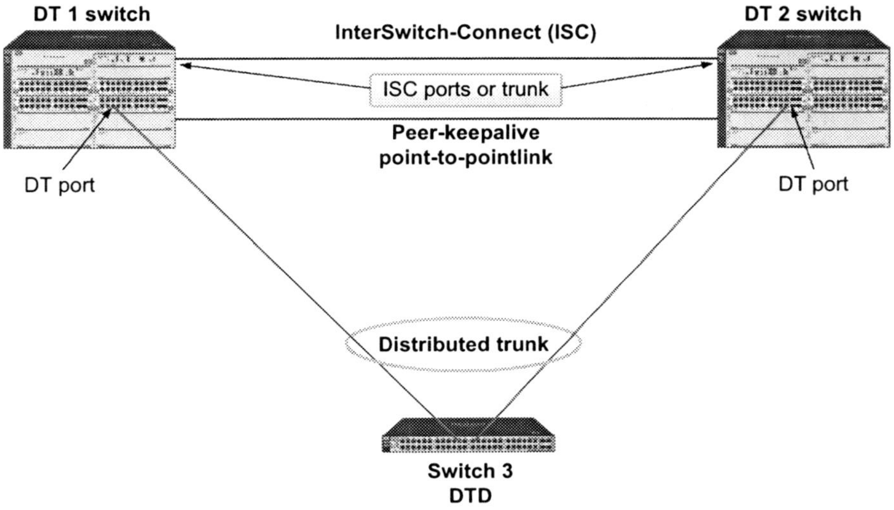

Figure 3-14: Switch-to-switch distributed trunking on ProVision switches

InterSwitch-Connect (ISC)

To appear as a single switch, the two DT switches must communicate frequently, exchanging and synchronizing information. The two DT switches communicate through the InterSwitch-Connect (ISC), which is a point-to-point link between the two switches as shown in Figure 3-15. The ISC can be a single Ethernet connection or a trunk, depending on the volume of traffic and the network topology.

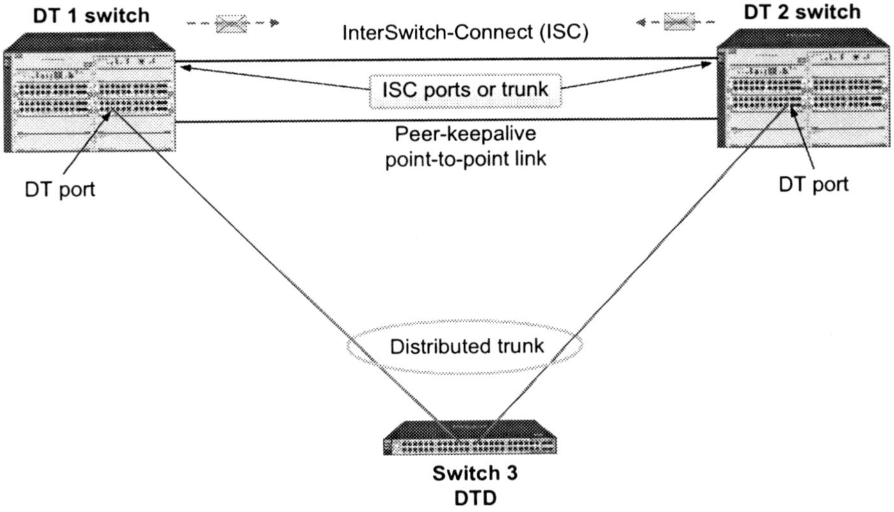

Figure 3-15: InterSwitch-Connect (ISC)

The DT switches use the Distributed Trunking Interconnect Protocol (DTIP) to exchange DT-related configuration information across the ISC. A DT switch sends this configuration information when:

- The distributed trunk is first established.
- The DT switch neighbor requests it.
- A change is made to the switch's DT configuration.

The DT settings on both switches should match.

If the distributed trunk is created through LACP, the DT switches also use the ISC to exchange information about LACP BPDUs that are received on DT ports.

In addition, the switches use the ISC to exchange information about loop-protection management, the Internet Group Management Protocol (IGMP) snooping database, DHCP snooping information, and IP routing—if these features are enabled on the switch. (IGMP snooping allows the switches to "listen" to IGMP traffic. The switch uses this information to determine which endpoints need specific multicast streams. You will learn more about IGMP snooping in "Chapter 11: IP Multicast." DHCP snooping protects against attackers who try to exploit DHCP.)

The DT switches also use the ISC to synchronize the MAC addresses that have been learned or have aged out. This allows each DT switch to forward traffic as efficiently as possible.

The ISC is not reserved solely for managing the distributed trunk, however. The DT switches also forward network traffic over the ISC—using the ISC as needed for efficient traffic flow.

Because the ISC is critical to maintaining the distributed trunk, the DT switches monitor it to ensure it remains up.

Peer-Keepalive link

To protect against an ISC or DT switch failure, you must set up a peer-keepalive link (see Figure 3-16). If such a failure occurs, the DT switches use this link to determine the type of failure that has occurred. For example, has the ISC failed or has one of the DT switches failed? The DT switches can then ensure that traffic to and from the DTD continues to be forwarded to its destination.

If the two DT switches cannot communicate over the ISC, each switch begins a hold timer. This hold timer is designed to reduce false positives (which might occur if the two switches mistakenly detect an ISC failure and immediately begin to send peer-keepalive messages). The hold timer also gives the network time to reconverge as needed if a switch has failed or becomes unavailable.

If the hold timer expires and the two DT switches still cannot communicate over the ISC, they enter the peer-keepalive timeout period. They begin to send peer-keepalive messages to determine if the failure occurred at the ISC or switch level. If a DT switch *receives* a peer-keepalive message during this peer-keepalive timeout, it determines that the failure occurred at the ISC level. If a DT switch does not receive a peer-keepalive message, it determines the other DT switch has failed.

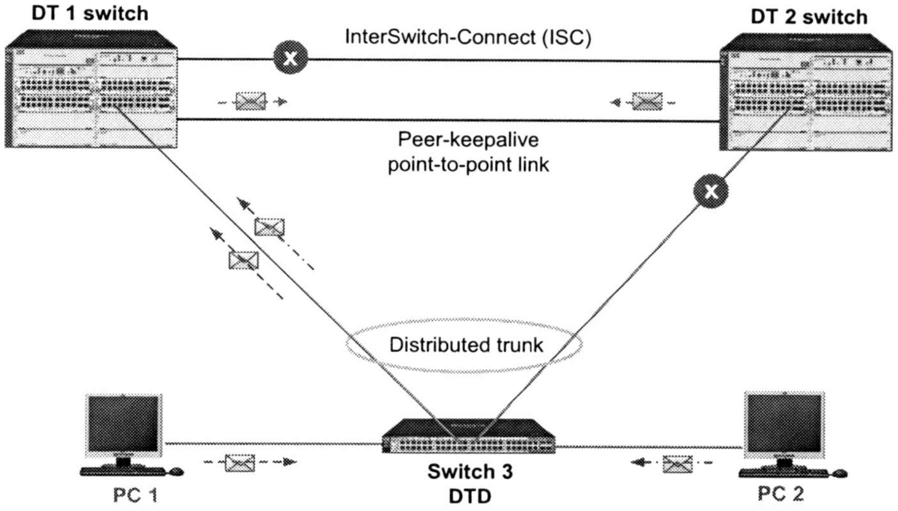

Figure 3-16: Peer-keepalive link

After the DT switch determines what type of failure has occurred, it can then take the appropriate action, based on whether it is the primary or secondary DT switch. Within a distributed trunk, the switch with the lowest MAC address is the primary switch; the other is the secondary switch.

If the primary DT switch is still operational, it assumes responsibility for forwarding traffic to and from the distributed trunking device no matter what type of failure has occurred. However, if the secondary DT switch receives a keepalive message—indicating that the primary DT switch is still operational—it disables its DT ports. It no longer handles traffic to and from the distributed trunking device. If the secondary DT switch does not receive a keepalive message—indicating that the primary DT switch is unavailable—the secondary DT switch continues to transmit traffic to and from the distributed trunking device.

It is very important that you reserve the peer-keepalive link for this purpose alone; it should not carry user traffic. You should also reserve a VLAN and subnet for this link, and if you are using dynamic routing, do not advertise this subnet in the dynamic routing protocol.

Updating switches running software versions K.14 to K.15.03

When HP first introduced distributed trunking (software version K.14 to K.15.02), it did not incorporate peer keepalive. HP introduced a shared peer keepalive with version K.15.03 and supported it in the K.15.04 software release. The dedicated peer keepalive was added in version K.15.05.

If a switch is configured to support distributed trunking but is running a software version that does not support dedicated peer keepalive, you must complete several setup steps when you update the software to K.15.05 and above. For example, if you are upgrading from K.14 to K.15.02, you must disable the ISC, update the software, and then set up the peer keepalive. Consult your switch documentation for specific steps.

Forwarding unicast traffic on the distributed trunk

You will now take a look at how traffic is forwarded on a distributed trunk, starting with unicast traffic. As you learned earlier, the DTD sees the distributed trunk as a normal aggregated link. It will, therefore, distribute traffic over the links in the distributed trunk using the same methodology it uses to distribute traffic over a normal aggregated link.

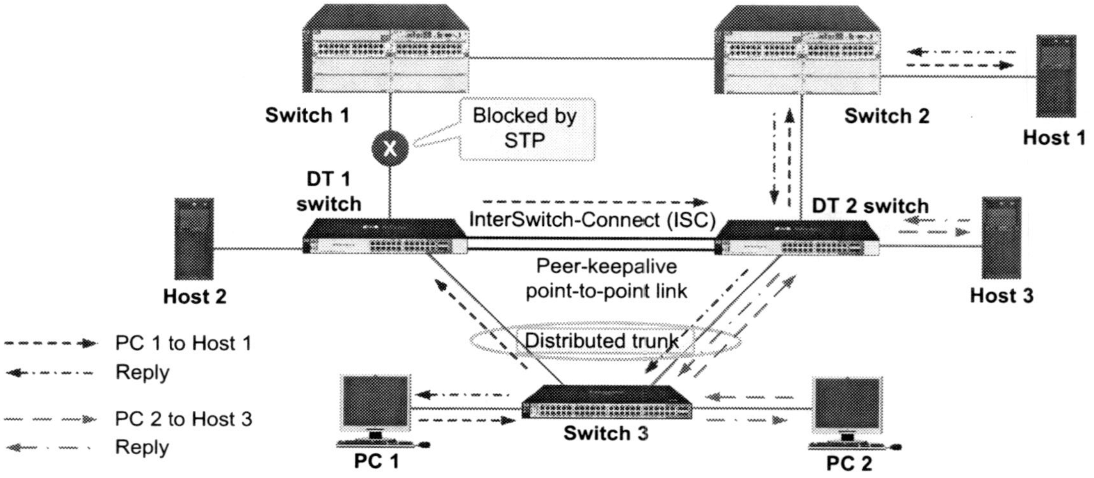

Figure 3-17: Forwarding unicast traffic on distributed trunks

If the DTD is a ProVision switch, it distributes traffic based on the source and destination address. That is, the switch identifies traffic sent from a particular source address to a particular destination address as a "conversation." It then uses a hash algorithm to distribute the conversations over the links in the distributed trunk. Consequently, the switch sends some traffic over the link (or links) to the DT 1 switch and some traffic over the link (or links) to the DT 2 switch. (You can configure ProVision switches that run the K.XX.XX and KA.XX.XX software to use Layer 4 information—UDP or TCP ports—to distribute traffic, but this discussion is based on the switch using the default setting for distributing traffic: source and destination address.)

When the traffic reaches one of the DT switches, the switch forwards or routes the traffic (if IP routing is enabled), as needed. Because the DT switches share the MAC addresses they learn and the MAC addresses that age out, they can forward traffic efficiently. For example, if the distributed trunking device sends a frame to the DT switch that is not closest to the destination device, the DT switch sends the frame over the ISC to reach the DT switch that is closest to the destination.

For example, if PC 1 in Figure 3-17 wants to access information on host 1, PC 1 sends a frame to its directly connected switch. For the purposes of this example, switch 3 sends the frame to the DT 1 switch. Because the DT 1 switch learned host 1's MAC address from the DT 2 switch, the DT 1 switch sends the frame across the ISC to the DT 2 switch. The DT 2 switch forwards the frame to the next switch in the path, which forwards the frame to its final destination, host 1.

The DT switches also handle the return traffic efficiently. When host 1 sends its reply to the DT 2 switch, the DT 2 switch forwards the frame directly to switch 3, which, in turn, forwards the frame to PC 1.

Now, imagine that PC 2 in Figure 3-17 sends a frame to host 3. This time, switch 3 sends the frame to the DT 2 switch. The DT 2 switch is directly connected to host 3, so it forwards the frame to host 3. When host 3 responds, the DT 2 switch sends the frame back to switch 3, which forwards the frame to PC 2.

Finally, suppose that host 2 sends a frame to host 1. When the DT 1 switch receives the frame, it sends the frame across the ISC to the DT 2 switch, which forwards the frame to host 1. Because the frame took the most efficient route to get to its destination, when host 1 responds, the frame follows the same route back to host 2.

Forwarding multicast and broadcast traffic

You will now examine how multicast and broadcast traffic is forwarded over a distributed trunk. When the distributed trunking device receives multicast or broadcast traffic, it handles this traffic just as it would handle multicast or broadcast traffic over a normal aggregated link. If the DTD is a ProVision switch, as shown in Figure 3-18, it forwards the traffic on one connection in the aggregated link, based on the traffic's source and destination address. Multicast traffic from PC 1, for example, might be sent to the DT 1 switch.

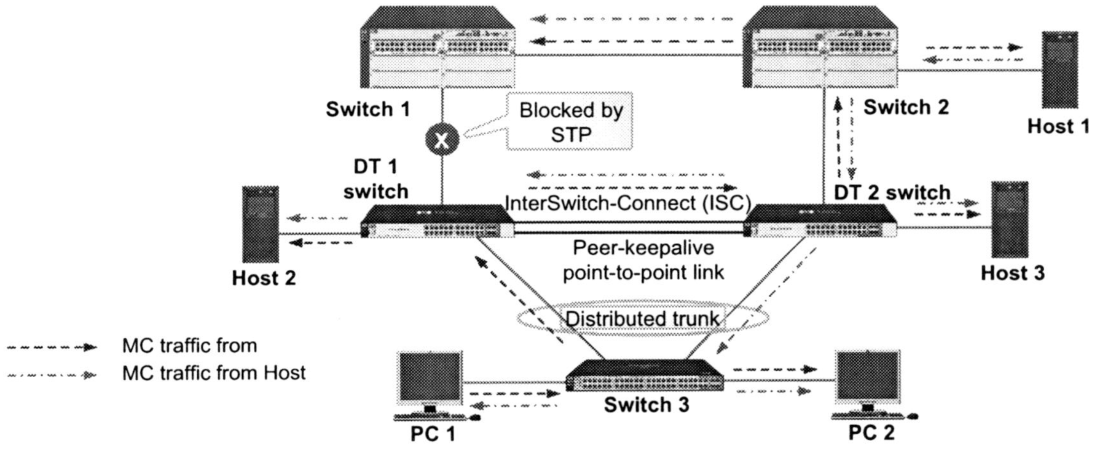

Figure 3-18: Forwarding multicast and broadcast traffic on distributed trunks

The receiving DT switch, in this case the DT 1 switch, forwards the traffic out all ports, *except* the distributed trunk (because it received the traffic on this trunk). When the second DT switch receives multicast traffic over the ISC, it forwards multicast traffic only on non-DT ports. One exception applies to this final rule: the second DT switch does forward multicasts on one link in a distributed trunk for which it knows that the peer has no active links.

If a DT switch receives multicast traffic on a non-DT port, it forwards that traffic on both DT and non-DT ports. For example, suppose switch 2 forwards multicast traffic to the DT 2 switch. The DT 2 switch forwards that traffic to all ports, including the ISC. Again, when the DT 1 switch receives the traffic over the ISC, it forwards that traffic only on non-DT ports. However, once again, the DT 1 switch does forward the traffic on a distributed trunk if it knows that its peer has no active links in that trunk.

Layer 3 routing with distributed trunking

Each DT switch in a distributed trunk has its own IP address in each active VLAN and can function as the default gateway for clients. In the example network shown in Figure 3-19, the DT 1 switch is the default gateway for PC 2. When PC 2 transmits traffic that must be routed, switch 3 forwards the traffic to the DT 2 switch, which, in turn, forwards it over the ISC to the default gateway, the DT 1 switch. The DT 1 switch then routes the traffic, sending it over the ISC to the DT 2 switch, which forwards it to the next switch in the path to its final destination.

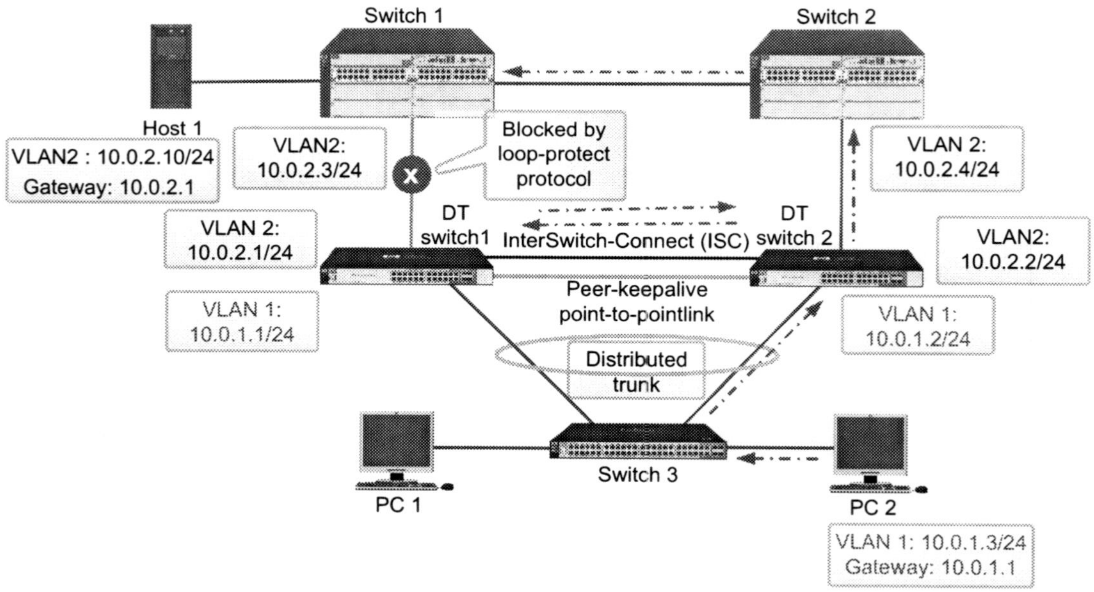

Figure 3-19: Distributed trunking and routing

When you determine which DT switch should function as the default gateway for certain clients, you should evaluate the traffic flow in your particular network environment. In the example network in Figure 3-19, you might want to use the DT 2 switch as the default gateway for PC 2.

Note
The K.15.05 software release added support for IP routing on DT switches. In previous versions of software that supported distributed trunking, DT switches could not route traffic.

Note
The ISC carries all VLANs on the DT switches except the keepalive VLAN. Connecting routing switches on links that are not dedicated to one VLAN can introduce issues with slower convergence. "Chapter 9: Advanced OSPF" explains more.

Distributed trunking and spanning tree

If you are using spanning tree on your network, you should plan how the distributed trunk will fit into your spanning tree topology. For example, because the distributed trunk must always be forwarding traffic to and from the DTD, spanning tree is disabled on all distributed trunking ports.

Likewise, the peer-keepalive link must always be available for the DT switches to exchange peer-keepalive messages (although it is never used to exchange data). Because the peer-keepalive link is isolated in its own VLAN, there is no reason for it to participate in a spanning tree network. With the K.15.05 software release, spanning tree is automatically disabled on the peer-keepalive link. (With the previous version of the software, spanning tree was supported on peer-keepalive links. You should check the release notes for these and other changes when you upgrade switch software.) The KA.XX.XX has always disabled spanning tree on the peer-keepalive link.

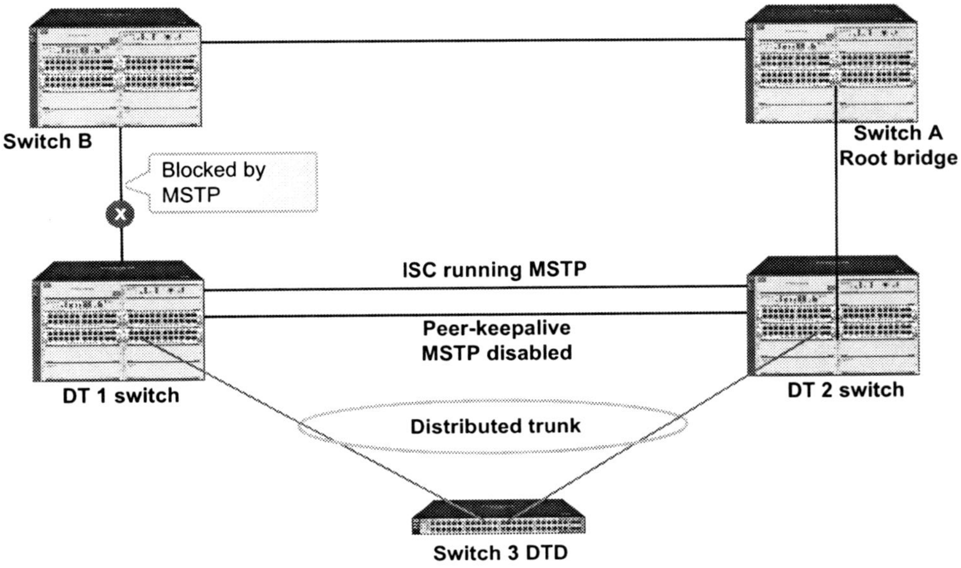

Figure 3-20: Distributed trunking and MSTP

You should also keep in mind that when you use **show spanning tree** commands, you will not be able to see that the STP BPDUs are filtered on these links.

Although spanning tree is not supported on the distributed trunk and the peer-keepalive link, it is supported on the ISC, as shown in Figure 3-20. In fact, the ISC should always be part of the preferred path in a spanning tree network, with the ISC ports or trunk operating in a forwarding state (for all instances).

Distributed trunking topologies

The distributed trunking examples you have seen thus far use a simple triangular topology, with the two DT switches being connected to one distributed trunking device.

Figure 3-21 shows a complex distributed trunking topology, with three distributed trunks each connecting to a different distributed trunking device. Notice that the two DT switches are still connected by one ISC.

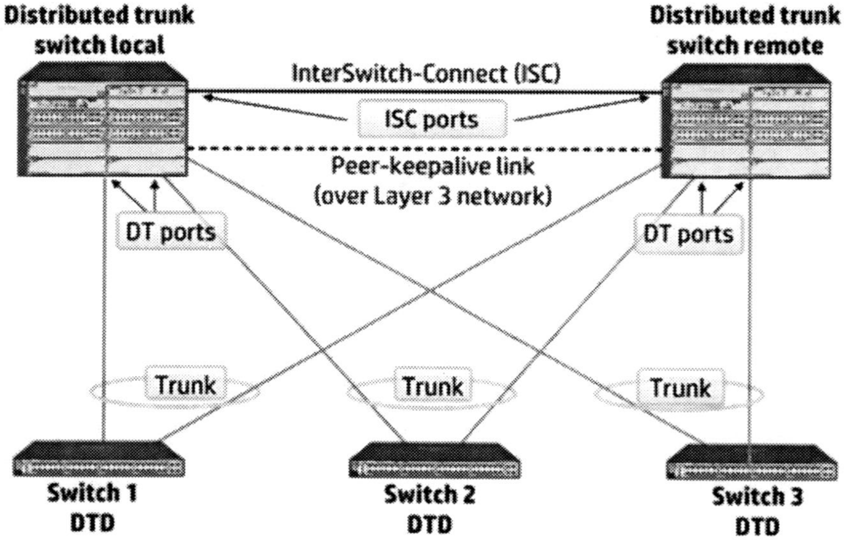

Figure 3-21: Complex triangular trunking topology

You can also configure distributed trunking in a square topology. In Figure 3-22, switches 2 and 3 are connected as DT switches with an ISC. Likewise, switches 4 and 5 are connected as DT switches with their own ISC. The two pairs of DT switches are then connected using distributed trunks.

Each DT pair also connects through another distributed trunk to a DTD. Switches 2 and 3 are connected to switch 1, and switches 4 and 5 are connected to switch 6. For these connections, a triangular topology is used.

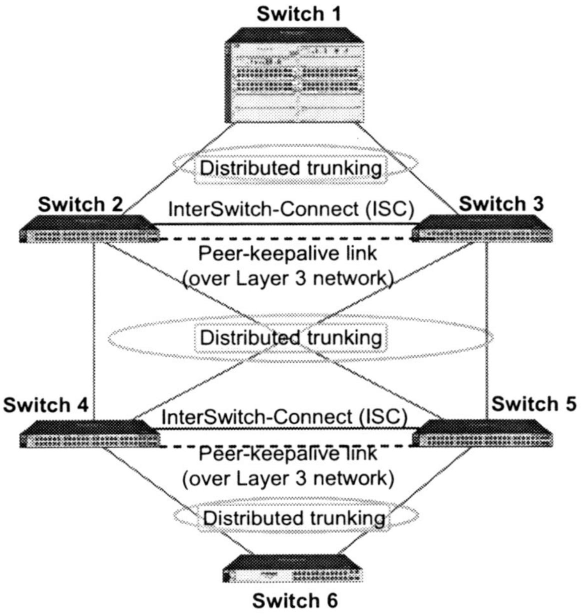

Figure 3-22: Square distributed trunking topology

Distributed trunking topologies and HP IRF

Any LACP link aggregation (or ProVision static link aggregation) can form the other side of the distributed trunk. Because an IRF virtual switch can form link aggregations that span multiple chassis, you can even connect two DT switches to multiple members of an IRF virtual switch. Figure 3-23 shows two DT switches and an IRF virtual switch with two members. Each DT switch connects to each IRF member, but all four links are part of the same link aggregation, providing a high degree of resiliency and availability.

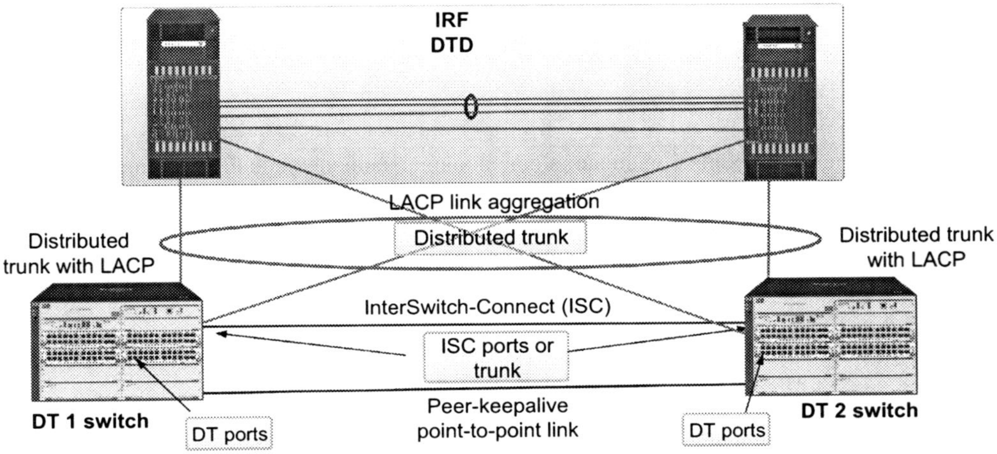

Figure 3-23: Distributed trunking and IRF

Distributed trunking with a non-HP switch

Now that you know how to set up a distributed trunk, you will examine a distributed trunk that is established between two DT switches and a third-party switch. In this example, a company is connecting two DT switches through a distributed trunk to a Cisco virtual switching system (VSS), which allows two switches to be linked together as a single network device.

At this company, the VSS provides the network core, and the two DT switches are functioning at the distribution layer as shown in Figure 3-24. The distributed trunk will provide the much needed bandwidth and redundancy between the core and distribution layers.

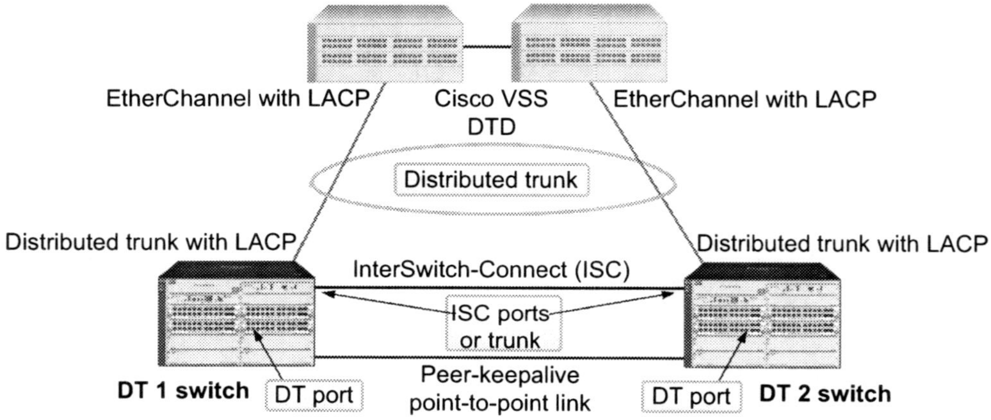

Figure 3-24: Distributed trunking with a non-HP switch

Because both the HP DT switches and the Cisco switches support LACP, the trunk on all switches is configured to use this industry standard. On the DT switches, the network administrator simply configures a distributed trunk with the **dt-lacp** option. On the Cisco VSS, the network administrator configures a multichassis EtherChannel using static LACP. (EtherChannel is the term used for aggregated links on Cisco switches.) When the switches are connected, they use LACP to establish the aggregated link. To the DT switches, it appears that the LACP aggregated link terminates on a single device. Likewise, on the Cisco VSS, it appears that the aggregated link terminates on a single device.

Although this example shows a Cisco VSS, the distributed trunking device could be *any* third-party switch or server that supports LACP or can establish a trunk through HP port trunking. In addition, this example shows a simple setup, but keep in mind that each switch could have multiple connections to the other switch. For example, the DT 1 switch could connect to both Cisco switches, and vice versa.

It is important for you to consider the load-balancing algorithm used on the DTD switch (the Cisco VSS switches, in this case). If this switch sends frames for the same session over different links, the session might experience an unacceptable number of out-of-order frames (because some of the frames cross the ISC and others do not). Therefore, you should avoid round-robin load-balancing mechanisms in favor of the ones that send all traffic for a session over the same link.

Configuring distributed trunking

This section outlines the general process for preparing for and configuring distributed trunking.

Preparation

Before you begin configuring distributed trunking, you should ensure that both DT switches are using the same software. You should also ensure that both DT switches support the same VLANs and the DT trunk ports on each switch support the same VLANs. You must manually configure these VLAN memberships, just as you would for any other trunk.

In addition, you should make sure the IGMP, DHCP snooping, and loop-protection settings on both DT switches match. Later in this training, you will learn how to compare these configuration settings using a simple **show** command.

Finally, you should keep in mind that all versions of spanning tree—STP, RSTP, MSTP, and RPVST+—are disabled on DT and peer-keepalive ports.

Note
Refer to the management and configuration guide for your switches for a complete list of restrictions.

Guidelines for ISC

As mentioned earlier, the DT switches support only one ISC link, which must be a direct connection between the two switches. You can use a single port or a trunk, which is established through static LACP or port trunking. Dynamic LACP is not supported.

Once configured, the ISC port is automatically made a member of each VLAN created on the switch, except the dedicated VLAN you create for the peer-keepalive link. If you make sure that the DT switches support the same VLANs, the two switches' ISC ports automatically support the same VLANs.

Guidelines for the peer-keepalive link

Like the ISC, the peer-keepalive link should be a point-to-point connection between the two DT switches.

In addition, you must create a dedicated VLAN for the peer-keepalive link and then make the peer-keepalive port a member of this VLAN. The DT switch will allow you to make only one port a member of this VLAN. It will further ensure that the ISC is not a member of this VLAN. If you are using a routing protocol, do not advertise the peer-keepalive subnet in that protocol.

Finally, you should try to separate the port that a switch uses for the peer-keepalive link from the port used for the ISC; choose ports on different modules or port banks. This practice increases the chance the peer-keepalive link remains up and unaffected by any problem that affects the ISC—such problems, of course, create the precise situation in which the peer-keepalive link must be up.

If spanning tree is enabled on the DT switches, the switch automatically disables it for the peer-keepalive port. This port cannot be blocked; it must always be able to send peer-keepalive messages. (When you use the **show spanning-tree** commands to view information about the spanning tree network, you will not be able to see that the BPDUs are being filtered on the peer-keepalive link.)

Configuration steps

There are three main steps for configuring distributed trunking:

1. Configure the ISC.
2. Configure the peer-keepalive link.
3. Configure the distributed trunk.

Configuring the ISC. To configure the ISC, you enter the **switch-interconnect** command, specifying a port or a static trunk (which you have configured through LACP or port trunking):

```
Switch(config)# switch-interconnect [port-number | trkx]
```

Because each switch supports only one ISC, you can override any existing ISC configuration by entering this command with new values.

To remove the ISC, enter the **no** form of the **switch-interconnect** command.

Configuring the peer-keepalive link. To configure the peer-keepalive link, you first create a dedicated peer-keepalive VLAN on each DT switch and assign each switch an IP address in that VLAN. For example, you could create VLAN 50 on each DT switch and assign the primary DT switch the IP address 10.1.50.1/24 and the secondary DT switch the IP address 10.1.50.2/24.

You must then specify that this is the peer-keepalive VLAN:

```
Switch(config)# distributed peer-keepalive vlan <vlan_ID>
```

You must also configure the peer-keepalive destination address—the peer-keepalive IP address on the other DT switch:

```
Switch(config)# distributed peer-keepalive destination <ip_address>
```

You can also customize the peer-keepalive parameters for your environment. For example, the peer-keepalive timeout is, by default, 5 seconds, but you can assign it a value between 3 and 20 seconds. You can change the default interval for sending peer-keepalive messages, which is 1000 milliseconds, to a value between 400 and 10,000. You can change the default peer-keepalive timeout (which is 5 seconds), specifying 3 to 20 seconds. And finally you can change the UDP port the DT switches use for the peer-keepalive messages. By default, the DT switches use port 1024.

To configure the peer-keepalive parameters, use the following command:

```
Switch(config)# distributed peer-keepalive [hold-timer <3-10> | interval
<400-10000> | timeout <3-20> | udp-port <1024-49151>]
```

Configuring the distributed trunk. After you configure the ISC and peer-keepalive link, you can configure the distributed trunk itself. If you have experience configuring trunks on ProVision switches, you are already familiar with the **trunk** command. Options have been added to allow you to configure a distributed trunk:

```
Switch(config)# trunk <port-list> <trkx> [trunk | lacp | dt-lacp |
dt-trunk]
```

Just as you would for any static trunk, you specify the trunk name, using trk and adding a number. For example, you might specify trk1. For distributed trunks, you must use the same name, as well as the same trunk type, on both DT switches.

You then specify the type of trunk. For distributed trunking, you use **dt-lacp** or **dt-trunk**. Specify **dt-lacp** if you are establishing a trunk with an LACP-compatible device. On the distributed trunking device, configure a static LACP aggregated link. (The distributed trunking device does not have to support distributed trunking.) If you want to establish a nonprotocol trunk, specify **dt-trunk**.

Maximums and limitations

The ProVision switches that support distributed trunking support a maximum of 144 aggregated links. This maximum applies to distributed trunks as well.

In addition, a maximum of two switches can form a distributed trunk, and each DT switch can be connected to the DTD with four ports. The two DT switches are always connected by one ISC link.

Distributed trunking has some limitations as well. When setting up distributed trunking, you cannot use dynamic LACP. If you want to use LACP, only static LACP is supported.

In addition, distributed trunking cannot be used with some other features:

- ARP protection and dynamic IP lockdown are not supported on distributed trunks.
- Meshing and distributed trunking are mutually exclusive.
- QinQ-mixed VLAN mode and distributed trunking are mutually exclusive.
- Multicast routing and distributed trunking are mutually exclusive.
- IPv6 routing and distributed trunking are mutually exclusive.

Note

Refer to the management and configuration guide for your switches for a complete list of restrictions.

HP IMC: Monitoring data link layer redundancy

HP Intelligent Management Center (IMC) allows you to monitor your network topology, helping you to maintain redundant links, understand the current topology, and troubleshoot any issues. Among other tasks, you can:

- Deploy MSTP settings as a batch operation

  Note that you would usually first configure individual settings on the root switches because they require different priorities in each instance. You then run the batch operation to deploy standard settings (using the default priority in each instance) to the rest of the switches.

- View the spanning tree topology

  HP IMC makes it easy for you to trace the spanning tree that your network has actually created. As shown in Figure 3-25, you simply click the MSTP icon in the network topology and select the CIST or an instance to see which ports are forwarding and which are blocking. You can further explore by right-clicking switches and viewing their MSTP settings.

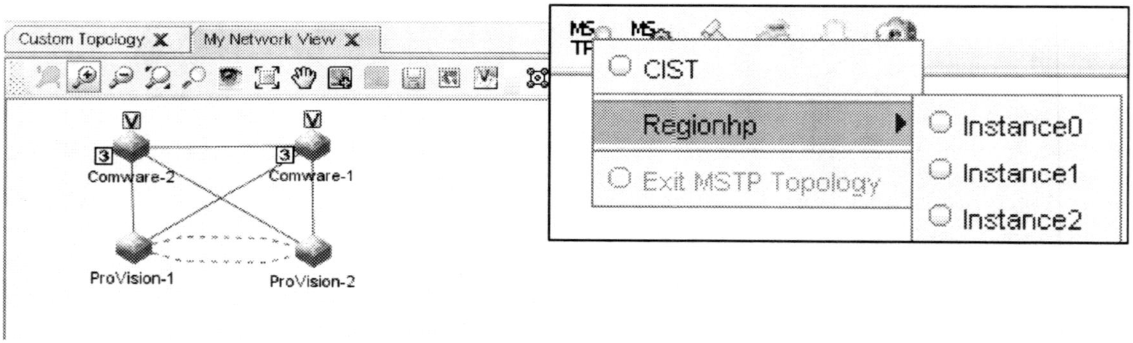

Figure 3-25: Using IMC to monitor data link Layer redundancy

HP also helps you to monitor link aggregations, including distributed trunks.

HP IMC displays the link aggregation in its network topologies as a line with a shaded circle. You can right-click the link and see a variety of information about the protocol, the ports involved, and the switches involved. For a distributed trunk, note that the topology shows the link connecting to just one of the DT switches (see Figure 3-26). However, if you right-click and view the detailed information, you can see the names of both DT switches with ports in the trunk.

With these features, you can troubleshoot problems and resolve design flaws.

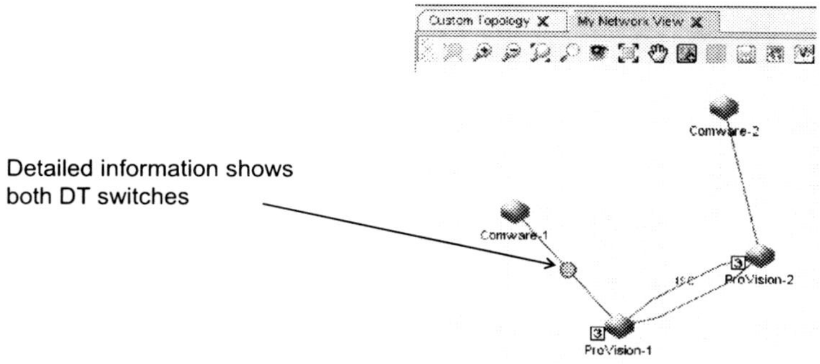

Figure 3-26: HP IMC

Summary

This chapter has built on your knowledge of STP, RSTP, and MSTP, showing you how these protocols form just a part of your plan for data link layer redundancy. You have learned about RPVST+, and you know how to deploy it on HP switches as required for interoperability with PVST+/RPVST+-based deployments. You learned how UDLD on ProVision switches and DLDP on Comware switches protect unidirectional Ethernet links, a problem that might occur on fiber links, from negatively affecting network connectivity.

Then you learned how to implement distributed trunks on pairs of ProVision switches, improving bandwidth utilization and failover time over similar logical topologies that use MSTP instead.

Finally, you saw how IMC helps you to manage and monitor your complete data link layer redundancy solution.

Learning check

Answer these questions:

1. What are some of the similarities and differences between MSTP, PVST+, and RPVST+?

2. How does Device Link Detection Protocol (DLDP) detect and handle unidirectional links?

3. How do DT switches that are part of a distrusted trunk use ISC?

4. What steps should you take before configuring a distributed trunk?

Answers

1. What are some of the similarities and differences between MSTP, PVST+, and RPVST+?

   All of these technologies can establish multiple spanning trees such that a link can forward traffic in some VLANs and block traffic in other VLANs.

   With PVST+ and RPVST+, each VLAN has a root bridge and a spanning tree rooted at that bridge. Although multiple links forward traffic, each VLAN (broadcast domain) remains free of loops. PVST± and RPVST+ must always create one spanning tree for each *VLAN*, even if many VLANs use the same paths. Therefore, the switches need to process more BPDUs and maintain more spanning trees, creating a higher CPU utilization that can cause issues.

   MSTP creates a spanning tree for each instance. Multiple VLANs can be mapped to the same instance. MSTP also uses a single BPDU for all instances. Therefore, MSTP imposes less of a burden on switches' CPU.

   When PVST+ and RPVST+ establish a spanning tree for a VLAN, the spanning tree only uses links that can actually carry traffic for that VLAN. MSTP, on the other hand, establishes the spanning tree for an instance without regard to the VLANs actually allowed on the link. Therefore, with MSTP, you must be careful to permit all VLANs on each switch-to-switch link to avoid connectivity issues. You can do this more safely than with PVST+ and RPVST+ because each VLAN does *not* introduce another BPDU and spanning tree instance.

   RPVST+ offers faster convergence for PVST+ without the need for the proprietary PortFast, UplinkFast, or BackboneFast features. Basically, RPVST+ is RSTP operating per-VLAN in a single Layer 2 domain. MSTP also bases its operations for each spanning tree on RSTP.

2. How does Comware Device Link Detection Protocol (DLDP) detect and handle unidirectional links?

   Comware switches use DLDP to identify a neighbor device and monitor a link. If the link becomes unidirectional, DLDP shuts down the port.

   DLDP works with the physical layer protocols to determine if the ports at both ends of the link are operating properly at that layer. If they are, DLDP further verifies the link is operating properly at the data link layer.

   In normal mode, DLDP can detect fiber links that are cross-connected, resulting in unidirectional traffic. In addition to cross-connected fibers, DLDP in enhanced mode can detect fiber-optic connections that have one broken fiber or one disconnected fiber.

3. How do DT switches that are part of a distrusted trunk use the InterSwitch-Connect (ISC)?

   The two DT switches communicate through the ISC, which is a point-to-point link between the two switches. The switches also use the ISC to exchange information about loop-protection management, the Internet Group Management Protocol (IGMP) snooping database, DHCP snooping information, and IP routing.

The DT switches also use the ISC to synchronize the MAC addresses that they have learned or aged out.

The DT switches can also forward network traffic over the ISC as necessary.

4. What steps should you take before configuring a distributed trunk?

   You should ensure that both DT switches are using the same software. You should also ensure that both DT switches support the same VLANs, which will be automatically added to the ISC.

   In addition, you should make sure the IGMP, DHCP snooping, and loop-protection settings on both DT switches match.

   You should keep in mind that all versions of spanning tree—STP, RSTP, MSTP, RPVST+—are disabled on DT and peer-keepalive ports.

   You should create the ISC, which is a dedicated link between the DT switches. Each side of the link can be either a single Ethernet port or a link aggregation.

   You should establish the peer-keepalive link, which has a VLAN dedicated to it alone. You should assign each DT switch an IP address on this VLAN; do *not* add the subnet to the routing solution. You should establish the peer-keepalive settings.

4 Virtual Router Redundancy Protocol (VRRP)

EXAM OBJECTIVES

In this chapter, you learn to:

- ✓ Ensure redundancy for a network's default gateway by configuring Virtual Router Redundancy Protocol (VRRP) on HP ProVision and HP Comware switches.
- ✓ Establish load balancing of active routing in several different ways.
- ✓ Implement VRRP in topologies that use various data link layer redundancy mechanisms.

INTRODUCTION

The previous chapter covered various technologies that ensure link redundancy. You not only reviewed Multiple Spanning Tree Protocol (MSTP) and Rapid Per-VLAN Spanning Tree Plus (RPVST+), you also learned how each access layer switch can connect to two redundant core (or distribution layer) switches that are implementing distributed trunking. These technologies ensure that endpoints across the network can reach each other even if a link or sometimes an entire device fails. Distributed trunking speeds failover and provides a more efficient use of bandwidth. However, neither MSTP nor distributed trunking on its own can provide redundancy for the default gateway role—without additional measures, only one core switch can route traffic for endpoints in each subnet, and if that switch fails, users cannot reach services in other subnets.

To establish redundancy for the default gateway, you can add a second router to a subnet with relative ease. The challenge arises in ensuring that endpoints understand when they need to use that default gateway. IPv4 does not provide a gateway discovery mechanism. Often endpoints learn their IP settings, including their default gateway address from a Dynamic Host Configuration Protocol (DHCP) server; however, DHCP does not help the endpoints realize when one gateway is unavailable and they must switch to a new one. Clearly endpoints with static IP settings have an even worse time responding. It would be difficult to deploy and maintain a solution for discovering an available gateway across all endpoints. Therefore, to provide the fastest, most seamless failover, a gateway redundancy protocol should ensure that:

- Endpoints can continue to use the same default gateway IP address during a failover situation. They can receive this address as usual from static settings or a DHCP server; you do not need to deploy a complicated, difficult-to-manage solution to all of the endpoints.
- Endpoints can continue to use the same default gateway MAC address during a failover situation. Address Resolution Protocol (ARP) table entries do not need to time out for failover to occur.

This chapter explains how you can implement VRRP to meet these needs. VRRP allows multiple core (or distribution) switches to act as default gateway that appears to endpoints as a single device. The protocol ensures that endpoints and other devices within a network can still reach other networks if a default gateway fails. VRRP can be an important part of a strategy for improving network reliability and ensuring that network services are always available to users.

You should use VRRP as the default gateway redundancy protocol when ProVision switches provide the default gateway services. If Comware switches provide the default gateway services, you should typically use HP Intelligent Resilient Framework (IRF), discussed in the next chapter, instead; IRF provides redundancy for the default gateway among other benefits.

This chapter focuses on VRRP, which allows you to implement redundancy for default gateways.

VRRP overview

VRRP is a standards-based protocol that provides automatic failover for gateways.

Note

The current version of VRRP, VRRP v3, described in RFC 5798, adds VRRP support for IPv6 subnets, providing a faster transition of the active routing role than IPv6's native Neighbor Discovery mechanisms. Comware switches support VRRPv3. Check the HP website to see whether your ProVision switch model supports VRRPv3. For IPv4 subnets, VRRP v3 operates in essentially the same manner as VRRP v2. This chapter focuses on the IPv4 functions.

With VRRP, two or more routers are grouped together to form a virtual router. Using a single virtual IP address, the group of routers appears as a single default gateway to the endpoints in the subnet.

As Figure 4-1 shows, one router in the group acts as the master and routes all traffic to other networks. The other routers in the group function as backups. If the master fails, VRRP enables a backup to become the new master.

Note

Comware documentation refers to the virtual router as a VRRP group.

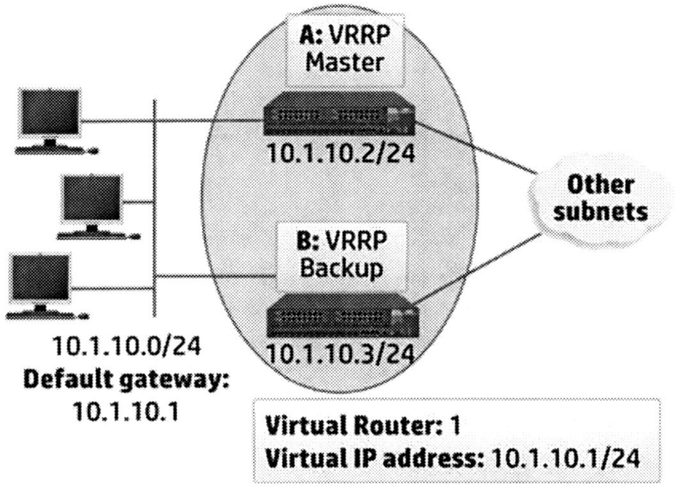

Figure 4-1: VRRP overview

Controlling VRRP states

Members of a virtual router hold an election to determine which one will be the master. The router with the highest priority is elected master.

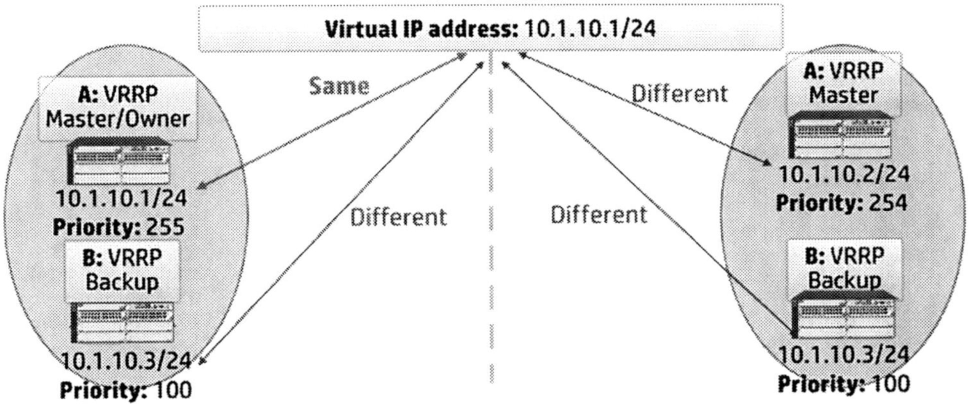

Figure 4-2: VRRP states

Controlling the states with a VRRP owner

A virtual router can have an owner, which is a member that has the same actual IP address as the virtual IP address (see Figure 4-2). The owner is always the master when it is available, automatically having a priority of 255, the highest possible value, without you assigning the priority. The 255 value, in fact, is reserved for VRRP owners.

To configure a device as the virtual router owner, simply set the virtual IP address to match the device's actual IP address. However, with some ProVision software, you enter the **owner** command to set the role *before* you configure the virtual IP address.

Controlling the roles without a VRRP owner

You can also configure a virtual IP address that is the VRRP's virtual address alone—none of the routing devices within the group owns it. You configure the routers' priority to determine which router becomes master. The default value is 100, and you can set the value between 1 and 254. (Priority 0 is reserved for a master that has released its master responsibilities.)

With some ProVision software, you must enter the **backup** command before you configure a VRRP virtual IP address that does not match the router's actual IP address.

In Figure 4-2, router A has a higher priority, so it is elected as the master.

If all the routers have the same priority number, the router with the highest IP address becomes the master. It is usually a good idea to set the priority manually, however. You can then predict and control which router becomes master during normal operation, giving you better knowledge of your network and control over the load-balancing strategies that you will examine in the next section.

VRRP basic operation

Figure 4-3 shows a simple network with VRRP enabled. Router A and router B are configured to form a virtual router that acts as the default gateway. They share a virtual router identifier (VRID), a virtual IP address, and a virtual MAC address.

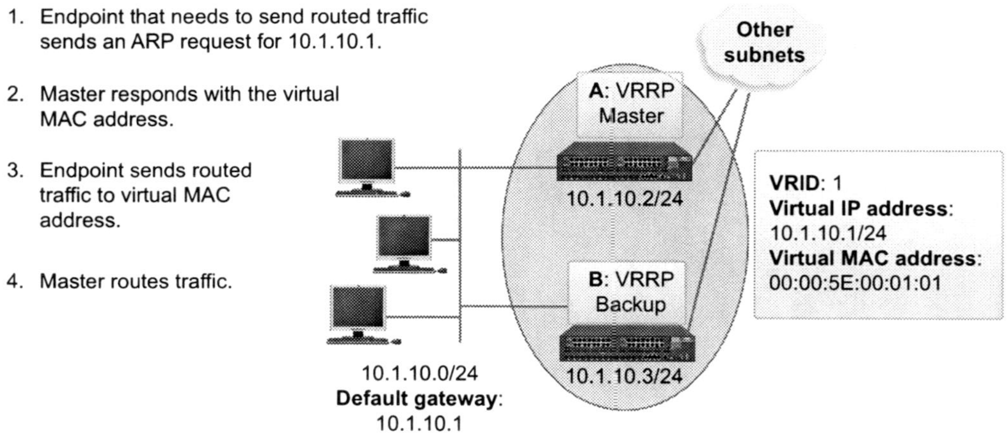

Figure 4-3: VRRP basic operation

With the virtual IP address and virtual MAC address, the virtual router always presents a consistent front to the endpoints for which it provides routing services—no matter which particular router is actually handling the services at the moment. Endpoints know only one IP address for their default gateway (configured statically or, more often, learned through DHCP). When an endpoint needs to learn the

MAC address for its default gateway, it always sends an ARP request to that consistent virtual IP address. Similarly, it always addresses packets that need to be routed to the same destination MAC address.

All routers in the virtual router know the virtual IP address and virtual MAC address. However, only the master responds to ARP requests for the virtual IP address (see Table 4-1). Similarly, only the master accepts frames destined to the virtual MAC address. Just like any router, it decapsulates these frames and routes them.

Table 4-1: Master and backup responsibilities

Virtual router state	Responds to ARP requests for the virtual IP address	Receives and routes frames destined to the virtual MAC address
Master	X	X
Backup		

Choosing the virtual router settings

You configure the VRID and virtual IP address when you set up VRRP—make sure that they match on all routers. The routers automatically calculate the virtual MAC address based on the VRID, as follows:

```
00:00:5E:00:01:VRID
```

Note

The virtual MAC address for IPv6 subnets is 00:00:5E:00:02:VRID.

As for choosing the VRID, you can choose any value between 1 and 255 that you like—as long as no other virtual router on the subnet is using it. You can, however, set the same VRID on different virtual routers on different subnets.

You might correlate the VRID with the virtual router's host address. For example, if the virtual IP address is 10.1.10.1, you might set the VRID to 1. Or if you plan to set up only one virtual router (you will learn more about using multiple virtual routers a bit later), you can correlate the VRID with the third octet of a/24 subnet.

Of course, the VRID does not have to provide any meaning, and you can simply choose any value that matches on all VRRP devices in the subnet.

VRRP failover

In addition to handling routing, the VRRP master is responsible for transmitting VRRP advertisements, which it sends every advertisement interval to multicast address 224.0.0.18. The default advertisement interval for VRRPv2 is 1 second, but you can raise this interval. By introducing *shorter* advertisement intervals, as low as 1 centisecond, VRRPv3 allows much faster failover. This change is one of the few changes for IPv4 in VRRPv3.

The virtual router's backup routers listen for the advertisements to determine whether the master is still up. If they miss three advertisements, the backup router with the next highest priority assumes the master role (see Figure 4-4).

The new master starts sending VRRP advertisements. It also begins responding to ARP requests to the virtual IP address. And it accepts frames destined to the virtual MAC address, which it then decapsulates and routes appropriately—in short, it assumes all the functions of a default router and the virtual router master. The new master also sends a gratuitous ARP response which allows other switches to associate the virtual MAC address with the port they use to reach the new master (see Table 4-2).

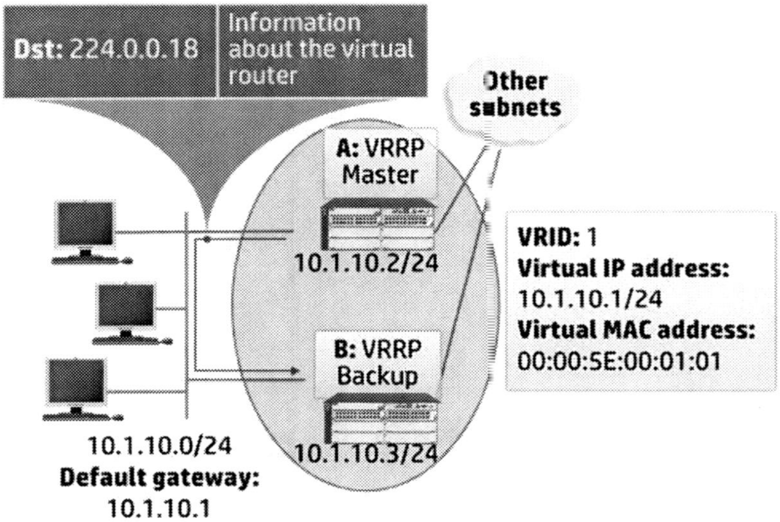

Figure 4-4: VRRP failover

As far as the actual endpoints are concerned, nothing has changed; they continue to send traffic they need routed to the same MAC address.

Table 4-2: Master and backup responsibilities

Virtual router state	Responds to ARP requests for the virtual IP address	Receives and routes frames destined to the virtual MAC address	Transmits virtual router advertisements	Listens for virtual router advertisements	Becomes master when the master down interval expires
Master	X	X	X		
Backup				X	X

Supplemental information on electing the new master

A skew time ensures that the remaining member with the highest priority takes the master role. A VRRP backup router includes a very short skew time, which is inversely based on its own priority, in its master down interval. Therefore, the member with the highest priority has a slightly shorter interval. It assumes the master role and starts sending VRRP advertisements first. Other backup routers hear the advertisements and so do not compete in taking over as master.

Often the virtual router has only two members with a single backup router to take the master role. But if you do have a virtual router with more than two members, you should set the priority values relatively far apart, even if you do not particularly care which member becomes master, to prevent brief periods in which backups compete for the master role.

Supplemental information on controlling the failover time

You can roughly determine the master down interval, and the failover time, by multiplying the master advertisement interval by three. At default settings, a virtual router recovers from a master failure in 3 seconds, and a VRRPv3 virtual router can recover within a small fraction of a second.

You can lower the advertisement interval as required to meet your failover needs, taking into account network congestion and the processing burden on the routers that need to send and listen for advertisements. Also be careful to set the same interval on all routers participating in the virtual router; otherwise, the backup router might try to take over the role prematurely.

If you see that a router frequently changes between the backup and master role, you might have set too short of an advertisement interval; try raising the interval.

VRRP preempt and nonpreempt mode

VRRP failover provides continuity of services during the emergency situation of a total failure of the active router. But eventually the network returns to the normal operational topology with the failed device once again able to provide routing services. Depending on the type of problem that caused the original master to fail, the router might come back on its own, you might restore it by fixing failed links or power supplies, or you might replace the failed master with new hardware running the failed master's configuration.

You should understand how the restored master affects the VRRP operations, which in turn depends on whether the group operates in nonpreemptive or preemptive mode. Note that the same processes occur when a new device joins the virtual router (which is, in fact, what is happening when you add new hardware running a failed device's configuration):

- **Preempt mode**—When a router operates in preempt mode, it will take over the virtual router master role whenever it has a higher priority than the current master. The router sees the advertisements with a lower priority and sends a preempt master message telling the device with the lower priority to become a backup router. It also begins sending its own advertisements.

The preempt mode applies when you add a new router with a higher priority to the virtual router. It also applies when a failed master is restored. That router assumes the master role again because it has the higher priority.

You can configure a preemption delay, which allows the router with the higher priority to take over the master role but forces it to wait a certain amount of time after initializing its VRRP functions to do so. This delay allows the router to build up its routing table *before* taking over as the active router, preventing disruptions in service—you will learn more in "Chapter 9: Advanced OSPF."

- **Nonpreempt mode**—If a router is using nonpreempt mode, the router does not attempt to take over the master role from an active master even if it has a higher priority. This behavior holds true whether the backup router with a higher priority is a new device or a former master that failed.

The preempt mode determines whether the router attempts to preempt the master role itself—not whether it accepts preemption from another router (see Figure 4-5). Whether or not preempt mode is enabled on a router's VRRP instance, a master will change to a backup router in response to preempt master messages from a router with higher priority. Therefore, even if routers' preempt settings do not match, they can agree on the master. However, you should generally match the preempt mode on the routers in the same VRID.

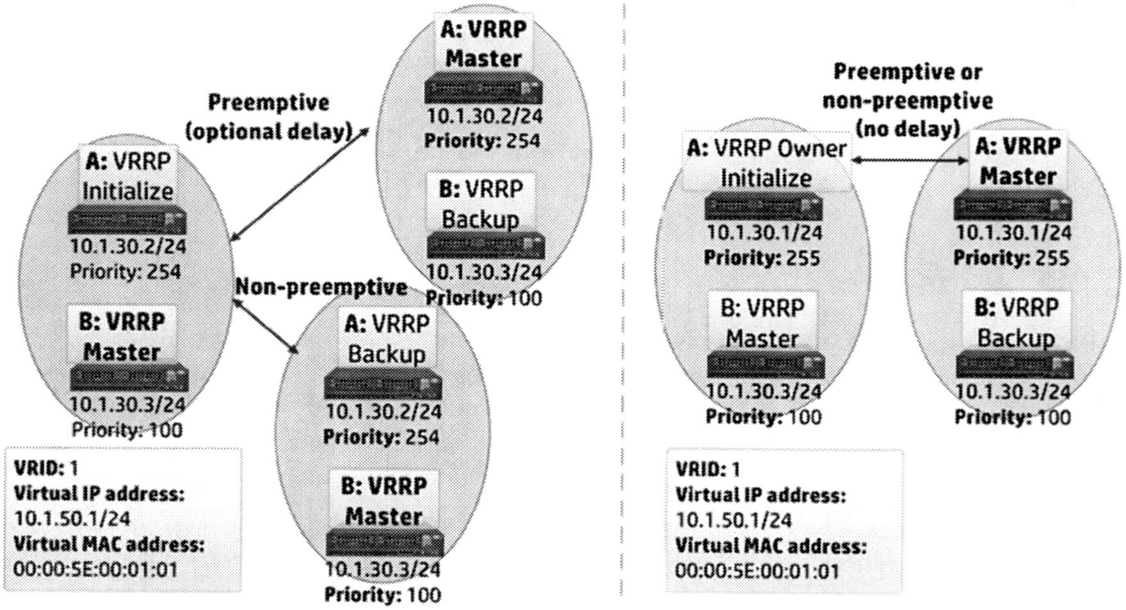

Figure 4-5: VRRP preempt and nonpreempt mode

By default, Comware and ProVision switches operate in preempt mode with no delay. You can add a delay or disable preemption entirely with settings configurable per VRID.

Effect of owner role on preemption

When one of the VRRP routers owns the virtual IP address, it *always* preempts the master role *despite what you configure for the VRID's preempt mode*. In addition, any preemption delay that you configure does *not* take effect.

Load-balancing routing within a subnet

A company might not want one router standing idle while another router bears the burden of routing all traffic for the network segment's subnet or subnets. VRRP provides several ways for you to distribute the routing functions across multiple devices.

The VRRP standard allows you to create multiple virtual routers on the same subnet, identifying each router with a different VRID. Because VRRP supports 255 VRIDs, you can theoretically create that many virtual routers on a subnet. Practically, you only need to create, at most, one virtual router per router participating in VRRP. A different router acts as master in each virtual router while providing backup for the other virtual routers.

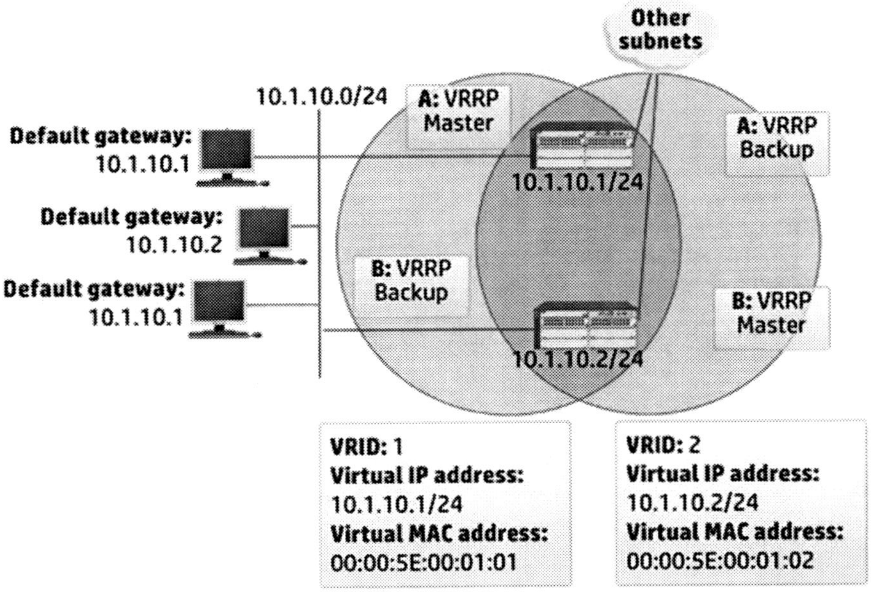

Figure 4-6: VRRP: Load balancing within a subnet

In the example shown in Figure 4-6, router A is the master for virtual router 1 but the backup for virtual router 2. Router B, on the other hand, is the master for virtual router 2 and the backup for virtual router 1.

The primary drawback of this load-balancing strategy is that it features multiple virtual IP addresses. The load balancing depends on different endpoints in the same subnet using different virtual IP addresses for their default gateway. You either need to configure the default gateway addresses manually or the DHCP administrator needs to create different scopes, which adds complexity.

Load-balancing routing across different subnets

If your VRRP routers support multiple subnets, you can take a simpler approach to load balancing the routing functions. You configure a single VRID for each subnet, but you alternate the master role on different subnets.

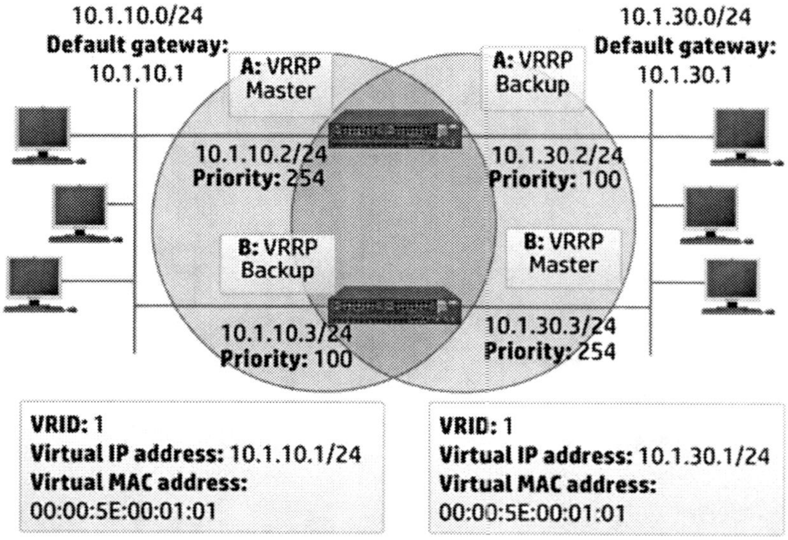

Figure 4-7: VRRP: Load balancing across subnets

In the example shown in Figure 4-7, router A is the master for virtual router 1 on VLAN 1 but the backup for virtual router 1 on VLAN 10. Router B, on the other hand, is the master for virtual router 1 on VLAN 10 and the backup for virtual router 1 on VLAN 1. Neither router owns the virtual IP address, but you have set priorities to control which router enters the master state during normal operation. (If necessary, you could also create a similar solution in which you alternated the owner role between subnets.)

This strategy might offer less fine-grained load balancing than using multiple virtual routers on the same subnet, but you do not have to worry about establishing more than one default gateway address per subnet. Later in this chapter, you will also see how you can create an efficient flow for traffic using this strategy.

HP Comware: Load balancing within a subnet with one virtual router

You have learned how you can balance the active routing role *across* VLANs by configuring a higher priority on different VRRP routers in different VLANs. You have also learned how you can balance this rule *within* a VLAN using multiple virtual routers.

HP Comware switches also provide a load-balancing mode, which enables you to load-balance the routing role within a subnet using a *single* virtual router and virtual IP address—thus simplifying the configuration of endpoints' default gateway. (This feature is not provided in the VRRP standard. It is proprietary to HP.) This mode of VRRP uses messages similar in format to standard VRRP packets; however, an option field is appended to the packets.

In this mode, *multiple* virtual MAC addresses are associated with a virtual router's virtual IP address, as shown in Figure 4-8. Each virtual MAC address represents a virtual forwarder (VF), which can actively route traffic. Although a load-balancing virtual router still has just one master, every backup router can act as a VF.

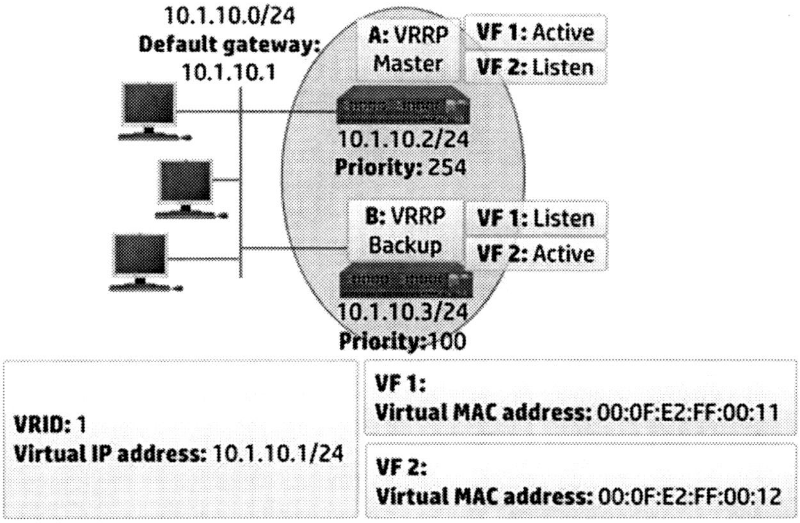

Figure 4-8: Comware: Load balancing within a subnet with one virtual router

Creating the VFs

When a backup router that supports load balancing initializes VRRP, it sends a request to the master. The master responds by assigning the backup router a virtual MAC address. The master also assigns itself a virtual MAC address.

Each router in the virtual router creates a VF associated with its virtual MAC address. It advertises this VF information to other members of the virtual router group using a similar advertisement packet to the master's. The advertisement indicates the router's priority for the virtual MAC address, which is typically 255 for the owner.

During normal operation, each backup router is in the active state for the VF that it originally created (because it has the highest priority). Other members of the virtual router hear the VF advertisements and create the VF in the listening state.

An active VF behaves similarly to a master router in standard VRRP. It accepts and routes frames destined to the virtual MAC address associated with the VF.

However, the virtual router *master* remains solely responsible for responding to ARP requests from endpoints. It uses a hashing algorithm to choose the virtual MAC address for its response to each new ARP request. In this way, each endpoint retains a stable MAC address for its default gateway. However, different endpoints in the subnet know different MAC addresses for the default gateway, distributing the routing load across the members of the virtual router.

Failing over for VF states

Just as every backup router listens for the master's advertisements and stands ready to become master, every member of the virtual router that is not active for a VF is in a listening state for that VF. If the active VF becomes unavailable and stops transmitting advertisements, a router in the listening state becomes active, taking on the responsibility for routing traffic destined to the associated MAC address.

Often you only have two routers in the virtual group, which means, of course, that the other router becomes the new active VF. In virtual routers with more than two members, the remaining members' priority for the VF comes into play. The priority derives from the router's weight, which indicates the router's forwarding capability. The higher the weight, the greater the amount of traffic the device can route. A router's weight must exceed a certain lower limit; if it does, the router's priority for a VF is:

```
Weight/(Number of VFs in the virtual router + 1)
```

The only reason for a virtual router to maintain a VF owned by a failed member is to provide continuity of services to the endpoints that were using that router as their default gateway. Therefore, the virtual router gradually phases out the failed VF, as follows:

- When a VF that owns a virtual MAC address stops sending advertisements, the master continues responding to ARP requests with that VF's virtual MAC address over the redirect timer period. This period grants the VF a brief grace period to return without the danger of disrupting connectivity because the new active VF routes the traffic. After the redirect time period expires, the master stops responding to ARP requests with the virtual MAC address.
- Endpoints now no longer learn the virtual MAC address of the failed router. However, the VF remains up, with the new active VF routing the traffic, over the timeout time, giving endpoints that had already learned this MAC address time to send new ARP requests and learn a different VF's MAC address. After the timeout time expires—at which point, presumably endpoints are no longer sending packets to the VF's MAC address—the router now responsible for the VF deletes the VF, sending a release message to notify other members of the virtual router.

VFs always operate in preempt mode, so if a router with higher priority for the VF initializes, it takes over as the active VF. For example, if a failed router is restored, it takes over the active state for the VF that it owns.

Summary of load-balancing VRRP functions

The standard VRRP functions, such as master election and preemption, continue to apply in load-balancing virtual routers, as Table 4-3 and Table 4-4 summarize.

Note

Just like you can create tracks to lower a router's priority for being master, you can create tracks to lower a router's priority for a VF.

Table 4-3: Virtual router responsibilities (Comware load-balancing mode)

Virtual router state	Responds to ARP requests for the virtual IP address	Transmits virtual router advertisements	Listens for virtual router advertisements	Becomes master when the master down interval expires
Master	X	X		
Backup			X	X

Table 4-4: VF responsibilities (Comware load-balancing mode)

VF state	Receives and routes frames destined to the virtual MAC address	Transmits VF advertisements	Listens for VF advertisements	Becomes active when the down interval expires
Active	X	X		
Listening			X	X

HP ProVision: Combining VRRP and distributed trunking

The next section explains how to implement VRRP together with a data link layer redundancy protocol to provide a high level of network redundancy.

When you have two HP ProVision switches, such as HP 8200 zl Series switches, acting as redundant default gateways at the network core and distribution layer, you should typically combine VRRP with distributed trunking. Distributed trunking provides redundant load-balancing pathways between access layer switches and the default gateway with a bit more stability and faster failover than MSTP. VRRP, for its part, ensures redundancy for the gateway services.

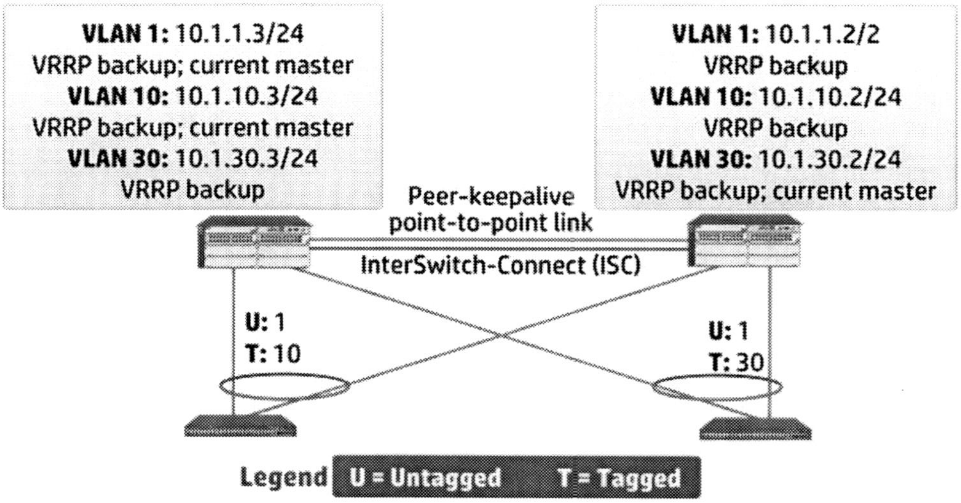

Figure 4-9: ProVision: Combining VRRP and distributed trunking

Figure 4-9 illustrates the traffic flow. The access layer switch learns the virtual MAC address on the link aggregation rather than a specific link, so it might forward the traffic on the link to either the master or the backup router. If the backup router happens to receive the traffic, it forwards the traffic over the interswitch connection (ISC), which means that this link might need to handle quite a bit of traffic. Plan appropriately—for example, use a link aggregation (trunk) of two or three 10G links for the ISC.

Note that traffic flowing from a VRRP router to an access layer switch on a distributed trunk takes a more efficient path. Both DT switches/VRRP routers learn the source MAC address on the distributed trunk, so the VRRP router forwards the return traffic on its own link to the access layer switch. (In "Chapter 9: Advanced OSPF," you will learn strategies for constructing an efficient traffic flow for traffic routed to the DT switches/VRRP routers from upstream routers *not* connected on distributed trunks.)

As you have learned, you can load-balance routing duties on the ProVision VRRP devices in two ways: by establishing multiple VRIDs with different masters on the same subnet or by using one VRID per subnet but alternating the master role on different subnets. These designs reduce the routing load on each VRRP router, but it only reduces the traffic load on the ISC somewhat. The access layer switches' choice for the link on which to forward traffic depends entirely on their link aggregation load-sharing mechanism and not on which DT switch is routing the traffic. Therefore, if the DT switch that receives the traffic does not happen to be the master for the virtual router to which the traffic is destined, it forwards the traffic on the ISC.

As an alternative to combining VRRP with distributed trunking, you can implement VRRP within a topology that uses spanning tree to provide data link redundancy.

Combining VRRP and MSTP

When you implement VRRP in a spanning tree topology, you should generally use MSTP. The benefits of MSTP, as opposed to RSTP, include:

- **Higher bandwidth (ability to utilize links more fully)**—Switch can use redundant links since each MSTP instance has a different active path. When a port belongs to multiple VLANs, it may be blocked in one MSTP instance, but forwarding in another instance.

  Note that distributed trunking offers the same benefit (with even better distribution of the load across multiple links).

- **Load balancing of the master role without overutilizing the link between the VRRP routers**—You might want to lower the burden on a single router by configuring different routers as the master in different subnets. In an RSTP topology, only one of the VRRP routers can be the root bridge with the active link to the downstream switches. That router must pass all traffic being routed in VLANs for which it plays the backup role over the link to the other VRRP router. That link can potentially become a bottleneck.

  When you use MSTP, on the other hand, you can create a topology in which downstream switches always forward traffic on the link to the master. As Figure 4-10 shows, you simply correlate the MSTP primary root bridge role in an instance with the VRRP master role for all VLANs mapped to that instance.

  Note that as far as minimizing traffic on the link between the VRRP routers is concerned, MSTP has the advantage over distributed trunking.

  Also note that using the Comware load-balancing mode somewhat reduces this advantage. You can still see some reduction in the traffic on the link by configuring different masters/root bridges

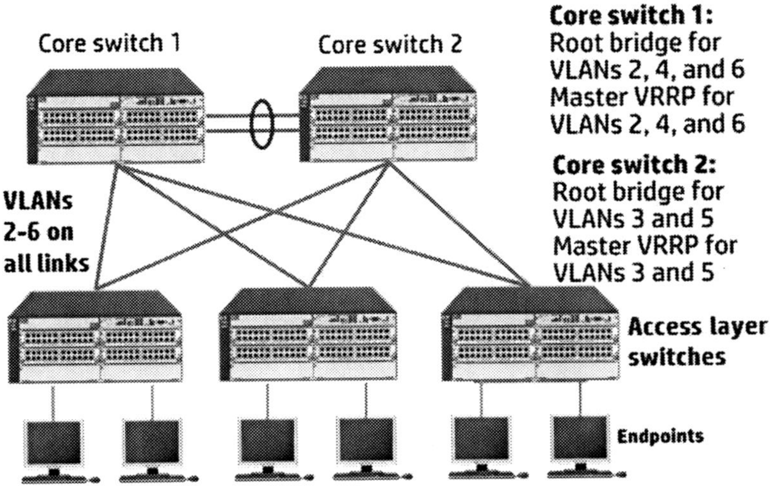

Figure 4-10: Combining VRRP and MSTP

in different VLANs, but the VRRP master/MSTP root bridge now passes roughly half of traffic to be routed to the other VRRP router. You need to decide whether load-balancing routing within a subnet is more important or reducing traffic on the VRRP routers' direct link.

Now you will learn the basic steps for setting up a VRRP and MSTP solution on your network.

Important

Whether you are using MSTP or RSTP, you should create a direct link that carries all VLANs between the VRRP routers, ensuring that the routers receive each other's VRRP advertisements and preventing disruptive situations in which both routers attempt to route traffic. This link must provide high reliability; for example, you might use a link aggregation. If you cannot guarantee that this link always remains up, you must extend all VLANs on all switch-to-switch links. This strategy ensures that even if the link between the VRRP routers goes down, the routers maintain connectivity in every VLAN no matter which blocked port transitions to the forwarding state.

Configure MSTP

- Create an MST instance for each redundant path. Because a virtual router often consists of two routing switches, you generally create two instances (as well as instance 0).
- Make sure that each core switch acts as the primary root bridge in one instance and the secondary root in another:
  - Configure the bridge priority for each MST instance on the core switches to make the correct primary and secondary root assignments.
  - Leave the bridge priority at its default setting on access switches to ensure that the core switches have higher priority.
- Map a set of VLANs to each instance. You can simply map half of the VLANs to each instance, or you can look at the traffic load on each VLAN and attempt to create two sets of VLANs, each of which supports half of the traffic load.

Configure VRRP

- Align the VRRP master state with the MSTP primary root role:
  - Each core switch should be the VRRP master for the VLANs in the instance for which it is the root bridge. On each VLAN that is mapped to the instance for which the switch is the primary root, set a high VRRP priority such as 254.
  - Likewise, the VRRP backup state should correspond to the MSTP secondary root bridge. You can keep the default priority of 100 on the VLANs mapped to the instance for which the switch is the secondary root.

- This configuration ensures that each access layer switch transmits traffic to be routed in each VLAN directly to the master. Also, if an uplink between an access switch and the VRRP master switch fails, the access switch can begin sending traffic to the backup VRRP switch.

Table 4-5 and Table 4-6 show how you might distribute MSTP root roles and VRRP master and backup states between the core switches.

Table 4-5: MSTP instances

MSTP instance	Mapped VLANs	Core switch 1 role (normal operation)	Core switch 2 role (normal operation)
1	2,4,6	Primary root (priority 0)	Secondary root (priority 1 or 4096)
2	3,5	Secondary root (priority 1 or 4096)	Primary root (priority 0)
0	1	Secondary root (priority 1 or 4096)	Primary root (priority 0)

Table 4-6: VRRP virtual routers

VLAN	VRID	Core switch 1 state (normal operation)	Core switch 2 state(normal operation)
1	2,4,6	Master (priority 254)	Backup (priority 100)
2	3,5	Backup (priority 100)	Master (priority 254)
0	1	Backup (priority 100)	Master (priority 254)

Monitoring VRRP with HP IMC

HP Intelligent Management Center (IMC) collects information about the VRRP configuration for devices that it is managing. As shown in Figure 4-11, the topology indicates devices implementing VRRP with a V. You can right-click those devices and select **VRRP Information**. A window shows each virtual router configured on that switch, the VRID, the virtual IP address, the master IP address, and backup router IP addresses.

You can monitor VRRP settings:

- **For managed devices**
- **In the IMC network topology**

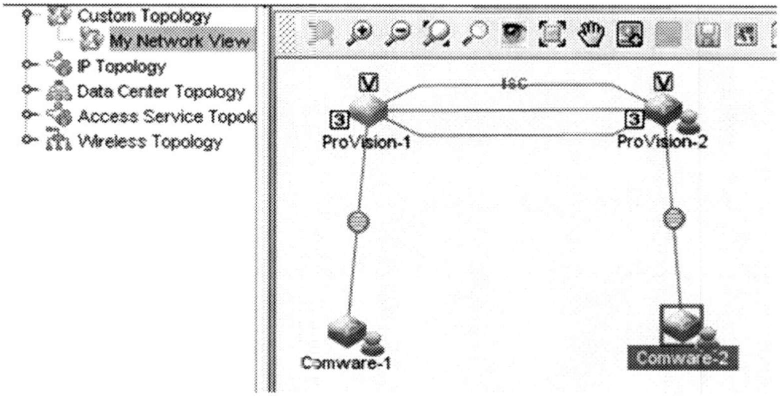

Figure 4-11: Monitoring VRRP with IMC

Summary

This chapter introduced you to VRRP. This redundancy protocol, when implemented correctly and combined with data link layer redundancy protocols, increases fault tolerance on a network, allowing users to continue sending traffic, even if network links or devices fail.

Learning check

Answer the following questions:

1. What are some of the responsibilities of the master router in a VRRP-enabled network?

2. What is a drawback to using the standard VRRP strategy for load-balancing routing within a subnet?

   a. Different endpoints in the same subnet use different IP addresses for their default gateway.

   b. This strategy usually offers less fine-grained load balancing than a strategy in which routers load-balance routing between subnets.

   c. This strategy does not permit companies to use DHCP to assign IP settings to endpoints.

   d. This strategy requires you to associate multiple virtual MAC addresses with a virtual router's virtual IP address.

3. What are guidelines for assigning VRRP states (or roles) in a network that uses MSTP?

Answers

1. What are some of the responsibilities of the master router in a VRRP-enabled network?

   The master, like all routers in the virtual router, knows the virtual IP address and virtual MAC address. However, only the master responds to ARP requests for the virtual IP address. Similarly, only the master accepts frames destined to the virtual MAC address. Just like any router, it decapsulates these frames and routes them.

   In addition to handling routing, the VRRP master is responsible for transmitting VRRP advertisements, which it sends every advertisement interval to multicast address 224.0.0.18.

2. What is a drawback to using the standard VRRP strategy for load-balancing routing within a subnet?

   a. Different endpoints in the same subnet use different IP addresses for their default gateway.

   b. This strategy usually offers less fine-grained load balancing than a strategy in which routers load-balance routing between subnets.

   c. This strategy does not permit companies to use DHCP to assign IP settings to endpoints.

   d. This strategy requires you to associate multiple virtual MAC addresses with a virtual router's virtual IP address.

3. What are guidelines for assigning VRRP states (or roles) in a network that uses MSTP?

   You should align the VRRP master state with the MSTP primary root role. On each VLAN in the MSTP instance for which the switch is the primary root, set a high VRRP priority such as 254. (Or configure the switch as the owner of that VLAN's virtual router, if you are using that approach.)

   Likewise, the VRRP backup state should correspond to the MSTP secondary root bridge. You can keep the default priority of 100 on the VLANs in the MSTP instance for which the switch is the secondary root.

   This configuration ensures that each access layer switch transmits traffic to be routed in each VLAN on its link to the master. But if an uplink between an access switch and the VRRP master switch fails, the access switch can begin sending traffic to the backup VRRP switch.

5 HP Intelligent Resilient Framework and Backplane Stacking

EXAM OBJECTIVES

In this chapter, you learn to:

- ✓ Design an HP Intelligent Resilient Framework (IRF) virtual device for the core, distribution, or access layer.
- ✓ Implement and manage an IRF virtual device.
- ✓ Describe what a split IRF stack is and configure the mechanisms designed to detect and remedy this problem.
- ✓ Describe the three topologies supported with backplane stacking and the roles members play in the stack.
- ✓ Explain how backplane stacking handles stack fragments.

ASSUMED KNOWLEDGE

Before studying this chapter, you should be familiar with IRF and know how it is implemented. You should be able to list the main benefits of using IRF and describe the roles of IRF members, IRF ports, member IDs, and IRF topologies.

In this chapter, you will build on this knowledge, learning how to implement IRF at each tier in a one-, two-, or three-tier architecture. You will also learn how to manage an IRF virtual device, avoid a split stack, and monitor IRF through HP Intelligent Management Center (IMC).

INTRODUCTION

This chapter focuses on two HP technologies that help you simplify your network architecture, operations, and management while providing a high level of redundancy and scalability.

- IRF
- HP backplane stacking

Available on many HP Comware switches, IRF can be used at the core, distribution, or access layer. Backplane stacking, which is available on the HP 3800 and 2920 Switch Series, is typically used at the access layer. As you will learn in this chapter, both technologies enable a two-tier architecture, which is designed to better meet the needs of organizations today.

Note

You should check the HP networking website to determine which switch series currently support IRF and backplane stacking.

Implementing HP IRF at the access, distribution, or core layer

When you plan an HP IRF virtual device, you should factor in where the IRF virtual device will be deployed in the network. Organizations have traditionally implemented a three-tier architecture, which includes the core, distribution, and access layers as shown in Figure 5-1.

Because networks have changed, HP recommends that you design a two-tier topology whenever possible. This topology provides the greatest simplicity and best performance for many network environments. Traffic flows directly from the edge to the core, reducing latency. One of the key technologies for implementing a two-tier architecture is IRF.

In the sections that follow, you will review the advantages of implementing IRF at each layer. Keep in mind that for each layer in both a two-tier and three-tier architecture, you will consider many of the same IRF benefits. However, you will consider how the benefits apply to each network layer.

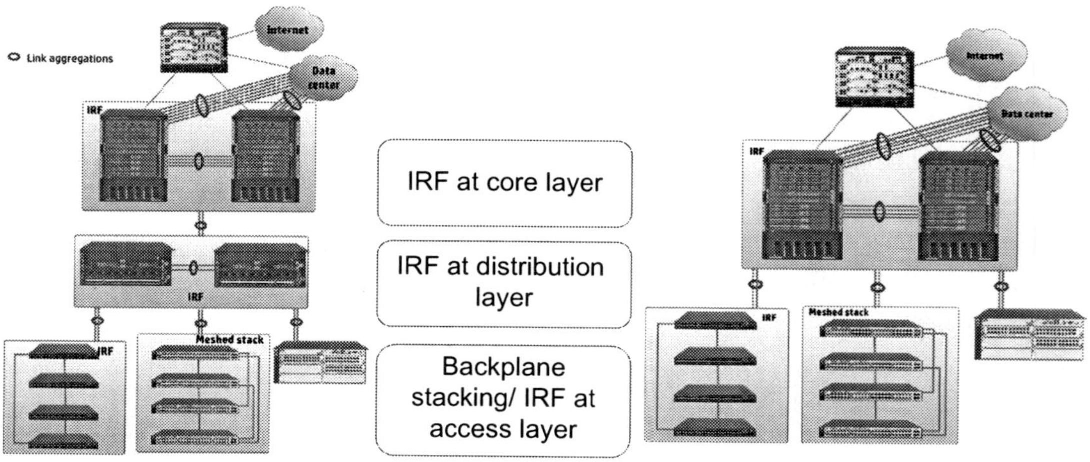

Figure 5-1: Implementing IRF at the access, distribution, or core

Advantages of implementing IRF at the access layer

Implementing IRF at the access layer provides a number of benefits. Keep in mind that backplane stacking, which you will learn more about later in this article, provides the same benefits.

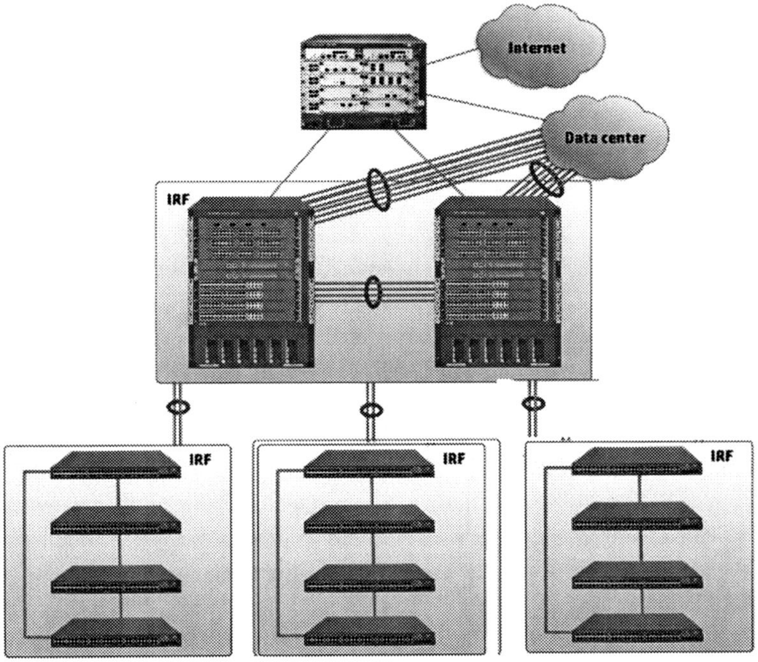

Figure 5-2: Implementing IRF at the access layer

- **Two-tier architecture**—IRF, whether implemented at the core or access layer or both, helps you to implement a two-tier architecture. For example, consider the access layer IRF virtual devices shown in Figure 5-2. Assume that these switches are deployed in a different building from the core, and you need to connect them to the core. It is not feasible to extend enough fiber to connect every access layer switch to a core switch—or even worse, if redundancy is required, to both core switches. Therefore, you have two choices. You can implement a distribution layer to aggregate the access layer connections into higher-bandwidth connections between the distribution switches and the core. Or you can connect several access layer switches to each other and have these switches share a couple of higher-bandwidth connections to the core. Traditionally, the distribution layer option provided the better choice; if you used the second option, MSTP or another spanning tree protocol blocked some of the links between the access layer switches. The design became error-prone and inefficient. On the other hand, when the access layer switches that share uplinks are configured as an IRF virtual device and are connected with high-speed unblocked IRF links, this design becomes much more efficient and resilient. Eliminating the distribution layer and saving the cost of that equipment becomes a real possibility.

- **Redundancy**—IRF provides redundancy. Although organizations typically understand the need to provide redundancy at the core layer, they should not ignore the access layer. In some environments, servers or even users connecting at the access layer provide mission-critical or even life-critical services. Services are considered life critical if downtime could potentially cause physical harm or loss of life. For such services, the organization must, of course, ensure 100% availability. Services are considered mission critical if downtime will disrupt core business operations, result in revenue loss, decrease productivity, or harm the organization's reputation. For mission-critical services, organizations must ensure at least 99.99% availability.

  Most organizations require N + 1 link redundancy—in other words, at least two uplinks. Keep in mind that if one of the links fails, the bandwidth will drop, which might decrease quality. (The fewer the number of uplinks, the greater the effect.)

  Organizations also need device-level redundancy.

  Because the IRF virtual device acts like a single switch, it becomes easy to design both types of redundancy for the mission- or life-critical servers. Whatever the NIC teaming solution used by a particular server, the server can connect to multiple physical chassis within the IRF virtual device; yet, to the server, the connection appears transparently as a link to a single upstream device. If one of the switches in an IRF virtual device fails, the other switches seamlessly take over the failed switch's functions.

- **Resiliency**—IRF provides resiliency. Closely related to redundancy, resiliency refers to the network's the ability to quickly adapt to change and to recover from errors such as link and hardware failures. IRF has rapid failover, as shown in Table 5-1.

Table 5-1: Failover in an IRF virtual device

Scenario	Failover time
Link aggregation: port removal/insertion	2 ms/0.7 ms
Link aggregation: board removal/insertion	2 ms/1 ms
Chassis off/on	2 ms/0.14 ms
Software upgrade	2 ms

- **Port density**—IRF increases the number of available edge ports without adding management burden. For example, in a data center you could group two 5800AF-48G switches, giving you a total of 96 10/100/1000 ports and 12 1000/10000 SFP+ ports.

- **Simplified management**—You can manage all the switches in an IRF virtual device as a single virtual switch. IRF also eliminates the need for protocols such as VRRP and MSTP, making it easier to manage the network as a whole.
- **Load balancing across multiple links**—You can connect the IRF virtual device to other switches using aggregated links. An aggregated link load balances traffic more efficiently than redundant links that implement MSTP. Uneven load balancing means that you cannot simply sum the bandwidth on the uplinks and assume that all of that bandwidth will be available. With MSTP, for example, these imbalances can be significant.
- **Elimination of MSTP and other network protocols**—You can eliminate MSTP, making the network more responsive while providing a deeper level of redundancy and faster failover times. IRF fails over in 2 ms or less. By comparison, MSTP fails over in 1 to 6 seconds. In addition, MSTP can be difficult to troubleshoot, adding to organizations' management burden. IRF also eliminates the need for other network protocols such as VRRP, further reducing complexity.
- **Access layer routing**—By grouping multiple access layer devices into a single virtual device, IRF makes it simpler to implement access layer routing solutions. You simply have fewer routers to support. In this way, IRF promotes efficient handling of traffic, in which access layer IRF virtual devices route traffic as it enters the network, rather than forwarding it to the core or distribution layer.

  Within the virtual device itself, IRF adopts an efficient, distributed forwarding strategy for Layer 2 and Layer 3 packet forwarding, making use of the processing capability of each member. Each member has complete Layer 2 and Layer 3 forwarding capabilities. For example, when a member receives packets to route at Layer 3, it scans its Layer 3 routing table to identify the forwarding egress port as well as the next hop and then sends the packets to the appropriate egress port, which might be on the local member or another member in the IRF virtual device. Because the IRF virtual device functions as a single virtual device, forwarding the packet to a port on any member is an internal action, which is not noticeable to outside devices. For Layer 3 packets, the hop number increases only by one no matter how many IRF members handle the packets as they are forwarded through the IRF virtual device. To the outside network, the packets travel one hop—as if one device routed the packets.

Best practices for implementing IRF at the access layer

Whenever you implement IRF, you should follow certain guidelines. For example, you should implement a ring topology because this topology provides the greatest level of redundancy.

It also provides the best protection against an IRF split system. (The IRF split system might also be called a fragment.) You will examine split IRF systems later in this chapter. For now, however, consider the added protection a ring topology provides against such a split. If any one link between two switches in the IRF virtual device fails, the IRF virtual device will remain intact and continue functioning. By comparison, if such a failure happened in a chain topology, a split system would be created.

As a secondary safeguard against IRF split systems, you should implement Multi-Active Detection (MAD). Although MAD will not prevent a split system, it will help the IRF virtual device to detect the split.

You should also configure a different IRF domain ID for each IRF virtual device. When you have two IRF virtual devices on the same network, assigning each one a unique domain ID prevents the systems from interfering with each other in case of a misconfiguration or incorrectly connected cable.

In addition, you should distribute uplinks across members in the IRF virtual device. This increases redundancy and improves performance.

You also need to determine the optimal number of switches to include in each IRF virtual device. For an IRF virtual device that is deployed at the access layer, you should plan for a maximum of four or five switches. This increases port density but allows the IRF virtual device to function efficiently. When you limit the number of switches to four or five and implement them in a ring topology, it reduces the number of hops between any two members in the IRF virtual device to one or two. This, in turn, minimizes member-to-member traffic on the IRF links and keeps the IRF virtual device performing well.

Advantages of implementing IRF at the distribution layer

Implementing IRF at the distribution layer optimizes the three-tier architecture (see Figure 5-3). The IRF virtual device provides the much needed redundancy for the distribution layer without introducing complexity. In addition, the IRF virtual device load shares traffic over the links that connect the IRF virtual device itself and the uplinks to upstream and downstream switches.

Further, the IRF virtual device simplifies management, allowing you to manage all the switches in the IRF virtual device through a single IP address.

The IRF virtual device also provides router redundancy. The switches within the IRF virtual device share routing information; if a switch becomes unavailable, all the other switches in the IRF virtual device have the routing information and will continue to route traffic. IRF provides a less complex routing redundancy solution than VRRP.

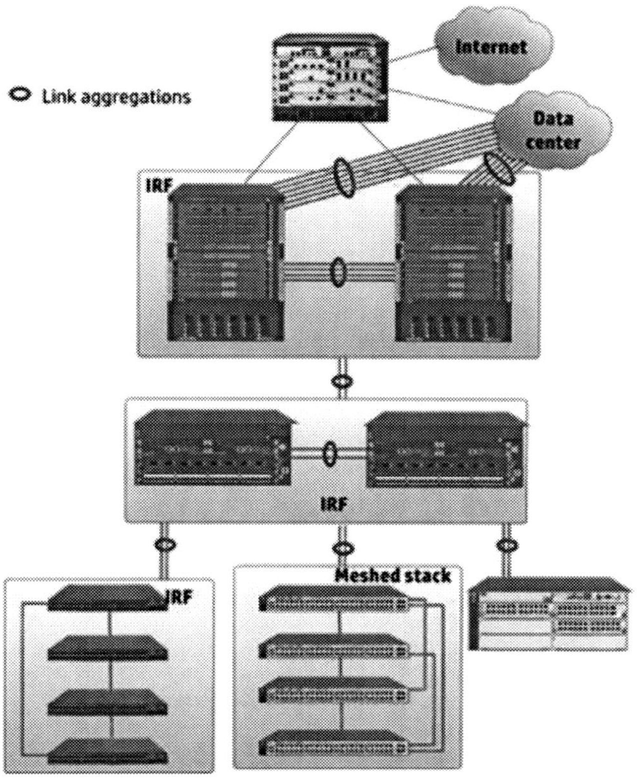

Figure 5-3: Advantages of implementing IRF at the distribution layer

Best practices for implementing IRF at the distribution layer

You should follow the general guidelines for implementing IRF virtual devices:

- Implement a ring topology.
- Protect the IRF virtual device from splits and detect if a split occurs:
    - Primary defense—ring topology
    - Secondary defense—MAD
- Configure a unique domain ID for every IRF virtual device.
- Distribute uplinks across members.

Other guidelines for implementing IRF at the distribution layer are a bit different from those for access layer switches. For an IRF virtual device at the distribution layer, you should plan for two members.

Because switches at the distribution layer handle so much traffic, you should connect the two switches in the IRF virtual device with multiple links. In general, try to keep the utilization on each IRF port under 70 or 80%, adding links if necessary to reduce the utilization.

You should also use aggregated links to connect the IRF virtual device to the core layer.

Advantages of implementing IRF at the core layer

For a two-tier solution, the core must fulfill the distribution layer switch roles such as routing and controlling the traffic (see Figure 5-4). For a three-tier solution, the core is simply routing and forwarding traffic as quickly as possible.

Redundancy for the vital core routing switches is a must in most environments, and IRF usually provides the simplest, most effective way to deliver redundancy. Once again, the advantages of using IRF include:

- Increased resiliency through the creation of multiple load-balancing links and elimination of all single point of failures at the network core
- Simplified topologies based on link aggregations rather than complex MSTP solutions
- Simplified management through a single management IP address for the virtual switch
- Simplified router redundancy using a single IP address rather than complex VRRP solutions

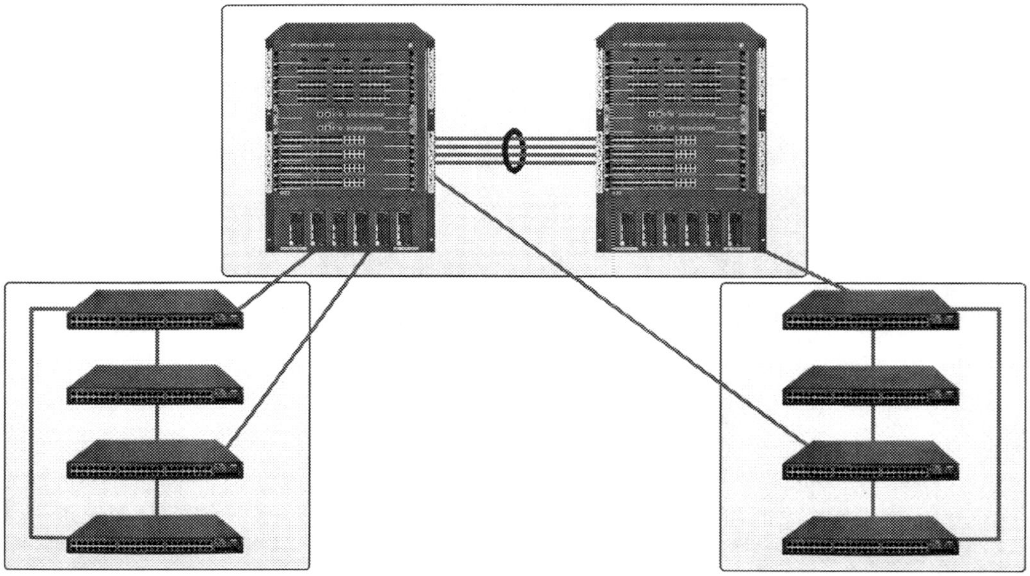

Figure 5-4: Advantages of implementing IRF at the core layer

Best practices for implementing IRF at core layer

As for the access layer and distribution layer, you should follow best practices for implementing an IRF virtual device. Although the modular Comware switches support up to four members in an IRF virtual device, they often include just two members for the core layer.

As you would for the distribution layer, you should connect the two switches in the IRF virtual device with multiple physical links (which are bound to an IRF port). For the core layer, you might begin with two to four links between the core switches. Again, check the utilization and try to keep the utilization on each IRF port under 70 or 80%. Add links if necessary to reduce the utilization.

For an optimized three-tier architecture, connect the distribution IRF virtual device to the core with aggregated links.

Configuring IRF: Option 1

After you plan your IRF implementation, you can turn your attention to the configuration process. This chapter presents two options for configuring the switches that will form an IRF virtual device.

Both options are designed to help you completely configure the IRF settings and save those settings before you allow the switches to connect and begin forming the IRF virtual device. If the two switches try to form an IRF virtual device before all the configuration settings are entered and saved, problems can occur.

This first option is designed to protect against configuration errors (see Figure 5-5). It ensures that the switches are physically unable to form a connection and begin establishing an IRF virtual device until after you finish configuring each one and save the configuration settings.

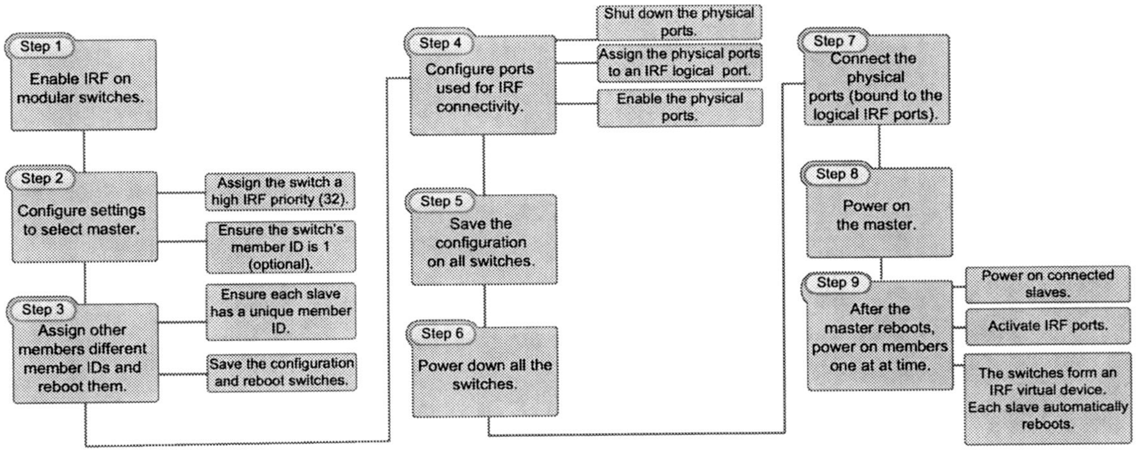

Figure 5-5: Configuration option 1

Before you begin configuring the switches, ensure they are physically disconnected.

1. If you are configuring a modular switch, you must enable IRF:

   ```
   [Switch] chassis convert mode irf
   ```

   The switch automatically reboots in IRF mode.

Note

The commands used to configure IRF vary slightly among the different switch models that support this feature. Please check the documentation for the switch model you are using.

---

2. Configure the master.

   When you implement IRF, you should determine which switch you want to function as the master. On this switch, configure a high IRF priority. To ensure this switch is always elected master, you should set the IRF priority to its highest number, which is 32:

   ```
   [Switch] irf member 1 priority 32
   ```

   HP networking also recommends that you ensure this switch has a member ID of 1, which is the default setting when IRF is enabled. This is not required, but using this convention will simplify the process of managing and troubleshooting an IRF virtual device.

3. Assign all other members a different member ID, save the configuration, and reboot members. If you do not assign each member a unique member ID, the IRF virtual device cannot be established:

   ```
   [Switch] irf member <member-id> renumber <new-member-id>
   [Switch] save
   [Switch] quit
   <Switch> reboot
   ```

   When you change the member ID of a switch, you must reboot the switch:

   - If you do not reboot the switch, the original member ID remains in effect, and all physical resources are identified by the original member ID. If another member is using the same member ID, the IRF virtual device will not be established.
   - If you save the current configuration and reboot the switch, the new member ID takes effect, and all physical resources are identified by the new member ID.

Note

Modifying a member ID may change the switch configuration, resulting in the loss in some configuration settings. When configuring an IRF virtual device, however, it is best practice to return slaves to factory default settings. Only the master should have a configuration (and you should save a copy of the configuration to a safe location on your network).

---

4. Configure the ports used for IRF connectivity on all switches:
   a. Shut down the physical ports that you want to bind to the logical IRF ports.
   b. Assign the physical ports to an IRF logical port. If the physical ports are not shut down, the switch will not let you complete this step:

```
[Switch] irf-port memberid/port-number
[Switch-irf-port 2/2] port group interface <interface-type interface-number>
```

   c. Enable the physical ports. To follow best practices, do not physically connect the ports yet.
5. Save the configuration on all the switches.
6. Power down all the switches.
7. Physically connect the 10 GbE ports on the switches. When connecting ports, remember to connect the physical ports that are bound to IRF port 1 on one switch to the physical ports bound to IRF port 2 on another switch.
8. Power on the switch you configured to be the master.
9. After the master powers up, power on the switches that are connected to the master and add them to the IRF virtual device:
   a. On all of the switches, enable the IRF port configuration (if this is an option on your switches):

```
[Switch] irf-port-configuration active
```

   b. As the switches activate their IRF ports, they join the IRF virtual device and recognize that it has a master. They will automatically reboot.
   c. You can then power on and add other members, if any.

Configuring IRF: Option 2

The second option for configuring IRF is designed to require the minimum number of reboots for the switches that will be slaves (see Figure 5-6). It is also designed for network administrators who must configure the switches from a remote location and cannot physically disconnect the switches, power them down, and power them up.

Because this option provides fewer safeguards for configuration errors, you must be careful to perform the steps in the order listed. In particular, you must enable the ports that connect to other switches in the IRF virtual device in the order outlined below so that you can save the configuration changes before each switch tries to join the IRF virtual device and reboot.

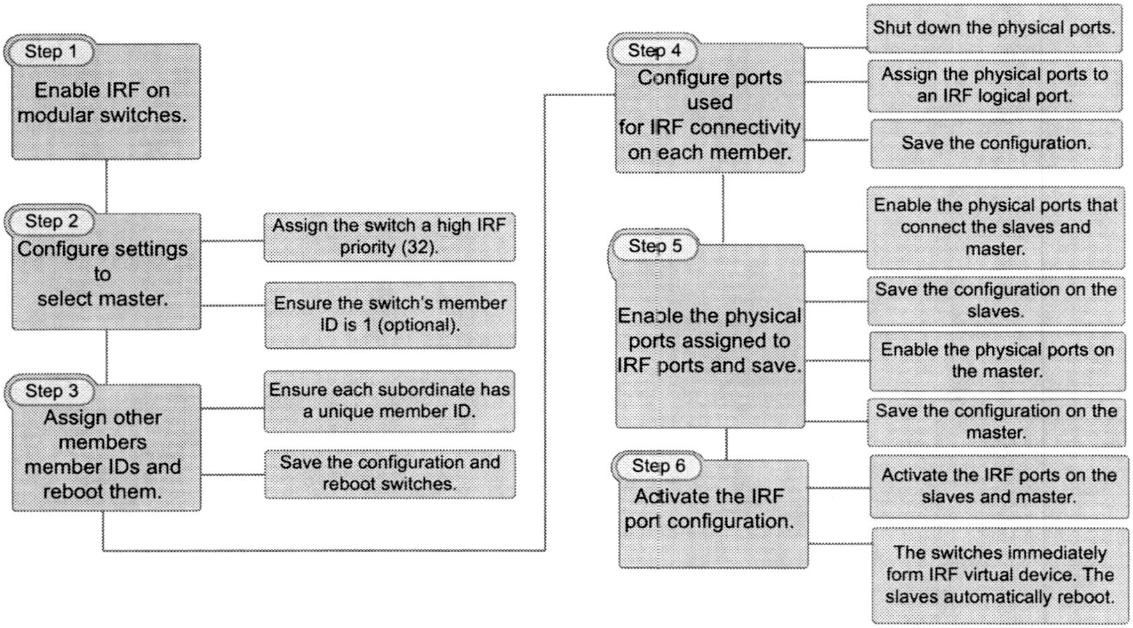

Figure 5-6: Configuration option 2

The first three steps are exactly the same as the first three steps for option 1:

1. If you are configuring a chassis-based switch, you must enable IRF:

```
[Switch] chassis convert mode irf
```

The switch automatically reboots in IRF mode.

Note

The commands used to configure IRF vary slightly among the different switch models that support this feature. Please check the documentation for the switch model you are using.

---

2. Configure the master:
   a. Set the IRF priority to its highest number, which is 32:

```
[Switch] irf member 1 priority 32
```

   b. Ensure this switch has a member ID of 1, which is the default setting when IRF is enabled.
3. Assign all other members a different member ID, save the configuration, and reboot members:

```
[Switch] irf member <member-id> renumber <new-member-id>
[Switch] save
```

```
[Switch] quit
<Switch> reboot
```

4. Configure the ports used for IRF connectivity on all switches but do not enable the physical ports yet:
   a. Shut down the physical ports that you want to bind to the logical IRF ports.
   b. Assign those ports to an IRF logical port. If the physical ports are not shut down, the switch will not let you complete this step:

```
[Switch] irf-port member_id/port-number
[Switch-irf-port 2/2] port group interface <interface-type interface-number>
```

   c. Save the configuration on all the switches.
5. Finish configuring the physical ports used for IRF connectivity.
   a. Enable the physical ports associated with the IRF ports on slaves that connect to the master.
   b. Save the configuration on slaves.
   c. Enable the physical ports on the master.
   d. Save the configuration on the master.

Note

Make sure to save the configuration as soon as you enable the ports on each device. If you forget to save, a device might join the IRF virtual device, reboot with its IRF physical ports disabled, and be unable to join the device again.

Although the slaves will not form the IRF virtual device until you activate the IRF port configuration, it can be a good idea to enable the ports in a particular order. Then you can be doubly sure that you can save this configuration change on each slave before it forms an IRF virtual device and reboots. You can also avoid forming IRF virtual devices between slaves without the master. Follow the order outlined below, based on the number of switches in the IRF virtual device:

- **Two-member IRF virtual device**—If you are implementing a two-member IRF virtual device (with a master and one slave), you can simply enable the ports on the slave and save the changes.
- **Three-member IRF virtual device**—If you are implementing a three-member IRF virtual device (with a master and two slaves), you should enable the physical ports that connect to the master and save the changes.

If you are implementing an IRF virtual device with a ring topology, do not enable the physical ports that connect the two slaves. If you mistakenly do this and activate the IRF port configuration too soon, the two slaves will form an IRF virtual device. Then when you try to connect the intended master, it will detect an existing IRF virtual device with an existing master. In this case, what will happen?

- **Four-member IRF virtual device**—If you are implementing a four-member IRF virtual device (with a master and two slaves), enable only the physical ports on the two switches that are directly connected to the master. Again, if the IRF virtual device is set up in a ring topology, enable only the physical ports on the switches that connect to the master. Save the changes.

6. Activate the IRF port configuration on each of these switches.

   As soon as you activate the IRF port configuration, the switches begin to exchange IRF messages. They elect the switch with the highest priority as the master and form an IRF virtual device. The slaves then reboot.

You now have an operational IRF virtual device.

Remaining configuration steps for a three-member IRF virtual device

7. If you have a three-member IRF virtual device that is connected in a ring topology, you can now access the IRF virtual device's CLI and enable the physical ports that connect the two slaves.

8. Save the changes.

Remaining configuration steps for a four-member IRF virtual device

9. If you have a four-member IRF virtual device, you must add member 4 to the IRF virtual device:

   a. Access the IRF virtual device's CLI and ensure that the ports that connect to member 4 are shut down.

      In the daisy chain example shown in Figure 5-7, you would ensure that the member 3 ports that connect to member 4 are shut down.

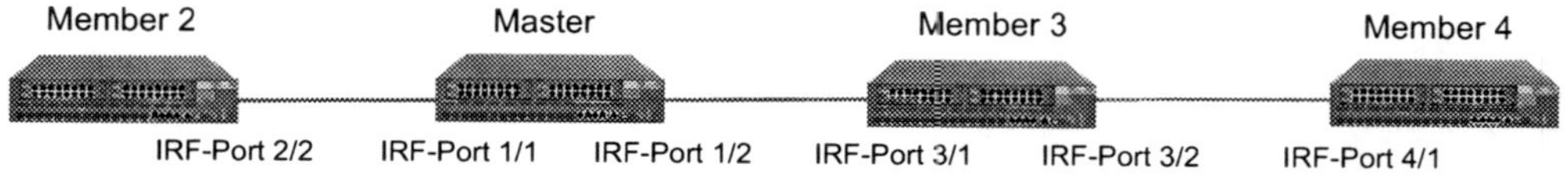

Figure 5-7: IRF daisy chain topology

      In the ring topology example in Figure 5-8, you would ensure that member 2 and member 3 ports that connect to member 4 are shut down.

   b. Then, access member 4's CLI and enable the physical ports that connect to the IRF virtual device. Save the changes.

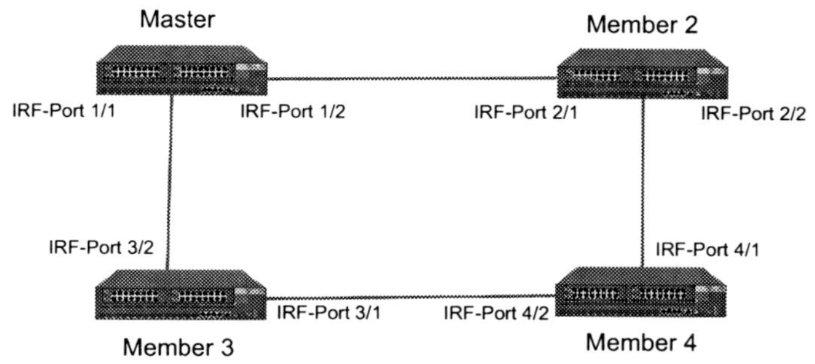

Figure 5-8: IRF ring topology

c. Access the IRF virtual device's CLI and enable the physical ports that connect to member 4.

d. Activate the IRF port configuration on member 4. Member 4 should merge with the IRF virtual device and immediately reboot.

e. Save the configuration changes on the IRF virtual device.

Synchronizing the configuration files

To function as a single virtual device, the members in an IRF virtual device must use the same configuration file. When you begin to set up the IRF virtual device, you should ensure that the switches that you want to function as slaves have a minimal configuration. As the switches begin to establish the IRF virtual device, the slaves automatically identify the master, synchronize their configuration file with the master's configuration file, and then execute it. If the master switch fails, all slaves have the same configuration and can continue to operate without interruption of services.

When you make a configuration change to the master's configuration file, those changes are synchronized to each switch in the IRF virtual device (see Figure 5-9). If you enter the **Save** command to save the current configuration to the startup configuration file on the master, all slaves complete the same action.

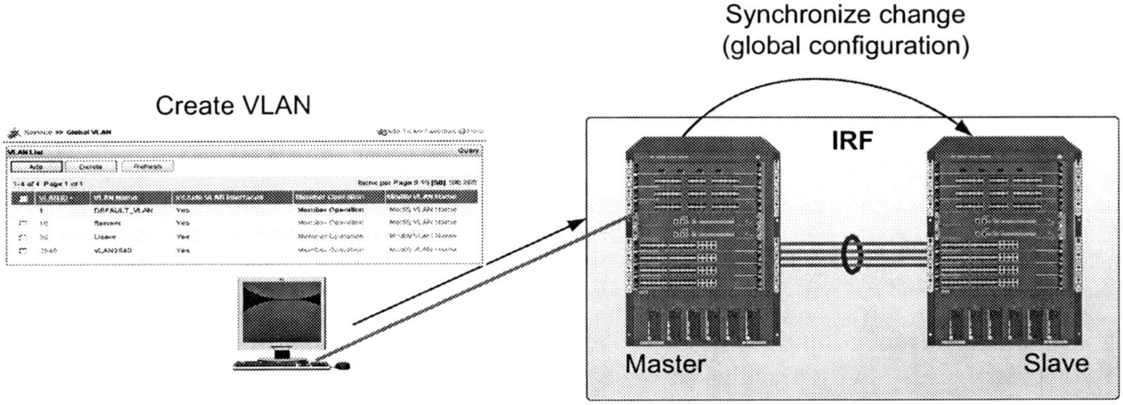

Figure 5-9: Synchronizing the configuration files in an IRF virtual device

The configuration file can be divided into two parts: global configuration and port configuration. When a slave applies these two kinds of configurations, it deals with them in different ways:

- **Global configuration**—All slaves apply the master's current global configuration.
- **Port configuration**—When a slave receives port configurations, it applies the configuration related to its own ports. For example, if a slave has a member ID of 3, it applies the configuration related to the GigabitEthernet 3/0/x port on the master. If a port configuration is not related to it, the slave uses the null configuration.

Load sharing across physical links bound to IRF ports

If you bind multiple physical links to one virtual IRF port, the switch load shares traffic across those physical links. You can configure the switch to load share traffic based on the following:

- Source IP address
- Destination IP address
- Source MAC address
- Destination MAC address
- Combination of source and destination IP addresses
- Combination of source and destination MAC addresses

Each Comware switch uses a particular criterion to load share the traffic by default. Most Comware switches use a different criterion for Layer 2, Layer 3, or Layer 4 traffic. For example, the 12500 Switch Series uses the following criteria by default:

- At Layer 2, a 12500 switch uses source and destination MAC addresses.
- At Layer 3, a 12500 switch uses source and destination IP addresses.
- At Layer 4, a 12500 switch uses source and destination ports.

However, some Comware switches may use a different criterion. For example, some of the 5800 switches (the 5800-48G Switch with 1 Interface Slot, 5800-48G TAA Switch with 1 Interface Slot, 5800-48G-PoE+Switch with 1 Interface Slot, and 5800-48G-PoE+TAA Switch with 1 Interface Slot) load share traffic based on the port number on which the traffic is received. Other switches in the HP 5800 and 5820X Switch Series load share Layer 2 traffic based on the source and destination MAC addresses and Layer 3 traffic based on the source and destination IP addresses. (Check the documentation for the particular Comware switches you are using to form the IRF virtual device to determine how these switches load share across the physical links bound to IRF ports.)

You can configure load-sharing criteria for a specific IRF port or globally for the entire switch. The switch will apply the specific IRF port setting before the global setting. Note that if a criteria combination is not supported, the IRF virtual device displays an error message.

To configure the load-sharing criteria globally on the IRF virtual device, use the following command:

```
[Switch] irf-port load-sharing mode [destination-ip | destination-mac |
source-ip | source-mac]
```

To configure the load-sharing criteria for a particular IRF port, you must move to the IRF port view:

```
[Switch] irf-port <member-id/port-number>

[Switch-irf-port 1/1] irf-port load-sharing mode [destination-ip |
destination-mac | source-ip | source-mac]
```

Managing the bridge MAC address for the IRF virtual device

The IRF virtual device uses the master's bridge MAC address as its MAC address (see Figure 5-10). For example, in Layer 2 protocols such as MSTP and Link Aggregation Control Protocol (LACP), this is the MAC address associated with the IRF virtual device.

Because this bridge MAC address must be unique, you must determine how you want the IRF virtual device to handle the MAC address if the master leaves the IRF virtual device. By default, the IRF virtual device maintains the existing bridge MAC address for 6 minutes after the master leaves the IRF virtual device. If the master does not rejoin the IRF stack before this timer expires, the IRF

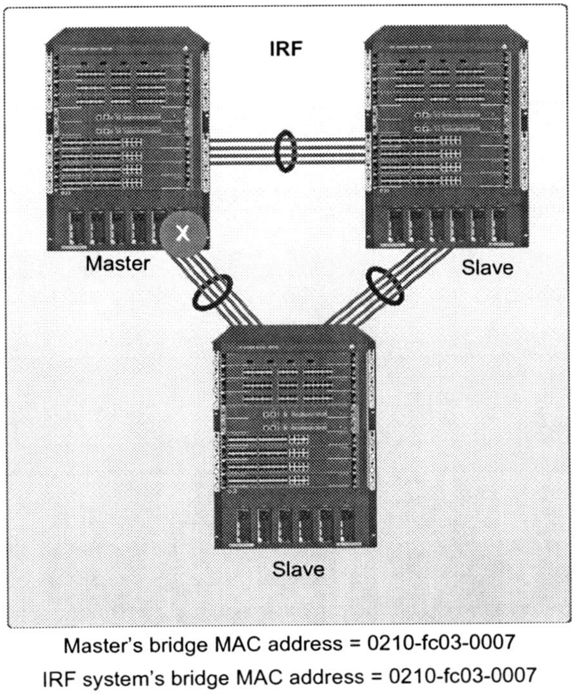

Figure 5-10: Managing the bridge MAC address for the IRF virtual device

virtual device begins using the bridge MAC address of the newly elected master as its bridge MAC address. This option avoids unnecessary switching of bridge MAC address due to a device reboot or transient link failure.

Rather than use the default setting, you can configure the IRF virtual device to maintain the existing MAC address or to immediately change its bridge MAC address after the master leaves. To configure this setting, use the following commands:

```
[Switch] irf mac-address persistent always
```

This command configures the IRF virtual device to use the current bridge MAC address even after the master leaves:

```
[Switch] undo irf mac-address persistent
```

This command configures the IRF virtual device to use the bridge MAC address of the newly elected master to replace the current address as soon as the former master leaves:

```
[Switch] irf mac-address persistent timer
```

This command configures the IRF virtual device to use the current bridge MAC address for 6 minutes after the master leaves the IRF virtual device.

ISSU upgrade process

When you need to upgrade an IRF virtual device, you should use the In-Service Software Upgrade (ISSU) process, which allows you to upgrade the IRF virtual device without disrupting the services it provides. You can also use ISSU for switches that have two main processing units (MPUs) and are operating in stand-alone mode.

HP provides two ISSU upgrade processes:

- **MPU-based ISSU**—With this process, you upgrade the master's MPUs one at a time. You can use the MPU-based ISSU upgrade method if:
  - The master in the IRF virtual device has two MPUs.

    The IRF virtual device has one active MPU, which manages the entire IRF virtual device. The master's MPU functions as the active MPU. If the master has two MPUs, one functions as a standby MPU.
  - The software version the switches are running is compatible with the new software version.
- **Member switch-based ISSU**—With this process, you upgrade members one at a time, starting with a slave. As with the MPU-based ISSU upgrade process, the switches must be running a software version that is compatible with the new software version.

In addition to preventing service disruption, ISSU supports version rollback. You can go back to the previous software version if the upgrade is not completed successfully or if you want to revert to the previous software version for any reason.

You will now take a more in-depth look at the ISSU process.

ISSU: Compatibility check

Before starting any upgrade process, you should review the release notes—not only to learn about fixes but also to find out known issues and workarounds. In particular, you should review the commands in the documentation to learn how they function and whether or not there are any restrictions. Avoiding an issue is much better than trying to fix issues that might occur during the actual upgrade process.

In addition to comparing the software running on the MPUs, the compatibility check examines the Line Processing Units (LPUs) and switch fabric modules and determines how they can be updated:

- **HotFix**—The LPU or switching fabric module can be updated through a hotfix. Because a reboot is not required, the flow of traffic is not interrupted.
- **Reboot**—The LPU or switching fabric module must be rebooted. The flow of traffic will be briefly interrupted during the reboot process.

It is important that you perform the compatibility check because the results determine whether or not you can use the ISSU process to upgrade the switch and which ISSU method you can use.

To determine the compatibility of the two software versions, use the following command:

```
<Switch> display version comp-matrix [ file upgrading-filename ]

[| { begin | exclude | include } regular-expression ]
```

You will receive one of the following responses:

- **Compatible**—The software version the switches are running is compatible with the new software. If the versions are compatible, you can use either ISSU upgrade method.
- **Incompatible**—The software version the switches are running is *not* compatible with the new software. You can continue with the upgrade, but you should use the member-based upgrade method.
- **Same**—The switch is already running the software version to which you want to upgrade.
- **Unknown**—The software version the switch is running is not included in the version compatibility database. You cannot use ISSU to update the switch software.

For example, if the new software version is named version2.bin, you would enter the following command:

```
<Switch> display version comp-matrix file version2.bin

Number of Matrices in Table = 1

Matrix for HP 10504

Running Version: 1104

Version Compatibility List:

6606L08 (Compatible)

chassis 0 slot 2: Hotfix

chassis 0 slot 5: Hotfix

chassis 0 slot 6: Hotfix
```

As you can see from the output, the new software in this example is compatible with the previous version. In addition, the output shows that the LPUs and switching fabric modules can be updated with a hotfix—which is part of the ISSU upgrade process.

ISSU prerequisite checklist

Before performing an ISSU upgrade, you should complete this checklist, which will help ensure that the upgrade is completed successfully.

- Save the current configuration, using the **save** command.
- If you want to use the MPU-based ISSU upgrade process:
  - Ensure that the master and the slaves have two MPUs and all interface cards operate normally.
  - Verify the running status and software version of each card. To check this information, use the **display device** command. Ensure that the active MPU and the standby MPU are running the same software version. The software files for both MPUs will have the same file path, name, and content. Use the **display boot-loader** command to verify this information.
  - Determine the active or standby state of each MPU and ensure that the standby MPU is in a real-time backup state. Use the **display switchover state** command to verify this information.
- Copy the new system software to the root directory of the storage medium for both active and standby MPUs. If you are using the member-based ISSU process, copy the new system software to the master and the slaves. (You can use FTP, TFTP, or other methods to copy the software.)

- Use the **dir** command to verify that:
  - The new system software and the old system software exist on both active and standby MPUs or on all members.
  - The file paths and file names are the same on these MPUs or members.

MPU-based ISSU upgrade

As you learned earlier, the MPU-based ISSU upgrade method has two requirements:

- The master in the IRF virtual device must have two MPUs.
- The switches must be running software versions that are compatible with the new software version.

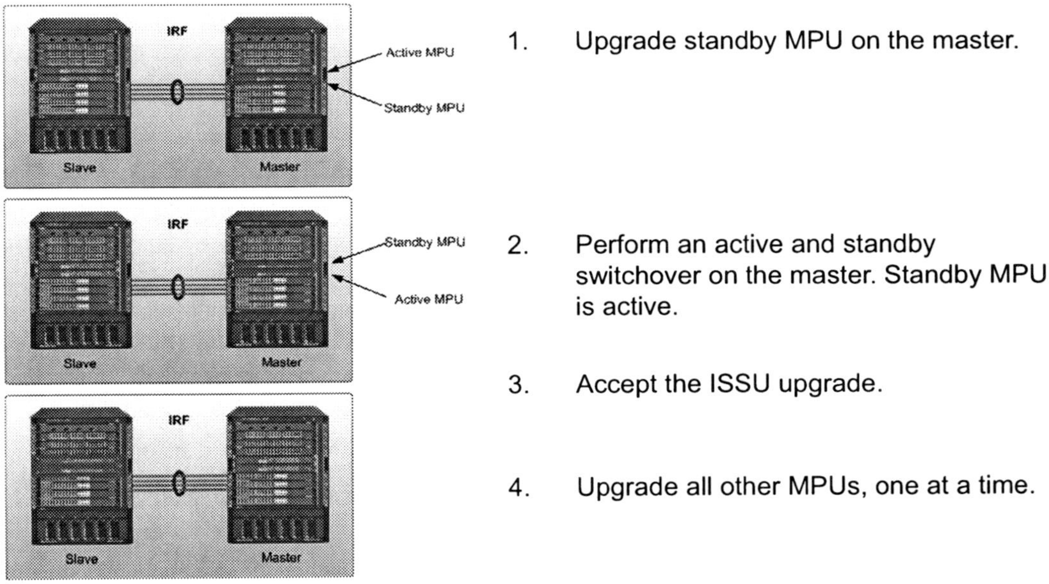

Figure 5-11: MPU-based ISSU upgrade

If your IRF virtual device meets these requirements, you can follow these steps to perform an MPU-based ISSU upgrade (see Figure 5-11):

1. Upgrade the standby MPU on the master. This upgrade will not affect the active MPU, allowing it—and the IRF virtual device—to operate normally:

   ```
   [Switch] issu load file <filename> chassis <chassis-number> slot
   <slot-number>
   ```

   The standby MPU will reboot. After the reboot is completed, you can move to the next step.

2. Perform an active and standby switchover to make the newly updated standby MPU the active MPU:

```
[Switch] issu run switchover chassis <chassis-number> slot <slot-number>
```

   The upgraded standby MPU will become the active MPU. The former active MPU will reboot and come back as the standby MPU.

3. Accept the ISSU upgrade (optional):

```
[Switch] issu accept chassis <chassis-number >
```

4. Upgrade all the other MPUs, one by one.

This upgrade process provides nonstop service because the IRF virtual device ensures the data and state of all MPUs remain synchronized. The interface cards are upgraded by installing a hotfix and do not require a reboot.

Version rollback

ISSU allows you to roll back the upgrade to the previous version. For example, you may want to cancel an upgrade if you discover the software image is not compatible with your current configuration settings.

Version rollback is triggered by either of the following events:

- **Autorollback**—When the **issu load** command is entered, the IRF virtual device automatically creates a version rollback timer. If you do not execute the **issu accept** or **issu commit** command before the rollback timer expires, the IRF virtual device automatically stops the ISSU process and reverts to the previous software version. If you enter the **issu accept** or **issu commit** command, the IRF virtual device deletes the rollback timer and does not allow a version rollback.
- **Manual rollback**—When your switch is in the load, switchover, or accept state, you can use the **issu rollback** command to reboot the upgraded cards with the previous version of the system software image or to delete the upgrade hotfixes. The system reverts to the previous version and the MPUs resume their former roles.

  For MPU-based ISSU upgrades, enter:

```
[Switch]issu rollback chassis <chassis-number> slot <slot-number>
```

  For member switch-based ISSU upgrades, enter:

```
[Switch] issu rollback chassis <chassis-number>
```

During the ISSU upgrade, a switch may be in one of the states listed in Table 5-2.

To view the ISSU state, use the **display issue state** command.

Table 5-2: ISSU states

States	Failover time
Init	ISSU upgrade process has not been started, or the process has not been completed
Load	Standby MPU is being upgraded or has been upgraded. The rollback timer has been started
Switchover	Upgraded standby MPU has become the new active MPU, and LPUs and switching fabric modules are being upgraded
Accept	Upgrade of the new active MPU has been accepted, and the rollback timer has been deleted
Commit	Multiple standby MPUs are available, and at least one standby MPU has not been upgraded to a new version. In this state, neither the manual nor the autorollback can be performed

Member-based ISSU upgrade

Rather than upgrading each main board one by one, you can upgrade each of the member switches one by one. This method does not require upgrade hotfixes in the boot file. However, each member switch is upgraded through a switch reboot, which might cause a very brief interruption to the traffic passing through the member switches. However, the other members continue handling traffic.

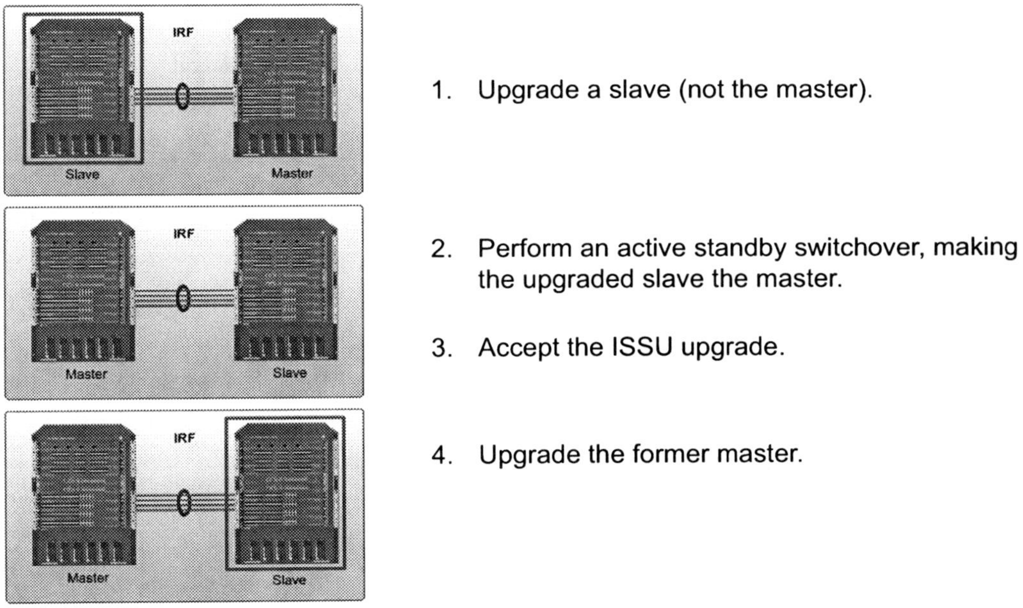

Figure 5-12: Member-based ISSU upgrade

For this upgrade process, you complete these steps (see Figure 5-12):

1. Upgrade a slave in the IRF virtual device. This upgrade will not affect the master, which will continue to operate normally. (If you have more than two members in the IRF virtual device, upgrade all of the slaves, except the master, at this time.)

```
[Switch] issu load file <filename> chassis <chassis-number>
```

2. Perform an active and standby switchover, making an upgraded member become the new master:

```
[Switch] issu run switchover chassis chassis-number
```

3. Accept the ISSU upgrade (optional):

```
[Switch] issu accept chassis <chassis-number >
```

4. Upgrade the former master:

```
[Switch] issu commit chassis chassis-number
```

   Entering this command reboots the former master.

IRF split stack

If an IRF link failure occurs and members in an IRF virtual device cannot communicate, it could cause an IRF split stack, as shown in Figure 5-13. If this occurs, two separate IRF virtual devices are formed. Because both virtual devices are unaware of the other, each will elect a master and use the IP addresses and configuration settings assigned to the original IRF virtual device.

You can immediately see the issues such a split stack would cause. Address collisions would occur, creating havoc on the network.

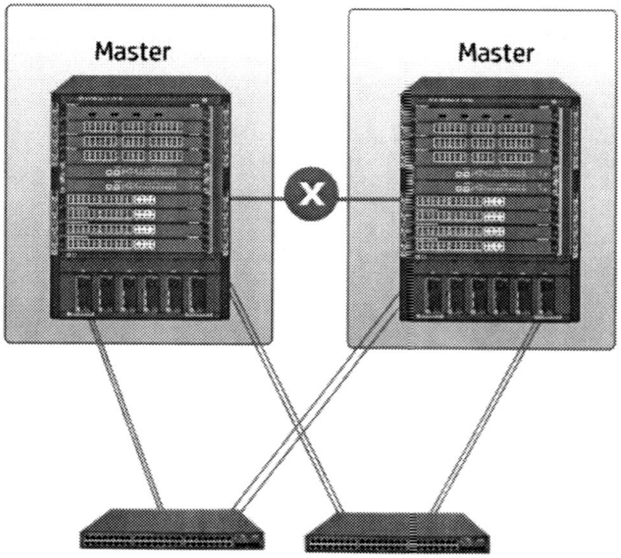

Figure 5-13: IRF split stack

IRF includes a mechanism, called multiactive detection (MAD), to quickly discover IRF split stacks. MAD mitigates the effect of the split stack on the network. It first prevents address conflicts by permitting only one IRF virtual device to be active. Once this problem is resolved, MAD tries to reestablish the failed link and reestablish the IRF virtual device.

You can implement three types of MAD:

- Link Aggregation Control Protocol (LACP) MAD
- Bidirectional Forwarding Detection (BFD) MAD
- Address Resolution Protocol (ARP) MAD

You will now learn more about each one.

Detecting IRF split stacks with LACP MAD

With LACP MAD, you connect each IRF member to an intermediate switch, which must be a Comware switch. The intermediate switch is connected to the IRF virtual device with dynamic aggregated links.

After you configure the IRF virtual device to support LACP MAD, the IRF members send extended LACP data units (LACPDUs) that include a type length value (TLV) (see Figure 5-14). The IRF members use the TLV to specify the active ID of the IRF virtual device and the domain ID (if one has been configured). The active ID matches the member ID of the current master.

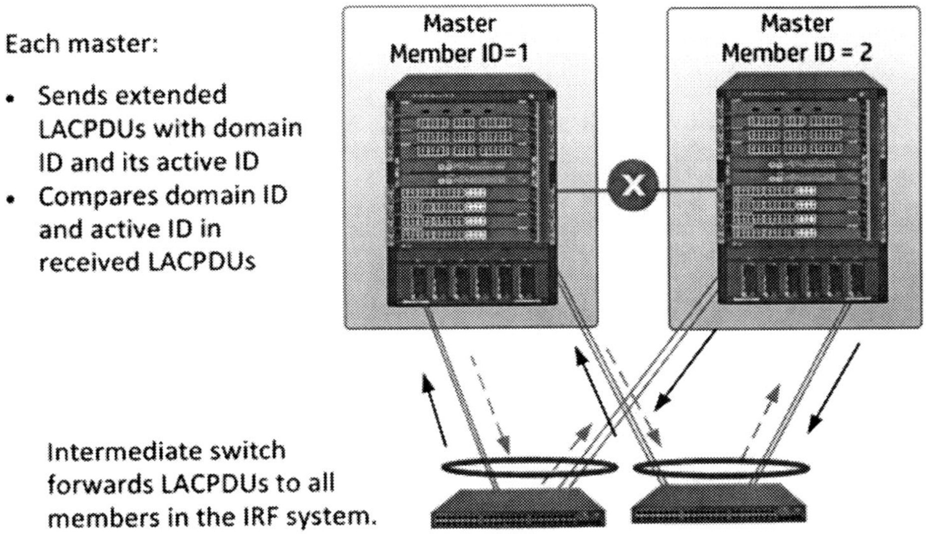

Figure 5-14: Detecting IRF split stacks with LACP MAD

Because the intermediate switch runs Comware, it can interpret the LACPDUs, which include the TLV, and transparently forward the extended LACPDUs received from each IRF member to all other member switches.

When an IRF member switch receives an LACPDU, it compares the domain ID and the active ID with its own domain ID and active ID:

- If the domain ID in the received LACPDU is different than its own domain ID, the member switch recognizes that a switch in another IRF virtual device sent the LACPDU and quits processing it.
- If the domain ID in the received LACPDU matches its own domain ID, the member switch compares the active ID to its own active ID:
  - If the active IDs do not match, the member switch recognizes that the IRF virtual device has split. This triggers MAD to take action to prevent address conflicts. You will learn more about the actions MAD takes later in this chapter.
  - If the active IDs are the same, the member switch recognizes that the IRF virtual device is functioning properly.

To set up LACP MAD, you need an intermediate switch that runs Comware software. Comware switches can properly handle the extended LACPDUs, which include the TLV containing the domain ID and active ID.

You then create an aggregated link between the IRF virtual device and the intermediate switch as shown in the figure. Because the switches must exchange LACPDUs, the aggregated link must run in dynamic mode.

To set up LACP MAD, you complete the following steps. (Note that you should have already configured the IRF virtual device.)

1. Ensure that the IRF virtual device has a domain ID. If it does not, enter:

   ```
   [Switch] irf domain <domain-ID>
   ```

2. Create a dynamic aggregated link:

   ```
   [Switch] interface bridge-aggregation <interface-number>
   [Switch-Bridge-Aggregation1] link-aggregation mode dynamic
   ```

3. Enable LACP MAD:

   ```
   [Switch-Bridge-Aggregation1] mad enable
   [Switch-Bridge-Aggregation1] quit
   ```

4. Add ports to the aggregation interface:

   ```
   [Switch] interface gigabitethernet 1/0/1
   ```

```
[Switch-GigabitEthernet1/0/1] port link-aggregation group 1
[Switch-GigabitEthernet1/0/1] quit
[Switch] interface gigabitethernet 2/0/1
[Switch-GigabitEthernet2/0/1] port link-aggregation group 1
[Switch-GigabitEthernet2/0/1]quit
```

On the intermediate switch, you complete the following steps:

1. Create a dynamic aggregate interface:

```
[Sysname] interface bridge-aggregation 1
[Sysname-Bridge-Aggregation1] link-aggregation mode dynamic
```

2. Assign ports to the aggregate interface:

```
[Sysname] interface gigabitethernet 1/0/4
[Sysname-GigabitEthernet1/0/1] port link-aggregation group 1
[Sysname-GigabitEthernet1/0/1] quit
[Sysname] interface gigabitethernet 1/0/3
[Sysname-GigabitEthernet2/0/1] port link-aggregation group 1
[Sysname-GigabitEthernet2/0/1] quit
```

Excluding a port from being shut down by MAD

If MAD detects a split stack, it shuts down all ports except the IRF ports when the IRF virtual device moves to the recovery state. If a port must always be in the up state for some reason (such as the port supports a Telnet connection), you can "exclude" it, preventing IRF from shutting it down.

To exclude ports from being shut down if MAD detects a split stack, enter:

```
[Switch] mad exclude interface <interface-type interface-number>
```

Detecting IRF split stacks with BFD MAD

BFD MAD requires you to connect the IRF members with a link that is dedicated to running the Bidirectional Forwarding Detection protocol to detect multiactive IRF virtual devices. No other services should run on this link. Because spanning tree must not shut down this link, it is disabled on it.

As Figure 5-15 shows, BFD MAD also requires you to configure an IP address for the BFD interface on each IRF member. The IP address for the BFD interface on all switches must be in the same subnet and VLAN (which cannot be VLAN 1). The IP address identifies the IRF member during BFD MAD detection. Note that you must not assign the BFD VLAN to any links except the BFD MAD link.

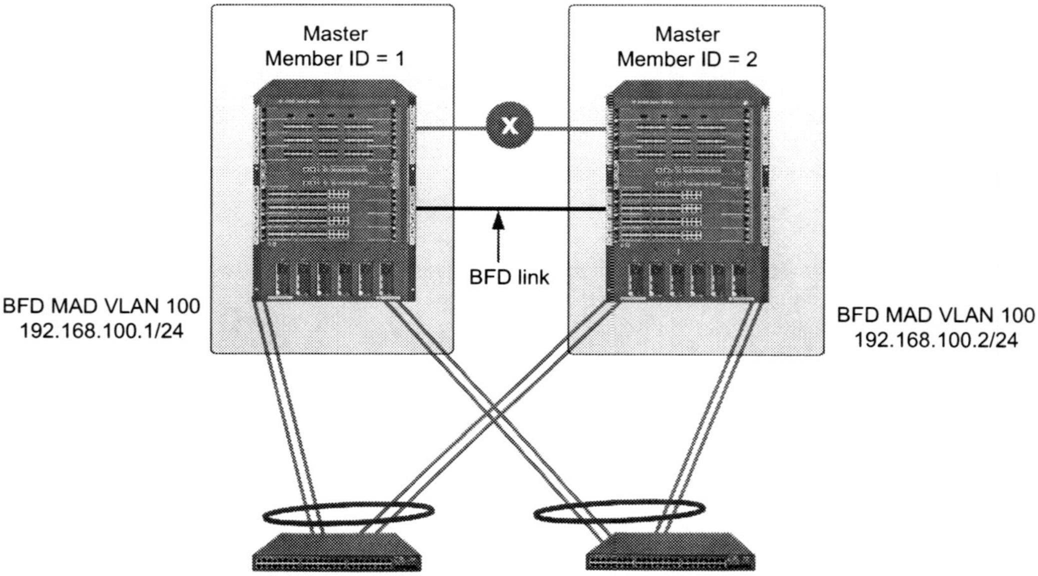

Figure 5-15: Detecting IRF split stacks with BFD MAD

When you enable BFD MAD, the master tries to establish a BFD session with the other IRF members. If the IRF virtual device is operating normally, only the MAD IP address of the master is active. If a failure occurs, creating an IRF split stack, the MAD IP addresses of both masters in the separate IRF virtual devices become active. This triggers MAD to take action to avoid address conflicts caused by the split stack.

Note

You cannot enable spanning tree on the BFD link.

To configure BFD MAD, you complete these steps:

1. Create the BFD MAD VLAN. You cannot use VLAN 1 for BFD VLAN:

```
[Switch] vlan <vlan-ID>

[Switch??] quit
```

2. Configure the VLAN settings for the ports that establish the BFD link. Note that BFD MAD does not require the ports to be an access port or a trunk port. (You must configure these steps for each port.)

```
[Switch] interface <interface-type interfacenumber>

[Switch-GigabitEthernet<ID>]port access vlan <vlan-ID>

[Switch-GigabitEthernet<ID>] quit
```

3. Enable BFD MAD:

```
[Switch] interface vlan-interface <interfacenumber>
[Switch-Vlan-interface100] mad bfd enable
```

4. Configure a BFD MAD IP address for both members:

```
[Switch-Vlan-interface100] mad ip address ip-address { mask | mask-
length } member <member-id>
```

If a port must always be in an up state (for whatever reason), you can exclude it from being shut down. See the "Excluding a port from being shut down by MAD" section.

Detecting IRF splits with ARP MAD

With ARP MAD, IRF virtual devices detect a split stack by sending extended gratuitous ARP packets. Each extended gratuitous ARP packet includes the active IDs of the IRF virtual devices.

To implement ARP MAD, you enable ARP MAD between IRF members or between each IRF member and an intermediate switch. The latter setup, as shown in Figure 5-16, is the more common implementation. This intermediate switch could be a switch that handles production traffic, but often you should dedicate a switch to this role, as shown in the exhibit.

When you use an intermediate switch, you enable MSTP on the IRF virtual device and the intermediate switch, and MSTP blocks the redundant links. During normal operation, each member of the IRF virtual device sends extended gratuitous ARP packets but does not receive other member's extended gratuitous ARP packets.

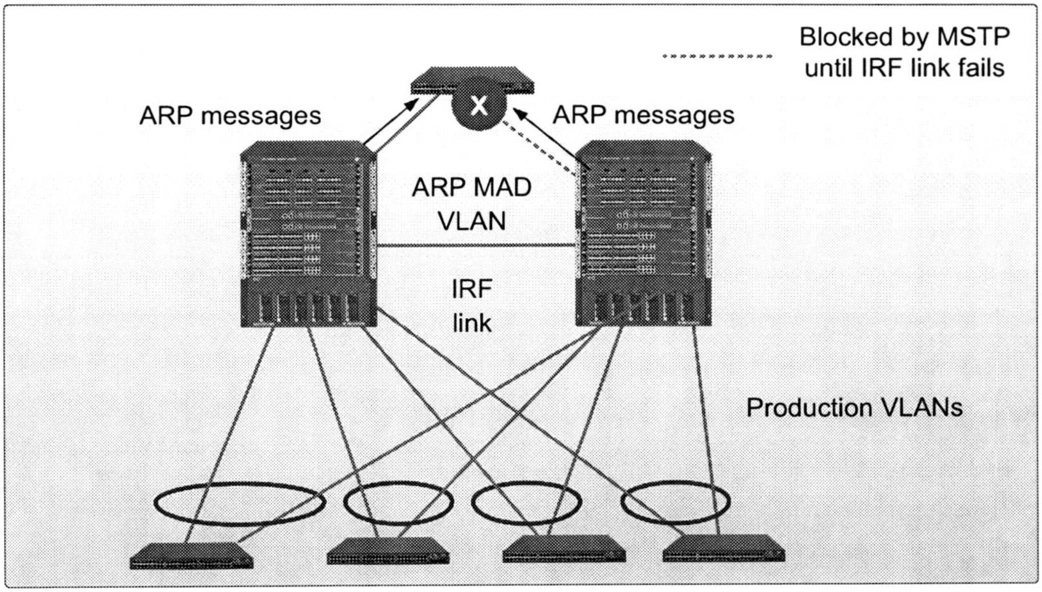

Figure 5-16: Detecting IRF splits with ARP MAD

If the IRF link between the members in the IRF virtual device fails, creating a split stack, MSTP unblocks redundant links. Members of the IRF virtual device will then begin to receive extended gratuitous ARP packets from other members. Because the extended gratuitous ARP packets use the same IP addresses but different MAC addresses, MAD will be activated to resolve the address conflicts. Each IRF member will begin to compare its active ID to the active IDs in the extended gratuitous ARP packets it receives from other members. You will learn exactly how MAD uses this information to resolve the split stack in the next section.

LACP MAD lets you use the link aggregations that you would implement between an IRF virtual device and other switches in any case. ARP MAD, on the other hand, requires a dedicated VLAN and MSTP, which would complicate the topology if you selected an existing switch as the intermediate switch. Therefore, it is less often implemented than LACP MAD. The split stack detection time also depends on MSTP failover times, so it is also implemented less frequently than BFD MAD.

To configure ARP MAD for an implementation that includes an intermediate switch, complete these steps:

1. Identify the links on which the IRF members will send gratuitous ARP packets.
2. Enable MSTP on the IRF virtual device.
3. Configure the IRF virtual device to change its bridge MAC address immediately if the master leaves the IRF virtual device:

   ```
   [Switch] undo irf mac-address persistent
   ```

4. Create a VLAN that is dedicated to ARP MAD:

   ```
   [Switch] vlan <vlan_id>
   [Switch-vlan3] quit
   ```

5. Assign the ports on which the IRF members will send the gratuitous ARP packets to this VLAN. Note that ARP MAD works on access ports or trunk ports, but the commands below are for an access port. (You must complete these steps for each port.)

   ```
   [Switch] interface <interface-type interfacenumber>
   [Switch-GigabitEthernet<ID>]port access vlan <vlan-ID>
   [Switch-GigabitEthernet<ID>] quit
   ```

6. Create the VLAN interface and assign it an IP address:

   ```
   [Switch] interface vlan-interface <vlan_id>
   [Switch-Vlan-interface100] ip address <ip_address>
   ```

7. Enable ARP MAD on the VLAN interface:

   ```
   [Switch-Vlan-interface100] mad arp enable
   ```

8. Configure MSTP on the intermediate switch.
9. Configure the VLAN that is dedicated to ARP MAD on the intermediate switch.
10. Connect the intermediate switch to each member of the IRF virtual device and add the ARP MAD VLAN to the links.

If a port must always be in an up state (for whatever reason), you can exclude it from being shut down. See the "Excluding a port from being shut down by MAD" section.

MAD: Detecting and resolving the split stack

When MAD detects the split stack, it initiates an election between the two IRF virtual devices. Specifically, MAD determines which of the two IRF virtual devices has the master with the smaller member ID. For example, suppose one IRF virtual device has a master with a member ID of 1, and the other virtual device has a master with a member ID of 4. In this example, the IRF virtual device that has a master with the member ID of 1 wins the election. MAD allows this IRF virtual device to operate normally, using the IP addresses and the other configuration settings assigned to the original IRF virtual device.

MAD then places the other IRF virtual device into a recovery state, which shuts down all of its ports except the IRF ones and those excluded from MAD. This action prevents the second IRF virtual device from using the IP addresses configured for the original IRF virtual device, thereby eliminating address conflicts on the network.

Note

You can exclude a port from MAD manually. You might do this when the connecting device uses the port for a purpose such as a Telnet connection.

After placing one device in recovery mode, MAD tries to automatically repair the failed IRF links. If the repair operation fails, however, you will need to manually repair the failed links. When the link is recovered, the IRF virtual device that is in recovery state automatically reboots, and the IRF virtual devices both in active state and in recovery state automatically merge into one—just as they were before the IRF split stack occurred.

Typically, recovery occurs automatically in this way. You would only need to implement recovery manually with the mad restore command in the rare event that MAD places one IRF device in recovery mode and the active device then fails. In that case, entering mad restore activates the IRF device shutdown by MAD recovery.

You realize the most benefit from MAD (and IRF) when you design the topology properly. Whenever possible, devices should have multiple link-aggregated connections to different chassis within the IRF virtual device. Therefore, when one IRF member is shut down by MAD, the connecting devices seamlessly fail over to the other links in their link aggregation group—just as they fail over seamlessly if one of the IRF members fails.

Using HP IMC to monitor the IRF virtual device

You can use IMC to monitor your IRF virtual device. HP IMC recognizes each IRF virtual device and displays it in the topology map as a single virtual switch as shown in Figure 5-17. This allows you to see the IRF virtual device in the bigger context of your network.

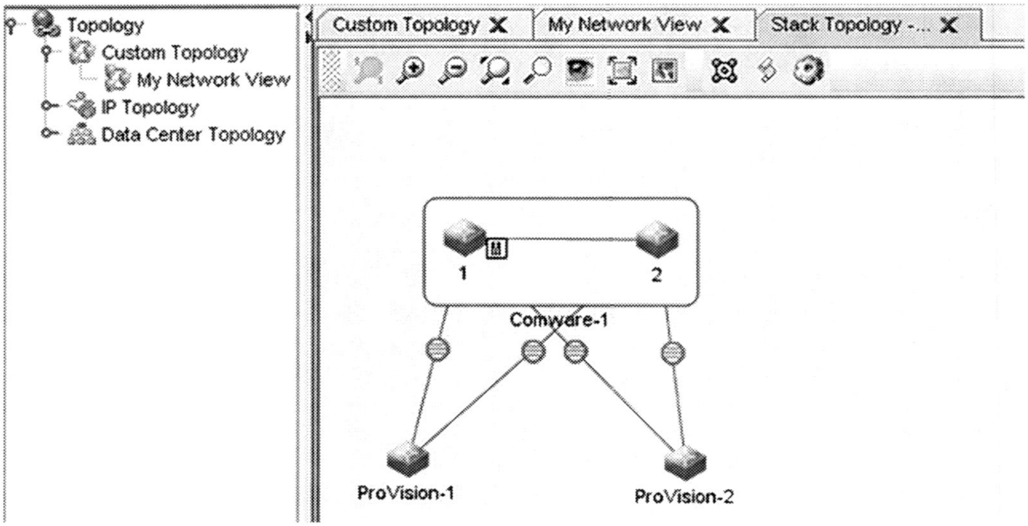

Figure 5-17: Using IMC to monitor the IRF virtual device

HP IMC allows you to view all the switches comprising the IRF virtual device. You can also view and monitor the links connecting the IRF virtual device to other devices on the network. You can easily check to ensure that each member has a link to an upstream or downstream switch.

Backplane stacking on HP 3800 and 2920 Switch Series (HP ProVision)

You will now examine a similar feature, backplane stacking, which is available on the HP 3800 and 2920 Series Switch (see Figure 5-18). The 3800 and 2920 Switch Series run the HP ProVision software.

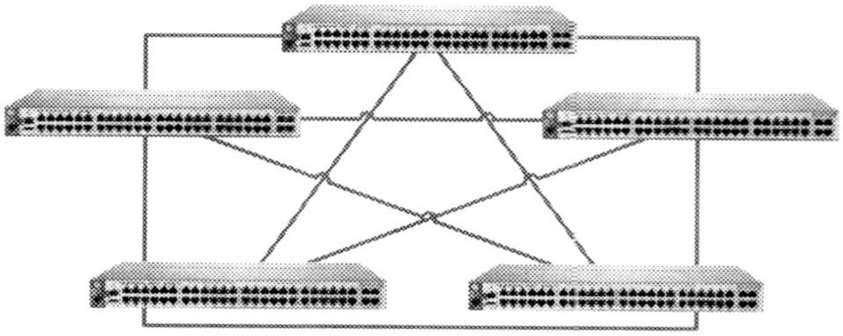

Figure 5-18: Backplane stacking on HP 3800 and 2920 Switch Series (ProVision)

Like IRF, the backplane stacking feature on the HP 3800 and 2920 Switch Series allows you to combine multiple switches to create a single logical virtual device. The stack operates as if it were a single switch with all the member switches providing ports.

A backplane stack provides the following benefits:

- Simplified management, allowing you to manage the stack from a single IP address
- Scalability, enabling the network to handle increasing demands for faster processing and higher bandwidth
- Simplified deployment, allowing you to add switches to the network without configuring IP addresses and completing other tasks
- Resiliency and redundancy, protecting the organization from link, power, or device failure

Backplane stacking topologies

When you implement backplane stacking, you can choose from the three topologies shown in Figure 5-19, depending on how many switches you want to combine in a backplane stack.

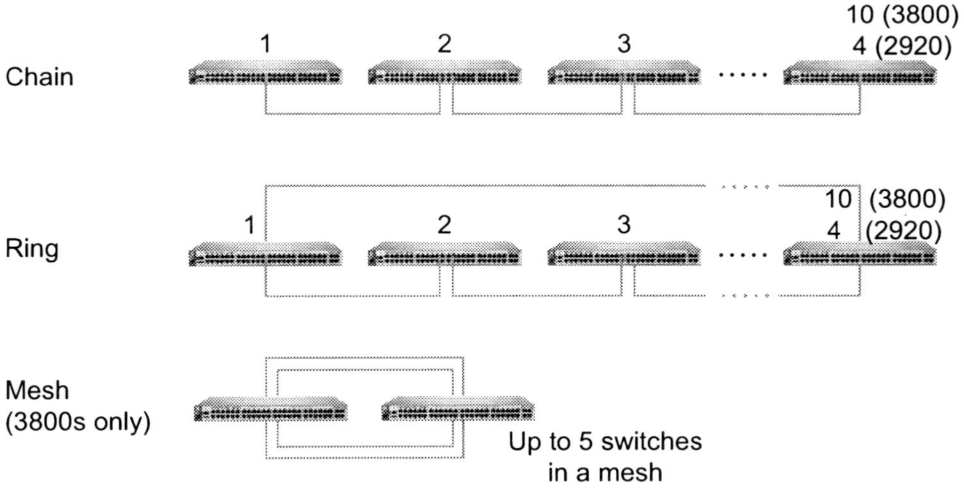

Figure 5-19: Backplane stacking topologies

- **Chain topology**—In a chain topology, each switch in the backplane stack is connected to another through a single cable, as shown in Figure 5-19. As with IRF virtual devices, chain topologies are generally not recommended because a single failure (whether it is a switch, stacking module, or cable failure) disrupts communication among the switches in the backplane stack. Unless the failure occurs at the end of the chain, two stack fragments are created. One of the fragments becomes inactive to ensure that two virtual switches are not created. On the inactive fragment, the switches remain powered on, but all of the switch ports in the inactive stack fragment are disabled and do not transmit network traffic.

You can combine up to ten 3800 Series switches in a chain topology or up to four 2920 Series switches.

- **Ring topology**—As the name suggests, each switch in a ring topology is connected to two switches, thereby forming a ring as shown in Figure 5-19. The ring topology provides a higher level of redundancy than the chain topology. A single failure in the switch, stacking module, or cable does not sever communications within the backplane stack. If such a failure occurs, the ring topology becomes a chain topology.

  The ring topology supports the same number of members as the chain topology: ten 3800 Series switches or four 2920 Series switches.

- **Mesh topology**—In a mesh topology, every switch in the stack is connected to every other switch. As a result, the mesh topology provides the highest level of redundancy because the stack can continue to function even if multiple failures occur.

  The 2920 Series switches do not support a meshed topology, and for the 3800 Series switches, the mesh topology supports fewer members. You can connect two to five switches in this topology.

Roles in the backplane stack

Within the backplane stack, each switch fulfills one of the following roles:

- Commander
- Standby member
- Member

At a minimum, each stack must have a commander and a standby member. Depending on the switch series and topology you use, you can combine up to 10 switches in a stack.

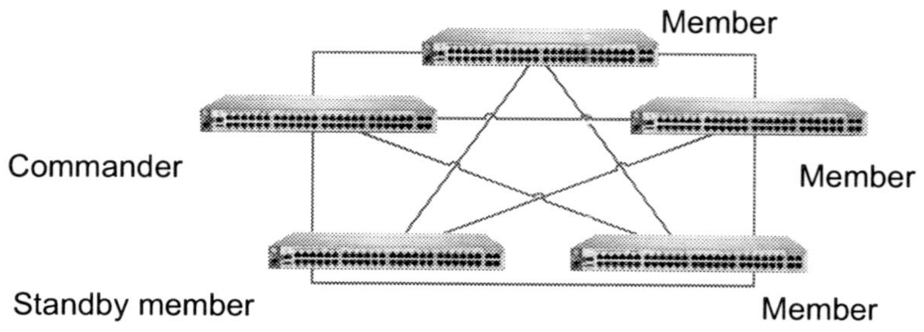

Figure 5-20: Roles in the backplane stack

Commander

The commander is responsible for:

- Managing the stack
- Responding to management requests for the stack
- Running all the routing and switching protocols for the entire stack
- Managing its own local subsystem and ports

Only one Commander is elected at any given time (see Figure 5-20).

The commander's MAC address and out-of-band management (OOBM) IP address are used to access and manage the stack.

Standby member

The standby member takes over management of the stack if the commander fails or if the network administrator forces a failover. The standby member also manages its own local subsystem and ports.

Only one standby member is elected at any given time (see Figure 5-20). If the standby member fails or becomes the commander (because the commander failed), the member switches in the backplane stack elect a new standby.

Members

Members do not participate in managing the stack. Each member must manage its local subsystems and ports to operate correctly as part of the stack.

Requirements

To set up the 3800 switches in a stack, you must purchase and install the HP 3800 4-port Stacking Module (see Figure 5-21). You will also need one of the following cables:

- HP 3800 0.5m Stacking Cable (J9578A)
- HP 3800 1m Stacking Cable (J9665A)
- HP 3800 3m Stacking Cable (J9579A)

To set up the 2920 switches in a stack, you must purchase and install the HP 2920 2-Port Stacking Module:

- HP 2920 0.5m Stacking Cable (J9734A)
- HP 2920 1.0m Stacking Cable (J9735A)
- HP 2920 3.0m Stacking Cable (J9736A)

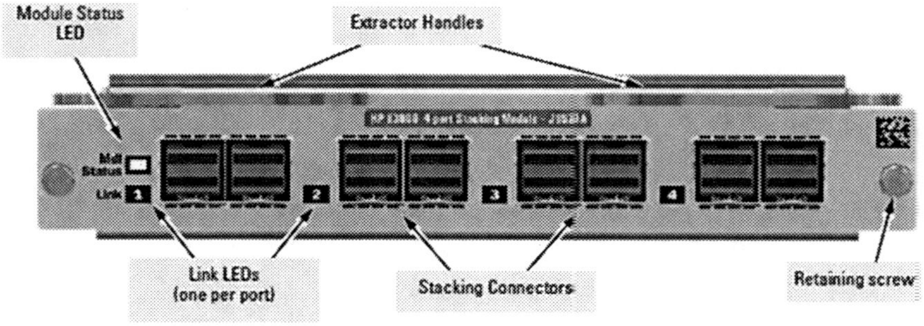

Figure 5-21: Hardware requirements for backplane stacking

Configuring a backplane stack

Backplane stacking is enabled on 3800 and 2920 switches by default. When a switch boots initially, it checks to see if a stacking module is installed. If the switch does not detect a stacking module, it disables the backplane stack feature. If you later install a stacking module, you must manually enable the feature.

There are two configuration options:

- **Deterministic**—Use this option if you want to control which switches function as the commander and standby member by configuring priorities as shown in Figure 5-22.
- **Plug-and-go**—Use this option if it does not matter which switches function as the commander and standby member.

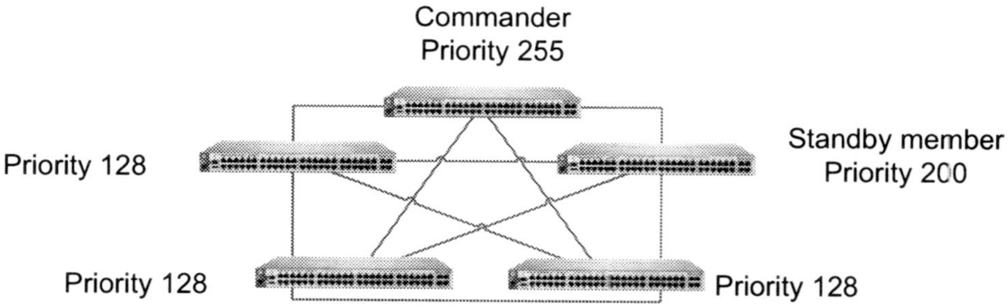

Figure 5-22: Configuring a priority for a backplane stack

Deterministic installation method

To configure the switches so that you determine which switches function as the commander and standby member, complete these steps:

1. Power off all switches that will be part of the backplane stack.
2. Install the stacking modules in the switches and connect the stacking cables to create your stacking topology.
3. Configure the switch that you want to function as the commander:
   a. Power on the switch that you want to function as the commander.
   b. Ensure that stacking is enabled on the switch:

```
Switch# show stacking
```

   c. If necessary, enable stacking:

```
Switch(config)# stacking enable
```

The switch automatically reboots.

   d. Display the switch's member ID:

```
Switch# show stacking
```

   e. If you want this switch to always function as the commander, configure the stacking priority. If all switches are booted simultaneously, the switch with the highest priority is selected as commander:

```
Switch(config)stacking member 1 priority 255
```

Note

If you add a switch to an existing backplane stack, that switch will not be elected commander or standby, no matter what priority you assign it. The current commander and standby member will continue to function, thereby preventing a system disruption and maintaining the stability of the backplane stack. In general terms, this means higher priority devices do not preempt switches that are currently acting as the commander or standby.

1. Repeat steps 3a to 3e on the switch that you want to function as the standby. When setting the priority, use a value that is lower than the priority you assigned the commander.
2. Power on the members, one at a time, and ensure that stacking is enabled on each switch.

   The member ID assigned to the switches is determined by the sequence in which they are booted.

Plug-and-go installation method

If you do not care which switches function as the commander and standby member, complete these steps:

1. Power off all switches that will be part of the backplane stack.

2. Install the stacking modules in the switches and connect the stacking cables to create your stacking topology.
3. Power on the switches, one at a time. The order in which you power on the switches will determine which switch becomes the commander and the standby member. If you want to use the order in which switches are booted to determine which switch becomes commander and standby, complete these steps:
   a. First, power on the switch that you want to function as the commander.
   b. Second, power on the switch that you want to function as the standby.
   c. Power on the remaining switches, one at a time. The switches member ID will be determined by the order in which they were booted and joined the stack.
4. Ensure that stacking is enabled on each switch:
   ```
   Switch# show stacking
   ```
5. If necessary, enable stacking:
   ```
   Switch(config)# stacking enable
   ```
   The switch automatically reboots.
6. Determine the switch's member ID role in the stack:
   ```
   Switch# show stacking
   ```

Election of the commander and standby member

If the stack is rebooted simultaneously, such as during a building power failure recovery, the members will undergo an election process. An election timer is set for 60 seconds and if no new switches are discovered during that timeout period, the switches in the stack enter the election phase.

During the election process, for each group of switches that has the same stack ID, the following criteria are used for electing the commander:

- The switches with the highest stack revision are discovered.
- The switch with the highest configured priority is selected as the commander.
- If more than one switch has the same high priority, the switch that was the previous commander is selected.
- If none of the switches were previously the commander, the switch that was the standby member is selected.
- If none of the above conditions apply, the switch with the lowest MAC address is selected as the commander.

The commander then selects the standby member, using the same rules that were used to elect the commander.

Stack operations

When the backplane stack is formed, the commander creates a stack ID, using its own MAC address hashed with time, and sends this stack ID to all the members of the stack. If the commander fails or if you manually force a failover and a new commander manages the stack, the stack ID does not change. The commander also assigns each switch a unique member ID.

Once the stack is formed, all members' consoles are redirected to the commander. Each member's OOBM IP address is redirected to the commander as well.

The commander uses a stack revision to track the current stack topology. When the stack is created, the stack revision is stored in the configuration. Whenever a change is made to the stack—such as a new member is added—the stack revision is incremented and stored in the configuration.

To view information about the stack, you can enter the **show stacking** command. You will see output similar to the following:

```
HP Stack# show stacking

Stack ID : 00031cc1-de4d48c0

MAC Address : 1cc1de-4d48c9

Stack Topology : Mesh

Stack Status : Active

Uptime : 1d 2h 35m

Software Version : KA.15.05

Mbr

ID Mac Address    Model                                Pri Status
--- ------------  ------------------------------------ --- ---------
1 1cc1de-4d48c0   HP J9574A 3800-48G-PoE+-4SFP+ Switch 250 Commander

2 1cc1de-4d8680   HP J9573A 3800-24G-PoE+-2SFP+ Switch 230 Standby

3 1cc1de-4e3180   HP J9574A 3800-48G-PoE+-4SFP+ Switch 128 Member

4 78acc0-3c2180   HP J9576A 3800-48G-4SFP+ Switch      128 Member
```

Stack fragments

A backplane stack has built-in mechanisms for handling splits. You do not have to configure a protection mechanism.

Figure 5-23: Stack fragments

If a link or a switch failure occurs, separating a stack into two groups, the smaller group, or fragment, becomes inactive, as shown in Figure 5-23. All of the user ports on the switches in this stack fragment will be disabled. However, the OOBM and stack ports will be active, allowing you to manage the switches.

The stack fragment that includes more switches will become the active stack and will inherit the MAC address and IP addresses assigned to the stack. This allows more of the user ports to remain operational.

If the two stack fragments have the same number of switches, the fragment that includes the original commander will be the active fragment.

When the two fragments are reconnected, the commander and standby of the active fragment retain those roles for the recombined stack. If the original commander was not in that fragment, then the stack will have a new commander when the stack is reformed. The switches in the inactive fragment reboot and assume their new roles in the stack.

Summary

In this chapter, you expanded your knowledge of IRF. You considered the advantages it provides at the access, distribution, and core layers, and you examined best practices for implementing IRF at each layer. You also learned about ISSU, which allows you to upgrade an IRF virtual device without interrupting services. In addition, you learned how to implement MAD to detect and handle split stacks.

Finally, you learned about backplane stacking, a feature that provides similar features on HP 3800 and 2900 Series switches.

Learning check

Answer these questions:

1. What are some of the advantages to implementing IRF at the access layer?

2. What is a split IRF stack?

   a. It is the creation of two separate IRF virtual devices, each unaware of the other, in the event that members in an IRF virtual device cannot communicate.

   b. It is a mechanism that prevents address conflicts by permitting only one IRF virtual device to be active and tries to reestablish both the failed link and the IRF virtual device.

   c. It is a piece of software that allows an administrator to upgrade the IRF virtual device without disrupting the services it provides.

   d. It is a process that supports version rollback, which allows a user to go back to the previous software version.

3. What role do members play in a mesh topology implemented with backplane stacking?

   a. Members manage their local subsystems and ports to operate correctly as part of the stack.

   b. Members are responsible for responding to management requests for the stack.

   c. Members take over management of the stack if the commander fails or if the network administrator forces a failover.

   d. Members run all the routing and switching protocols for the entire stack.

4. How does backplane stacking handle stack fragments in the event of a link or switch failure?

Answers

1. What are some of the advantages to implementing IRF at the access layer?

   IRF helps you to implement a two-tier architecture, provides redundancy and resiliency, increases the number of available edge ports without adding management burden, and allows you to manage all the switches in an IRF virtual device as a single virtual switch. With IRF, you can also connect the IRF virtual device to other switches using aggregated links, eliminate MSTP, and group multiple access layer devices into a single virtual device.

2. What is a split IRF stack?

   a. It is the creation of two separate IRF virtual devices, each unaware of the other, in the event that members in an IRF virtual device cannot communicate.

   b. It is a mechanism that prevents address conflicts by permitting only one IRF virtual device to be active and tries to reestablish both the failed link and the IRF virtual device.

   c. It is a piece of software that allows an administrator to upgrade the IRF virtual device without disrupting the services it provides.

   d. It is a process that supports version rollback, which allows a user to go back to the previous software version.

3. What role do members play in a mesh topology implemented with backplane stacking?

   a. Members manage their local subsystems and ports to operate correctly as part of the stack.

   b. Members are responsible for responding to management requests for the stack.

   c. Members take over management of the stack if the commander fails or if the network administrator forces a failover.

   d. Members run all the routing and switching protocols for the entire stack.

4. How does backplane stacking handle stack fragments in the event of a link or switch failure?

   The smaller fragment becomes inactive, making it so all of the user ports on the switches in that stack fragment are disabled.

   The stack fragment that includes more switches will become the active stack and will inherit the MAC address and IP addresses assigned to the stack. If the two stack fragments have the same number of switches, the fragment that includes the original commander will be the active fragment.

   When the two fragments are reconnected, the commander and standby of the active fragment retain those roles for the recombined stack. If the original commander was not in that fragment, then the stack will have a new commander when the stack is reformed. The switches in the inactive fragment reboot and assume their new roles in the stack.

6 Access Control Lists (ACLs)

EXAM OBJECTIVES

In this chapter, you learn to:

- ✓ Define access control list (ACL) and identify the criteria by which ACLs select traffic.
- ✓ Configure ACLs on HP Comware and HP ProVision switches to select given traffic.
- ✓ Apply static ACLs to interfaces to meet the needs of a particular scenario.
- ✓ Examine an ACL configuration and determine the action taken on specific packets.

ASSUMED KNOWLEDGE

Before studying this chapter, you should understand how virtual LANs (VLANs) are implemented on ProVision and Comware switches.

INTRODUCTION

In this chapter, you learn about ACLs, one of the basic mechanisms for controlling traffic on your network. You will find out which types of ACLs are available on Comware and ProVision switches and which services these ACLs provide. You will also learn how to configure ACLs and apply them to interfaces on both Comware and ProVision switches.

What is an ACL?

An ACL is simply a list of rules against which a network infrastructure device, such as a routing switch, checks packet headers. Each rule consists of specific criteria such as source IP address, destination IP address, and TCP or UDP port number. As Figure 6-1 shows, an ACL also consists of an action:

- Permit (select matching packets)
- Deny (do not select matching packets)

What the routing switch does with selected packets depends on how you apply the ACL. This chapter focuses on packet-filtering ACLs—ACLs that are applied to routing switch interfaces to filter incoming and outgoing traffic. For packet-filtering ACLs, the permit action allows packets that match the rule, and the deny action discards packets that match the rule.

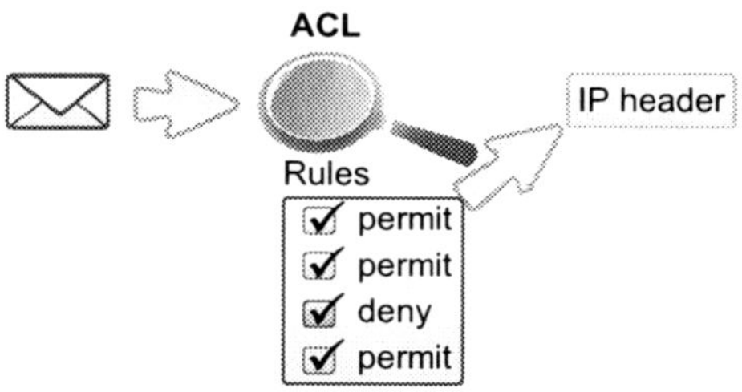

Figure 6-1: What is an ACL?

Note
Rules are sometimes referred to as access control entries (ACEs).

Reasons to implement ACLs

This chapter focuses on applying ACLs to filter traffic. Without such packet-filtering ACLs, each routing switch interface accepts all packets and forwards that traffic based on its forwarding tables. ACLs enable you to implement more control, improving network performance and enhancing security. As you will see, you can apply packet-filtering ACLs to filter traffic to or from an endpoint, a group of endpoints, or entire subnets.

You might implement packet-filtering ACLs for reasons such as these:

- At the network edge, ACLs can prevent unwanted IP packets from entering the network infrastructure. You can take restrictive approaches in which you permit the required traffic and deny all other traffic. Or you can take a permissive approach in which you deny specific traffic and permit all other traffic. By combining permit and deny rules, you can create highly flexible policies.
- Implementing ACLs at the network edge can also improve network performance and reduce switch buffer and CPU utilization by reducing the volume of packets that upstream switches and routers handle.

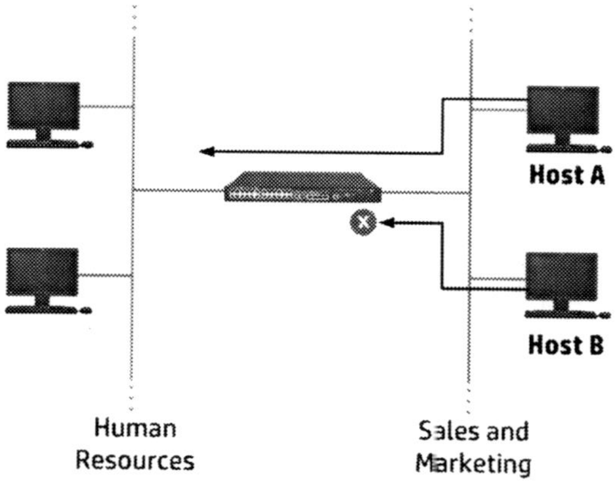

Figure 6-2: Implementing ACLs at the access layer

- At the network core and distribution layer, ACLs can similarly improve performance and security. For example, you can use ACLs to ensure that various collections of clients only have access to selected destinations (see Figure 6-2). These destinations may be specific hosts, entire subnets, or even particular applications. For example, you may want to ensure that all hosts and servers in a given VLAN are only allowed to communicate within that VLAN or with a limited number of other specific VLANs.
- You can also use ACLs to block IP traffic that has the switch as the destination address. For example, ACLs can deny unauthorized clients attempting to access the switch using functions such as Telnet, Secure Shell (SSH), or a web browser.
- In addition to securing the network infrastructure, you can also use ACLs to control a performance problem quickly by limiting IP traffic generated by a specific device, group of devices, or a subnet. This gives you an opportunity to troubleshoot without sacrificing performance for users outside of the problem area.

Note

Generally, HP switches that support more advanced IP routing services also support the use of ACLs. Switches that provide limited IP routing services, such as IP static routes only, typically do not support the use of ACLs.

Types of ACLs

There are three types of ACLs:

- Basic or standard ACLs
- Advanced or extended ACLs
- Ethernet frame header ACLs

Basic/standard ACLs

On Comware switches, you would configure a basic ACL to permit or deny IP traffic based on source IP address only. On ProVision switches, the same ACL is known as a standard ACL. You should use basic/standard ACLs whenever you want to select traffic based on the endpoint, subnet, or range of subnets from which it derives.

Note

You can create IPv4 or IPv6 basic or standard ACLs; however, this study guide covers only IPv4. You should keep the following in mind when configuring your switches: for an IPv4 basic or advanced ACLs, its ACL number and name must be unique in IPv4, and for an IPv6 basic or advanced ACL, its ACL number and name must be unique in IPv6.

Advanced/extended ACLs

Often you want to select traffic according to more specific criteria than the IP source address. For example, you want a particular set of endpoints to be able to communicate with one server but not another. Or you want the endpoints to send traffic to a particular set of TCP/UDP ports but not others.

In this case, you would create an advanced ACL on Comware switches or an extended ACL on ProVision switches.

Advanced/extended ACLs support flexible filtering with criteria such as these:

- Source and/or destination IP addresses
- IP protocol type (such as Generic Routing Encapsulation [GRE], Internet Control Messaging Protocol [ICMP], and Encapsulating Security Payload [ESP])
- Source and/or destination TCP/UDP port
- TCP flag
- ICMP or Internet Group Management Protocol (IGMP) type
- Fragmented packet
- Virtual private network (VPN) instance
- Quality of Service (QoS) mark (IP precedence, Differentiated Services Code Point [DSCP], or type of service [ToS] value)

Note
You will learn more about QoS marks in Chapter 8.

The TCP/UDP destination port option can play an important role because it enables you to permit or deny IP connections using TCP for applications such as Telnet, HTTP, FTP, and others. For example, TCP destination port 21 is for FTP traffic, and TCP destination port 80 is for HTTP traffic.

Note
Not all HP switches support all criteria for advanced ACLs. For details regarding the features your switch supports, refer to your switch configuration guide containing details of ACL configuration.

Note
You can also create IPv4 or IPv6 advanced or extended ACLs, but only IPv4 ACLs are covered in this study guide.

Ethernet frame header ACLs

Comware switches also support Ethernet ACLs, which select traffic according to fields in the MAC header. For example, you can permit or deny traffic based on its source MAC address.

Note

Some Comware switches support user-defined ACLs, which let you filter packets according to any content. A rule includes specific hexadecimal characters with a mask that might make some of those characters variable. The rule also includes an offset, indicating where in the packet the switch begins looking for a match. However, these ACLs are beyond the scope of this study guide.

Creating valid IDs for basic/standard ACLs

When you first create a basic/standard ACL, you assign it a valid ACL identifier (ID)—a number and/or an alphanumeric name. This ID groups together all the rules in the ACL, allowing you to view the ACL, add rules to the ACL, and otherwise edit the ACL.

Assigning a name (as opposed to a number or a number only) can make it easier to identify the ACL's intended purpose. Figure 6-3 summarizes basic guidelines for creating basic/standard ACLs on Comware and ProVision switches.

Category	Comware ACL ID	ProVision ACL ID
Basic/standard ACLs	Always use a number: 2000–2999 Optionally, add an alphanumeric name (< 64 characters, including spaces).	Use a number: 1–99 OR Use an alphanumeric name (< 64 characters, including spaces).

Figure 6-3: Creating valid IDs for basic/standard ACLs

Creating and identifying an HP Comware basic ACL

When creating a basic ACL on an HP Comware switch, you must assign it a number in the range of 2000 to 2999. You can also assign it a name.

From the system view, use this command to create the ACL and assign the ACL its number:

```
acl number <2000-2999>
```

If you want to assign the ACL a name as well, you *must* do so when you first create it. Use this command:

```
acl number <2000-2999> name <string>
```

The name can be up to 64 case-sensitive alphanumeric characters in length, including spaces. To include spaces in the string you must enclose the string in single or double quotes (" ").

Creating the ACL moves you to a configuration view for that ACL from which you can add rules.

Creating and identifying an HP ProVision standard ACL

HP ProVision switches use *either* a number or a name to identify an ACL. Enter this command from the global configuration context to create a named standard ACL:

```
ip access-list standard <string | 1-99>
```

If you choose to use a number, you must select one in the correct range: 1 to 99. The format for a name follows the same rules as the format for a Comware name.

As for Comware, creating the ACL with this command moves you to configuration context for adding rules and editing the list.

As an alternative, which is valid only for ACLs identified by numbers, you can create the ACL with this command:

```
access-list <1-99> <rule>
```

As you see, when you enter this command, you must also add options to configure a rule for the ACL. (The next section shows the format for such a rule.)

Important

Even if you used the **access-list** command to create a numbered ACL, you can use the **ip access-list <standard | extended>** command later to edit the list.

Creating rules for basic/standard ACLs

An ACL rule consists of three basic components:

- ID
- Action
- Matching criteria

ID

The ID identifies the rule within the list, enabling you to identify and, if necessary, remove it. The rule also tells the switch when to process this rule in relation to other rules. You will examine this process in more detail in a moment.

Action

A permit action tells the ACL to select any packet that matches the criteria. The deny action tells the ACL not to select any packet that matches the criteria.

When you are using the ACL to filter packets, permitted packets are forwarded and denied packets are dropped. However, you can use ACLs for other purposes, in which case permitted packets are selected for the particular function and denied packets are not. You will look at several of these specific cases in other chapters throughout this study guide.

Matching criteria

A basic/standard ACL simply selects traffic basic on its source IP address, so the rules are quite simple. A rule can select one IP address, any IP address, or multiple IP addresses. Figure 6-4 shows the syntax for matching one IP address or any IP address. To match multiple IP addresses, you must use wildcard bits or prefix lengths.

Options for basic/standard matching criteria	HP Comware syntax	HP ProVision syntax
Any source IP address	source any	any
One source IP address	source *<IP address>* 0	host *<IP address>*
A range of source IP addresses	source *<IP address>* *<wildcard bits>*	*<network address/prefix>* | *<IP address>* *<wildcard bits>*

Figure 6-4: Creating rules for basic/standard ACLs

Supplemental information about rule options

In basic/standard ACLs, the source IP address is the primary matching criteria. However, you can set additional options for special circumstances.

Comware options

In addition to the source address options, options for a basic ACL rule on a Comware switch include:

- **counting**—This option configures the switch to record the number of packets that it has matched to the rule. To enable counting for rule matches performed in hardware, configure the **hardware-count enable** command for the ACL. (This command does not take effect for ACLs referenced by QoS policies.)
- **fragment**—Without this keyword, the rule applies to all fragments and nonfragments.

  Traditional packet filtering matches only first fragments of packets and allows all subsequent non-first fragments to pass through. This is because the Layer 4 information that is used to filter TCP and UDP packets is only present in the initial fragment. Thus, attackers can fabricate non-first fragments to attack networks. To avoid the risks, Comware switches filter all fragments by default, including non-first fragments, which are evaluated solely on the Layer 3 portion of any configured rule. You can, however, also set matching criteria—to filter only non-first fragments, for example.
- **logging**—In addition to counting packets that are matched to the rule, the switch can log the matches. The switch stores logs and transmits them to its information center at the ACL logging interval, which you set with the **acl logging frequence *<minutes>*** command. Refer to your switch's documentation to learn how to configure the information center to forward the logs to other devices such as a syslog server. You could use this option for "deny" rules, for example, to audit potential security risks.
- **time range**—On Comware switches, you can implement ACL rules based on the time of day by creating time definitions globally and then applying a time range to a rule. A time-based ACL rule takes effect only in any time periods specified by the time range.
- **vpn instance**—This option enables you to specify a vpn instance as the match criteria.

ProVision options: Logging

In addition to the source address options, a standard ACL rule on a ProVision switch can include the **log** option. This option can be used after the source address to generate an Event Log message. The log is created if the action is "deny" *or* "permit," if there is a match, and if you have enabled ACL logging.

To enable ACL logging to a syslog server, you must:

1. Specify the syslog server IP address:

   ```
   ProVision(config)# logging <IP address>
   ```

2. Configure the switch to send log messages to the syslog server:

   ```
   ProVision(config)# logging facility syslog
   ```

3. Set the debug destination for ACLs to logging:

   ```
   ProVision(config)# debug destination logging
   ```

4. Enable ACL debug messages:

   ```
   ProVision(config)# debug acl
   ```

The first time that a packet matches the rule with the log option, the switch sends a log to the syslog server. It then tracks the activity over the logtimer interval and sends the syslog server a summary at the end of this interval, which you set with this command (timer set in seconds):

```
ProVision(config)# ip access-list logtimer <30-300>
```

Specifying IP address ranges with wildcard bits

You will now learn how to specify a range of IP addresses on Comware and ProVision switches:

- Comware—Use wildcard bits (also called an inverse mask).
- ProVision—Use either wildcard bits or a prefix length.

To understand how wildcard bits works, you will review subnet masks.

Common IP addressing mask

In common IP addressing, a network (or subnet) mask defines which part of the IP address to use for the network (or subnet) address and which part to use for the hosts on the network:

- The bits set to 1 define the part of the IP address to use for the network address.
- The bits set to 0 define the part of the IP address to use for the host address.

In Figure 6-5, the subnet mask is 255.255.255.0, or all 1s within the first three octets. Therefore, 10.1.10.0 is the network address. Valid host numbers in the fourth octet are between 1 and 254. Therefore, valid IP addresses that could be assigned to devices are from 10.1.10.1 through 10.1.10.254. The IP addresses 10.1.10.0 and 10.1.10.255 are reserved for identifying the subnet and broadcast addresses, respectively.

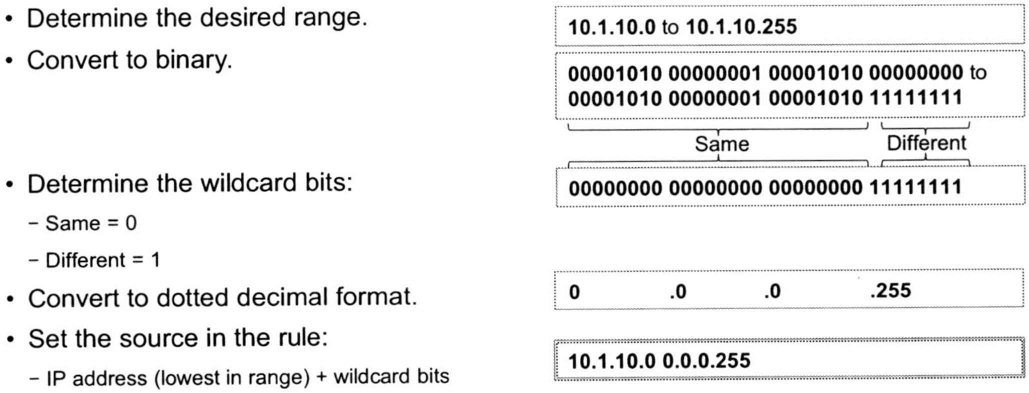

Figure 6-5: Using wildcard bits

Wildcard bits

A standard network mask defines which bits belong to the network address and which to the host address. Wildcard bits in an ACL rule define which bits in a packet's IP address must match the corresponding bits in the rule's IP address and which bits can be wildcards (any value).

- A mask-bit setting of 0 ("off") requires a match between the corresponding bits in the packet's address and in the rule's address. Thus, if a bit in the rule's address is set to 1, the rule only selects packets that have the same bit set to 1. Similarly, if the bit in the rule's address is set to 0, the bit in the packet's address must be 0.
- A mask-bit setting of 1 ("on") means the corresponding bits in the packet's address and in the rule's address do not have to be the same. As a matter of standard format, the rule specifies 0 bits for the segment of the IP address that corresponds to 1s in the wildcard mask, as you see in Figure 6-5. The corresponding bits in a matching packet can be either 1 or 0.

Note

Although most of the time ACL masks are continuous zeroes followed by continuous ones, it is possible to have any combination of ones and zeroes. Managing such masks, however, is difficult, and they are not recommended.

Example

Return to the example in Figure 6-5 to see how to apply wildcard masks in rules. Assume you want a rule that selects any host within the 10.1.10.0/24 subnet. You would specify 10.1.10.0 0.0.0.255. As you can see in Figure 6-5, zeroes comprise the first three octets of the wildcard mask—which means that the corresponding part of the IP address, 10.1.10, must match exactly. However, the packet can have any host address in that subnet because the last octet of the wildcard mask consists of all 1s.

You can also now understand why you use 0.0.0.0 or 0 to specify a single host in a rule on a Comware switch. Because the mask is all zeroes, the rule policy dictates that a match occurs only when the source address on packets is identical to the address configured in the rule.

Invert the subnet or subtract the first address from the last address

As the example in Figure 6-5 shows, you can think of a wildcard mask as essentially the inverse of a subnet mask. Whenever you want to create a rule to select multiple IP addresses, follow these guidelines:

- Determine the subnet that contains the range of IP addresses that you want to select.
- Determine the subnet mask.
- Invert the mask.

Figure 6-6 shows the binary logic behind inverting the mask. However, a simple way to invert the mask is to subtract the value for each octet in the mask from 255. For example, the subnet mask 255.255.255.0 becomes 0.0.0.255. The subnet mask 255.255.255.252 becomes 0.0.0.3.

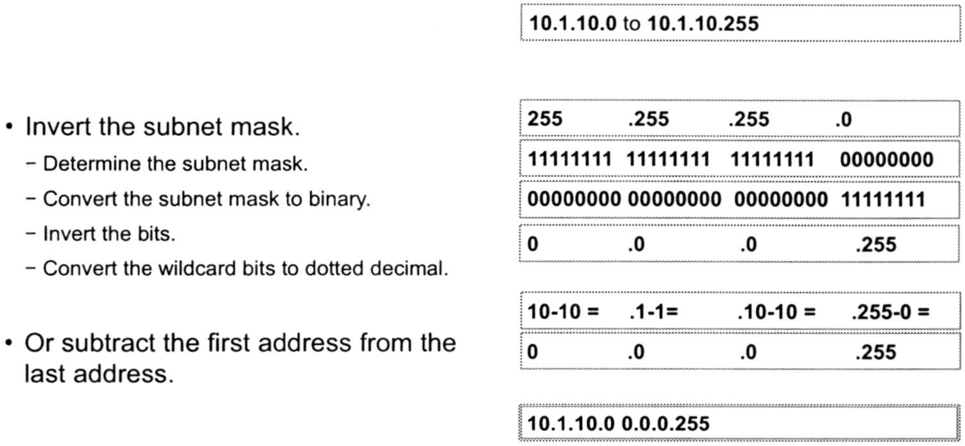

Figure 6-6: Invert the mask

In fact, a very simple way to determine the wildcard bits is to subtract the first IP address in the range from the last address in the range. The result is the wildcard bits. This approach works when the range could be a subnet. You must include the broadcast and network address although, of course, actual devices do not use these addresses.

ProVision: Specifying IP address ranges with prefix length

Classless Inter-Domain Routing (CIDR) notation uses a prefix length instead of a subnet mask. In other words, the prefix length specifies the number of bits set to 1 in the subnet mask. For example, the subnet mask in Figure 6-7 has 24 1s and defines a/24 subnet.

You double the size of a subnet when you subtract one from the prefix length and halve the size when you add one.

On some switches, such as ProVision switches, you can use a network address and prefix length to define a range of IP addresses in a rule rather than an IP address and wildcard mask. The switch automatically converts the prefix length to the correct wildcard mask to select all IP addresses within the subnet.

Example

Specifying **10.1.10.0/24** within a rule selects the same range of IP addresses as **10.1.10.0 0.0.0.255**. The switch first converts prefix length/24 to subnet mask 255.255.255.0 and then inverts the subnet mask to wildcard mask 0.0.0.255, as shown in Figure 6-7.

- If the range is a subnet, determine the prefix length.

10.1.10.0 to **10.1.10.255 =**
10.1.10.0 255.255.255.0 (24 bits set) =
10.1.10.0/24

- Set the source in the rule:
  - Network address/prefix length

10.1.10.0/24

Figure 6-7: Specifying IP address ranges with prefix length on ProVision switches

Processing order and implicit permit or deny

When applying an ACL, the switch compares a packet against each rule's matching criteria in sequence until it finds a match. By default, Comware and ProVision switches process ACL rules according to each rule's ID (lowest to highest). When the switch finds a match, it takes the action indicated by the matching rule (deny or permit) and stops processing the ACL (see Figure 6-8).

By combining permit and deny rules, you can create highly flexible policies. For example, you can deny one device within a subnet (such as 10.1.10.10) and permit the complete subnet (10.1.10.0/24). Or with advanced/extended ACLs, which you will learn about momentarily, you can select traffic associated with a particular application on a particular server; first permit traffic destined to a specific IP address and range of TCP or UDP ports and then deny all other traffic to the IP address.

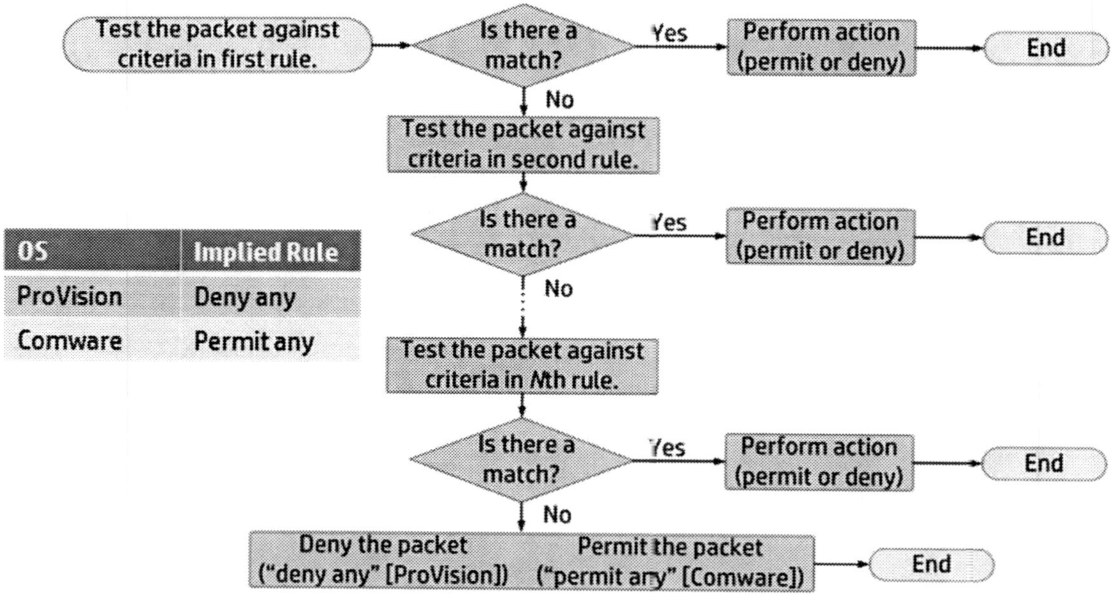

Figure 6-8: Processing order and implicit permit or deny

These scenarios highlight the importance of the order in which the switch processes the rules. For example, in the first scenario, the switch must match packets against the rule that denies a specific device, 10.1.10.10, before the rule that permits the device's subnet, 10.1.10.0/24. If the switch processed the rule that permits the entire subnet first, the switch would permit the packet from 10.1.10.10 and stop processing the ACL. The deny rule would never take effect.

The next sections present guidelines for configuring rules with the correct order.

Implicit action

First, however, you also need to understand how your switch treats packets that do not match any of the rules defined in the ACL. You can then avoid blocking or permitting the wrong traffic (see Figure 6-8).

Important

ProVision switches use an implicit "deny any" for ACLs. Comware switches use an implicit "permit any" when applying ACLs as packet filters.

ACLs on ProVision switches have an implicit "deny any" at the end: if the switch finds no matches between an inbound packet and the configured criteria in an applicable ACL, the packet is not selected by the ACL. If you apply the ACL to an interface, the switch denies (drops) the packet.

Understanding the functionality for Comware switch ACLs is a bit more complicated. When you apply the ACL as a packet-filtering ACL on an interface, packets that do not match any of the explicit rules are *forwarded*. Therefore, the ACL functions as if it has an implicit "permit any" at the end. On the other hand, ACLs applied to select traffic for other functions might *not* select traffic that does not match an explicit rule, as if the ACL has an implicit "deny any" as you will learn in Chapter 8.

You should understand the default behavior that you want—implicitly deny or implicitly permit. If your switch's implicit behavior matches the type you want, you do not need to explicitly define that rule. However, on both switches, the implicit final rule is not displayed when viewing an ACL. Therefore, you may want to explicitly define the implicit rule at the end of an ACL to clarify the behavior. Make sure to assign the rule a very high ID.

You can also override the default settings. On a ProVision switch, for example, you can add **permit any** as the last rule in the basic ACL. This directs the ACL to permit (forward) packets that do not have a match with any earlier rule and prevents these packets from being filtered by the implicit "deny any."

Conversely, you can add a **deny any** rule at the end of a Comware basic ACL to tell the switch to drop any packet that does not match any earlier rules.

Specifying rule IDs

The previous section highlighted the importance of assigning the correct rule IDs so that the ACL selects packets as you expect. You can specify the rule ID manually, or you can enter the rules in the order in which you want them processed.

In any case, you must plan the ACL carefully and configure rules for processing in the correct order, generally entering more specific rules with a lower ID than the ID for less specific rules.

Automatic numbering

If you do not enter a manual ID, the switch automatically adds rule IDs in order, leaving spaces. New rules are added to the *end* of the list, even if spaces exist earlier in the list.

For example, if the first rule that you enter does not have a manual ID, a ProVision switch numbers the rule as 10. Subsequent rules without rule numbers are numbered in increments of 10: 10, 20, 30, and so on. The switch always adds a rule without a manual ID after the *final* explicit rule. If this rule has an ID that is not divisible by 10, such as 45, the switch adds the new rule with the next ID that is divisible by 10, such as 50.

Comware switches follow very similar rules when you add rules without explicit rule IDs. However, by default, they add rules in increments of 5s and begin at rule 0. You can change the rule numbering step. In the ACL view, use the command **step <step value>**. The wider the numbering step, the more rules you can insert between two rules.

Manual numbering

When you enter rules manually, it is best practice for you to space rule IDs, preferably by 10s, so that you can insert additional rules at a later point, if necessary. Then, when you do not want to add a rule to the end of the list, you can choose a rule ID between existing rules to insert the rule in the correct location.

Sequence renumbering on ProVision switches

Sometimes, despite your best planning, you need to change rule IDs to place rules in the correct order.

From the Named Access Control List (NACL) context (accessed with the **ip access-list <extended | standard** ***<name>*** command), you can also easily remove a rule using the **no <sequence-number>** command. In addition, you can redefine the sequence numbers of all rules in a given ACL using the **ip access-list resequence <start-sequence> <increment>** command, in which **<start-sequence>** is the sequence number you want to assign to the first rule and **<increment>** is the incrementing value by which subsequent rules will be numbered.

The **resequence** command leaves the rules in the same relative order. However, with more spaces between the rules, you can remove rules and add them in the appropriate location as necessary.

Sequence renumbering on Comware switches

You can follow similar strategies on the Comware switches.

Remove a rule with the **undo rule** *<rule ID>* command in the ACL view (accessed with the **acl <number** *<number>* **| name** *<name>*>command).

Whenever you change the step value, the rules are renumbered, starting from 0. For example, if there are eight rules numbered 0, 5, 6, 7, 10, 13, 14, and 15, changing the step from 5 to 10 causes the rules to be renumbered 0, 10, 20, 30, 40, 50, 60, and 70. You can use the **step** command whenever you need to create additional space between your rules so that you can easily add new rules in the correct location.

Comware: Matching based on logical order

Comware switches also provide a useful alternative method of matching rules, called "auto," in which they intelligently select the order in which to process rules regardless of the rule ID and configuration order. This feature reduces the chance of misconfiguration and also saves you time if you need to insert multiple rules at a later date and have not left enough free IDs of the correct value for the new rules. With the auto match order method, you do not need to reconfigure the entire list to add the rules with the correct IDs; instead, the switch automatically matches traffic against the new rule in the most logical order.

ACL category	Sequence of tie breakers
IPv4 basic ACL	1. VPN instance 2. More 0s in the source IP address wildcard (more 0s means a narrower IP address range) 3. Smaller rule ID

Figure 6-9: Comware: matching based on logical order

This method is also known as "depth-first" order, because it ensures that any subset of a rule is always matched before the rule. Figure 6-9 illustrates how this method is applied to basic IPv4 ACLs.

1. Any rule associated with a VPN instance is processed before other rules. However, this study guide does not deal with VPNs, so you do not need to consider such rules.
2. The switch sorts rules by source IP address wildcard, listing the rules in order of the number of zeroes in the wildcard bits. The more zeroes, the more specific the entry. Therefore, the switch compares packets against the rule with the most zeroes in the source IP address wildcard, then against the rule with the next highest number of zeroes, and so on.

For example, the rule with the source IP address wildcard 0.0.0.255 is compared prior to the rule with the source IP address wildcard 0.0.255.255.

3. If two rules are present with the same number of zeroes in their source IP address wildcards, the one with the smaller rule ID takes precedence.

You specify the match order method when you create the ACL:

```
acl number <number> [name <name>] [match-order [config | auto]]
```

If you do not enter a specific match order, the ACL uses the config method (it uses the rule ID).

Even if you do not want to use the auto match order (or if it is not an option on your switches), you should follow similar rules to specify the correct sequence for rules. More specific rules (more zeroes in the wildcard bits) should come first.

Note

You can assign manual IDs to rules in an ACL that uses the auto match order. If you do not specify a manual ID, the switch automatically creates an ID in the same way that you learned for an ACL using the config match order. However, *neither* the manual nor the automatic rule ID affects the processing order except as a final tie breaker. When you view the ACL, the output displays the rules in the processing (depth-first) order rather than in the order of the rule ID.

Creating valid IDs for advanced/extended ACLs

You will now take a closer look at advanced/extended ACLs, which permit you to create more sophisticated rules. Start with how you would configure a valid ACL ID.

Creating and identifying a Comware advanced ACL

A Comware advanced ACL must have a number between 3000 and 3999 and can also have a name (see Figure 6-10). If you want to use a name, remember to specify it when you first create the list.

From the system view, use the command:

```
acl number <3000-3999> [name <string>]
```

Creating the ACL moves you to a configuration view for the advanced ACL from which you can add more complicated rules than for basic ACLs.

Creating and identifying a ProVision standard ACL

Again, ProVision switches use *either* a number or a name to identify an ACL. On ProVision switches, an extended ACL can have a number in the range of 100 to 199 (see Figure 6–10).

Enter this command from the global configuration context to create an extended ACL:

```
ip access-list extended <string | 100-199>
```

The alternative (**access-list**) command is also available for extended ACLs. Remember that the ACL must be identified by a number, and you must configure a rule as part of the command.

Important

Even if you used the **access-list** command to create a numbered ACL, you can use the **ip access-list <standard | extended> <*number*>** command later to edit the list.

Category	Comware ACL ID	ProVision ACL ID
Basic/standard ACLs	Always use a number: 2000–2999 Optionally, add an alphanumeric name (< 64 characters, including spaces).	Use a number: 1–99 OR Use an alphanumeric name (< 64 characters, including spaces).
Advanced/extended ACLs	Alwaysuse a number: 3000–3999 Optionally, add an alphanumeric name (< 64 characters, including spaces).	Use a number: 100–199 OR Use an alphanumeric name (< 64 characters, including spaces)

Figure 6-10: Creating valid IDs for advanced/extended ACLs

Creating rules for advanced/extended ACLs

Rules for advanced/extended ACLs include the familiar three components:

- Rule ID, which you add manually or is added automatically according to the process you examined in the "Processing order and implicit permit or deny" section
- Action (permit or deny)
- Matching criteria

It is the final component that differs in advanced/extended ACLs.

Basic/standard ACLs match only a packet's source IP address to a rule. In an advanced/extended ACL, you can specify a variety of packet header fields against which the switch matches packets. The components of a rule depend on the protocol that you specify for the rule. For this study guide, you are examining IPv4 ACLs.

Basic components of a rule:

Protocol	Source (IP address)	Destination (IP address)
ip	Any, one, or range	Any, one, or range

Basic components of a rule:

Protocol	Source (IP address)	Source port	Destination (IP address)	Destination port
tcp	Any, one, or range	One, range, any except one	Any, one, or range	One, range, any except one
udp	Any, one, or range	One, range, any except one	Any, one, or range	One, range, any except one

Figure 6-11: Creating rules for advanced/extended ACLs

With an advanced/extended ACL, you can select packets based on their destination IP addresses instead of, or in addition to, their source IP addresses. The first table in Figure 6-11 shows the components of a rule that meets this basic need:

1. You specify **ip** for the protocol keyword, which matches packets that use any protocol in the IP suite.
2. You specify the source IP address, which can be any address, a single address, or a range of addresses as defined by wildcard bits or prefix length, just as in a basic list.
3. You specify the destination IP address, following the same rules.

You can also create more complicated rules for TCP and UDP traffic, enabling you to open and close specific ports:

1. You specify **tcp** or **udp** for the protocol, which matches only packets that use that Layer 4 protocol.
2. You specify the source IP address and destination IP addresses as in rules that specify IP for the protocol. As always, you can use the **any** option if you do not want to define a specific component.
3. You can also specify the source port and the destination port. Defining the destination port is most typical because it enables you to define the permitted applications. Table 6-1 indicates some of the keywords for defining single ports and ranges of ports.

Table 6-1: Specifying ports

Need	Keyword	Example
Specify one port	eq	eq 80 Any traffic with port 80 is selected.
Specify a range of ports	range	range 50000–50100 Any traffic with port 50000–50100 is selected (including 50000 and 50100)
Specify any port with a smaller value than that listed	lt	lt 20 Any traffic with port 1–19 (including 1 and 19) is selected
Specify the listed port and any port with a greater value	gt	gt 49999 Any traffic with port 50000–65535 (including 50000 and 65535) is selected
Specify all ports except for one (the rule, whether permit or deny, applies to all ports except the listed port)	neq	neq 80 Any traffic with port 1–79 and 81–65535 is selected (including 1, 79, 81, and 65535)

The precise syntax for defining rules depends on the switch model. With Comware switches, you specify a keyword for a particular component of the rule (such as **destination**) and then the match criteria (such as **10.1.10.0 0.0.0.255**). If you do not specify a component, the switch sets that component to **any** automatically.

For example, you enter this command:

```
[Comware-acl-adv-<number>] rule permit ip destination 10.1.1.5 0
```

This rule permits traffic destined to 10.1.1.5 from any source IP address because you did not specify the **source** option.

On ProVision switches, on the other hand, you enter the match criteria without keywords, but you must specify each component in the order indicated in Figure 6-11. (However, you skip the source port or destination port component if you want to allow any number.)

For example, you could enter this command to permit HTTP requests to 10.1.1.5 from any source IP address:

```
ProVision(config-ext-nacl)# permit tcp any host 10.1.1.5 eq http
```

As you see, you must specify the **any** keyword for the source IP address.

You will see more examples of command syntax as you complete the labs.

IP packets can carry a variety of protocols, and you can configure rules that select packets based on those protocols, filtering out unnecessary protocols that pose a risk. Consider ICMP packets, which an attacker can use to map the network and implement certain exploits and denial of service (DoS attacks). For example, in the Smurf attack, the attacker sends an ICMP echo request with a spoofed source address to a broadcast address. Devices then flood the spoofed source with replies. You might want to protect areas of your network by filtering out ICMP packets that cross from a less trusted network segment to a trusted subnet.

Because you can create rules that select ICMP traffic by type, you can protect the network while still maintaining normal functions. You could perhaps filter ICMP echo packets entering a protected subnet but allow ICMP echo reply packets, enabling your own servers to send pings and receive replies, as necessary.

Figure 6-12 indicates the components of a rule that selects ICMP packets:

- Protocol set to ICMP
- Source IP address (any, one, or a range)
- Destination IP address (any, one, or a range)
- ICMP type, specified by number or name (see the specific options on your switch model)

Figure 6-12 also indicates some of the other IP protocols that you can specify for rules. For most of the protocols, you simply specify the protocol and the source and destination IP address without additional options. On some switch models, you might be able to define specific packet types for the protocol—for example, for IGMP.

Basic components of a rule:

Protocol	Source (IP address)	Destination (IP address)	ICMP type
icmp	Any, one, or range	Any, one, or range	Any or specific type (name or number)

Basic components of a rule:

Protocol	Source	Destination	X type
igmp pim ospf esp ah ...	Any, one, or range	Any, one, or range	Valid for some protocols

Figure 6-12: Creating rules for advanced/extended ACLs

Supplemental information

As a final note, you should understand that you can add options to any of these types of rules to further limit the selected traffic. For example, you can specify a QoS mark so that the rule selects only traffic that meets the other criteria and is also marked for a specific type of service. You will learn about QoS marks in Chapter 8. On some Comware devices, you can associate rules with VPN instances and so on.

The following sections summarize all the available options for advanced/extended rules.

Comware

This list summarizes all the available options for advanced ACLs on Comware devices:

- **rule <*rule ID*>**—Specifies a rule ID, which ranges from 0 to 65534. If no rule ID is provided when you create an ACL rule, the system automatically assigns it a rule ID. This rule ID takes the nearest higher multiple of the numbering step to the current highest rule ID, starting from 0. For example, if the rule numbering step is 5 and the current highest rule ID is 28, the rule is numbered 30.
- **deny | permit**—These keywords specify whether the rule denies or permits a packet matching the criteria in the ACE.
- **protocol**—To filter by protocol you can specify by protocol name or code. (The ack, fin, psh, rst, syn, and urg keywords specify TCP flags.)
- **established**—For the TCP protocol, you can optionally include this keyword, which is used to control TCP connection traffic. It blocks in one direction on a VLAN the synchronizing packets associated with establishing a TCP connection, while allowing all other IP traffic for the same type of connection in the opposite direction.

  For example, a Telnet connection request requires TCP traffic to move both ways between a host and the target device. Simply applying a deny action to inbound Telnet traffic on a VLAN would prevent Telnet sessions in either direction, because responses to outbound requests would be blocked. By using the **established** keyword, inbound Telnet traffic arriving in response to outbound Telnet requests would be permitted, but inbound Telnet traffic trying to establish a connection would be denied.
- **counting**—This option configures the switch to record the number of packets that it has matched to the rule. To enable counting for rule matches performed in hardware, configure the **hardware-count enable** command for the ACL.
- **destination {<*dest-addr dest-wildcard*> | any}**—This option defines the destination IP address a packet must carry for a match with the rule.
- **any**—This option allows IP packets from any destination IP address.
- ***<dest-addr dest-wildcard>* | any**—This option specifies the specific destination address(es) and wildcard mask.
- **dscp**—You would use this keyword to filter by DiffServ codepoint.
- **fragment**—Without this keyword, the rule applies to all fragments and nonfragments.

  Traditional packet filtering matches only first fragments of packets and allows all subsequent non-first fragments to pass through. This is because the Layer 4 information that is used to filter TCP and UDP packets is only present in the initial fragment. Thus, attackers can fabricate non-first fragments to attack networks. To avoid the risks, Comware switches filter all fragments by default, including non-first fragments, which are evaluated solely on the Layer 3 portion of any configured rule. You can, however, also set matching criteria—to filter only non-first fragments, for example.

- **icmp**—This option enables you to enter packet-type criteria for ICMP traffic. IP hosts rely heavily on the Internet Control Messaging Protocol (ICMP), and it would likely have a negative impact on network communication if you were to deny all ICMP traffic.
- **logging**—In addition to counting packets that are matched to the rule, the switch can log the matches. The switch stores logs and transmits them to the information center at the ACL logging interval, which you set with the **acl logging frequence <*minutes*>** command.
- **precedence**—This option enables you to enter IP precedence criteria.
- **reflective**—The reflective keyword is not supported in the current software version, and is reserved for future support.
- **source <*source-address source-wildcard*> | any**—This option defines the source IP address (SA) a packet must carry for a match with the rule.
- **any**—This option allows IP packets from any source address.
- **<*source-address source-wildcard*>**—This option specifies the specific source address(es) and wildcard mask.
- **source-port operator**—This option specifies one or more UDP or TCP source ports. The operator argument can be lt (lower than), gt (greater than), eq (equal to), neq (not equal to), or range (inclusive range).
- **time range**—On Comware switches, you can implement ACL rules based on the time of day by applying a time range to them. A time-based ACL rule takes effect only in any time periods specified by the time range.
- **tos**—This option enables you to enter ToS criteria.
- **vpn instance**—This option enables you to specify a vpn instance as the match criteria.

ProVision

This list summarizes all the available options for extended ACL rules on ProVision switches:

- **1-2147483647**—You can specify a manual ID for the rule.
- **deny | permit**—These keywords specify whether the rule denies or permits a packet matching the criteria in the rule.
- **ip | *ip-protocol-name* | *ip-protocol-number*>**—This option is used after deny or permit to specify the packet protocol type required for a match. "ip-protocol-nbr" is the IP protocol number of an IP packet type, such as "8" for Exterior Gateway Protocol or 121 for Simple Message Protocol. As you learned, specify **ip** to select all IP protocols.

Note

For details about the specific protocols your switch supports, refer to the *HP Switch Software Access Security Guide* (or *HP Switch Software Advanced Traffic Management Guide*) for the switch model.

- **any | host < *source IP address*> | <*source IP address* > < *mask* > | < *source IP address/prefix-length* >**—This defines the source IP address a packet must carry for a match with the rule.
  - **any**—Allows IP packets from any source IP address.
  - **host < *source IP address* >**—Use this criterion when you want to match only the IP packets from a single source address.
  - **<*source IP address*> < *mask* > or <*source IP address/prefix-length*>**—Specifies packets received from a source address where the source address is either a subnet or a group of IP addresses. The mask can be in either dotted-decimal format or CIDR format (number of significant bits). The mask is applied to the source address in the ACL to define which bits in a packet's source address must exactly match the source address configured in the ACL and which bits need not match.
- **any | host < *destination IP address* > | <*destination IP address*> <mask> | <*destination IP address/prefix-length*>**—This defines the destination IP address that a packet must carry in order to have a match with the rule.
- **precedence < 0–7 | *precedence-name* >**—This option can be used after the destination IP address to cause the rule to match packets with the specified IP precedence value.
- **tos < *tos-setting* >**—This option can be used after the destination IP address to cause the rule to match packets with the specified IP ToS setting.
- **log**—This option can be used after the destination IP address to generate an Event Log message when a packet matches this rule. Refer to the "Creating rules for basic/standard ACLs" section for more details about logging.
- **established** – For rules that use the TCP protocol, you can optionally include the **established** keyword, which is used to control TCP connection traffic. It blocks in one direction on a VLAN the synchronizing packets associated with establishing a TCP connection while allowing all other IP traffic for the same type of connection in the opposite direction.

Comware: Matching based on logical order

Remember, as you learned earlier, that you must specify the **match-order auto** option when you create the ACL if you want the Comware switch to automatically match packets against rules in a logical order.

ACL category	Sequence of tie breakers
IPv4 Advanced ACL	1. VPN instance (not used in this course) 2. Specific protocol type rather than IP (IP represents any protocol carried by IP) 3. More 0s in the source IP address wildcard mask 4. More 0s in the destination IP address wildcard 5. Narrower TCP/UDP service port number range 6. Smaller rule ID

Figure 6-13: Matching based on logical order on Comware switches

Figure 6-13 illustrates how this method is applied to advanced IPv4 ACLs.

1. As for basic ACLs, the Comware device places rules associated with specific VPN instances first in order. However, this study guide does not cover VPNs.
2. The switch executes a rule configured with a specific protocol prior to a rule with the protocol type set to IP.
3. Next the switch sorts rules by source IP address wildcard bits, as you learned for basic ACLs. That is, the switch compares packets against the rule with the most zeroes in the source IP address wildcard, then against the rule with the next highest number of zeroes, and so on.
4. If two rules are present with the same number of zeroes in their source IP address wildcards, the switch examines the destination IP address wildcards, comparing packets first against the rules with more zeroes in the wildcard bits.
5. If the number of zeroes in the destination IP address wildcards is the same, the switch looks for the narrower TCP/UDP service port number range.
6. If further differentiation is necessary, the switch compares the rule IDs. The packet with the smaller rule ID takes precedence.

Creating valid IDs for Ethernet frame header ACLs

Currently, ProVision switches do not support Ethernet ACLs. Comware Ethernet ACLs have numbers between 4000 and 4999. As for basic and extended IPv4 ACLs, you can add an alphanumeric name to help yourself and your fellow administrators to identify the ACL (see Figure 6-14).

Creating valid IDs for Ethernet frame header ACLs

Category	Comware ACL ID
Ethernet ACLs	Always use a number: 4000–4999 Optionally, add an alphanumeric name (< 64 characters, including spaces).

Figure 6-14: Creating valid IDs for Ethernet frame header ACLs

Enter this command from the system view to create the ACL:

```
[Comware]acl number <4000-4999> [name <ACL name>] [match-order {auto |
config}]
```

Creating rules for Ethernet frame header ACLs

An Ethernet frame header ACL rule has the same basic structure as IP ACLs (see Figure 6-15):

```
[Comware-acl-number-<4000-4999>]rule [ID] <permit | deny> [source-mac <MAC
address> <mask>] [dest-mac <MAC address> <mask>] [lsap <0-ffff> <lsap mask>]
[cos <0-7>] [time-range <time>][counting]
```

Match criteria:

Source MAC address and mask	Destination MAC address	Link layer protocol	CoS (802.1p value)
Any, one, or range	Any, one, or range	Any, one type, or range of types	0–7

In masks:

F = Match hex character

0 = Don't match hex character

Figure 6-15: Creating rules for Ethernet frame header rules

Rule ID

As with IP ACLs, you can specify the rule ID manually. Or if you do not specify an ID, the switch adds the rule to the end of the list, setting the ID to the ID of the rule with the highest ID plus the step value.

Action

The rule can permit matching frames or deny them. When you apply the ACL as a packet filter, the switch forwards permitted frames and drops denied frames. You can also apply Ethernet frame header ACLs for other features such as QoS policies, covered in the next chapters.

Match criteria

The basic match criteria include source MAC address and destination MAC address. For either type of address, you must specify both the MAC address and the MAC address mask, which indicates whether the rule matches an exact MAC address or a range of address.

You can also add a few other criteria, described below.

Whenever you do not specify a particular criterion in a rule, the switch automatically allows any match. For example, if you do not specify a destination MAC address, the rule permits all destinations.

MAC addresses and MAC address masks

The mask follows the typical mask logic (not wildcard logic). Each f indicates that a frame's source or MAC address must match the corresponding hexadecimal character in the address in the rule. If the mask has a 0, on the other hand, the switch ignores the corresponding character in frames' addresses.

For example, **source-mac 0050-5600-0001 ffff-ffff-ffff** matches only frames from a device with that precise MAC address. But to match 0050-5600-0000 ffff-ff00-000, a frame can have any source MAC address that begins with 0050–56.

You can create more complicated rules using different hexadecimal characters in the mask. Determine the exact bits with source or destination MAC addresses that you want the switch to check or ignore. Set checked bits to 1s and ignored bits to 0; then convert to hexadecimal. For example, you enter **dest-mac 0050-5600-0000 ffff-ffff-fffc**. This mask tells the switch to check all bits for all hexadecimal characters until the last character. At that point, the switch checks the first two bits but ignores the last two bits (1100 = hexadecimal c). This rule selects packets with destination MAC addresses 0050-5600-0000 through 0050-5600-0003. However, complicated rules such as these are usually unnecessary.

Additional criteria

Other criteria include:

- The link layer protocol type (**lsap**)

- The class of service (CoS), or 802.1p, value (which you will examine in "Chapter 8: QoS")
- The time

Additional options

You can also add options such as counting to the rule. The counting option configures the switch to record matches to the rule.

Note that the switch does not typically record matches for frames that are processed against the ACL in the hardware. However, you can enable this function with this command:

```
[Comware-acl-ethernetframe-<number>] hardware-count enable
```

You should take care when enabling hardware counting. Also note that the switch ignores this command when you specify the ACL in a QoS; in this case, it never records hardware matches.

Auto match order

When you create an Ethernet frame header ACL, you can choose the auto match order. The switch automatically processes frames against the ACL in this order:

- First against a rule with the more specific source MAC address (more 1s in the mask)
- To break ties, against a rule with the more specific destination MAC address (more 1s in the mask)
- To break ties, against a rule with a smaller rule ID

Applying ACLs to switch interfaces

After you create an ACL, you can apply it to one or more interfaces. Until you actually assign an ACL to a VLAN interface or physical interface, it is present in the configuration but not used to filter traffic. When you apply an ACL to an interface using a configuration command, the ACL is called a static ACL. Once you apply the static ACL, it remains on the interface permanently until you modify or remove it.

Note

HP switches can also apply dynamic port ACLs as part of an authentication, authorization, and accounting (AAA) solution. This chapter focuses on static ACLs. You will examine the dynamic port ACLs in Chapter 8.

You can apply three types of static ACLs:

- Routed ACLs (RACLs)
- Port-based ACLs (PACLs)
- VLAN-based ACLs (VACLs) on ProVision switches only

Type of ACL	HP Comware	HP ProVision
Layer 2 VLAN (VACL)	No	Yes, in
Layer 3 VLAN (RACL)	Yes, in/out	Yes, in/out
Static port ACL	Yes, in/out	Yes, in

IMPORTANT: Applying an ACL ID for an ACL that does not exist or has no entries permits all traffic.

Figure 6-16: Applying ACLs to switch interfaces

As Figure 6-16 shows, for RACLs and port ACLs, you can also choose whether the ACL applies to inbound or outbound traffic.

These choices together determine which packets and frames the switch processes against the ACL.

Note

The extent of ACL support varies among the HP switch families. For example, some switches may support RACLs, VACLs, and both static and dynamic port ACLs, whereas other switches may support static port ACLs only. Refer to your switch documentation for more information.

Supplemental tips

Keep the following points in mind when applying ACLs:

- Both ProVision and Comware switches allow you to assign a nonexistent ACL name or number to an interface; they might provide a warning that the ACL has no rules. When you apply a nonexistent or empty ACL to an interface, the interface permits all traffic until you create an ACL that uses that name or number *and* add a rule to the ACL. The ProVision implicit "deny any" does not take effect until at least one rule is defined. In short, you should be careful to specify the correct ID.
- On ProVision switches, if you delete an assigned ACL from the switch without subsequently removing the ACL from an interface to which it is assigned, the ACL assignment remains on the interface. The empty ACL does not affect traffic (all traffic is permitted). However, the switch will automatically begin applying any new ACL you create with the same identifier (name or number). You should clean up your configurations to ensure that ACLs are applied when and where you expect.

RACLs

You would configure an RACL to filter *routed* traffic between subnets. RACLs apply to Layer 3 VLAN interfaces, and you should apply them on a routing switch that handles routing for that VLAN. For example, your core routing switch is the default gateway for VLAN 2, which is also configured on several access layer switches. If you want to apply an RACL to control traffic routed to or from VLAN 2, apply the ACL to VLAN 2 on the core routing switch.

On ProVision switches, you apply the ACL to the VLAN and specify the direction (**ip access-group <*name*> <in | out>**). On Comware switches, you apply the ACL to the Layer 3 VLAN interface as a packet filter and specify the direction (**packet-filter <*number* | name *name*> <inbound | outbound>**).

An inbound RACL filters the following traffic:

- **Traffic received on this Layer 3 interface and routed out another subnet**—For example, you have applied the RACL inbound to VLAN 2, which uses the 10.1.2.0/24 subnet. The ACL filters packets with sources in the 10.1.2.0/24 subnet and destinations in other subnets. Routing can be between different VLANs or between different subnets in the same multinetted VLAN. Because the ACL applies to traffic that *this* network infrastructure device routes out of this interface, the device must be routing the traffic. You must enable routing on a ProVision switch.
- **Traffic received on this Layer 3 interface and destined to this network infrastructure device**—An RACL does not filter switched IP traffic unless the switch's own IP address on the interface is the destination. On a ProVision switch, you do not need to enable routing on the switch for the ACL to perform this service. (Of course, as long as the device is routing, it also filters packets destined to its own address on other Layer 3 interfaces.)

Note
The RACL does *not* filter traffic generated on the interface by the network infrastructure device itself.

Direction	Filters	Does not filter
in	Traffic received on the VLAN and: • Routed to any other subnet, including another subnet on the same multinetted VLAN • Switched to the switch's own IP address on the subnet	• Traffic received on the VLAN and switched on the same subnet (except to the switch itself) • Traffic received on other VLANs and routed out this VLAN
out	Traffic routed to this VLAN (received on any other subnet, including another subnet on the same multinetted VLAN)	• Traffic generated by the switch and routed out this VLAN • Traffic received on this VLAN and switched out the same subnet • Traffic received on this VLAN and routed out another subnet

Figure 6-17: Filters for RACLs

An outbound RACL filters the following traffic:

- **Traffic received from any subnet and routed to a subnet on this Layer 3 interface**—For example, you have applied an RACL outbound to a switch's VLAN 2 interface, which uses the 10.1.2.0/24 subnet. The switch receives a packet destined to 10.1.2.100 on VLAN interface 10. This ACL filters that packet because the switch routes the packet to VLAN 2. If, however, the switch receives a packet destined to 10.1.2.100 from 10.1.2.10, it does not route the packet or apply the ACL. Again, routing can be between different VLANs, or between different subnets in the same multinetted VLAN.

Note

The RACL does *not* filter traffic generated on the network infrastructure device itself and routed out this interface.

On both types of switches, you can apply only one RACL per interface to inbound traffic and one RACL per interface to outbound traffic. In the case of a multinetted VLAN, this rule implies the following:

- IP traffic received on different subnets in the same VLAN is screened by the same inbound RACL.
- IP traffic routed out different subnets is screened by the same outbound RACL.

Figure 6-17 summarizes the traffic RACLs filter when applied inbound or outbound.

You can use the same ACL for both inbound and outbound RACL applications. Of course, you would need to ensure that the ACL has the correct entries for filtering both inbound and outbound traffic.

You can apply standard (basic) or extended (advanced) IP ACLs as RACLs, as well as Ethernet frame header ACLs on Comware switches.

Port ACLs

On a ProVision switch, you would configure a static port ACL on a port or link aggregation to filter IP traffic inbound on that physical interface. The switch filters *all* inbound traffic regardless of whether it is routed or switched. The port ACL also applies to traffic inbound on the interface with a destination on the switch itself.

On a Comware switch, you also apply the static PACL to a port, but you can apply the port ACL in either direction (inbound or outbound). The inbound port ACL filters all packets inbound on the physical interface regardless of whether they are routed, switched, or destined for the network infrastructure device itself. Similarly, the outbound port ACL filters all packets outbound on the physical interface regardless of whether they were routed, switched, or generated by the network infrastructure device itself. A PACL can be either an IP ACL or an Ethernet ACL.

Comware routers and some Comware switches support Layer 3 physical interfaces, which support IP and routing functions. In this case, the port ACL applies to routed traffic because, of course, all traffic received and forwarded on the interface is routed.

Figure 6-18 summarizes how port ACLs can be applied and what traffic they filter.

You apply the port ACL with the same command as an RACL, but you enter the command from the physical interface context rather than the VLAN interface.

Direction	Port type	Filters
in	Layer 2	All inbound traffic, routed and switched
	Layer 3 (some Comware only)	All inbound traffic (routed by definition)
out (Comware only)	Layer 2	All outbound traffic, routed and switched
	Layer 3	All outbound traffic (routed by definition)

Figure 6-18: Applying port ACLs

Applying port ACLs to link aggregations (ProVision)

You must apply port ACLs to the link aggregation (trunk) rather than to individual ports within the link aggregation. You should also remove any ACLs from a port before adding it to a link aggregation.

When you apply the port ACL to the link aggregation, it automatically applies to the member ports. (Similarly, removing the port ACL from the link aggregation removes it from member ports.) Removing a port from the link aggregation removes the port ACL from that port.

Applying port ACLs to link aggregation (Comware)

On Comware switches, you *must* apply the port ACL to *individual* ports within a link aggregation group.

Important
Make sure to apply the same ACL to all member ports to avoid inconsistencies.

ProVision: VACLs

As you learned, RACLs applied to VLAN interfaces do not filter traffic that flows within the *same* subnet. When you want to control traffic switched within a subnet, you must apply a different type of ACL.

A PACL filters both routed and switched traffic and meets this need. However, when you want to control all devices in a VLAN in the same way, a VACL is the most efficient mechanism. One VACL applies to all ports that are members of the VLAN. The VACL filters all IP traffic entering a switch on the specific VLAN interface. On a multinetted VLAN, this includes inbound IP traffic from any subnet.

The VACL filters routed traffic like an inbound RACL but also filters switched traffic. In specific, the VACL filters this traffic:

- Traffic received on this VLAN interface and routed out another subnet
- Traffic received on this VLAN interface and forwarded in the same subnet
- Traffic received on this VLAN interface and destined to this network infrastructure device (on any interface)

The VACL does *not* filter traffic received on another interface and routed to this VLAN interface. (However, the return traffic from a device in this VLAN would be filtered.) In addition, the VACL does not filter traffic generated by the switch itself using its IP address on this VLAN.

Figure 6-19 summarizes the traffic that VACLs filter and do not filter.

Keyword for applying	Filters	Does not filter
vlan	All traffic received on the VLAN: • Switched on the same subnet on which it was received • Routed out a subnet on another VLAN • Routed out a different subnet on the multinetted VLAN	• Traffic received on other VLANs and routed out this VLAN • Traffic generated by this switch on this VLAN

Figure 6-19: ProVision VACLs

When you configure VACLs on ProVision switches, you apply the ACL to the VLAN but specify the **vlan** option instead of a direction (**ip access-group <*name*> vlan**).

You can assign the same ACL to multiple VLANs. HP switches, however, allow you to configure just one VACL per VLAN. (This is in addition to any other ACLs you assigned to the VLAN or to ports in the VLAN.)

Comware: Alternative method for filtering traffic

You have learned about using the packet-filter command to apply ACLs to specific interfaces on Comware switches. You can alternatively use a QoS policy, which applies to all traffic routed on the switch.

If possible, you should typically use the packet-filter command to apply ACLs to interfaces. This method easily allows for a mixed set of rules that both permit and deny traffic and allows for dynamic reconfiguration of the ACL. To modify an ACL applied with a QoS policy, the policy must first be removed. The initial setup might also be more complicated.

However, you might decide that a global QoS policy best meets your needs. First, decide whether you want the filter to allow all traffic that is not explicitly denied or to discard all traffic that is not explicitly permitted.

Whitelist global packet filter

The left side of Figure 6-20 illustrates a policy that takes the latter approach, allowing only traffic from devices in the 10.1.0.0/16 and denying all other traffic.

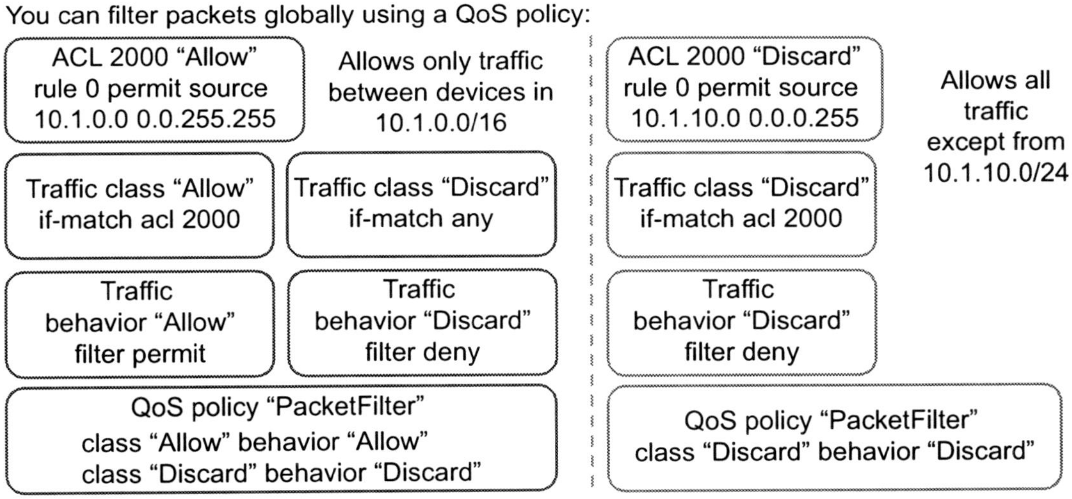

Figure 6-20: Alternative method for filtering traffic on Comware switches

To create a policy such as this, you follow these steps:

1. Create an ACL.

   Rather than allow and discard traffic itself, this ACL selects traffic for a traffic class. That is, it selects any traffic that matches a permit rule, but does not select any traffic that matches a deny rule. It also does *not* select any traffic that does not match any rule—unlike what you might expect from the behavior of a packet-filtering ACL.

The ACL must select all the traffic that you want to allow. Create permit rules that match that traffic.

The ACL might also have deny rules that filter out some traffic that would otherwise match a permit rule. For example, consider a different scenario in which you want to allow traffic from all IP addresses in the 10.1.0.0/20 range (10.1.0.0 to 10.1.15.255) except 10.1.8.0/24. In that case, you would create a deny rule for 10.1.8.0/24 (10.1.0.0 0.0.0.255) and a permit rule for 10.1.0.0/20 (10.1.0.0 0.0.15.255).

2. Create a traffic classifier that matches the ACL.
3. Create a traffic behavior that specifies the "filter permit" action.
4. Create a traffic classifier that matches all traffic.
5. Create a traffic behavior that specifies the "filter deny" action.
6. Create a QoS policy. First, link the traffic classifier associated with the ACL and the "filter permit" behavior. Then link the traffic classifier that matches all traffic and the "filter deny" behavior.

   You must create the second statement due to how the QoS policy processes traffic. Any traffic that does not match one of its classes, it simply passes on without altering. Therefore, you need a traffic class that selects all remaining traffic and applies the "filter deny" behavior.
7. Assign the QoS policy to the switch. On most Comware switches, only global QoS policies support the traffic filtering behaviors. The QoS policy acts like an RACL in that it affects all routed (but not switched) traffic.

Remember that the global inbound policy applies to *all* routed traffic that arrives on the switch, and the global outbound policy applies to *all* routed traffic that leaves the switch. Remember to permit return traffic as well.

For example, with this policy, the switch would allow traffic from 10.1.1.1 to 10.2.2.2; however, when 10.2.2.2 sent traffic back to 10.1.1.1, it would drop the traffic. Therefore, the policy only allows traffic between devices in the 10.1.0.0/16 range.

Blacklist global packet filter

As shown on the right side of Figure 6-20, you can take a different approach to the QoS policy-based packet filtering. You create an ACL that select selects the traffic that you want to discard. That ACL composes the match criteria for a traffic class, which you map to the "filter deny" behavior in a QoS policy. You apply the QoS policy globally to the switch.

Make sure that you understand that the policy discards all routed traffic that matches *permit* rules in the ACL. (The permit rules select the traffic for the traffic class, and the QoS policy applies the filter deny behavior to this class. In a way, the "permit" rules become "deny" rules.)

Example basic/standard ACL implementation

For a more solid understanding of the concepts in this chapter, explore a basic scenario. Figure 6-21 shows a very simple network with an access layer switch that supports two VLANs:

- VLAN 10, 10.1.10.0/24
- VLAN 20, 10.1.20.0/24

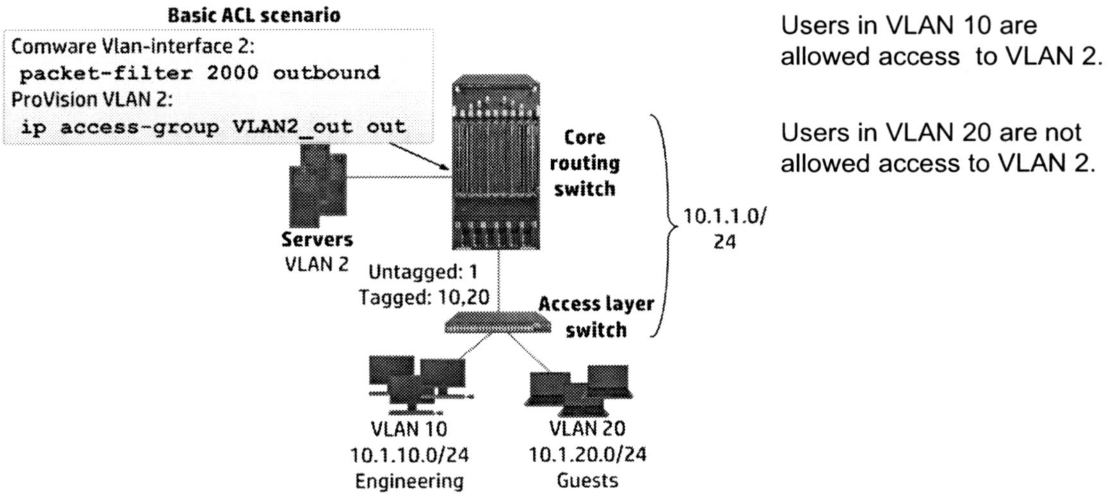

Figure 6-21: Example of basic/standard ACL

The core routing switch supports VLAN 2, which connects to servers. (It is more likely that this switch connects to another access layer switch, which in turn connects to servers, but the infrastructure is simplified for this example.)

Both switches shown have their IP addresses on 10.1.1.0/24.

For this scenario, the network infrastructure and trusted users on VLAN 10 are allowed access to the servers in VLAN 2. In addition, devices in VLAN 2 can communicate with each other. Users on the guest VLAN (VLAN 20), however, are not allowed access to VLAN 2.

Whenever possible, you should try to enforce desired filtering with the fewest number of ACLs. In this example, VLAN 2 is the common factor, so you should apply the ACL to VLAN 2. You want to control traffic routed *to* VLAN 2, 10.1.2.0/24, from other subnets, so you should apply the ACL as an *outbound* RACL on the VLAN 2 interface.

You need to control the traffic based on the subnet of the source device, so a basic ACL will work. The ACL must permit traffic with source IP addresses in 10.1.1.0/24 and 10.1.10.0/24 and deny traffic with source IP addresses in 10.1.20.0/24. You do not need to create rules for allowing communications between devices within the 10.1.2.0/24 subnet because RACLs do not filter switched traffic.

You will first consider Comware switches. Assume that you have chosen the number "2000" and the name "VLAN2_out" for the ACL ID. In that case, the syntax is as follows:

```
acl number 2000 name VLAN2_out
```

```
rule 10 permit source 10.1.1.0 0.0.0.255
```

(This rule allows the network infrastructure access to VLAN 2.)

```
rule 20 permit source 10.1.10.0 0.0.0.255
```

(This rule allows trusted users on VLAN 10 access to VLAN 2.)

```
rule 65534 deny
```

(Remember that the Comware switches have an implicit "permit all" when you use them to filter traffic on an interface. That is why you need a rule to deny all other traffic. Assign the rule the highest ID so that you can insert other permit rules as necessary.)

```
interface vlan-interface 2
```

(Access the Layer 3 VLAN 2 interface.)

```
packet-filter 2000 outbound
```

(Apply the ACL as an outbound RACL.)

The ProVision syntax for the same setup is as follows:

```
ip access-list standard VLAN2_out
```

(Remember that on ProVision switches, a standard ACL can have either a name or a number in the range of 1 to 99. In this example, you have assigned it the name "VLAN2_out.")

```
10 permit 10.1.1.0 0.0.0.255
```

(This rule allows the network infrastructure access to VLAN 2.)

```
20 permit 10.1.10.0 0.0.0.255
```

(This rule allows trusted users on VLAN 10 access to VLAN 2.)

```
2147483647 deny any
```

(This rule is not necessary. ProVision switches have an implicit "deny any," so the switch will drop any packets that do not match the rules you created, including all traffic from VLAN 20 outbound to VLAN 2. However, if you want, you can make this entry explicit to clarify the behavior.)

```
vlan 2
```

(Access VLAN 2 to apply the ACL.)

```
ip access-group VLAN2_out out
```

(Apply the ACL as an outbound RACL.)

Of course, instead of permitting specific source addresses and denying all others, you could choose to deny access to VLAN 2 to users in VLAN 20, while permitting all other sources.

The Comware ACL for that approach is as follows:

```
acl number 2000 name VLAN2_out
```

```
rule 10 deny 10.1.20.0 0.0.0.255
```

(This rule denies users in VLAN 2 access to VLAN 20.)

```
Rule 20 permit any
```

(This rule is not necessary. Because Comware switches have an implicit "permit any" when you use them to filter traffic on an interface, all traffic from VLAN 10 outbound to VLAN 2 will be allowed access to VLAN 2. Devices in VLAN 2 are also able to communicate with each other.)

The ProVision syntax is as follows:

```
ip access-list standard VLAN2_out
```

```
10 deny 10.1.20.0 0.0.0.255
```

(This rule denies users in VLAN 2 access to VLAN 20.)

```
20 permit any
```

(Because ProVision switches have an implicit "deny any," you need rule 20 in order to permit access to all traffic that you have not explicitly configured the switch to deny.)

You apply the ACLs in the same way. Whether you choose the white-list approach (permit explicitly and then deny all) or blacklist approach (deny explicitly and then permit all) depends on how many addresses you want to permit versus deny, the security needs of the environment, and your own preference.

Note

In this scenario, you are just protecting VLAN 2. VLAN 20 can still reach VLAN 10 and the network infrastructure addresses. If you want to protect those VLANs, you would need to set up other lists for them.

Example advanced/extended ACL implementation

If you want more control over specific types of traffic, you must create advanced/extended ACLs for the network. As before, the access layer switch supports two VLANs:

- VLAN 10, 10.1.10.0/24
- VLAN 20, 10.1.20.0/24

The core routing switch supports VLAN 2, which connects to servers. Both switches shown have their IP addresses on 10.1.1.0/24.

In the first scenario, the guests in VLAN 20 could not reach any company services in VLAN 2. In this scenario, you want to deny VLAN 20 access to most of VLAN 2—and actually most of the network. However, the untrusted VLAN 20 users can reach a device that will give them DHCP addresses, answer DNS requests, and act as a proxy web server.

In this scenario, you want to control the guest VLAN 20. Other VLANs are assumed to be trusted and allowed to communicate with each other. Because you want to control all traffic originating from VLAN 20, you will create one ACL to apply as an *inbound* RACL on VLAN 20. As Figure 6-22 shows, this ACL will affect all traffic routed out of 10.1.20.0/24, no matter what its destination.

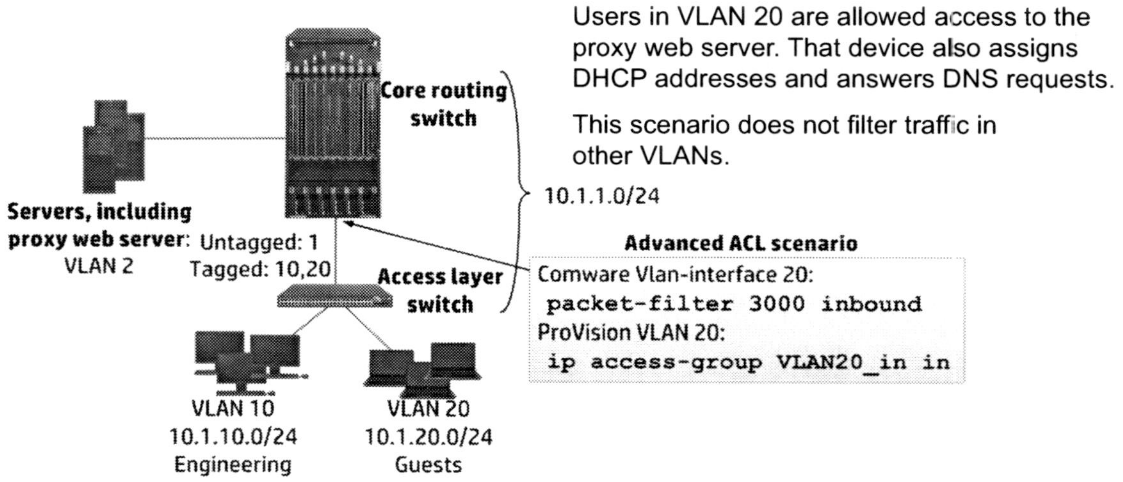

Figure 6-22: Example of an advanced/extended ACL

Begin with the Comware setup. Assume that you have chosen the number "3000" and the name "VLAN20_in" for the ACL ID. In that case, the syntax is as follows:

```
acl number 3000 VLAN20_in
rule 0 permit udp destination 10.1.2.100 0 destination-port eq bootps
rule 5 permit udp destination 10.1.2.100 0 destination-port eq dns
rule 10 permit tcp destination 10.1.2.100 0 destination-port eq 8080
rule 65534 deny ip
```

Remember that on Comware switches, you can use a single zero to identify an all-zeroes mask. The **eq** stands for "equal to" and specifies a match for a single port number or well-known service name.

After you create the rule, you can apply it:

```
interface vlan-interface 20
packet-filter 3000 inbound
```

The ProVision syntax for the same setup is as follows:

```
ip access-list extended VLAN20_in
10 permit udp 0.0.0.0 255.255.255.255 10.1.2.100 0.0.0.0 eq 68
20 permit udp 0.0.0.0 255.255.255.255 10.1.2.100 0.0.0.0 eq 53
30 permit tcp 0.0.0.0 255.255.255.255 10.1.2.100 0.0.0.0 eq 8080
2147483647 deny any
```

(You could add a rule that explicitly denies all other traffic, or leave this rule out, as you choose.)

```
vlan 20
ip access-group VLAN20_in in
```

If you want to prevent guests from sending traffic to each other, you could alternatively apply this ACL as a VACL. Note that the VACL would only affect traffic switched through the core, not at the access layer.

HP IMC ACL Management

You will now learn how HP Intelligent Management Center (IMC) helps you to create and manage ACLs on your network infrastructure devices. This section guides you through the basic components of the ACL Manager. (It does not show you all the pages and options nor give you step-by-step instructions for configuration.)

The ACL Management service helps you to configure ACLs in an intuitive manner and to apply them to the appropriate switches and interfaces with efficient batch operations. You can create all the types of ACLs that you have examined in this chapter as well as user-defined ACLs that filter packets based on content within the packet itself. (User-defined ACLs are beyond the scope of this study guide.)

In the following sections, you will see how you can centralize and standardize your ACL configurations across multiple devices using ACL resources, templates, and other defined objects. ACL Management lets you create rules in the way you choose—whether using wildcard bits or subnet masks. You can also define variables and other dynamic parameters to configure ACLs more quickly.

IMC brings disparate ACL configurations under its management so that you can standardize them and clone already configured ACLs for use elsewhere. ACL Management also helps you to manage the rules. For example, when you edit a rule set, you can easily sort the rules to fix their order, as necessary.

Creating rules for ACL resources manually

IMC allows you to define ACL resources, which are essentially equivalent to ACLs. Once you have defined an ACL resource, you can deploy it to any number of devices.

The ACL resource includes these parameters:

- **Type**—The types are basic IP, advanced IP, link (Ethernet frame header), or user-defined.
- **Identifier**—ACL Manager assigns this ID to the ACL when you deploy it to the switch. You can choose a number *or* name. Comware switches require numbers (and can add optional names), so you should select that option for them.
- **ACL resource name**—The ACL Manager uses this name to identify the resource, but it does not affect the device configuration.

You must also add a rule set to the resource. The rule set defines the rules for the ACL. You can configure a rule set in several ways.

First, you can simply add a rule set, defining its name and other basic settings such as the match order (config or auto) and whether the ACL uses time ranges. You then add rules, configured in a window such as that shown in the figure.

Rather than type syntax in a CLI, you select options from pulldown menus and type values in fields; however, you have all the same options for configuring a rule that you examined earlier. In Figure 6-23, you see a rule set with three defined rules, each configured in a page such as the one you see in the figure (note that the figure does not show the complete page).

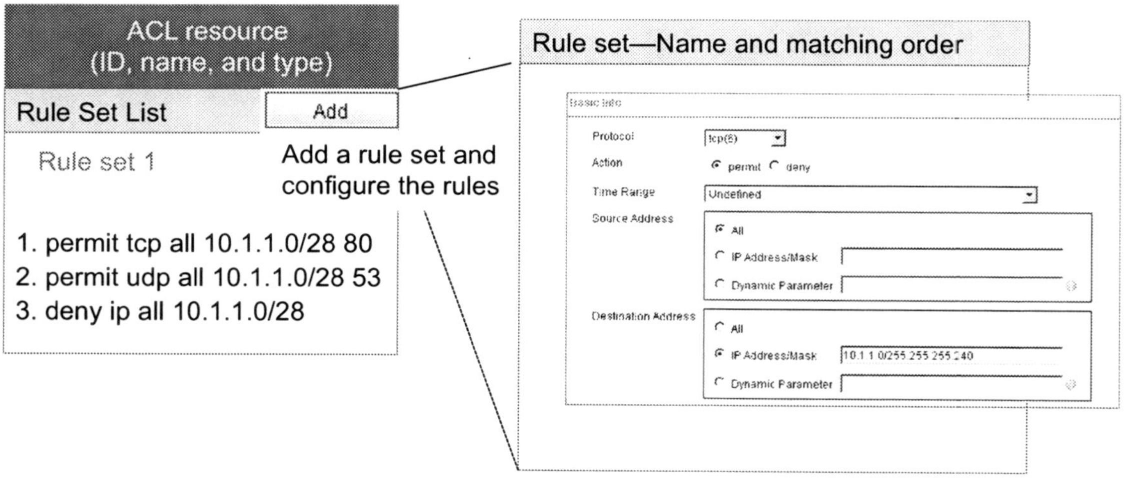

Figure 6-23: Creating rules for ACL management in IMC

Note

As you configure IP address ranges for rules, you must specify the correct subnet mask or wildcard bits, as you learned how to do earlier in this chapter. The ACL Manager lets you select either wildcard bits or subnet masks, as you prefer, in the global options. You must then configure every rule using the setting that you chose (you can specify a subnet mask or prefix length when you select the subnet mask option). The ACL Manager automatically adjusts rules to use the correct syntax when it applies the rules to a device.

For example, you select subnet mask as the global option. You then specify rules with source and destination IP addresses such as 10.1.1.0/24. When the ACL Manager deploys these rules to a Comware switch, it inverts the mask to the wildcard bits recognized by that device.

Importing rules from templates with static rules

You can also create a rule set by importing rules from an ACL template (see Figure 6-24). When you choose the import option, ACL Management presents all the ACL templates that have been defined with the same type as this resource (such as basic or standard).

An ACL template consists of a set of rules, defined much like rules in a rule set. In fact, you can configure the rules exactly like a rule set. You can then import the template into various ACL resources as needed.

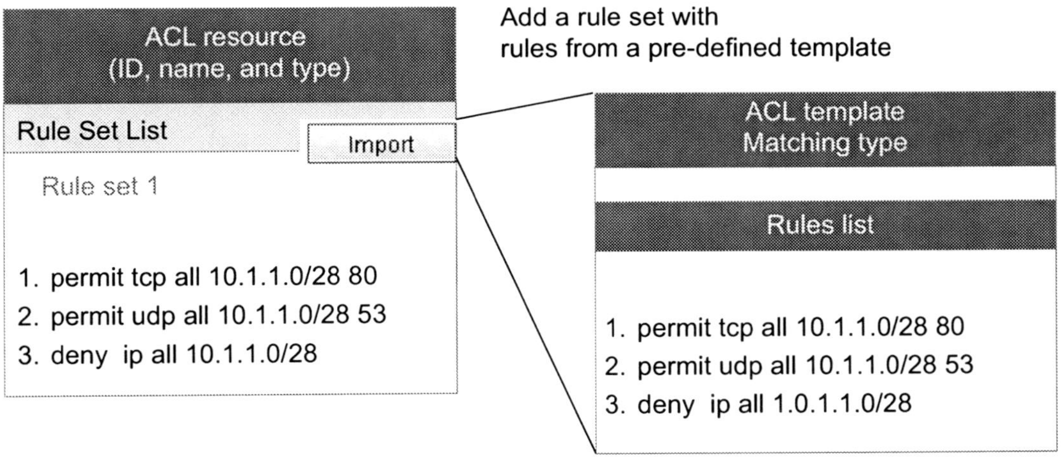

Figure 6-24: Importing rules from templates with static rules

Importing rules from templates with variables

However, the ACL template provides additional options for creating ACLs in an efficient and flexible manner. In an ACL template, rather than type a specific value for a particular parameter, you can choose to define a variable. For example, your network requires similar ACLs to control traffic received on the teacher VLAN and on the student VLAN. Teachers and students require the same services, but they are allowed access to different servers.

Figure 6-25 shows how you can create a standard ACL template that works for both ACLs. In this template, rather than define a specific range of destination addresses, you select Variable Address and type Servers. (Note that IMC automatically adds the % to the variable. When you specify the variable, you simply type the string, such as "Servers.")

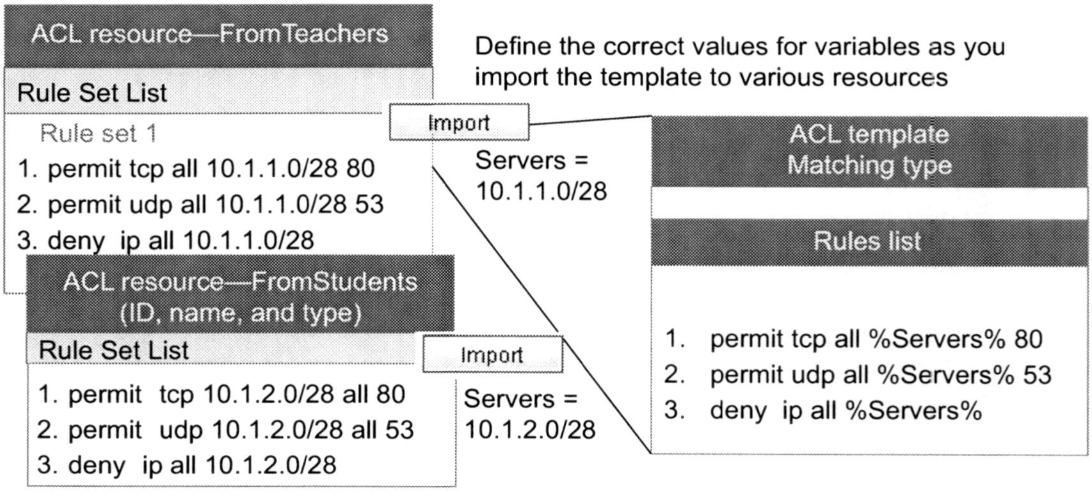

Figure 6-25: Importing rules from templates with variables

You can specify the same variable in other rules within that template. You can also create other variables with other names.

Then, when you import the template as a rule set within an ACL resource, you specify the value for your variables. In the example scenario, you set the Servers variable to 10.1.1.0/28 (or 10.1.1.0 0.0.0.15 if you are using wildcard bits) for the FromTeachers ACL resource. Variables are significant across the rule set. In other words, if you define the Servers variable in three rules within the template, you define the variable *once* when you import the template as a resource rule set. That same value is then applied to all three rules.

When you import the same template into the FromStudents ACL, you set the Servers variable to 10.1.2.0/28, thus granting students access to the correct servers. Of course, this is a simplified scenario, but you can see how you can create sophisticated ACLs in this way.

You have another option for specifying the values for variables when you import a template as a rule set: you can select a predefined object. The ACL Manager then calls up predefined objects that correspond to the type of variable:

- Network address groups for source IP address and destination IP address variables
- Services for TCP or UDP ports

These objects make it easier for you to create rules quickly and intuitively. You define the object once and then call it up by an intuitive name whenever required.

For example, you have defined a network address group named TeacherServers that includes all the IP addresses for servers to which teachers are allowed access (see Figure 6-26). Now you import the same template that you saw previously. However, rather than specifying 10.1.1.0/28, you select the TeacherServers network address group.

You can sometimes even select multiple network address groups for the variable. However, whether you are allowed to select multiple objects depends on factors such as whether multiple rules in the template use that variable and whether you have defined multiple objects for other variables.

In addition to the services and network address groups that you define yourself, the ACL Manager includes many well-known services such as Telnet, FTP, and DNS by default.

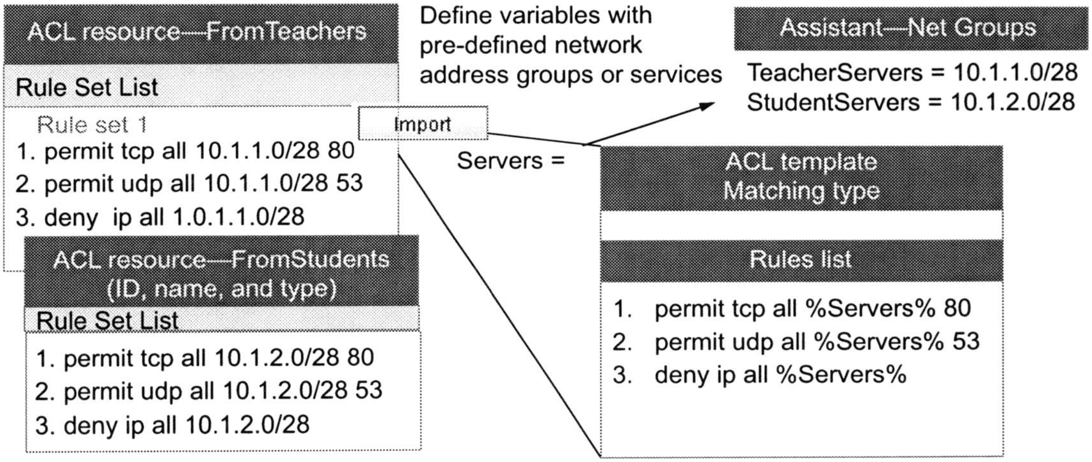

Figure 6-26: Importing rules from templates with variables

Creating standard rules for devices controlling different networks

Sometimes you want to configure a standard ACL for devices across the network.

For example, your company might have 50 branch offices. Each branch office uses standard, but different addressing schemes. Branch office 1 uses 10.1.0.0/16, branch office 2 uses 10.2.0.0/16, and so on. User, guest, and server subnets have similar addresses—for example, 10.X.1.0/28 for one set of servers. The company wants to implement standard ACLs on the local routing switch at each branch office, but configuring 50 ACLs would be tedious.

As Figure 6-27 shows, you can add a rule set with rules that include dynamic parameters. The dynamic parameter refers to values in the aclDynamicPara.csv file that you have saved in the {imc_home}\client\conf\acl\DynamicParaFile folder (you might need to create the acl and DynamicParaFile folders).

This .csv file always includes a first column for which the first row is Device_IP, which references the IP addresses of the managed devices to which you will deploy this ACL resource rule set. The first row of the next column specifies the name of a dynamic parameter—LocalServers, in this example. You then specify the appropriate match criteria in each. You can use dynamic parameters for a source or destination IP address or a source or destination MAC address. In the .csv file, define addresses with the same format that you would use in the GUI. Make sure to save the file in ANSI format.

Now, when you deploy the ACL resource rule set to a particular branch routing switch, IMC ACL Management sends the rules with IP addresses at that branch.

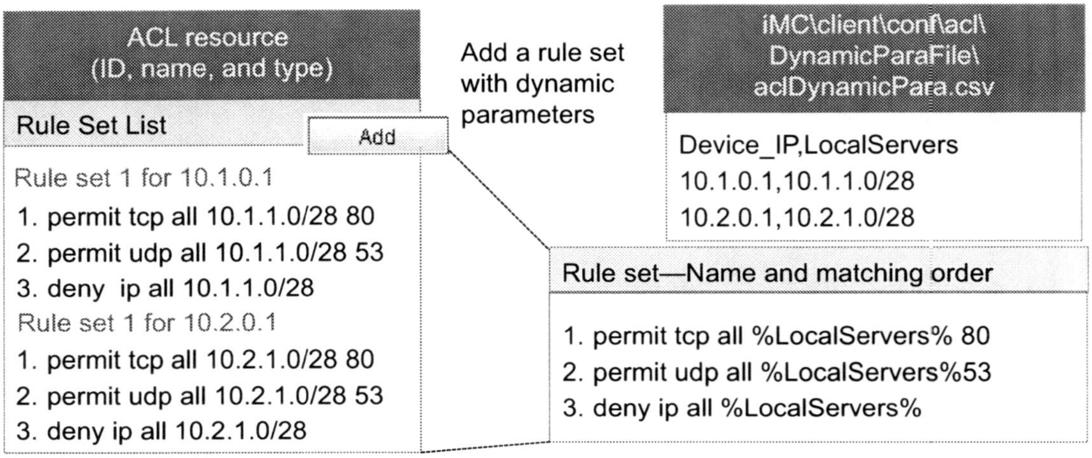

Figure 6-27: Creating standard rules for devices controlling different networks

Discovering and importing ACLs

The ACL Manager monitors the devices managed by IMC. It probes devices that are ACL capable and discovers any ACLs configured on them. You can then manage those ACLs from IMC, as shown in Figure 6-28. You might want to bring ACLs already configured on the devices within the ACL Manager's structure. You can then duplicate and adjust the ACL for other devices.

You can import an ACL discovered on a device as:

- An ACL rule set for an existing ACL resource

  If you want to define a new ACL resource for the discovered ACL, simply create that resource before you import the ACL.

- An ACL template, which you can then import to ACL resources for deployment to other devices

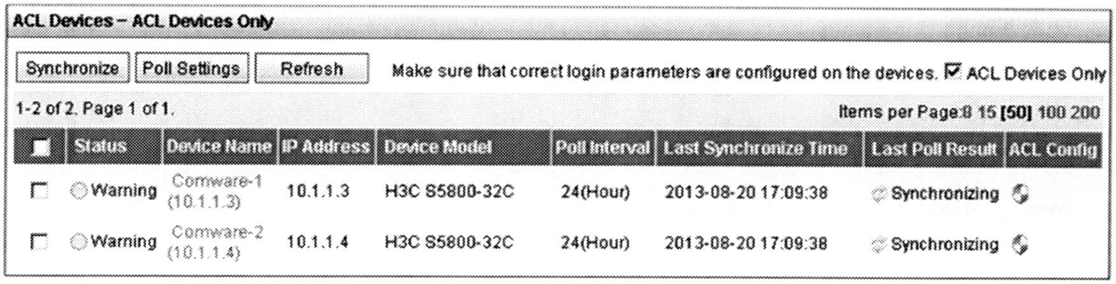

Figure 6-28: Discovering and importing ACLs

Deploying ACLs to managed devices

After you create a rule set for the ACL resource, you are ready to apply the ACL to the appropriate network infrastructure devices. You must complete two steps:

1. Deploy the resource and rule set, which configures an ACL on the targeted devices (see Figure 6-29).

   As you deploy the resource, you make choices such as what to do if the device already has an ACL with this ID.

   Note that an ACL resource can include multiple rule sets, which you might have created because you want to apply slightly different rules to different devices or because you imported multiple templates to the resource. You select one rule set to deploy. If you want to deploy multiple rule sets in the same resource to the same device, you must complete the deployment process multiple times; each time, choose to *append* the rules to the existing ACL rather than skip the deployment or replace (add) the ACL.

2. Add an ACL use, which applies an ACL that has already been deployed to one or more interfaces on the targeted devices.

For both steps, you can select an individual device or complete a batch operation with many targeted devices.

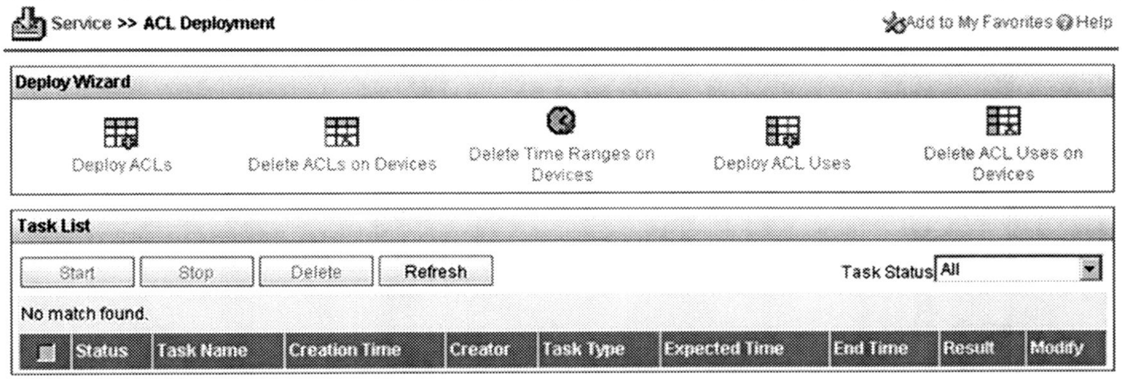

Figure 6-29: IMC ACL Management

Summary

This chapter provided an overview of the key concepts you need to understand to implement ACLs:

- ACLs are composed of one or more entries called rules and identified by a common ACL ID (a number, a name, or in some cases both).
- Rules specify the matching criteria the switch applies to determine whether the ACL selects a particular packet.
- You can configure two major types of ACLs:
    - You would use a basic/standard ACL when the source address is sufficient to filter the traffic of interest.
    - An advanced/extended ACL is useful for more complex filtering requirements.
- Once you create an ACL, you can assign it to a switch interface or to another function. Until you assign the ACL, it has no effect on traffic the switch processes.
- When you assign an ACL to a switch interface, it functions as a packet filter, permitting selected packets and dropping other packets.
- Depending on the switch model, you can assign an ACL to a VLAN interface as an inbound or outbound as RACL, to a VLAN interface (inbound) as a VACL, or to a physical port as an inbound or outbound port ACL. The type of ACL determines which packets are matched against the ACL.
- IMC ACL Management helps you to centralize and standardize your ACL deployments.

Learning check

Answer these questions:

1. Briefly describe ACLs. What components do they include, and what are some ways to apply them?

2. What are the three general components of an advanced/extended ACL rule? Give a brief description of each.

3. You want to control all inbound traffic on all interfaces in VLAN 3, regardless of whether the traffic is routed or switched. Which type of ACLs would you configure on a Comware switch?

   a. Port-based ACLs (PACLs)

   b. Routed ACLs (RACLs)

   c. VLAN-based ACLS (VACLs)

   d. ACLs for global QoS policies

4. What functionality does the HP IMC ACL Management service provide?

Answers

1. Briefly describe ACLs. What components do they include, and what are some ways to apply them?

   An ACL is a list of rules against which a network infrastructure device checks packet headers. Each rule consists of specific criteria such as source IP address, destination IP address, and TCP or UDP port number. It also consists of an action: permit (select matching packets) or deny (do not select matching packets). An ACL has a specific ID; valid IDs depend on the switch model and the type of ACL.

   You can apply an ACL in a variety of ways. This chapter focused on ACLs applied to network infrastructure device interfaces to filter packets. This type of ACL can be applied as RACL or a port ACL on inbound or outbound traffic. On ProVision switches, you can also apply a VACL.

   You can also apply ACLs to select traffic for features such as:

   - Traffic classes
   - Policy-based routing (PBR)
   - Traffic mirroring

2. What are the three general components of an advanced/extended ACL rule? Give a brief description of each.

   - Rule ID—The ID identifies the rule within the list, enabling you to identify and, if necessary, remove it. The rule also tells the switch when to process this rule in relation to other rules.
   - Action—A permit action tells the ACL to select any packet that matches the criteria. The deny action tells the ACL not to select any packet that matches the criteria. When you are using the ACL to filter packets, permitted packets are forwarded and denied packets are dropped.
   - Matching criteria—The matching criteria tell the switch how to select the traffic. In an extended/advanced ACL, the match criteria include IP header fields such as IP protocol type, source IP address, destination IP address, source TCP or UDP port, destination TCP or UDP port, and DSCP.

3. You want to control all inbound traffic on all interfaces in VLAN 3, regardless of whether the traffic is routed or switched. Which type of ACLs would you configure on a Comware switch?

   a. Port-based ACLs (PACLs)

   b. Routed ACLs (RACLs)

   c. VLAN-based ACLS (VACLs)

   d. ACLs for global QoS policies

4. What functionality does the HP IMC ACL Management service provide?

   It helps you configure ACLs in an intuitive manner and to apply them to the appropriate switches and interfaces with efficient batch operations. It also lets you create custom rules—whether using wildcard bits or subnet masks. You can also define variables and other dynamic parameters to configure ACLs more quickly.

   IMC brings disparate ACL configurations under its management so that you can standardize them and clone already configured ACLs for use elsewhere. ACL Management also helps you to manage the rules. For example, when you edit a rule set, you can easily sort the rules to fix their order, as necessary.

7 Port-Based Authentication

EXAM OBJECTIVES

In this chapter, you learn to:

- ✓ Implement 802.1X, MAC authentication (MAC-Auth), and web authentication (Web-Auth) on HP switch ports.
- ✓ Implement more complex port-based solutions with unauthenticated or guest VLANs, multiple clients, and multiple forms of authentication.
- ✓ Implement an HP User Access Manager (UAM) solution.
- ✓ Plan policies that apply dynamic VLAN assignments, access control lists (ACLs), quality of service (QoS) priorities, and rate limits.

ASSUMED KNOWLEDGE

Before studying this chapter, you should understand why companies need to apply additional security measures to protect their confidential data.

INTRODUCTION

This chapter introduces you to three port-based authentication technologies—802.1X, MAC-Auth, and Web-Auth—and teaches you how to implement them on HP switches.

Port-based authentication use models

Port-based authentication works within the authentication, authorization, and accounting (AAA) model. AAA provides a framework for implementing centralized policies to control access to particular services. For port-based authentication, network access is the service in question, and the AAA server is usually a RADIUS server. More specifically, AAA defines:

- **Authentication**—Users must prove their identity in some way to the authentication server. As you will see, for port-based authentication, the user usually submits a username and a password. The user might alternatively submit a digital certificate that includes a valid username and is signed by a trusted certificate authority (CA).
- **Authorization**—The authenticated user is authorized by a RADIUS server policy to obtain network access. The policy might authorize the user for particular port settings (such as VLAN and rate limit), which the switch port applies dynamically.
- **Accounting**—The switch submits information about the user's connection to the RADIUS accounting server, supporting user tracking and auditing functions.

Companies typically implement port-based authentication for one or more of these reasons:

- They need to prevent unauthorized users from connecting to open ports, whether because of their own security policies or to comply with government regulations.
- They want to provide guest access, ensuring that guests are placed in the proper guest VLAN and also forcing guests to log in for legal and tracking purposes.
- They have user groups with different access rights and want to ensure the correct port settings for each group. Users might connect to any port, so the IT staff would struggle to keep up. The authorization component of port-based authentication enables the port to adjust to the user dynamically.

For example, a school district might have a faculty/administrator VLAN and a student VLAN. Port-based authentication enables any user to connect to any port. Users submit their identity to the centralized authentication server, which directs the switch port to grant the user access in the correct VLAN.

As you will learn, port-based authentication can also dynamically provision ports with ACLs, rate limits, and other settings.

802.1X overview

You will first examine 802.1X, an open standard technology for controlling users' network access based on their identity. As Figure 7-1 shows, 802.1X forces users to authenticate as soon as the data link layer establishes an Ethernet connection between the switch port and the client. To restrict an unauthenticated user so that the client can send only authentication messages, 802.1X divides the port into two virtual ports:

- **Controlled port**—The controlled port allows all types of traffic, but it can be disabled. By default, the controlled port is disabled. Both sides of the connection control the port based on their assessment of the authentication state. In other words, no unauthorized traffic flows across the link.

- **Uncontrolled port**—The uncontrolled port is always active, but it can carry only the Extensible Authentication Protocol (EAP) packets used for authentication.

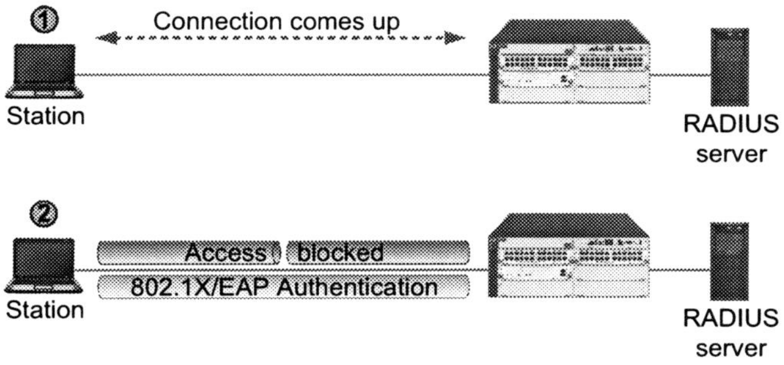

Figure 7-1: Overview of 802.1X authentication process

Without the uncontrolled port, this high level of security would shut out all traffic from users, preventing even authorized users from proving that that they are authorized.

Note, however, that the uncontrolled port does not carry Dynamic Host Configuration Protocol (DHCP) messages, Domain Name System (DNS) messages, and so on. The 802.1X authentication process is designed to allow the client to authenticate without an IP address.

802.1X roles

To understand 802.1X authentication, you must understand the devices involved in the process and the role each device plays:

- What does the device expect from the process?
- What responsibilities does the device assume?
- What protocols does the device use to communicate?

Keep these questions in mind as you review Figure 7-2 and learn about the three devices involved in the 802.1X authentication process:

- Supplicant
- Authenticator
- Authentication server

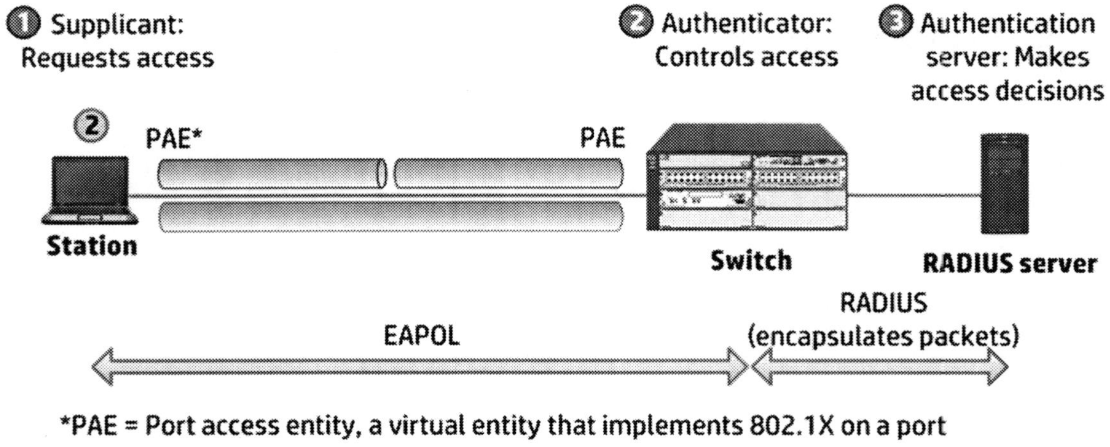

Figure 7-2: Devices that participate in the 802.1X process

Supplicant

On a wireless network, the station is the supplicant, or more precisely, the port-access entity (PAE), which implements 802.1X on the station.

The supplicant requests access to the network and proves that it deserves this access by authenticating—typically in response to a challenge from the far end of the connection (for example, the switch port).

In addition to responding to a challenge, the supplicant may also initiate the authentication process on its own behalf (by transmitting an EAP-Start packet). This mechanism protects supplicants that receive the challenge before the station has entirely booted, causing authentication to time out before users can submit their credentials.

Authenticator

The authenticator is the PAE on the far end of the supplicant's connection. In this study guide, the authenticator is the HP switch, which has as many PAEs as it has associations with clients.

The authenticator controls network access, forcing a supplicant to authenticate before it can send any non-EAP traffic over the connection. The authenticator initiates the authentication process but then relays authentication messages between the supplicant and the authentication server.

After the authentication is completed, the authenticator decides how to control the connection. If the authentication server has accepted the user's request, the authenticator activates the virtual port. In other words, the station is now completely connected to the network. If the authentication server rejects the user's request, the authenticator enforces this denial and keeps the virtual port closed.

Authentication server

The authentication server makes decisions about whether or not users can access the network. These decisions are based on whether or not the users:

- Can prove their identity (the users' credentials are correct)
- Are connecting at the proper time and location

For example, even legitimate guests might be prohibited from accessing the network after regular work hours.

The server also submits its own credentials to the supplicant. In essence, the supplicant and the server authenticate each other, although the authenticator always acts as a proxy in this process.

The authentication server can be any AAA server; however, it is typically a RADIUS server and will be referred to as such in this chapter.

The authenticator and authentication server use RADIUS messages to communicate. The authenticator encapsulates the supplicant's EAP messages in this protocol.

EAP framework

EAP, which the protocol 802.1X uses for authentication, defines a flexible framework into which you can fit authentication methods that meet your company's environment and security policies. Figure 7-3 shows this general framework.

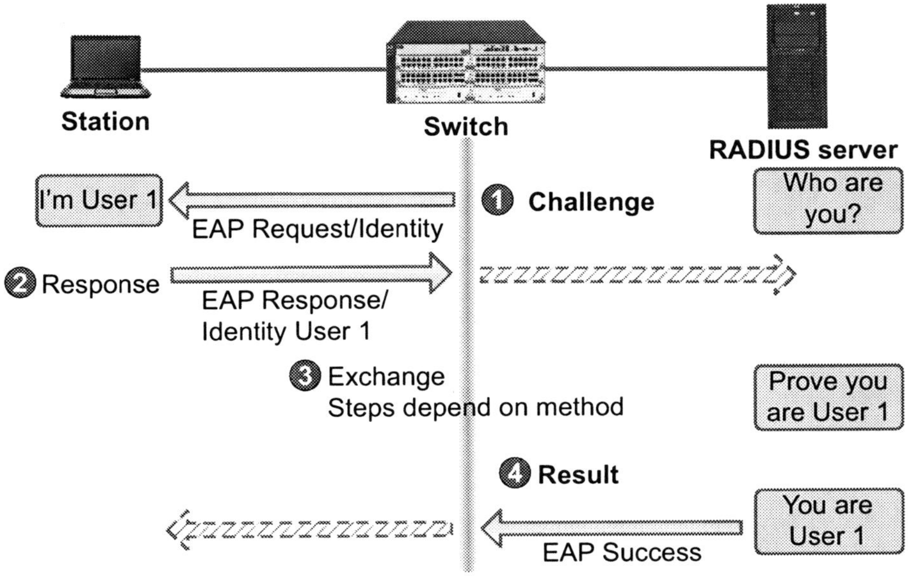

Figure 7-3: EAP framework

As you follow the process, remember that the authenticator (switch) relays all messages from the client to the RADIUS server, translating them as necessary. The vertical line underneath the switch shows the point at which the authenticator translates frames. The horizontal dashed lines show frames after they have been translated to the new format.

1. The client and switch port establish an Ethernet connection.

   The switch shuts down the controlled port, blocking all traffic except EAP, and issues a challenge. The challenge is an EAP Request/Identity packet. Basically, this frame initiates the authentication process and asks the client to identify itself but not to send any other information.

2. The client responds with an EAP Response/Identity packet, which typically includes a username.

   The authenticator includes the user's identity in all future frames so that the server can keep track of which EAP messages belong to which user. The user's identity also marks any accounting frames for the connection—which are important for companies with rigorous auditing requirements.

3. Depending on the EAP method, the client and RADIUS server exchange a particular series of messages, which might contain one of many types of authentication credentials.

4. Based on information received in step 3, the authentication server determines whether or not the client has authenticated successfully. The switch then either activates the connection and transmits an EAP-Success or leaves the connection deactivated and transmits an EAP-Failure.

Common EAP methods

Because EAP is, by definition, flexible, it supports different types of methods. The EAP method determines how the users prove their identity. As Figure 7-4 illustrates, different methods dictate different steps, so the station and the RADIUS server must agree upon the method.

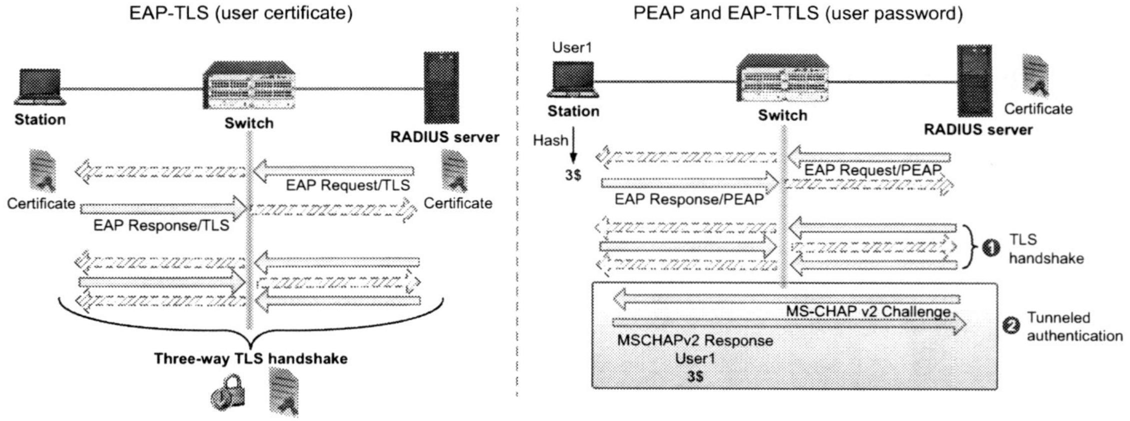

Figure 7-4: Common EAP methods

The first step in the exchange—the server's EAP Request/METHOD packet—both starts the process and indicates the method that the server requires. (Depending on your RADIUS server, you can program it to select methods according to conditions such as user identity and location.)

If the station supports the requested method, it continues the exchange. Otherwise, the station sends an EAP NAK packet, which can suggest a different method. The server, if it supports the alternative method, may then initiate the exchange with the new method.

EAP-Transport Layer Security (TLS)

Considered one of the most secure EAP methods, EAP-TLS uses a three-way TLS handshake to:

- Generate encryption keys that secure the rest of the authentication process.
- Exchange digital certificates.

The EAP Request/TLS and EAP Response/TLS packets include information such as:

- Digital certificates and certificate verifications
- Supported encryption suites
- Values for generating encryption keys (not the keys themselves)

Advantages

EAP-TLS is one of the most secure EAP methods because it provides mutual authentication with Public Key Infrastructure (PKI) digital certificates. Digital certificates rely on extremely strong asymmetric keys and trusted certificate authorities (CAs).

With TLS, authentication is based on a digital certificate (something the user has) rather than on a shared secret (something the user knows). The digital certificate is typically stored on a laptop or a smart card, providing stronger security. Although you can set up requirements to force users to create "stronger," and therefore more secure, passwords, you cannot prevent users from telling people their password or writing it on a paper displayed in plain sight. Generally, it is harder for users to leak certificates inappropriately; systems administrators might, for example, implement security measures to prevent users from copying certificates.

Hackers often run dictionary attacks on passwords, phish for passwords, or use other social engineering techniques to get users to leak their passwords. Although no credentials can be totally secure—malware might be able to steal certificates, for example—digital certificates are more resistant to many common attacks.

Someone who steals a laptop or smart card might be able to gain access to the certificates within, but they are usually protected by an additional layer of security such as a password. In addition, the user can immediately report the theft, allowing you to revoke the certificates.

Disadvantages

EAP-TLS is not vulnerable to eavesdropping and replay attacks that affect less-secure EAP methods such as EAP-MD5, but security comes at the cost of purchasing and managing the digital certificates—substantially more expensive than managing passwords. Maintaining a large number of certificates requires specialized software and trained IT staff.

Another barrier to adopting EAP-TLS is the requirement for digital certificates on all stations. In environments that have multiple operating systems, installing certificates on all clients can be difficult. In some environments, you might not even be able to install certificates on clients.

PEAP and EAP-Tunneled TLS (TTLS)

Tunneled Transport Layer Security (TTLS) and Protected EAP (PEAP) were developed to provide much of the security of EAP-TLS without forcing stations to have digital certificates—drastically reducing implementation costs. For this reason, these are among the most commonly used EAP methods. (TTLS was developed by Funk Software and Certicom; PEAP was developed by Microsoft, Cisco Systems, and RSA Security.)

TTLS and PEAP function in very similar ways. Both methods involve a two-step authentication process; in the first step, the outer method creates a secure tunnel in which the second step takes place.

Step 1: Outer method

Like EAP-TLS, TTLS and PEAP use a three-way TLS handshake to generate encryption keys and negotiate a tunnel secured by those keys. Only the server authenticates itself with a digital certificate during this exchange, but you must ensure that the clients trust the CA that signed the server certificate. (The encryption keys for the tunnel are derived from the public key in this certificate.)

Step 2: Inner method

The client authenticates itself in the second step; it uses a weaker—and so more easily implemented—authentication method, protected by the secure tunnel in which it takes place.

For the inner authentication method, TTLS can use a weaker EAP method, such as EAP-GTC, or a legacy RADIUS method, such as Challenge Handshake Authentication Protocol (CHAP), Password Authentication Protocol (PAP), or Microsoft CHAP variants (MS-CHAP v1 or MS-CHAP v2). PEAP supports methods such as MS-CHAP v2, EAP-GTC, and TLS. Because Windows native supplicants support PEAPv0 with MS-CHAP v2, this is by far the most prevalent EAP method.

Note

For 802.1X on wireless connections, whether implemented through EAP-TLS, EAP-TTLS, or PEAP, the authentication server sends the authenticator (AP) the encryption material. The wireless client and AP continue to use encryption to secure the connection. On Ethernet connections, however, the client and switch port to do not encrypt traffic in this way.

Advantages

Like EAP-TLS, EAP-TTLS and PEAP provide strong, mutual authentication and dynamic key distribution. Because TTLS and PEAP use encrypted tunnels to secure usernames and passwords (rather than require digital certificates on stations), you can implement these methods more easily than you can in TLS.

TTLS has one unique benefit: it always protects the username. Depending on how PEAP is implemented, the username might be transmitted in plain text, allowing a hacker to detect users' identity and possibly lock the users out of their account.

Disadvantages

As compared to EAP-TLS, EAP-TTLS and PEAP have the disadvantage of a weaker authentication method.

When choosing an EAP method, you must always consider if both the supplicant and the server support the method. EAP-TTLS tends to be less widely supported. For example, Microsoft Windows does not support TTLS natively; you must install additional software to provide this support. When HP switches act as supplicants (or in the rare cases that they act as servers), you also need to consider the methods that they support. Check the documentation for more details.

Complete process (EAP and RADIUS)

As you have seen, the client supplicant and the RADIUS server communicate during the EAP process. However, all of their communications are mediated by the authenticator (switch). You need to understand how the switch and the RADIUS server communicate so that you can implement the appropriate settings on the switch.

As Figure 7-5 illustrates, the authenticator encapsulates a supplicant's EAP message in a RADIUS Access-Request packet. It also hashes the packet with MD5, using the RADIUS shared secret as the key. It adds the hash to a message authenticator field. The authenticator then transmits the packet to the RADIUS server. Because the switch handles the actual delivery of the frames, the client does not need an IP address or any network access.

The RADIUS server checks the message authenticator field by hashing the packet with its shared secret. If the hashes match, it knows that a valid client sent the access request and that the packet was not altered in transit. The server then decapsulates the EAP messages and formulates the appropriate responses, which it encapsulates in RADIUS Challenge messages. These messages are also authenticated with the message authenticator field.

At the end of the process, the RADIUS server sends either:

- An Access-Accept message—Two requirements are met:
  - The client has successfully authenticated.
  - The policy that applies to this access request allows access.

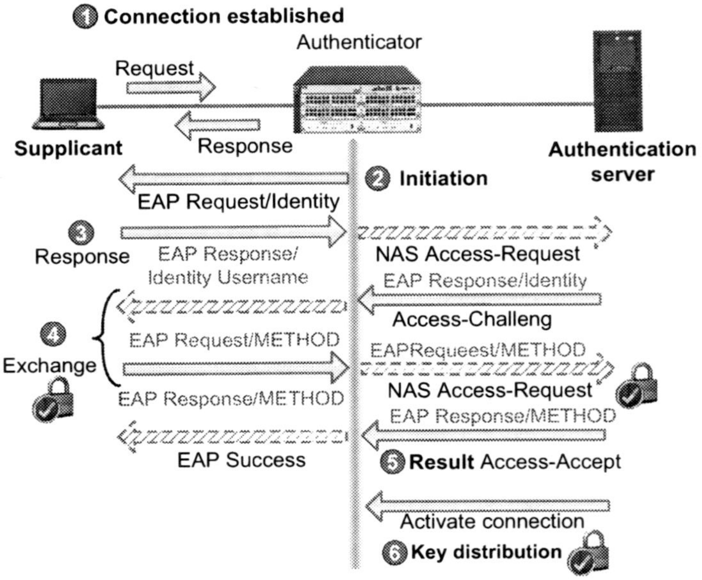

Figure 7-5: Overview of 802.1X process

- An Access-Reject message—One of these situations occurred:
  - The client could not complete authentication, whether because the RADIUS server does not trust the authenticator, because the client and server do not support compatible EAP methods, or because of some other problem.
  - The client does not trust the RADIUS server's certificate, so the server failed authentication.
  - The client failed authentication.
  - The policy that applies to this access request does not allow access.
  - No policy applies to this access request.

To ensure that the switch can participate in this process, you must set the correct RADIUS server address or addresses, the correct RADIUS server ports, and the correct RADIUS secret. You will look at the precise setup for RADIUS server and 802.1X settings in a moment.

Note

HP Comware switches distinguish between standard RADIUS servers and extended RADIUS servers, which support some additional HP options. (UAM is an extended server, but Microsoft Network Policy Server (NPS) is not.) You can choose whether the switch treats servers as standard or extended servers.

802.1X requirements

802.1X requires particular capabilities on clients, network servers, and switches.

First, the switch must support the 802.1X authentication standard. (Most of the HP managed switches support 802.1X. Check the documentation of the switch you are using to verify that it supports this standard.)

Your network also requires a RADIUS server (or other AAA servers) that supports EAPOL.

- HP provides a RADIUS server with a plug-in to HP Intelligent Management Center (IMC), called User Access Manager (UAM). UAM also provides a number of other capabilities that you will examine a bit later.
- You can alternatively use a third-party RADIUS server such as Microsoft Network Policy Server (NPS) or FreeRADIUS.

In addition, any device that needs an Ethernet connection must include an 802.1X supplicant. Windows and Mac OSs provide native supplicants, which you can use. If you are using UAM, you can install iNode on the devices. This application provides the 802.1X supplicant. It also integrates with UAM to allow UAM to apply more sophisticated centralized policies to clients.

Finally, you can choose from a number of third-party supplicants, including freely available ones.

802.1X certificate requirements

Several of the EAP methods about which you have learned use digital certificates as authentication credentials, as the source for the keys that secure an SSL tunnel between the client and the server, or both. Table 7-1 indicates common EAP methods and which components of the 802.1X process require certificates for that method. As you see, the RADIUS server requires a certificate for EAP-TLS, EAP-TTLS, and PEAP.

Table 7-1: EAP methods that require certificates

EAP method	RADIUS server certificate	Supplicant certificate
EAP-TLS	Yes	Yes
EAP-TTLS	Yes	No
PEAP	Yes	No

Defining the RADIUS server on the switch

An 802.1X solution requires a RADIUS server. You must configure your switches to communicate with this server, as described in the following sections.

HP ProVision switches

On ProVision switches, you can specify up to 15 RADIUS servers using this command:

```
ProVision(config)# radius-server host <IP address> [key secret] [auth-port number] [acct-port number]
```

You can set an individual key (shared secret). If you do not, the RADIUS server uses the global key. You can also change the authentication and accounting ports if your servers do not use the default ones (RADIUS standard 1812 for authentication and 1813 for accounting). Note that ProVision switches use the RADIUS server for both authentication and accounting.

You can add a number of other options, which are documented in the switch *Access Security Guide*.

The first three servers that you add are automatically added to the global RADIUS server group, which is called **radius**. The switch uses this group, by default, for all types of authentication for which you have enabled RADIUS.

The switch tries to access the RADIUS servers according to the order in which their IP addresses are listed when you enter the **show radius** or **show server-group radius** command. If the first server listed does not respond after a certain timeout and number of retries (both configurable), the switch will try to contact the next RADIUS server.

Although you can add up to 15 RADIUS servers, only the first three are automatically added to the **radius** server group.

You cannot edit the **radius** server group. If you want different servers to be part of this group, you must delete the servers that are currently part of the group. For example, you have entered four RADIUS servers, 10.1.10.11, 10.1.10.12, 10.1.10.13, and 10.1.10.14. The first three are in radius server group, but you want 10.1.10.14 to be in the group instead of 10.1.10.13. You must delete 10.1.10.13. You can then readd 10.1.10.13 and use it in different server groups.

If you want to use different RADIUS servers for different types of authentication, you can create new server groups, each of which can have up to three RADIUS servers.

You must first define the RADIUS server. You can then add it to a group.

The command for adding a RADIUS server to a new or existing server group is:

```
ProVision(config)# aaa server-group radius <group name> host <IP address>
```

Use the **no** form of the command to delete the server from the group.

A RADIUS server can be part of more than one group. If, for example, you add a server that is part of the **radius** server group to a new group, the server remains part of the **radius** group, as well.

To use the new RADIUS server group, you must choose the type of authentication to which it applies. Use this command to specify a server group for 802.1X authentication:

```
ProVision(config)# aaa authentication port-access <eap-radius | chap-radius> server-group <group name>
```

The **eap-radius** option is the usual choice for 802.1X.

You can specify the RADIUS server group for other purposes such as authenticating manager logins through the console port, Telnet, Secure Shell Protocol (SSH), or the web browser. However, this chapter focuses on port-based authentication.

As a final note, if you are using an endpoint integrity solution such as IMC's Endpoint Admission Defense (EAD), you must specify that server as a RADIUS server with the dyn-authorization option:

```
ProVision(config)# radius-server host <IP address> dyn-authorization
```

This setting allows the switch to accept disconnect messages (DMs) and change of authorization (CoA) messages from that server, allowing, for example, the server to move a client from an employee VLAN to a quarantine VLAN. By default, the switch accepts these messages on UDP port 3799, but you can change that setting:

```
ProVision(config)# radius-server dyn-autz-port <number>
```

HP Comware switches

HP Comware switches also allow you to set up particular RADIUS settings for particular types of authentication. A brief overview of these settings will help you to apply them in a variety of ways for different purposes.

A Comware RADIUS scheme groups the settings related to communicating with a particular set of RADIUS servers. You create a scheme to define these settings:

```
[Comware]radius scheme <name>
```

Within the scheme, you must specify the IP address of the primary RADIUS server:

```
[Comware-radius-<name>] primary authentication <IP address> [key shared_secret] [port]
```

As you see in the command syntax, you can specify an individual RADIUS shared secret for the server. You can also define the port on which the RADIUS server listens. If you do not define the port, the switch sends traffic to the default port, 1812.

It is recommended that you also set a secondary server:

```
[Comware-radius-<name>] secondary authentication <IP address> [key shared_secret] [port]
```

The Comware switch attempts to contact the primary server first. If attempts time out a certain number of times (retries and timeout are both configurable), the switch contacts the secondary server.

For Comware switches, you also set the server type to either standard (standard RADIUS) or extended (RADIUS with some HP extensions). Set the server type to extended if you are using UAM; otherwise, you should generally leave the type at standard, the default:

```
[Comware-radius-<name>] server-type <extended | standard>
```

For Comware, you define the RADIUS accounting servers separately:

```
[Comware-radius-<name>] <primary | secondary> authentication <IP address> [key shared_secret] [port]
```

If you do not specify a port for the accounting server, the switch uses the RADIUS default, 1813.

Rather than set the RADIUS shared secret for individual servers, you can set the secret for the scheme as a whole:

```
[Comware-radius-<name>]key <authentication| accounting> key <shared_secret>
```

In addition to configuring these settings, you might also need to configure the username-format setting. This setting configures the switch to submit the username in the RADIUS Access-Request as expected by the server:

```
[Comware-radius-<name>]username-format <with-name | without-name | keep-original>
```

- **with-name**—The switch always forwards the request with a domain name. It keeps the domain name submitted by the client or, if not present, it adds the name of this domain itself.
- **without-name**—The switch always forwards the request without a domain name (leaving the request without a domain name, if submitted in that manner, or stripping the submitted domain name).
- **keep-original**—The switch forwards the request with the precise username submitted, with or without the domain name.

Note

ProVision switches always forward the request in the original format, with or without the domain name. You must ensure that supplicants are configured to use the format expected by the server.

If you are using an endpoint integrity solution, such as IMC's Endpoint Admission Defense (EAD), you must set the IP address of the security policy server. Then the switch will accept authorization changes and settings from that server:

```
[Comware-radius-<name>]security-policy-server <IP address>
```

The scheme includes a number of other settings, which you can look up in your switch's *Security Configuration Guide*.

You can create multiple RADIUS schemes that apply to different types of authentication and different users. You apply RADIUS schemes within domains, which are described in the next section.

Comware RADIUS schemes and domains

Comware switches classify requests for services of all sorts in domains. For example, a client submits a user's authentication credentials as part of the 802.1X process. The username is bob@hp.com, so the switch can identify the request as belonging to the hp.com domain. Or—as you will learn about later—a manager might try to log in to the switch CLI using SSH. The manager submits username sara. Now, the switch does not have a domain name to use, but it still must classify the request in a domain.

It is important for you to understand the domain to which a particular request belongs because that domain determines how the switch handles the request.

As shown in Figure 7-6, a Comware switch uses the following process to assign requests to a domain:

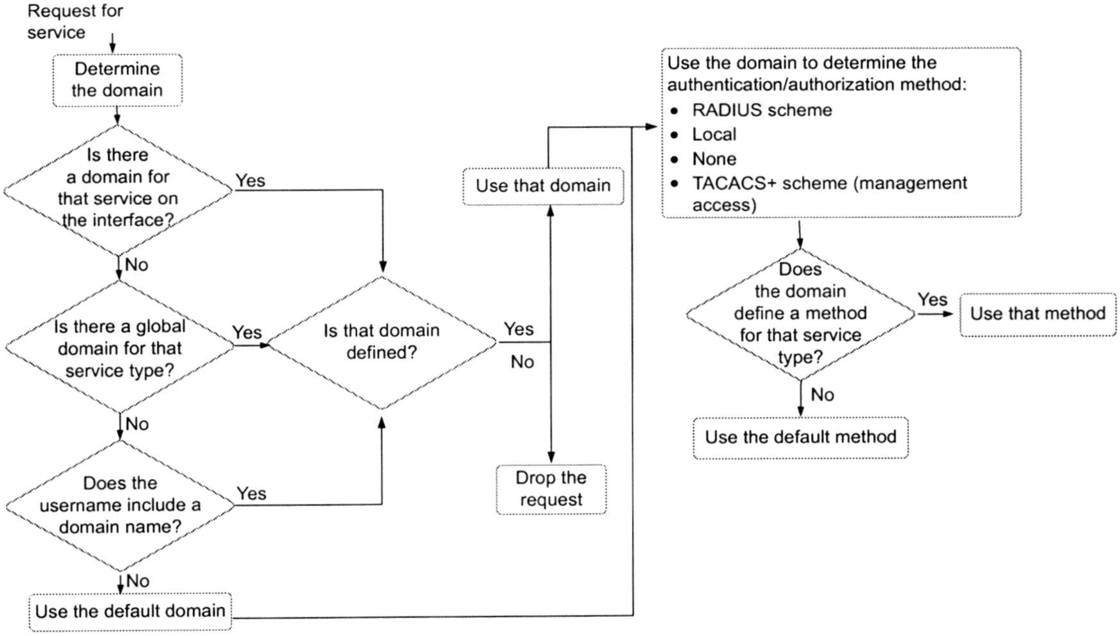

Figure 7-6: The process of assigning requests to domains on Comware switches

1. The switch first checks the interface associated with the service request. Is a domain specified for the service type? For example, you can specify the domain for 802.1X authentication on a particular physical interface. If this domain is specified, the switch always uses it.
2. In some cases, you can specify a domain for a service type globally. That domain applies if an interface configuration does not override it.
3. If no domain is specified for this service type, the switch examines the username submitted with the authentication request. It assigns the user to the domain name included with that username, if any.
4. If the request does not include a domain name, the switch assigns the request to the default domain.

   At factory settings, Comware switches have a default domain named "system." If you want, you can create a new domain and make that domain the default. Of course, only one domain can be the default. (Please note that the domain is specific to the switch; it is not bound to objects in a Lightweight Directory Access Protocol [LDAP]-compliant directory.)
5. After the switch has assigned the request to a domain, it checks whether that domain is defined on the switch. If it is, the switch can use that domain to determine how to handle authentication, authorization, and accounting for that request. If the request is assigned to a domain that is not defined, the switch does *not* use the default domain. Instead, it simply drops the request, causing the access attempt to fail.

   In other words, if you want to specify a particular domain for a particular access type, make sure to create a domain with a matching name. Similarly, if you expect users to send requests with a particular domain name, create a domain with a matching name.

Using the domain

The switch uses the domain to determine how to provide the three AAA functions for the access request:

- Authentication
- Authorization
- Accounting

For each of these functions, the switch can:

- Make the decision and fulfill the functions locally (**local**).
- Communicate with an external RADIUS server defined in a RADIUS scheme (**radius-scheme *scheme***).
- Not implement the function (**none**).

- For authentication, none means that the switch does not impose authentication for this form of access and permits all requests that match this domain. For accounting, none means that the switch does not submit accounting information for the session.

You can choose these options individually for several different types of access requests. For example, the lan-access type applies to 802.1X, and the login access type applies to management access to the switch. You could set up a RADIUS scheme for the one and local authentication for the other. You will learn about the various types of access throughout this chapter.

If you do not want that much granularity, you can simply define the default authentication, authorization, and accounting methods. The switch uses the default method for access types without a defined method. For example, you want to use the same RADIUS servers to authenticate 802.1X users, Web-Auth users, and switch management users assigned to this domain. You can just specify the RADIUS scheme as the default authentication and authorization method.

Similarly, although you *can* choose different methods for the different AAA functions, you often choose the same option for all three. In fact, if you specify a RADIUS scheme for authentication, you *must* specify the same RADIUS scheme for authorization. Often, you should also specify the same scheme for accounting. However, in some cases, you might need to direct accounting messages to a different set of RADIUS servers. In that case, you can create a RADIUS scheme for the authentication/authorization servers and a second scheme for the accounting servers. Then specify the correct scheme for each function in the domain.

Basic 802.1X configuration process

To set up 802.1X on Comware and ProVision switches in the CLI, you complete the following basic steps:

1. Specify the primary and secondary RADIUS server IP addresses and keys. On Comware switches, you complete this step by creating a RADIUS scheme, as you learned earlier. For both Comware and ProVision, you can configure more options, such as defining a nondefault port and setting up accounting services.

   The basic commands include:

   ```
   [Comware]radius scheme <name>
   [Comware-radius-<name>] [primary | secondary ] authentication <IP address>
   [Comware-radius-<name>] key authentication <secret>
   [Comware-radius-<name>] user-name-format [with-domain | without-domain]
   [Comware-radius-<name>] server-type [extended | standard]
   ```

   On ProVision switches, you specify global RADIUS servers or set up groups, as you learned earlier. The basic command for defining a server is:

```
ProVision(config)# radius-server host <IP address> key <secret>
```

2. This step only applies to Comware switches. You must configure the domain that matches the 802.1X user requests and apply the RADIUS scheme for the correct purposes. You explored cases in which you would use the default domain or create a new domain in the previous section.

   Remember, whether you are configuring the system domain or a new domain that you created, you must specify your RADIUS scheme for default authentication and default authorization. Or, if you plan to use different authentication methods for other types of access, specify the RADIUS scheme for the lan-access authentication and authorization methods:

   ```
   [Comware] domain <name>
   [Comware-isp-<name>] authentication lan-access radius-scheme <name>
   [Comware-isp-<name>]authorization lan-access radius-scheme <name>
   [Comware-isp-<name>]lan-access radius-scheme <name>
   ```

3. Configure the switch to use EAP for 802.1X authentication:

   ```
   [Comware] dot1x authentication-method eap
   ProVision(config)# aaa authentication port-access eap-radius
   ```

4. This step applies only to Comware switches. If you are not using iNode clients and UAM as the RADIUS server, you should disable the online handshaking feature.

   This feature enables the switch to exchange messages with the connected 802.1X supplicant (iNode) to determine whether the client is still online. If the client is not online, the switch deauthenticates the user, eliminating the potential security risk of an unauthorized user accessing the port activated by the authorized user. However, if the client does not support the handshake mechanism, the switch believe the client to be offline even when it is online, disrupting the users' status. Therefore, HP recommends that you disable this feature when you are not using iNode as the supplicant:

   ```
   [Comware] interface <interface ID>
   [Comware-<interface ID>] [undo] dot1x handshake
   ```

5. Enable 802.1X globally:

   ```
   [Comware] dot1x
   ProVision(config)# aaa port-access authenticator active
   ```

6. Enable 802.1X on individual ports:

   ```
   [Comware] dot1x interface <interface ID> [to] [interface ID]
   ProVision(config)# aaa port-access authenticator <port ID list>
   ```

Enabling 802.1X can interfere with user connections if their clients are not properly set up to complete authentication. Before implementing any form of port-based authentication, you should go through the proper change management processes for planning the solution, ensuring all components are in place, educating users, and so on.

This study guide focuses on the configuration steps, not on all of the planning and implementation processes.

HP IMC UAM does provide a testing option that can help. You configure the solution, but enable **User Authentication Test Mode** in the **Service >>User Access Manager >>Settings >>Parameters >>Settings** page. UAM then applies authentication, logging the success or failure of users' authentication attempts. However, it authorizes all users for access regardless of their success. You can then find any users failing authentication and resolve the problem. After resolving all problems, you disable test mode, and the 802.1X solution begins to secure your network without generating support calls.

HP IMC UAM overview

The switches must authenticate users to a RADIUS server, such as User Access Manager (UAM). UAM is a plug-in to IMC.

On all types of RADIUS servers, you generally complete at least these tasks:

- Specify RADIUS clients, which are also sometimes called network access servers (NASs) and which IMC calls access devices.

  For 802.1X, the authenticator plays this role, so the access devices are the switches to which authenticating users connect.

- Create policies that:
  - Select access requests.
  - Specify an authentication method for those guests.
  - Authorize users for particular rights and settings.
- Create user accounts or bind the RADIUS server to backend directory.

On UAM, you specify access devices, create service policies, and create access user accounts, assigned to at least one service policy.

In addition to providing these services, UAM also helps you to set up the access devices to work with the solution. In other words, you can use UAM to deploy the correct settings for 802.1X rather than configure them yourself in the switches' CLI.

UAM offers a wide variety of customizable settings. This study guide introduces you to a few, but many are beyond its scope.

Creating service policies

The service policy is the fundamental building block of UAM. In order for a user to log in, the user's account must be registered for a service. This service policy defines how users authenticate and for what services they are authorized when they log in. UAM supports service policies that are as simple or as powerful as you desire.

At its simplest, the policy refers to a single default rule, which defines:

- How users authenticate

  You must choose EAP for users who authenticate with 802.1X. Although this setting is labeled "Certificate Authentication," the users only require a certificate to log in when you select EAP-TLS. PEAP allows users to authenticate with passwords. The name derives from the fact that, unless you select EAP-MD5, which is deprecated, you must install a certificate on UAM to use EAP.

  If users or devices authenticate with MAC-Auth or Web-Auth, which are described later in this chapter, simply select None for Certificate Authentication.

- What customized settings the authenticator (switch port in this case) applies to the user's session

  These settings are discussed later in this chapter.

As Figure 7-7 shows, you can define multiple access rules in the service policy. The nondefault rules are bound to access conditions, which, like access rules, you configure in advance. Essentially, the access conditions set up scenarios for access—times, locations, and so on. The access rule then dictates how UAM handles the request for that particular scenario. The default access rule continues to apply for any requests that do not match a listed condition.

Note

UAM allows you to define service policies with complete sets of options on your own, or you can run the Service Quick Experience for help in quickly setting up common scenarios.

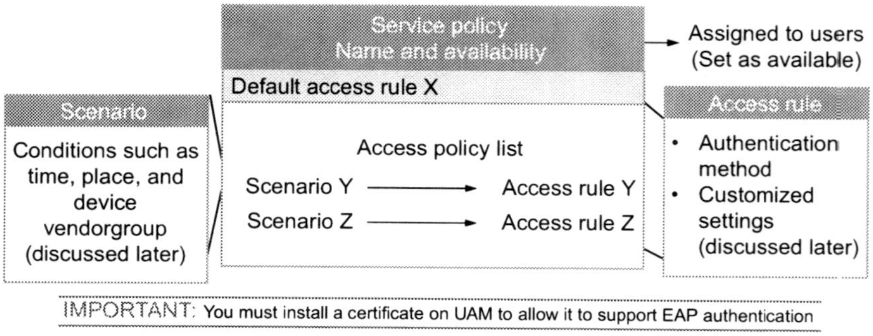

Figure 7-7: Service policies in UAM

Defining access users

IMC has its own user database. In order for UAM to authenticate a user, the user must have an account in IMC and also an access user account in UAM.

You can create UAM access user accounts manually, selecting an existing IMC user or creating the IMC user at the same time (see Figure 7-8).

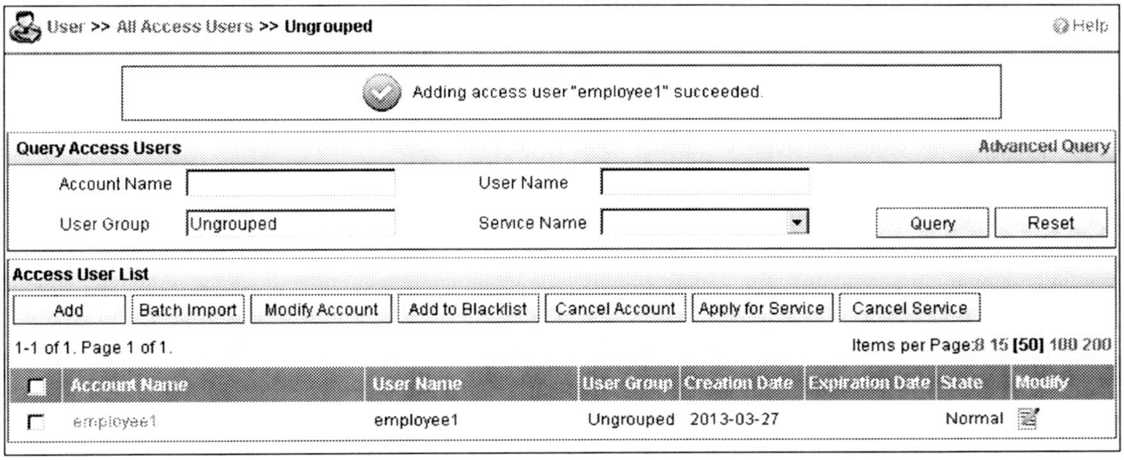

Figure 7-8: Defining access users in UAM

However, often a company has user accounts already defined in Windows Active Directory (AD) or another Lightweight Directory Access Protocol (LDAP) server. You can bind UAM to such a server and export user accounts from it, selecting the precise user groups and account attributes that you want to export. You can then import the spreadsheet into UAM and IMC at the same time, automatically creating the necessary user and access user accounts.

In fact you can import any spreadsheet file that you choose as long as it has the proper format (refer to the documentation).

For an even more seamless option, you can create synchronization policies that import user accounts from the LDAP server and also dynamically update the accounts if changes are made to the directory.

For UAM to authenticate the user and authorize the user for access, you must assign a service policy to the access user account.

Defining access devices

A switch that is discovered and managed by IMC is not necessarily allowed to send authentication requests to UAM. As Figure 7-9 shows, you must explicitly define the switches that act as NASs (RADIUS clients). As you do so, follow these guidelines:

- The secret must match the secret that you configure for the RADIUS server on the switch.

If you use UAM to deploy AAA settings to the switch, UAM automatically matches the secrets.

- You should select the correct device type. You will see later how doing so helps you to set up authorization attributes.

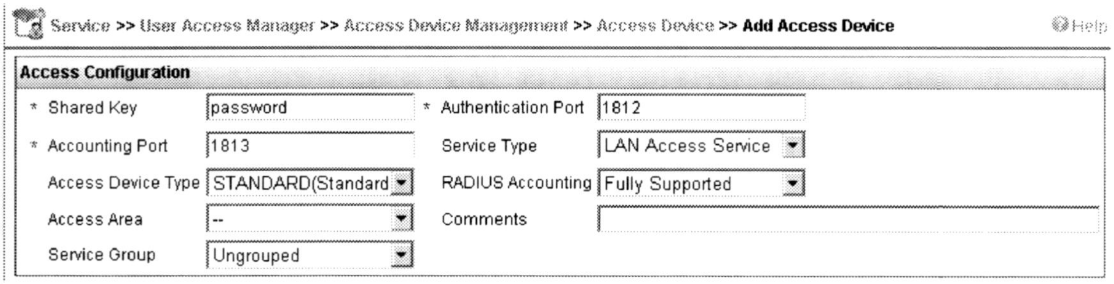

Figure 7-9: Defining access devices in UAM

Deploying AAA configurations to access devices

Earlier, you learned the procedure for configuring Comware and ProVision switches to implement 802.1X to a RADIUS server. When UAM acts as that RADIUS server, you can use UAM itself to configure these settings to supported access devices, as shown in Figure 7-10. (You can define any managed device as a UAM access device, but UAM supports deploying AAA configurations specifically to Comware switches, called H3C wired devices; ProVision switches, called HP wired devices; and HP MSM APs and controllers, called HP wireless devices.)

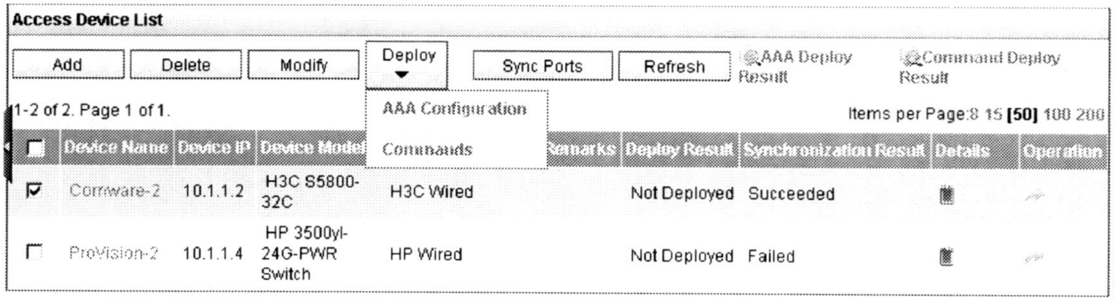

Figure 7-10: Deploying AAA configurations to access devices using UAM

The AAA deployment sets up the switch not only to support 802.1X but also MAC-Auth. To deploy settings for 802.1X, you simply need to make a couple of choices:

- For Comware switches (H3C wired), you must:
  - Select 802.1X.
  - Specify at least one domain.

Choose the settings for the domain following the same rules that you learned for configuring domains in the CLI. In brief:

- If users will log in with usernames that include domain name suffixes, you must set this domain name to match exactly. Keep in mind that certificate subject names might include domain names.
- If users will log in without domain names, set this as the default domain.
- If the RADIUS server—UAM itself, in this case—expects users to have domain names, whether or not users will, you must also use exactly the same domain name here. You set this domain suffix setting, or leave it blank, when you add the user account (or import it from LDAP).
- If the UAM user accounts do not have the domain suffix, choose to strip the name.

- UAM automatically sets itself as the authentication server, but you can add a secondary server if you want.

- For ProVision switches (HP wired), you must simply select 802.1X.

UAM handles converting your choices to the correct CLI commands. It also adds commands for various other required settings, as well as settings that help it to receive accounting information, and so on. When you finish the deployment, UAM logs in to targeted devices using the login type and credentials (Telnet or SSH) defined in the IMC resource management settings; it then implements the commands.

You can deploy settings to many devices at once; however, UAM must define all of them as the same basic configuration type.

Customizing authenticated users' access

Beyond the security benefits, port-based authentication such as 802.1X helps network administrators to provision the network edge more quickly. Traditionally, administrators had to configure each port with the appropriate settings for the user who connected to that port. If a user moved, the user might not receive the correct settings without a great deal of effort on the administrator's part.

With 802.1X, the RADIUS server, together with the switch acting as the NAS, has already determined the user's identity as well as other characteristics of the connection, such as the time and location. A RADIUS server policy can use this information to choose the appropriate settings for the connection. The network edge as a whole becomes more flexible and adaptable.

Examples of customized settings include:

- VLAN
- ACL

- QoS priority (you will examine QoS in more detail later in this study guide)
- Rate limits

Because the switch applies these settings to a user session automatically, as dictated by the RADIUS server, these settings are sometimes called RADIUS settings or dynamic settings. For example, a dynamic VLAN assignment is the VLAN assigned to a user session by the RADIUS server when the user authenticates to the network. A dynamic ACL is the (port-based) ACL applied to a user session by the RADIUS server when the user authenticates and so on.

Process for configuring customized settings

The RADIUS server has a policy that matches access requests. This policy defines the dynamic attributes that you considered previously. As Figure 7-11 shows, the server communicates the attributes to the switch acting as the NAS in attribute-value pairs (AVPs) within the RADIUS Accept message. The switch then applies the AVPs as settings on the authenticated user's port.

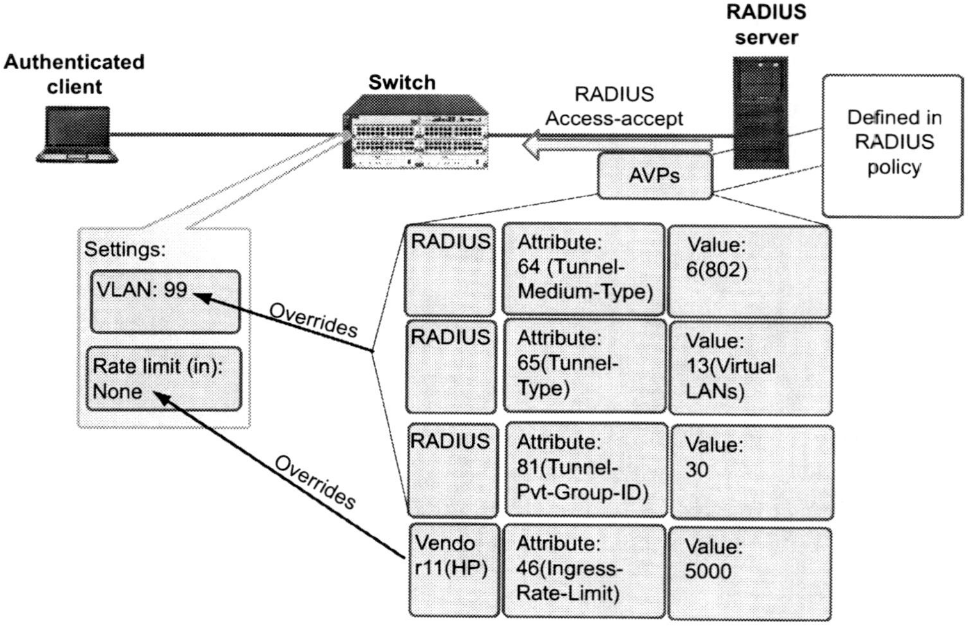

Figure 7-11: Configuring customized settings

As long as you understand the AVPs supported by your switch, you can work with the RADIUS administrator to set up whatever dynamic settings are required. Follow this basic process:

1. Find the attribute name or code for the parameter that you want to set. Table 7-2 shows the AVPs supported by Comware and ProVision switches as of the publication of this study guide; however, you should double-check the configuration guide for your own switches.

AVPs fall in several general categories: standard RADIUS, extended RADIUS, and vendor specific. When possible, you should use RADIUS attributes because they are supported across many vendors.

2. Determine the appropriate value for the AVP to set the parameter as required in your environment. The final column provides guidelines. For example, this column for the ACL parameter indicates the proper syntax for an ACL.
3. Provide the AVPs to the RADIUS administrator. For each AVP, include the attribute name (or code) and value.

   Make sure that the RADIUS server supports these AVPs. Also make sure that the administrators know to define the vendor-specific AVPs, when required.

Table 7-2: RADIUS attributes

<table>
<tr><th>Parameter</th><th>AVP type</th><th>Attribute name and code</th><th>Value</th></tr>
<tr><td rowspan="3">VLAN</td><td rowspan="3">Standard RADIUS</td><td>Tunnel-Medium-Type (64)</td><td>802 (6)</td></tr>
<tr><td>Tunnel-Type (65)</td><td>Virtual LANs (13)</td></tr>
<tr><td>Tunnel-Private-Group-ID (81)</td><td>VLAN ID string</td></tr>
<tr><td>ACL</td><td>Standard RADIUS</td><td>NAS-Filter-Rule (92)</td><td><permit | deny> <in | out> <ip | <i>protocol_number</i>> from <any | <i>IP_address</i> | <i>IP_address/PrefixLength</i>> to <any | <i>IP_address</i> | <i>IP_address/PrefixLength</i>> [<i>TCP/UDP port list</i> | <i>ICMPtype</i>] [<i>cnt</i>]</td></tr>
<tr><td>QoS priority (802.1p)</td><td>Standard RADIUS</td><td>User-Priority-Table (59)</td><td>$X_1 X_2 X_3 X_4 X_5 X_6 X_7$, in which X_n = priority to which traffic with n priority is mapped</td></tr>
<tr><td>Ingress rate limit</td><td>Vendor-specific: HP = 11</td><td>46</td><td>Kbps</td></tr>
<tr><td>Egress rate limit (most recent assignment applies to the port as a whole)</td><td>Vendor-specific: HP = 11</td><td>48</td><td>Kbps</td></tr>
</table>

Additional information on dynamic port ACLs

Dynamic port ACLs enable you to apply different access rights to users in the same VLAN based on their identity rather than port alone.

You can enhance network and switch management access security and traffic control by permitting or denying authenticated clients access to specific network resources: individual servers, entire subnets, and even the switch management interfaces. This includes preventing clients from using applications, such as Telnet, Secure Shell Protocol (SSH), web browsers, and Simple Network Management Protocol (SNMP), if you do not want their access privileges to include these capabilities.

As you learned, you configure users' dynamic port ACL as a series of AVPs in the RADIUS policy that authenticates those users. When the server authenticates the client, the ACL is assigned as an inbound ACL on the port. The ACL only applies to traffic that with that user's source MAC address. When the client session ends, the ACL is removed from the port. The switch allows as many dynamic port ACLs on a port as it allows authenticated clients.

Because every port must maintain its own ACLs, the switch's ACL limit might begin to fill up. You might want to apply most ACL rules in an ACL, a router ACL (RACL), or a VLAN ACL (VACL). Then you can apply particular rules with a dynamic port ACL.

Configuring customized settings using HP UAM

As you saw, it can be a bit complicated to define dynamic settings by manually adding AVPs to RADIUS policies. HP UAM helps you to define the settings in a more intuitive manner. You simply define a setting in an access rule defined within the appropriate service policy. For example, a service policy is using the employee access rule as its default access rule. You want to assign employees to VLAN 20 automatically, so you select the **Deploy VLAN** option in the access rule and type **20**. Figure 7-12 shows other options that you can define to customize authenticated users' sessions.

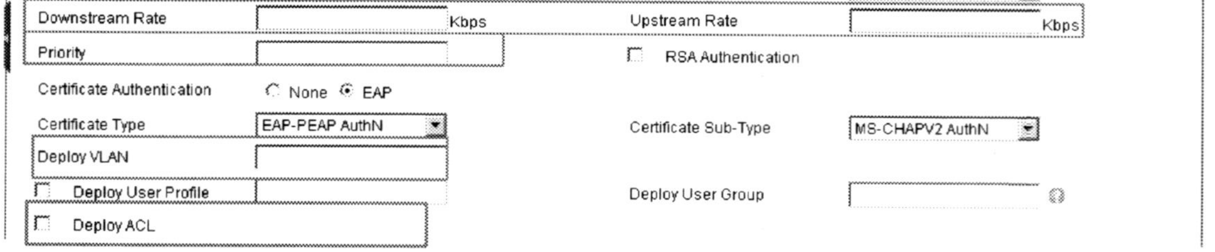

Figure 7-12: Configuring customized settings using UAM

When a user authenticates successfully, UAM converts these settings to the correct AVPs for the Access-Accept that it returns to the access device.

As you saw in Table 7-2, switches of various models might support vendor-specific attributes (VSAs) that define additional customized settings. These VSAs adjust the configuration on the port in many different ways, perhaps a phone's authentication prompts a VSA that adjusts the MAC address limit so that multiple users can authenticate on the port. Or perhaps a VSA applies a rate limit or QoS setting.

UAM comes with many VSAs that are predefined. When you configured a switch as an access device, you should have defined its type. UAM uses this type to associate the correct VSAs with devices. As Figure 7-13 shows, you can also add your own VSAs to the predefined device types or even create your own types and add VSAs to those. (As mentioned earlier, the vendor should document the values for these VSAs; you can find information about ProVision VSAs in the switches' *Access Security Guide*.)

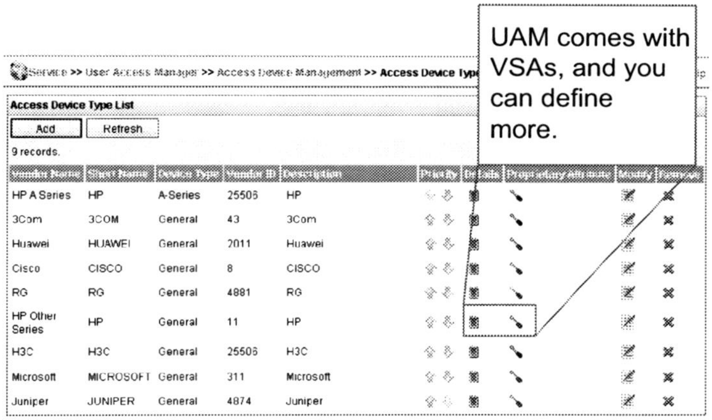

Figure 7-13: Preconfigured VSAs in UAM

After you have verified that the proper VSAs are defined in UAM, you can set up policies for applying the appropriate VSAs and values to users' sessions. Follow these general steps:

1. Create a named proprietary attribute assignment policy.

   A proprietary attribute assignment policy is to a VSA what an access rule is to a standard AVP. You use to specify VSAs to customize connections, and then you assign the proprietary attribute assignment policy in a service policy.

2. Add attributes to the policy. Select the vendor group and then choose the VSAs that you want to apply for particular users and access rights.

3. Define a value for each VSA. Usually, you choose values that the switch applies if the user authenticates successfully to customize the session. However, you can also define values for Access Rejects—customized settings for a failed authentication user.

4. Add the proprietary attribute assignment policy in the service policy for the user to which you want to apply these customized settings. As the simplest approach, you can define the policy as the default proprietary attribute assignment policy. In a moment, you will see how you can define this policy within a specialized scenario instead.

Creating specialized requirements for access

Until now, you have considered a user's identity as the sole requirement for access. However, the switch communicates additional information about the access request such as:

- Time
- Location (the NAS's IP address or MAC address, for example)
- Device (the MAC address of the device requesting access)

The RADIUS server can use this information make a user's access dependent on the user connecting at a particular time, at a particular place, or on a particular client. Or the server might assign different access rights to users based on these conditions. For example, the RADIUS server might have multiple policies that apply to the same users but at different locations. Each location has its own dynamic VLAN setting.

Once again, the switch and RADIUS server use AVPs to communicate the information. The switch includes these AVPs in the RADIUS Access-Request, and the RADIUS server uses them as inputs for a policy. That is, the server only uses a policy to reply to a request if the AVPs in the request match the AVPs defined in the policy. If you want to define special conditions for access requests then, you follow a similar process to defining the customized setting (policy outputs):

1. Find the attribute name or code for the parameter that you want to define. You can find information about the AVPs sent in requests by Comware and ProVision switches in their documentation.
2. Determine the appropriate value for the AVP to set the parameter as required in your environment.
3. Provide the AVPs to the RADIUS administrator. Make sure that the RADIUS server supports these AVPs. Also make sure that the administrators know how to define the vendor-specific AVPs, when required.

Defining specialized requirements and settings with UAM

Again, a solution such as UAM helps you to define the policies more easily. As you briefly reviewed before, each service policy can include an access policy list beyond the default access rule and default proprietary attribute policy.

An access policy binds a scenario to a different access rule. The scenario includes as many or as few of the settings as you want among those that you see listed in Figure 7-14. You define the parameters for these settings in the UAM Access Conditions pages. For example, you can set up access areas (defined by devices and ports), time ranges, vendor groups, and so on.

When you call up one of these settings within a scenario, UAM only uses the access policy if the request matches that setting. For example, if you define an access area, the user must connect to an access device in that area, for the access policy to apply. The access request must match all the various conditions within the scenario.

As Figure 7-14 shows, the policy also includes an access rule and, if you want, a proprietary attribute policy.

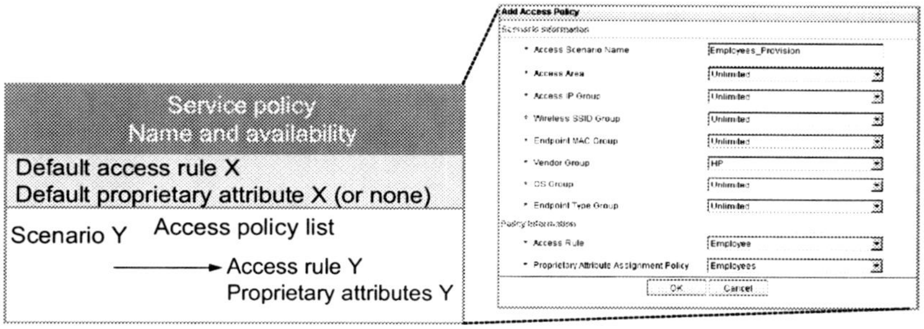

Figure 7-14: Defining specialized requirements and settings with UAM

In this way, you set the proper conditions for the type of access that your company wants to offer to users registered for this service.

If certain conditions require different customized settings, you can create multiple access policies. Each access policy features a scenario with different settings, and each scenario matches an access rule and optional proprietary attribute policy with settings specifically designed for that scenario.

If you want to prevent access for devices that do not match a specific scenario, remember to set the service policy's default access rule to denying access.

Permitting limited access for unauthorized clients

A company might want to provide limited access for guests and devices that cannot complete or that fail 802.1X authentication. In this scenario, the company wants to provide the limited access without forcing the users to log in. For example, the company might allow the unauthenticated guests to access the Internet, perhaps through a secondary connection or a proxy web server.

In a variation on this scenario, the company might want to help users who should undergo 802.1X authentication connect their devices for the first time. The devices receive limited access to a server from which the users can download an 802.1X supplicant.

You implement either scenario by configuring the switch to place devices that do not pass 802.1X authentication in a specific VLAN, rather than blocking their traffic.

The unauthorized devices fall into two categories. A device that never engages in the 802.1X authentication process, perhaps because it has no supplicant, differs from a device that submits invalid credentials. Table 7-3 shows the VLANs in which ProVision and Comware switches place particular devices.

Table 7-3: VLANs for unauthorized devices

Description	ProVision VLAN	Comware VLAN
Device never participates in the 802.1X process (no EAP messages)	Unauthorized-client VLAN	Guest VLAN
Device attempts to complete 802.1X authentication, but authentication fails for any of these reasons: • Invalid credentials • RADIUS policy does not permit access • RADIUS server timeout • Device supplicant rejects RADIUS server certificate	Unauthorized-client VLAN	Auth-Fail VLAN If no Auth-Fail VLAN, guest VLAN

Note

The switch assigns the unauthenticated (or failed authentication) devices' *untagged* traffic to the unauthorized-client VLAN indicated in Table 7-3. The switch drops any tagged traffic from these devices, even if the switch port has tagged memberships. Authenticated devices, on the other hand, can send tagged and untagged traffic. (You will learn more about MAC-based VLANs in a later section in this chapter.)

Guidelines for the network infrastructure setup

As a prerequisite for this scenario, you or another administrator must establish the VLAN (or VLANs) for the devices that do not pass 802.1X authentication. The default router for this VLAN often enforces ACLs that restrict the guests to access the Internet. The default router might be a guest solution that provides proxy web services, content filtering, and other such services—or it might redirect guest traffic to such a solution.

Of course, you must allow switch-to-switch links to carry the unauthorized-client/guest and Auth-Fail VLANs, typically as tagged.

Comware: Permitting limited access for unauthorized clients

By default, Comware switch ports implement 802.1X authentication in MAC-based mode, which enables them to distinguish between traffic from different devices (different source MAC addresses) and create a different authentication session for each (see Figure 7-15). This mode enables the switch port to support multiple clients but prevents guest and Auth-Fail VLANs from working on an access ports. (You can use guest and Auth-Fail VLANs with MAC-based mode, but you must use a hybrid port, as explained later in this chapter.)

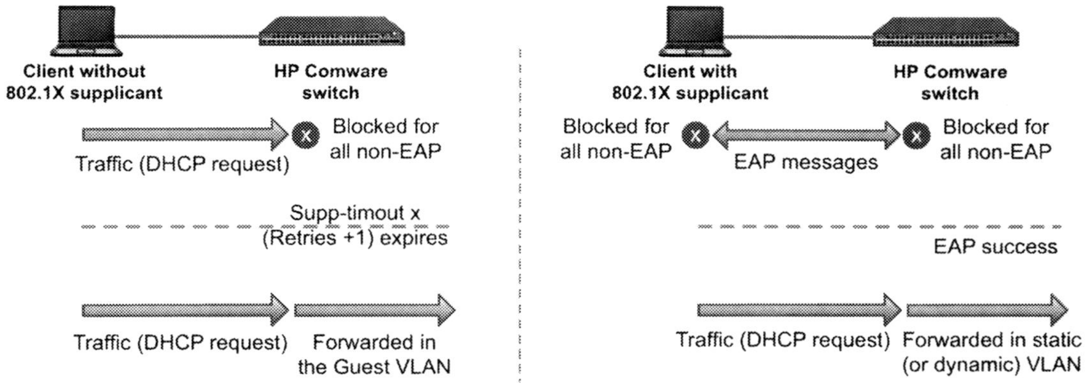

Figure 7-15: Comware: Permitting limited access for unauthorized clients

If you know that a single endpoint connects to the switch port, you should set the port to port-based mode, in which the switch applies a single authentication status for all traffic:

```
[Comware-<interface ID>] dot1x port-method portbased
```

You can then simply set the guest VLAN ID on the port, and the switch applies the VLAN to the port when the device connected to the port does not complete 802.1X authentication:

```
[Comware-<interface ID>] dot1x guest-vlan <ID>
```

Similarly, you can simply set the Auth-Fail VLAN:

```
[Comware-<interface ID>] dot1x auth-fail vlan <ID>
```

By default, the switch waits 90 seconds before it determines that a connected device does not support 802.1X and belongs in the guest VLAN. You can speed this process by changing the supp-timeout timer and 802.1X retry settings. The first timer determines how long the switch waits for a response to its authentication method. The second determines how many times the switch resends the message. To determine the amount of time that it takes for the switch to move a port that has just come up to the guest VLAN, multiply the supp-timeout timer by one plus the retry value. You establish these settings globally:

```
[Comware ] dot1x timer supp-timeout <seconds>

[Comware] dot1x retry <value>
```

Note, however, that lowering the timeout or retries too far could cause a device in the process of authenticating to be placed temporarily in the guest VLAN. As long as the client is also blocking traffic, it will not receive an IP address in the guest VLAN and should transition to its own VLAN after authentication seamlessly. However, in some cases, the client might send DHCP requests during the 802.1X authentication process and receive an IP address in the guest VLAN. After the device receives its new VLAN assignment, it must realize that its current IP address does not work and must obtain a new one. Consider these issues as you plan the supp-timeout timer setting.

As indicated in Table 7-3, Comware switches can distinguish between users who fail authentication and devices that never attempt authentication but can also place both types in a guest VLAN. You should usually start by defining the guest VLAN. Then, if you want to block access for users who actually *fail* authentication, set the Auth-Fail VLAN to a dead-end VLAN that exists only on the switch. Of course, you could also create an Auth-Fail VLAN that connects to the particular resources that you want to make available specifically to users who fail 802.1X authentication as opposed to devices that cannot perform 802.1X at all.

Note that, if you only set the guest VLAN, the guest VLAN applies to both devices that do not perform 802.1X and that fail 802.1X. However, if you set only the Auth-Fail VLAN, the VLAN applies only to devices that actually fail 802.1X. The switch would still block access for devices that cannot perform 802.1X authentication.

HP ProVision: Permitting limited access for unauthorized clients

HP ProVision switches also provide two modes of 802.1X access control (see Figure 7-16). User-based control, like MAC-based control on Comware switches, creates different authentication sessions for each connected device, while port-based control applies a single session. However, as long as only one endpoint connects to the port at the same time, you can use either mode with an unauthenticated-client VLAN without any special setup.

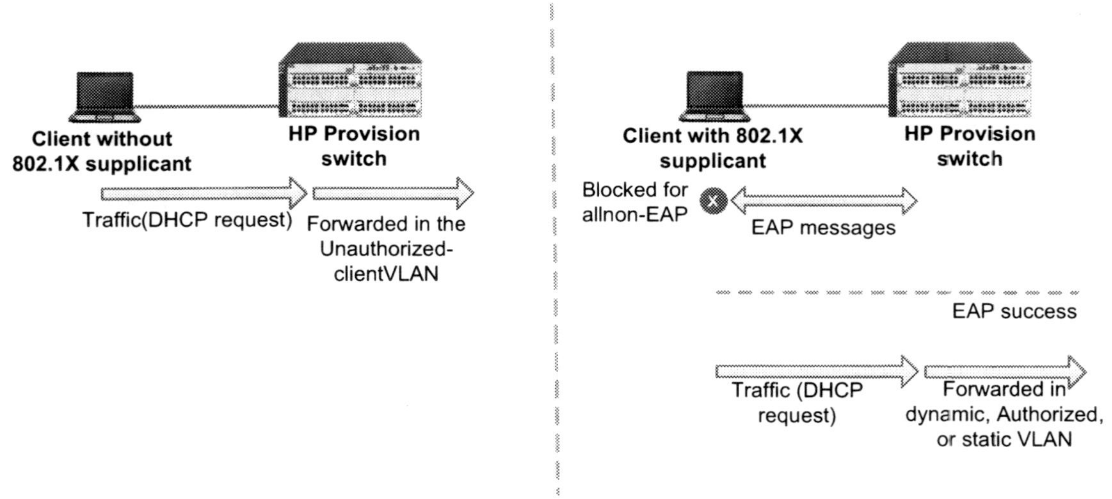

Figure 7-16: Permitting limited access for unauthorized clients on ProVision switches

To set an unauthenticated-client VLAN for one or more interfaces, use this command:

```
ProVision(config)# aaa port-access authenticator <port ID list> unauth-vid <ID>
```

On the ProVision switches, a switch port with a static VLAN ID and an unauthenticated-client VLAN ID is automatically part of the unauthenticated-client VLAN as soon as a device connects. If the device passes authentication, the port becomes an untagged member of the static VLAN.

This behavior helps guest devices and other devices without 802.1X supplicants to connect more quickly. However, it might cause issues for users who do use 802.1X to log in. If the user's device allows non-EAP traffic before authentication, it might receive a DHCP address in the unauthenticated-client VLAN, causing it to lose connectivity after the port moves to the VLAN for the authenticated user. Depending on its OS, the device usually drops the IP address after some delay and obtains a valid one.

You can use the unauth-period to prevent these connectivity delays for 802.1X devices:

```
ProVision(config)# aaa port-access authenticator <port ID list> unauth-
period <seconds>
```

When you set this period, the switch port delays placing a connected device in the unauthenticated-client VLAN until the client has failed to authenticate during that period. During this time, all traffic except EAP messages are blocked, giving the 802.1X device time to log in. In other words, the behavior becomes like that of the Comware switches. Fine-tune the unauth-period to balance delays for guest devices with delays for 802.1X devices.

Supporting multiple clients on the same port at the same time

You have assumed until now that you are implementing 802.1X on edge ports that support one endpoint at a time, which is true in many environments.

Sometimes, however, nonedge ports need to implement 802.1X. For example, a company might have legacy switches at the edge but still want to enforce 802.1X. You can implement 802.1X on an HP switch port that connects to the legacy switch. That port can then authenticate the devices connected to the legacy switch. For example, a switch might provide connectivity for the devices (such as company phones and laptops) and guests that connect to the network in a meeting room.

You might also encounter situations in which an edge port must support multiple devices at the same time. For example, the port might connect to a VoIP phone, which in turn connects to a computer.

You must understand how to set up the port to meet the particular needs of the scenario. First, you should be aware that this feature is supported only with IPv4. Second, you must consider the switch port's control mode.

Note

The legacy switch, if unmanaged, might not need to pass any traffic on its own behalf. If it does, however, it cannot complete 802.1X authentication. In that case, you might need to set up both MAC-Auth and 802.1X on the switch port. A later section in this chapter explains how to implement multiple authentication mechanisms on the same port.

Note

Theoretically, you could implement 802.1X on a switch port that connects to a legacy AP to authenticate the wireless clients. However, wireless clients usually use 802.1X authentication together with secure encryption, and the AP must support 802.1X for the solution to work. In practice, you will rarely encounter an enterprise AP that does not support 802.1X.

User-based/MAC-based versus port-based mode

You have two choices for how the port establishes authentication sessions. The port can authenticate *each* downstream device, as identified by source MAC address. Comware refers to this mode as MAC-based control, which you control with a command (it is the default setting):

```
[Comware-<interface ID>]dot1x port-method <macbased | portbased>
```

ProVision refers to this mode as user-based control, which is enabled when you set the client limit on the port to any value:

```
ProVision(config)# aaa port-access authenticator <port ID list> client-limit <1-32>
```

After you set the limit, the port begins tracking MAC addresses and defines the authorization status and settings for each separately. The limit value determines how many MAC addresses are allowed.

By default, the client limit is undefined. When then client limit is undefined (**no aaa port-access authenticator <*port ID list*> client-limit**), the interface operates in port-based mode.

In user-based/MAC-based mode, the port creates a separate session for each device, and each session has its own authorized status. In other words, the switch permits ingress traffic with a particular source MAC address only if that MAC address's session is authorized. Conversely, it permits egress traffic based on the destination MAC address's authorization status.

Each session also has its own dynamic attributes, such as VLAN and ACL. (However, you might need to implement some extra setup to ensure that the port can support multiple dynamic VLANs.)

With port-based control, on the other hand, the switch establishes a *single* session for all traffic sent and received on the port. As long as any one user passes authentication, the switch permits all traffic without referring to the source or destination MAC address. As Figure 7-17 shows, the most recently authorized user's dynamic settings determine the settings for the entire port. Therefore, port-based control introduces these potential issues:

- Unauthorized users gaining access to the network
- Inconsistent dynamic settings

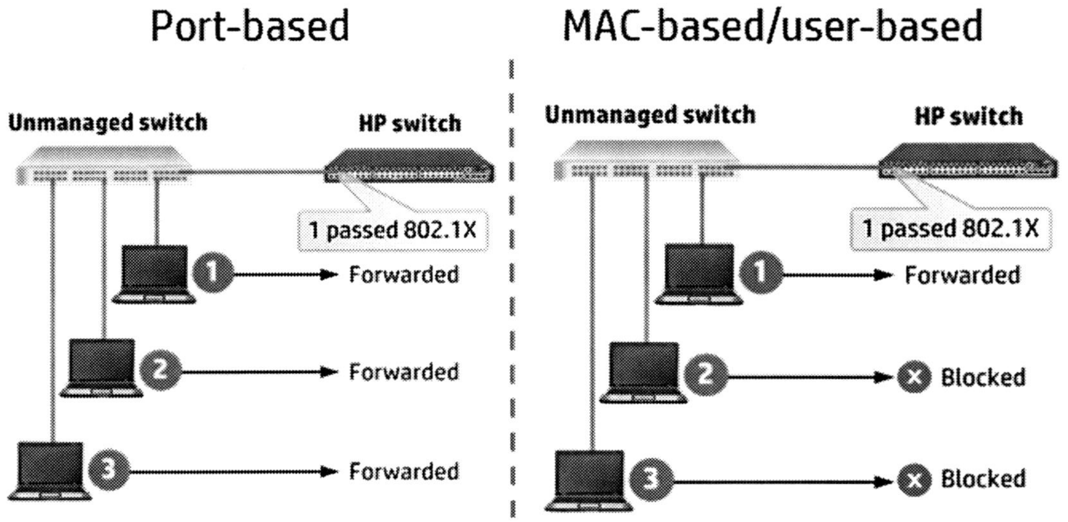

Figure 7-17: Port-based versus MAC-based or user-based mode

You might use port-based access when you want to use 802.1X to authenticate a physically insecure network infrastructure device such as an AP. However, that device itself implements authentication for its own endpoints.

Generally, user-based/MAC-based control is recommended for scenarios in which you want the switch port itself to authenticate and control multiple clients. In particular, it is recommended for cases in which the RADIUS server might implement different dynamic settings for different clients.

Multiple-client scenario: Unmanaged switch and users in different VLANs

In this scenario, users connect to an unmanaged switch, which then connects to an HP switch port. Multiple user devices might connect to the unmanaged switch, including employees in various user groups and guests. The company wants employees to log in to the network with 802.1X before they receive a VLAN assignment in the VLAN that gives them access to private resources. As you see, some employees are assigned to VLAN 10 and others to VLAN 20, based on their identity. Guest devices will be placed in a guest VLAN, VLAN 99, that lets the guests access the Internet.

You might encounter a situation such as this when a company wants cheap, nonenterprise switches at the edge but decide to implement some form of network access control using HP switches.

To implement the scenario, you set up 802.1X. The RADIUS server is configured with the appropriate VLAN assignments, and you specify the unauthorized-client/guest VLAN on the port, as shown in Figure 7-18.

You then enable the user-based/MAC-based control method. This method enables the port to authenticate each of the downstream devices separately and to apply the proper dynamic settings to each individual session. To enable the port to support more than one VLAN at a time, you *must* also implement MAC-based VLANs. This requirement applies whenever the port assigns concurrent users to multiple untagged VLANs, whether those VLANs include dynamic VLANs, unauthenticated-client/guest/Auth-Fail VLANs, or a combination, as in this scenario.

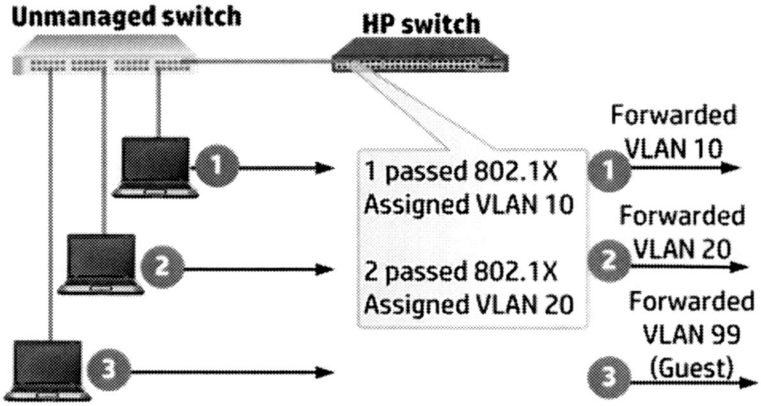

Figure 7-18: Multiple-client scenario: Unmanaged switch and users in different VLANs

Keep in mind that this solution has some disadvantages. For example, it is supported only with IPv4. With IPv6 advertisements, all the clients connecting through the same port will receive IP addresses in all the VLANs supported on that port.

In addition, you should not use MAC-based VLANs with multicast. If the port supports clients in two different VLANs and these clients request the same multicast stream, the switch will duplicate the stream and forward it to the port.

Using MAC-based VLANs to support multiple dynamic VLANs

On Comware switches, you must set up the MAC-based VLANs manually. You simply change the port to a hybrid port and enable MAC-based VLANs:

```
[Comware-<interface ID>] port link-type hybrid
[Comware-<interface ID>] mac-vlan enable
```

ProVision switches that support MAC-based VLANs automatically do so when necessary—in other words whenever multiple clients authenticate and receive different dynamic VLAN assignments. However, not all ProVision switch ports support this feature, so you need to check your switch model. As of the publication of this study guide, only ports on ProVision version 2 interface modules and some fixed-port switches such as the HP 3800 and 2920 Switch Series support MAC-based VLANs.

Behavior when the port does not support MAC-based VLANs

A Comware port in MAC-based mode that does not support MAC-based VLANs handles VLANs as follows:

- When the first client authenticates, the port applies that client's dynamic VLAN. If the client does not receive a dynamic VLAN assignment, the port applies the static untagged VLAN (only VLAN on an access port and port VLAN ID (PVID) on a trunk or hybrid port).
- When another client authenticates and does not receive a dynamic VLAN assignment, the port leaves the current VLAN assignment.
- When another client authenticates and receives a dynamic VLAN assignment, the port's behavior depends whether the new assignment matches the current VLAN.
    - If the new assignment matches, the port authorizes the new client.
    - If the new assignment is different, the port does not authorize the new client, even though it passed authentication.

A ProVision port in MAC-based mode that does not support MAC-based VLANs handles VLANs as follows:

- When the first client authenticates, the port applies that client's dynamic VLAN as the untagged VLAN. If the client does not receive a dynamic VLAN assignment, the port applies the static untagged VLAN (or authorized VLAN, if configured).
- When another client authenticates, the port leaves the current VLAN assignment whether or not the client receives a dynamic VLAN assignment.

These behaviors mean that you must ensure that all clients that might connect on the port use either:

- The static untagged (or authorized) VLAN
- The same dynamic VLAN assignment

Otherwise, you cannot predict whether clients will be able to connect or be placed in the correct VLAN.

Other scenarios with similar setup

The port might need to implement multiple dynamic VLANs if a VoIP phone and a computer connect on the same port and the VoIP phone sends *untagged* traffic. Dynamic VLAN assignments and MAC-based VLANs allow the port to place the phone's untagged traffic and the computer's untagged traffic in different VLANs.

In most cases, however, the phone sends tagged traffic.

Multiple-client scenario: Computer and VoIP phone

You will now look at the specifics for setting up another common scenario for multiple clients on an 802.1X port: a computer connects to a VoIP phone, which connects to the switch.

The phone should generally be connected at all times, whether or not someone is using the computer, so the phone requires an 802.1X supplicant that can authenticate to the RADIUS server. Usually, you want the user to authenticate separately; otherwise, you lose the security and provisioning benefits of 802.1X. Therefore, you should implement user-based/MAC-based control on the port.

Note

If the VoIP phone does not support an 802.1X supplicant, you would need to implement MAC-Auth on the port as well as 802.1X. These solutions are discussed later in the chapter.

You also need to consider the VLAN assignment for the computer and the phone. Usually, the computer sends untagged traffic in a user (or data) VLAN, and the VoIP phone sends tagged traffic in a voice VLAN.

As Figure 7-19 shows, you can create a static VLAN for the user, or the RADIUS server can send a dynamic VLAN assignment. Because the port assigns this VLAN as the untagged VLAN, the user's device does not need to be aware of the ID, as usual.

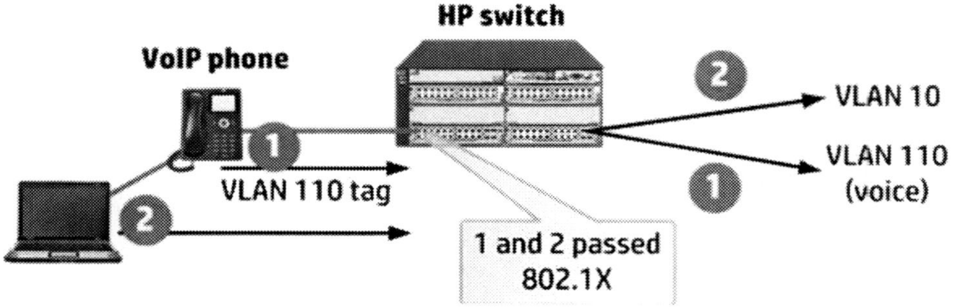

Figure 7-19: Multiple-client scenario: Computer and VOIP phone

The VoIP phone, on the other hand, needs to know the voice VLAN so that it can properly tag the traffic. It can learn the VLAN in several ways:

- Manual configuration
- Configuration file from a PBX
- From a DHCP server—The phone first sends untagged DHCP requests. The DHCP server sends a vendor-specific attribute that informs the phone of the correct voice VLAN ID.
- From LLDP-MED—The phone exchanges LLDP messages with the switch port. These messages include a variety of information related to the phone's settings, including power settings and information that help the phone register its location. They also include the voice VLAN ID.

You must ensure that the switch port supports tagged traffic in the voice VLAN. A port's tagged assignments do not change based on the dynamic VLAN assignments, and authorized clients can always send traffic tagged with these VLAN IDs. Because the port only requires one *untagged* VLAN at a time, you do not need to implement MAC-based VLANs.

The following sections provide more details about the proper setup for the first two scenarios on Comware and ProVision switches. You will learn more about the LLDP-MED scenario, as well as specifying the proper QoS settings for the phone's VLAN for all scenarios, in the next chapter.

Comware setup

You generally should not implement 802.1X for scenarios with VoIP phones and computers that connect to the same Comware port. Chapter 8 provides more details about setting up voice VLANs on Comware switches.

ProVision setup

The port implements 802.1X in user-based mode and supports at least two clients. The port has an untagged VLAN assignment in the user (data) VLAN. Or, if users receive dynamic VLAN assignments, the untagged VLAN can be any VLAN. The port has a tagged VLAN assignment in the voice VLAN.

For the DHCP scenario, make sure that the DHCP administrator creates the vendor-specific attribute with the voice VLAN ID in the scope for the PVID and every potential dynamic user VLAN.

MAC-Auth overview

MAC-Auth, which is one of the most basic security options available, imposes only minimal protection for network access.

Essentially, the switch uses a client's MAC address to authenticate the client. Because Ethernet clients always include their MAC address as the source of their Ethernet frames, the client does not need to be configured to submit credentials. Therefore, this authentication method provides an option for devices that do not have a user interface or support for 802.1X.

On the other hand, it is recommended that you use 802.1X when possible for more rigorous security. Often it is trivial for a hacker to discover MAC addresses used in your system (whether by eavesdropping or accessing a legitimate device). The hacker can then easily spoof an authorized MAC address. MAC authentication can also be more difficult to manage, requiring directory or domain administrators to create accounts for each MAC address on each device, whereas 802.1X can use the user accounts already in the directory or domain.

Nonetheless, MAC-Auth remains an option for enterprises that want to impose a degree of control over network access for devices without 802.1X supplicants.

Two types of MAC-Auth are available:

- **RADIUS MAC-Auth**—RADIUS MAC-Auth relies on a RADIUS server to approve or reject the authentication request.
- **Local MAC-Auth**—With local MAC-Auth, the switch checks the MAC address against a local list. (Check your switch documentation to determine whether the switch supports this option.)

RADIUS MAC-Auth

Figure 7-20 illustrates the basic process by which switches enforce RADIUS MAC-Auth.

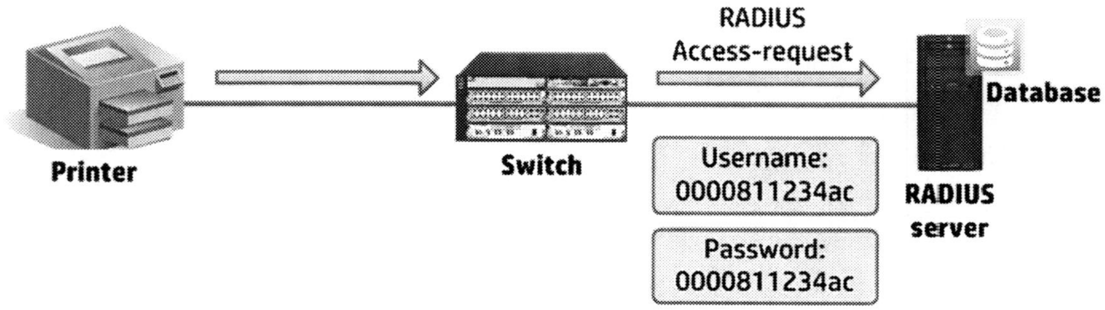

Figure 7-20: RADIUS MAC-Auth

1. A client is initially unauthenticated. The switch drops all traffic from it and also does not transmit traffic on the link to it.

   You can actually configure ProVision switches to transmit certain traffic. You will look at scenarios that require this configuration later in the chapter.

2. The switch submits credentials to a RADIUS server on the client's behalf.

   Usually, these credentials consist of the client's MAC address; however, you will look at some other options as well.

3. The RADIUS server authenticates the MAC address against a user account. Only the switch and the server are involved in the exchange of authentication messages, unlike in a typical 802.1X solution. The RADIUS server still uses policies to determine whether to grant access. It communicates its decision to the switch in a RADIUS Accept-Access or Reject-Access message.

4. The switch handles the client according to the RADIUS server's decision.

   - If the server accepted access, the switch allows all traffic to and from the client with that MAC address.
   - If the server rejected access, the client remains in the unauthenticated state, and the switch handles its traffic as indicated in step 1.

The switch continues to examine incoming frames. If a new client connects, the switch will authenticate that client. The switch also attempts to authenticate the client for which authentication failed again (called a silent client on Comware switches). By default, the switch suppresses authentication requests for such clients for 60 seconds. It then sends another authentication request. You can adjust that time period on both Comware and ProVision switches.

RADIUS MAC-Auth options

You need to make several choices related to RADIUS MAC-Auth.

Authentication protocol

ProVision switches can use PEAPv0/MS-CHAPv2 for MAC-Auth. In this case, the switch acts as the authentication client rather than the client itself, generating the EAP Request messages. PEAP often offers the best option because it secures the credentials within a tunnel between the switch and the RADIUS server. It requires the RADIUS server to have a certificate; however, most RADIUS servers already require a certificate for various functions.

Alternatively, on the ProVision switches, you can choose Challenge Handshake Authentication Protocol (CHAP) for the authentication protocol. RADIUS has specific messages for CHAP authentication and most RADIUS servers support this option. You should be aware that the mechanisms of CHAP require passwords to be stored with reversible encryption, which poses a potential security hazard and is not typical for most directories and domains. This limitation would mean, for example, that the directory, domain, or RADIUS administrator who creates the accounts for the clients would have to take special measures to store the passwords with reversal encryption. (For a Microsoft Windows domain, the administrators can edit the group policy object (GPO) and enable reversal encryption in the computer policies for account settings.)

Comware switches always use Password Authentication Protocol (PAP) with MAC-Auth. This less-secure protocol does not have all of the provisions of more advanced protocols such as PEAP. However, presumably, the network between the switch and RADIUS server is relatively trusted.

If you have both Comware and ProVision switches, make sure that the RADIUS policy allows an authentication method supported by both types of switches. (Alternatively, the RADIUS server could have a different policy for each type of switch; the policies could select access requests based on the called station identifier, which identifies the switch.)

Credentials

You need to decide on the precise credentials that switches will send so that the RADIUS, directory, or domain administrator can set up the accounts accordingly.

By default, both Comware and ProVision switches send the MAC address for both the username and password.

On a ProVision switch, you can set up a global password, which the switch uses for the password instead of the device's MAC address. The switch itself places the password in the authentication request. No configuration is required on the client. The switch continues to use the client's MAC address for the username, so every client requires a user account on the RADIUS server (or background directory).

On a Comware switch, you can create a single fixed account with a username and password that the switch submits when an unauthenticated client connects. The RADIUS server has a single matching account. The Comware switch *cannot* use a fixed password with individual clients' MAC addresses, as ProVision switches do. If you set a fixed password and no fixed username, the Comware switch uses the default fixed username, **mac**. Similarly, if you set a fixed username, you must set a fixed password. Otherwise, authentication fails for all devices. This Comware option does not provide any security because the switch automatically uses the account for all devices. You would use this option if you do not care about controlling devices' network access but you do want the switch and RADIUS servers to log the access attempts.

MAC address format

When you are using MAC addresses for the username, password, or both, you must ensure that the switch sends the credentials with the same format used in the user accounts. Table 7-4 shows the options and the syntax for configuring a specific format on Comware and on ProVision switches. The Comware switches do not support some of the options, in which case the Comware column shows a dash. Of course, if your environment includes switches of both types, you should select a format supported by both so that the solution can use a single set of accounts.

Note

Accounts in Microsoft Active Directory (AD) domains cannot use colons, so choose a format that does not include colons if you use AD.

Table 7-4: MAC address formats

Format	ProVision syntax	Comware syntax
Beginning of the command for specifying one of these formats:	aaa port-access mac-based addr-format	mac-authentication user-name-format mac-address
aabbccddeeff	no-delimiter	without-hyphen lowercase
aabbcc-ddeeff	single-dash	—
aa-bb-cc-dd-ee-ff	multi-dash	with-hyphen lowercase
aa:bb:cc:dd:ee:ff	multi-colon	—
AABBCCDDEEFF	no-delimiter-uppercase	without-hyphen uppercase
AABBCC-DDEEFF	single-dash-uppercase	—
AA-BB-CC-DD-EE-FF	multi-dash-uppercase	with-hyphen uppercase
AA:BB:CC:DD:EE:FF	multi-colon-uppercase	—

RADIUS MAC-Auth configuration process

The basic steps for setting up RADIUS MAC-Auth on both Comware and ProVision are outlined in the next sections.

Comware steps

1. Complete the same steps as for 802.1X to create a RADIUS scheme.
2. Also configure the domain much as you would for 802.1X authentication. You can often use the default domain. Exceptions include when you want to use the default domain for 802.1X authentication as well, but different RADIUS servers handle the 802.1X requests and the MAC-Auth requests.

   The lan-access service type (or, if this type is not defined, the default service type) applies to the MAC-Auth requests. Specify your RADIUS scheme for authentication and authorization and, optionally, accounting.
3. If you are not using the default domain, specify the new domain as the global MAC-Auth domain. You could also specify different domains on different physical ports; the port MAC-Auth domain overrides the global one:

```
[Comware]mac-authentication domain <name>
```

```
[Comware-<interface ID>] mac-authentication domain <name>
```

4. Configure the format for the MAC-Auth credentials so that they match the credentials configured on the RADIUS server (or its backend directory):
   - If you want to use that option, specify the global (fixed) password or the global (fixed) username and password:

```
[Comware]mac-authentication user-name format fixed [account <name>]
{password [simple | cipher] <password>}
```

   - If you are using MAC addresses, also set the format for the MAC address itself, indicating whether it is sent upper- or lowercase and with which delimiters:

```
[Comware]mac-authentication user-name format <format>
```

See Table 7-4 for formats.

5. Optionally, you can adjust other settings such as the timers for reauthenticating a client that fails authentication. You will look at some specific settings when you consider more advanced solutions at the end of this chapter. You can refer to your switches' *Security Configuration Guide* for more information as well.
6. Enable MAC-Auth on the individual ports that you want to enforce this type of authentication:

```
[Comware]mac-authentication interface <interface ID> [to] [interface ID]
```

7. Enable MAC-Auth globally:

```
[Comware]mac-authentication
```

ProVision steps

1. Follow the same steps as for 802.1X to specify the RADIUS servers on the ProVision switches.
2. Configure the format for the MAC-Auth credentials so that they match the credentials configured on the RADIUS server (or its backend directory):
   - If you want to use that option, specify the global (fixed) password or the global (fixed) username and password:

```
ProVision(config)# aaa port-access mac-based password <password>
```

   - If you are using MAC addresses, also set the format for the MAC address itself, indicating whether it is sent upper- or lowercase and with which delimiters:

```
ProVision(config)# aaa port-access mac-based addr-format <format>
```

See Table 7-4 for formats.

3. Choose between PEAP and CHAP for the authentication protocol. (Remember that CHAP requires passwords to be stored with reversible encryption, which is not the default for many directories.)

```
ProVision(config)# aaa authentication mac-based <chap-radius | peap-mschapv2
```

4. Optionally, you can adjust other settings such as the timers for reauthenticating a client that fails authentication. You will look at some specific settings when you consider more advanced solutions at the end of this chapter. You can refer to your switches' *Security Configuration Guide* for more information as well.
5. Enable MAC-Auth on the individual ports that you want to enforce this type of authentication:

```
ProVision(config)# aaa port-access mac-based <port ID list>
```

Comware: Local MAC-Auth

Local MAC-Auth proceeds in a similar manner as RADIUS MAC-Auth. In the domain used for MAC-Auth, you simply specify lan-access for local authentication for rather than specify a RADIUS scheme. As Figure 7-21 shows, you also configure the accounts for MAC-Auth as local-user accounts on each switch rather than use external RADIUS servers. For this reason, local MAC-Auth is less scalable than RADIUS MAC-Auth.

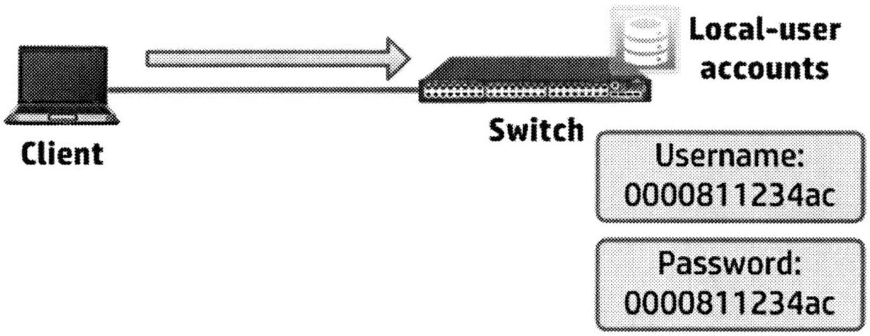

Figure 7-21: Comware: Local MAC-Auth

A local-user account on a Comware switch includes at least:

- A username
- A password, which is stored encrypted (cipher) or in plaintext (simple)
- Service type, which is comparable to the forms of access that you looked at for domains

For MAC-Auth, these settings are:

- Username—Client's MAC address (in the format that you have selected in the global MAC-Auth settings)
- Password—Client's MAC address (in the same format)
- Service type—lan-access

You would need one account for every client that might connect to this switch.

Of course, if you choose to use a fixed account simply so that you can log clients' access, you just need one account that matches the settings that you specify in the global MAC-Auth settings.

With these basic settings in place, the switch can authenticate the MAC-Auth clients.

The user account can also authorize a client for various settings much as AVPs specified a RADIUS policy can. For more information about local-user account settings, you can also refer to your switch's *Security Configuration Guide*.

Comware: Local MAC-Auth configuration process

Use these basic steps to configure local MAC-Auth on a Comware switch (commands are included only for new steps):

1. Configure the default domain or, if you do not want to use local authentication for lan-access in the default domain, create a new domain. Specify local for the lan-access authentication and authorization functions.

2. If you created a new domain, remember to specify the domain as the global MAC-Auth domain (or the domain on individual ports).
3. You still need to configure the MAC-Auth format.
4. Optionally, configure other settings, as for RADIUS MAC-Auth.
5. Create local-user accounts:
   - If you are using MAC addresses for the credentials, match the username and password to the MAC address format that you configured in step 3. Create an account for every authorized client that might connect to this switch.
   - If you are using a fixed account, match the fixed username and fixed password.
   - In either case, remember to set the service type to lan-access:

```
[Comware]local-user <MAC address>
[Comware-luser-<name>]password [simple | cipher] <MAC address>
[Comware-luser-<name>]service-type lan
```

6. Enable MAC-Auth on individual ports.
7. Enable MAC-Auth globally.

Web-Auth overview

You will now examine Web-Auth, another option for authenticating users and controlling their access to the network, typically used for guests. You will explore several options for this feature, including simple, switch-based solutions and more complicated centralized solutions. Note that Comware products refer to Web-Auth as portal authentication.

Today, customers and other guests request network services such as Internet access as a matter of course. If companies do not provide these network services to them, their competition will.

In these situations, you cannot be sure that all users' stations will support 802.1X or particular EAP methods. You cannot help the users configure their stations correctly to complete the authentication. On the other hand, you do not want to open your wireless and wired network to anyone with a wireless network interface card (NIC).

Web-Auth is one solution for this type of network access. As the name suggests, Web-Auth makes it easy for users to access the network through their familiar web browser interface. They can then connect to the Internet with a minimum of hassle.

Note

Comware switches refer to this form of authentication as "portal authentication."

Advantages of Web-Auth

Security solutions such as 802.1X require specific capabilities, but any client can authenticate to a switch that uses Web-Auth as long as the user has a legitimate username and password and a web browser.

Companies can use Web-Auth for employees; however, it is often a guest solution.

Web-Auth also lets companies control what users see before they receive network access. The company can present a customizable webpage that shows the guest information about the company providing the complimentary access. The company can also force the user to consent to terms and conditions, which can be a legal requirement in many regions.

Environments with external users who may benefit from Web-Auth include:

- Hospitals
- Universities
- Libraries, hotels, and other businesses that provide courtesy Ethernet access

Disadvantages of Web-Auth

Web-Auth does not provide as rigorous security as 802.1X. Users have a degree of network access before they authenticate. The credentials and authentication protocol might not be as secure as with 802.1X; however, you can improve the security by using HTTPS rather than HTTP.

Because Web-Auth requires interaction with the user; you cannot use it to authenticate devices without a user interface.

As you will learn, Web-Auth does not provide as much flexibility in terms of authenticating concurrent multiple clients on the same port as 802.1X. However, usually the port needs to support only one client at a time.

You should also note that the local switch Web-Auth solutions might not offer all of the options that wireless Web-Auth solutions typically do. As you will learn, though, the UAM portal solution does provide a great many options for supporting different types of clients and access.

Overview of Web-Auth process and options

Web-Auth is not a specific standard like 802.1X but rather a description of a common way to authenticate users. Implementations differ in terms of where they begin controlling user traffic, which device stores the webpages in which users enter their credentials, and so on. HP solutions themselves offer several options.

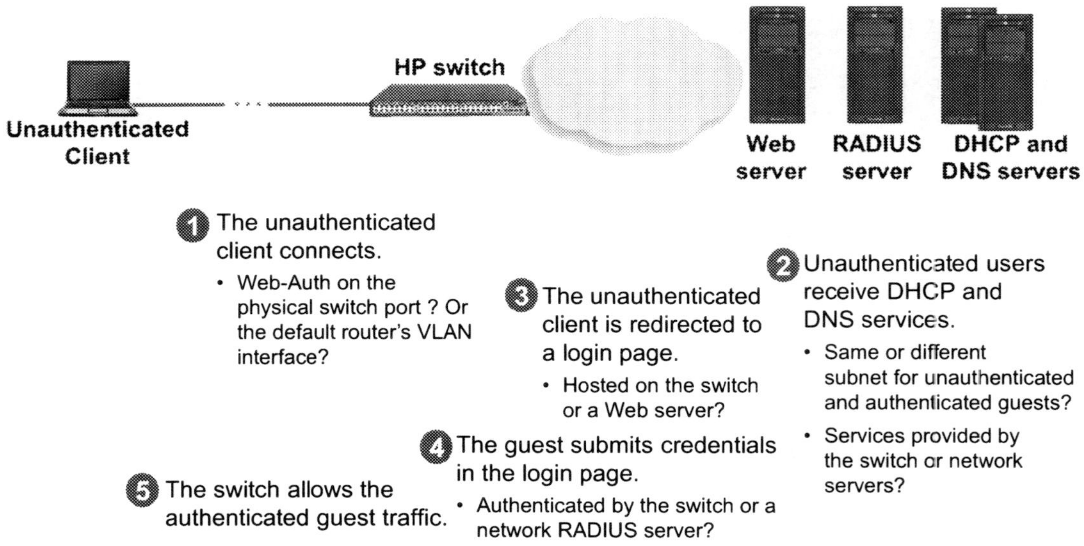

Figure 7-22: Web-Auth process and options

Figure 7-22 illustrates the general framework and points out some places in which you can choose among different options based on the needs of your environment:

1. The user's client establishes an Ethernet connection. The first decision for Web-Auth is the point at which the client's traffic is controlled. You can:
   - Apply Web-Auth at the physical port on which the client connects.
   - Let the client's traffic flow through the network until it reaches its default router, typically a routing switch, at which point that device implements Web-Auth on:
     - A Layer 3 interface (Comware Layer 3 switches only)
     - A physical port (Comware or ProVision)
   - Let the client's traffic flow through the VLAN until it reaches a specific choke point, at which point that device implements Web-Auth on the physical port. (This option is less common that the previous one since it usually makes sense to impose the control directly on the user's port or at the subnet boundary.)
2. The switch enforcing Web-Auth does not permit *most* traffic from unauthenticated clients to pass. However, unlike with 802.1X and MAC-Auth, the unauthorized client does receive a limited form of access:
   a. The client receives an address, usually through DHCP. The address might be:
   - An IP address in the same subnet that authenticated clients use
   - An IP address in a different subnet, specific for unauthenticated clients

The device is able to receive this address in one of these ways:

- The DHCP server exists within the unauthenticated clients' subnet before Web-Auth takes effect.
- The switch itself assigns the IP address.
- The switch enforcing Web-Auth allows the clients' DHCP messages to flow.

The precise mechanism depends on how you set up Web-Auth, as you will see.

b. The client also needs to send DNS requests so that it can attempt to browse to a webpage. Comware switches let clients' DNS messages pass. ProVision switches themselves act as the DNS server.

3. When the client attempts to send HTTP or HTTPS traffic, the switch recognizes the TCP port and redirects the client to a login webpage.

   As another decision, you can choose to use pages hosted on each switch enforcing Web-Auth or pages hosted on an external web server. You can also choose to use HTTP or HTTPS, the latter providing encryption for user credentials. You will examine the HTTPS options in the next chapter, which covers secure device management.

4. The user submits credentials to the login webpage. The webpage is configured to extract these credentials and send them back to the switch.

   You then have other choices as to how the switch verifies the credentials:

   - It can check them against local accounts.
   - It can submit them to a RADIUS server (Comware and ProVision). The protocols supported for the access request depends on the switch. The RADIUS server, as always, both authenticates the credentials and uses a matching policy to determine whether the user is allowed network access or not.

5. The switch enforces the decision, either continuing to block the client's traffic or now permitting all traffic from the authenticated client.

Comware: Local Web-Auth with a RADIUS server

Because you have so many options for Web-Auth, you will now examine how a few specific scenarios work.

In this scenario, you are:

- Using Comware switches
- Enforcing Web-Auth (portal authentication) directly at each user's physical Layer 2 port
- Using webpages hosted on the switch

On Comware switches, this is the only option for portal authentication implemented on a physical Layer 2 port.

- Assigning unauthenticated users IP addresses in the same subnet used for authenticated Web-Auth users

  On Comware switches, you cannot create a temporary subnet for unauthenticated Web-Auth users on the physical Layer 2 port.

Note

Comware switches *can* implement local Web-Auth solutions in which unauthenticated users receive temporary IP addresses in different subnets. However, you must set up the solution manually. That is, you set the PVID for the port to a VLAN reserved for this temporary subnet. You either set up DHCP services on the switch or on an external server, but you make sure that the scope has a very short lease. You then specify a dynamic VLAN assignment for guests so that authenticated guests receive a new IP address in that new VLAN.

To set a DHCP lease shorter than one minute on a Microsoft Windows DHCP, you must use NetShell (netsh) commands rather than the web-based management interface.

- Authenticating users to an external RADIUS server

How it works

In this scenario, Web-Auth proceeds as follows (see Figure 7-23):

1. The user's client establishes a connection with the switch port that enforces Web-Auth.
2. The switch drops all traffic from the unauthenticated client except:
   - DHCP discovery and request messages
   - DNS requests

   The switch also drops all traffic outbound on the port except:
   - DHCP inform and acknowledgment messages
   - DNS responses
   - Broadcasts
   - Traffic flooded because the destination MAC address is unknown
3. The user opens a browser and attempts to reach a webpage. The switch recognizes the TCP port and redirects the client to the login webpage hosted on itself.
4. The user submits credentials to the login webpage.
5. The switch extracts the credentials and submits them to the RADIUS server defined for portal authentication in the appropriate domain.

6. The RADIUS server both authenticates the credentials and uses a policy to determine whether the user is allowed network access or not.
7. The switch enforces the decision, either continuing to block the client's traffic or now permitting all traffic from the authenticated client.

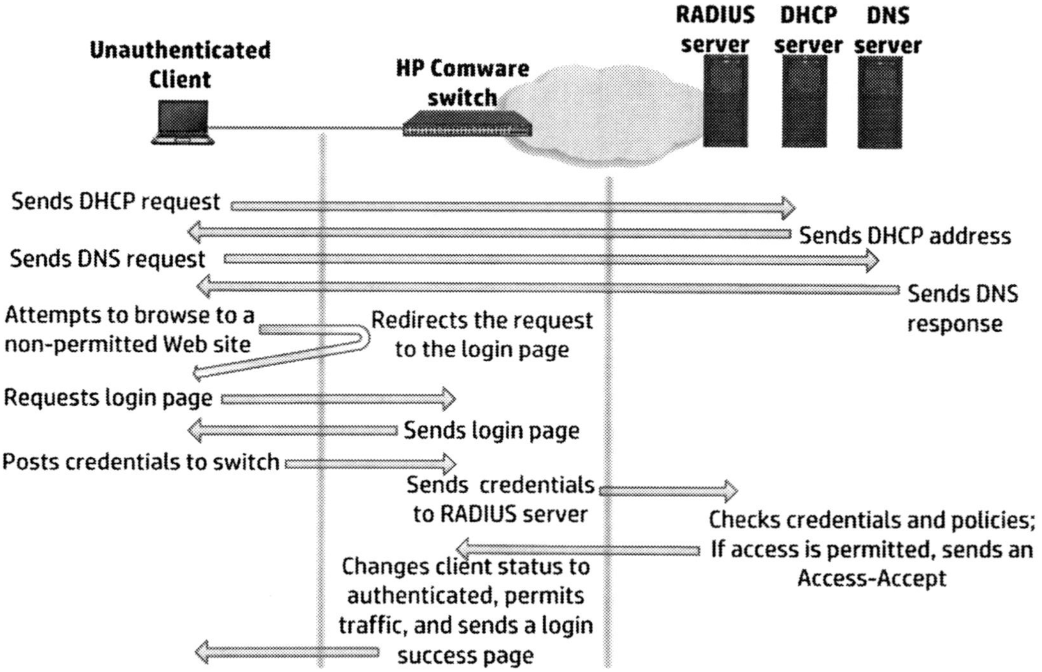

Figure 7-23: Local Web-Auth with a RADIUS server for a Comware switch

Comware: Set up local Web-Auth with a RADIUS server

To set up the local webpages, you must define the local portal server on the switch, specifying its IP address and protocol (HTTP or HTTPS). You can configure the switch to host the local login pages on any of its IP addresses. However, it is best practice to set up a loopback interface and use that interface's IP address. The address does not need to be routable within the network; the switch will just use it for Web-Auth communications with the directly connected hosts. For example, if your network uses an IP addressing scheme with 10.0.0.0/8 addresses, you might configure a loopback interface with IP address 192.168.1.1.

You must also define the RADIUS server that authenticates the Web-Auth users. As usual, you do so in a scheme.

You can create a new domain, which you then specify for portal authentication on the switch port. If you do not specify such a domain, the switch uses the default domain.

In either case, you must configure this domain to use your RADIUS scheme for portal authentication, authorization, or accounting. (Of course, you can also set the scheme for the default authentication, authorization, and accounting method when you do not need to distinguish portal services from others handled by the domain.)

Finally, you enable local portal authentication on the appropriate switch port or ports.

Comware: Local Web-Auth with local accounts

Next, consider a scenario identical to the previous one except that the switch checks the user credentials against local-user accounts rather than an external RADIUS server.

As you see in Figure 7-24, this Web-Auth solution works just like the previous one until the switch receives the guests' credentials. The switch then checks the credentials against its own local-user accounts and determines whether to provide access.

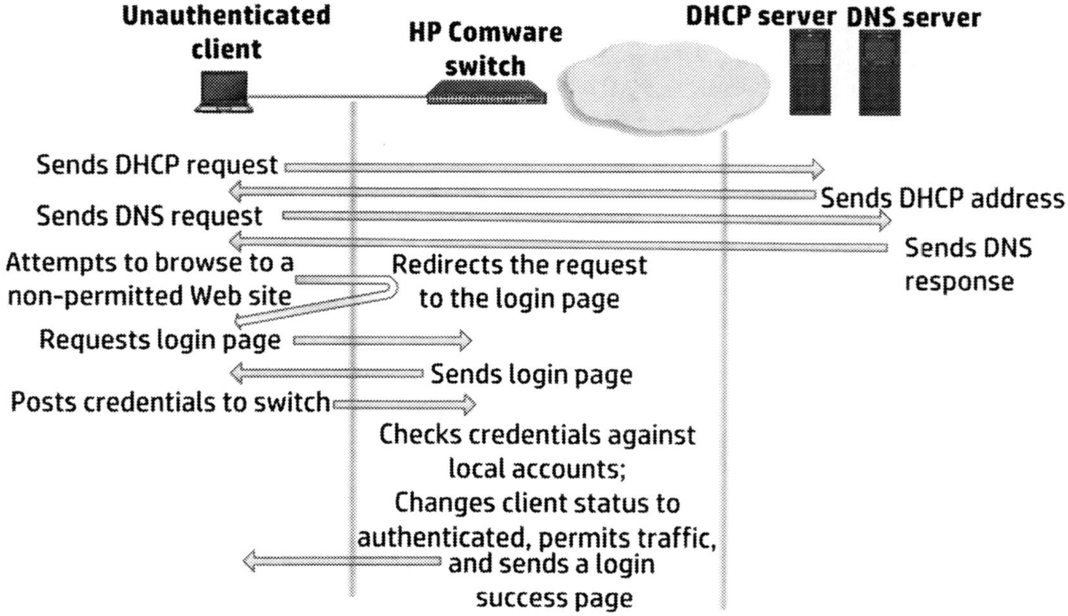

Figure 7-24: Local Web-Auth with local accounts for a Comware switch

Comware: Set up local Web-Auth with local accounts

For the most part, you set up this solution like the previous solution (see the "Comware: Set up local Web-Auth with a RADIUS server" section). You must set up local pages with:

- Loopback interface reserved for hosting the local pages
- Loopback address for the local portal server
- HTTP or HTTPS enabled

You must also create local-user accounts with at least these settings:

- Username—The desired username
- Password—The desired password
- Service type—Portal

You must then ensure that the domain that applies to the guests specifies local as the authentication and authorization method. It is highly unlikely that guests include domain information in their usernames. Therefore, you should take one of these approaches to assign the guests' access requests to a domain:

- If you want to apply local authentication to all guests, use the default domain. Within the default domain, set the portal authentication and authorization type to local. (Of course, you could also use the default type if you want the switch to use local-user accounts when another method is defined for a particular type of access.)
- If you want to apply local authentication to only some guests, configure a new domain. This domain uses local for the default or portal authentication and authorization method. Then configure that domain as the portal domain on specific physical interfaces.

You must also enable local portal authentication on the appropriate switch port or ports (just as you did in the "Comware: Set up local Web-Auth with a RADIUS server" section).

ProVision: Local Web-Auth to a RADIUS server

Next, you will examine a Web-Auth solution for ProVision switches. In this scenario, the switch is:

- Enforcing Web-Auth directly at each user's physical port

  This is the only option for ProVision switches.
- Using webpages hosted on the switch

  This is the default option. You will see what these pages look like in a moment. (You will also learn how to use webpages hosted on an external server a bit later in this chapter.)
- Authenticating Web-Auth users to an external RADIUS server

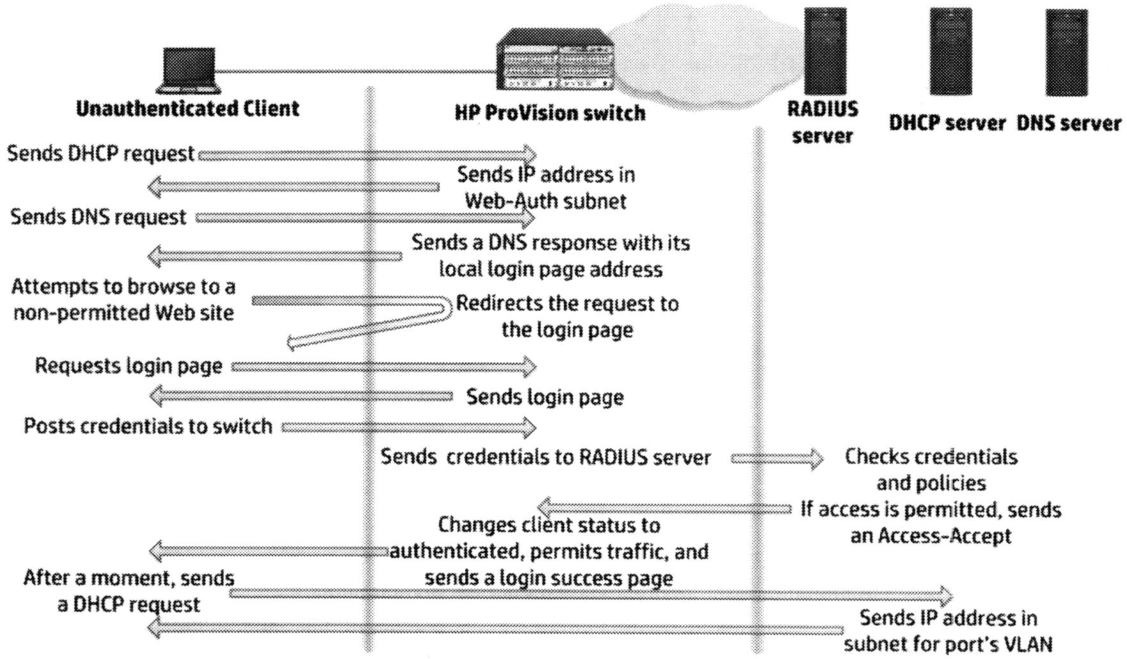

Figure 7-25: Local Web-Auth to a RADIUS server for a ProVision switch

In the ProVision solutions, the unauthenticated client receives an IP address in a subnet reserved for this purpose. It uses this IP address during the authentication process. You might not want unauthenticated users to have valid IP addresses for your network for several reasons:

- Unauthenticated users can gain information about your IP addressing scheme.
- When examining traffic, you cannot distinguish between unauthenticated and authenticated clients.
- Unauthenticated users use up some of the IP addresses that authenticated clients require. A hacker could attempt to obtain a lot of IP addresses to lock out legitimate Web-Auth users.
- Hackers might be able to use various exploits such as DNS tunneling to obtain unauthorized access.

How it works

This solution functions as follows (see Figure 7-25):

1. The user's client establishes a connection with the switch port that enforces Web-Auth.
2. The switch drops traffic from the unauthenticated client as explained previously. However, rather than forward the client's DHCP discovery messages, the switch responds itself with an IP address

in the Web-Auth subnet configured on it. The lease for this address is very short, 10 seconds, by default. (If the lease runs out before the user finishes authenticating, the switch allows the client to renew its same IP address.)

3. The user opens a browser and attempts to reach a webpage. The switch recognizes the TCP port and redirects the client to the login webpage hosted on itself. The switch uses the default gateway IP address that it assigns in the DHCP lease to host the webpage. The client only has access to this page; it cannot, for example, reach the switch's web browser management interface.

Steps 4 through 7—Proceed as in the first Comware scenario.

8. After the RADIUS server sends an Access-Accept message, the switch gives the client authenticated status. The client's DHCP lease for the temporary IP address rapidly expires. Now when the client sends a renew request, the switch ignores it.
9. The client then sends a DHCP discovery, which the switch forwards on the client's VLAN. The client receives a valid IP address for the network.

ProVision: Set up local Web-Auth to a RADIUS server

You specify the RADIUS server settings as you have already learned. Then enabling Web-Auth can be as simple as entering a command to enable it on the port.

However, it is recommended that you also configure a redirect URL. After a user successfully authenticates, the switch redirects the user's browser to this URL; you can choose any welcome webpage that you like. If you do not specify this URL, depending on their browsers, users might not see the success page and fail to understand that they have successfully logged in:

```
ProVision(config)# aaa port-access web-based <port ID list> redirect-url <URL>
```

Use standard URL format such as **http://server.example.com/welcome**.

If you want, you can also change the subnet for unauthenticated Web-Auth clients:

```
ProVision(config)# aaa port-access web-based dhcp-addr <network address>/<prefix length>
```

(You also have several more options, which you can read about in your switch's *Access Security Guide*.)

After specifying the options, enable Web-Auth on the ports:

```
ProVision(config)# aaa port-access web-based <port ID list>
```

Customizing local webpages

All HP switches provide default webpages hosted on the switch itself. These pages include the initial login page, pages that display when authentication succeeds, pages that display when authentication fails, and so on.

You can load new, customized pages on Comware switches. The pages simply need to include a few specific HTML forms and code, which are clearly defined in the switch documentation.

For ProVision switches, you can configure an authentication failure message that displays on the failed authentication page. If the company requires further customization, you should set up a solution that uses pages hosted on an external web server.

The next sections present these solutions.

ProVision: Remote Web-Auth

Next, you will examine several scenarios that feature login pages hosted on an external web server. An external web server provides benefits such as these:

- More advanced options for login page customization
- The ability to use HTTPS but the need for only one or two certificates, installed on the web servers rather than every switch
- A centralized, standardized experience for all guests who log in

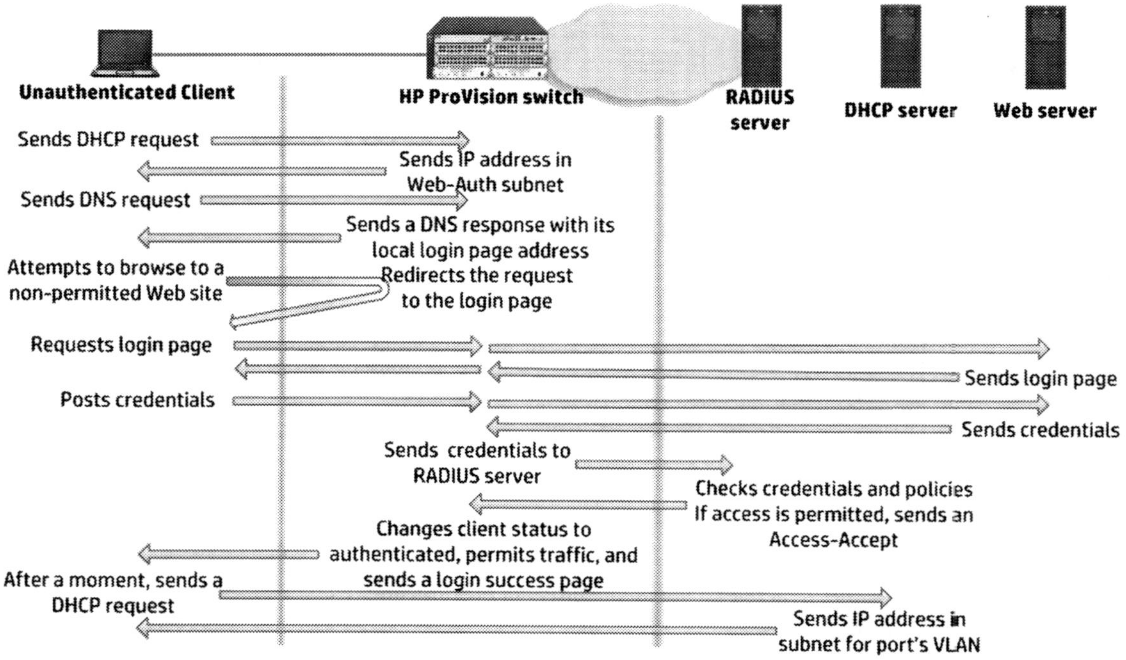

Figure 7-26: Remote Web-Auth for a ProVision switch

Scenario

First consider a ProVision scenario in which you are:

- Enforcing Web-Auth directly at each user's physical port
- Using webpages hosted on an external server
- Authenticating Web-Auth users to an external RADIUS server
- Assigning temporary IP addresses to unauthenticated users

How it works

As Figure 7-26 shows, the process for this form of Web-Auth does not differ from the others until the step at which the unauthenticated user attempts to browse to a webpage. Once again, the switch redirects the user to its local address on the Web-Auth subnet. However, instead of returning a local webpage, the switch initiates contact with the external web server URL, which you have configured in the switch's Web-Auth settings. The switch sends the webpage that this server sends to the client.

After the user submits the credentials, the web server returns them to the switch, and authentication proceeds as before.

At each point in the process, the switch requests a different webpage from the external web server. For example, the switch initially requests a page called index.html. While authenticating the user, the switch requests (and sends on to the client) a page called authen.html.

Note a couple of implementation considerations:

- The ProVision switch proxies HTTP traffic between the guest clients and the external web server. Therefore, you do not need to worry about extending the temporary subnet into the actual network. You can also use the same unauthorized subnet address on all switches.
- The switch also automatically permits unauthenticated users to send HTTP or HTTPS requests to the external web server.

ProVision: Set up remote Web-Auth

You set up this solution as the previous one. However, you must also specify the external web server's IP address or hostname, as well as the path to the login pages if the path is not the server's default one.

Use this CLI command:

```
ProVision(config)# aaa port-access web-based ewa-server <IP address
|hostname>[<page-path>]]
```

Because the ProVision switch requests specific pages at specific points in the process, the web server must have a precise set of pages:

- accept.html
- authen.html
- index.html
- reject_novlan.html
- reject_unauthvlan.html (if you are using an unauthenticated-client VLAN)
- retry_login.html
- sslredirect.html
- timeout.html

These webpages must also include a specific code for extracting users' credentials and sending them to the appropriate switch. It is recommended that you download template login pages provided on the download page for your switch's software. A company's web designers can customize these pages as much as they like as long as they do not change a few pieces of code, which are documented in your switches' *Access Security Guide.*

Web-Auth with UAM portal services

As the final Web-Auth solution, you will examine UAM portal services. The UAM solution provides many benefits beyond simple Web-Auth:

- Different login pages for different sets of users and for different OSs
- Centralized user tracking and logging
- Self-service page that enables guests to create their own credentials
- Integrated solution that helps you to create policies that restrict access according to specific rules
- Temporary IP addresses for unauthenticated guests (without UAM, this is an option for ProVision switches but not Comware ones)
- Endpoint integrity checks that ensure guest devices do not introduce malware and security threats

  The endpoint integrity features, which are not covered in this study guide, require an additional module, Endpoint Admission Defense (EAD), which integrates with UAM. Note that this module requires its own licensing. (Refer to the HP networking website for more information.)

For all of these reasons, this solution is the typically recommended enterprise guest solution.

UAM actually fulfills two roles in this solution: as a web server, it hosts the login pages, and, as a RADIUS server, it authenticates guest. The same server can fulfill both functions, or you can distribute the functions to different servers. Table 7-5 indicates the IMC components for each function.

Table 7-5: IMC components

IMC component	Role in the solution
User Access Manager—policy server	RADIUS server that authenticates guests
User Access Manager—portal server	Web server that hosts the login pages

UAM portal authentication: Process

To use the pages hosted on the UAM portal server, you must implement Web-Auth on a Comware switch's Layer 3 interface. That is, rather than enforce Web-Auth directly on the port to which the user connects, you allow the traffic to flow to the Web-Auth user's default router, which implements Web-Auth on the VLAN interface. In a network that uses access layer routing, each switch could be the default gateway for directly connected clients and implement Web-Auth. However, this study guide focuses on solutions that use a core routing switch (or, usually, an HP Intelligent Resilient Framework (IRF) virtual device) that is the default gateway for the entire network. Figure 7-27 illustrates this solution.

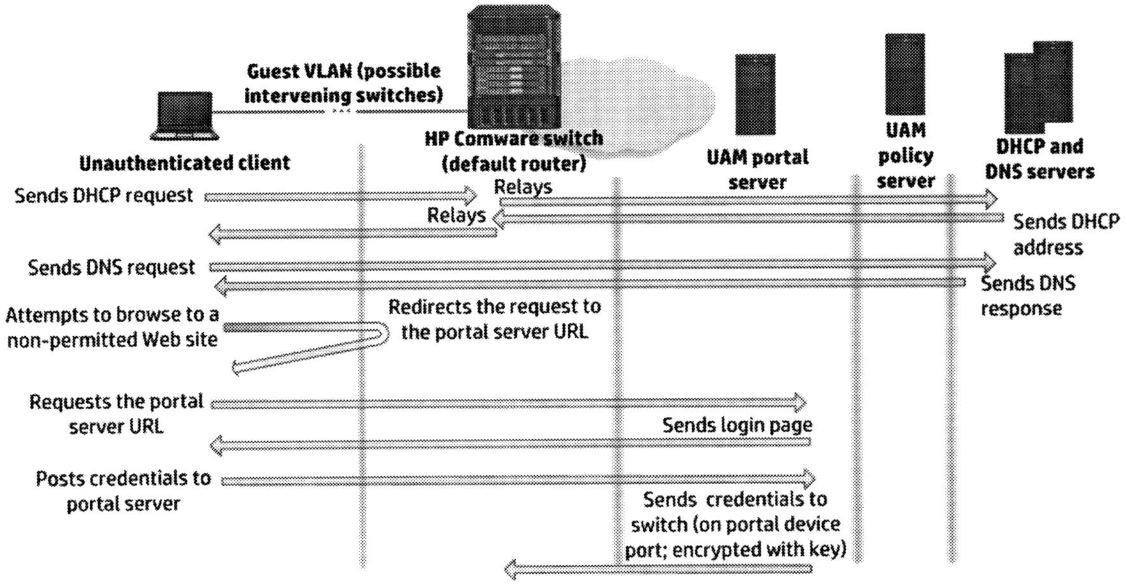

Figure 7-27: Configuring portal authentication using UAM

Note

Because this form of portal authentication operates on the Layer 3 interface, Layer 2 Comware switches do not support it.

First, the client must receive its IP address. When the user opens the browser, the Comware switch will redirect the HTTP or HTTPS traffic to any IP address to the portal server. But, as you see in Figure 7-27, before the client can even send HTTP traffic, it must send a DNS request and receive a DNS response. (Otherwise, it cannot resolve the desired URL's FQDN into an IP address.)

The Comware routing switch drops *all* unauthenticated traffic, so you must take care to ensure that the unauthenticated client can receive the proper DHCP and DNS services. The Comware routing switch will provide DHCP relay services to unauthenticated clients. However, to permit the DNS traffic, you must set up a portal free rule that permits traffic to the DNS server's IP address. (A portal free rule is a rule that selects unauthenticated clients' traffic and permits that traffic rather than drops it. You can create rules that select traffic by source IP address, destination IP address, or both.)

You might want to protect your DNS server by having the Comware switch act as a DNS proxy. To configure it as a DNS proxy, set the IP address of its DNS server, which should be a network server, enable DNS resolution, and enable DNS proxy. Then specify the switch's IP address as the DNS server in the client's DHCP scope. Even though this IP address belongs to the switch, you still need to create a portal free rule to the IP address that you specify as the client's DNS server.

With the appropriate DHCP and DNS services in place, the client can now send HTTP and HTTPS requests. The Comware routing switch, in its role of default router, receives the traffic. It redirects the HTTP and HTTPS traffic to a login URL, which, in this solution, is hosted on the UAM portal server.

When the user submits credentials in this login page, the portal server returns them to the user's default router, which is defined as a portal device. The credentials are secured by a key and sent on a specific TCP port.

Note

UAM's portal server is the supported external web server for Comware switches implementing Web-Auth.

The UAM portal solution is designed to work with Comware switches and routers. However, only the default router for the guest VLAN must be a Comware switch or router. Therefore, you might have ProVision switches functioning at the access layer.

As Figure 7-28 shows, the Comware default router looks up the proper method for authenticating the credentials using its domain and RADIUS scheme settings. The RADIUS server should be defined as the UAM policy server. The UAM server receives the credentials, authenticates them, and returns the Access-Accept to the switch. The user is now authenticated, redirected to a login success page, and allowed to send any traffic.

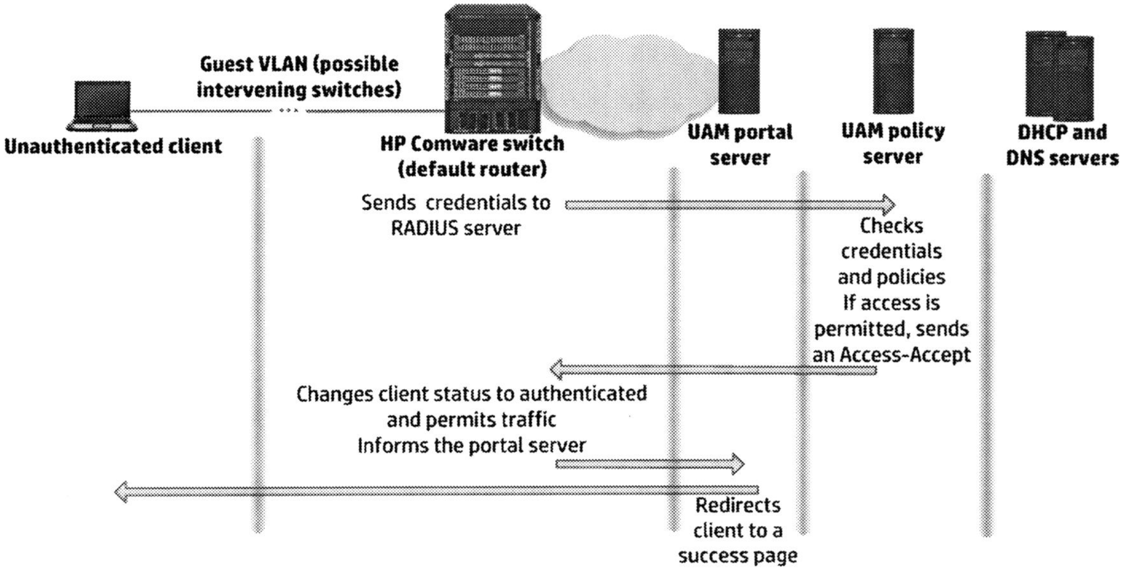

Figure 7-28: Configuring portal authentication using UAM (cont.)

UAM portal authentication: Set up UAM

When you implement Web-Auth on the VLAN interface, the Comware switch intercepts any HTTP/HTTPS requests that it is asked to route and redirects the device to the portal server URL.

The guest device then sends an HTTP or HTTPS request to the UAM portal server. When the portal server receives the request, it must determine the portal access device to which the guest connects. The portal server can then apply the correct policies and direct its communications to the correct switch. The server follows this process:

1. It checks the source IP address of the request and matches it to an IP group.
2. It looks up the portal access device that supports that IP group. This device should be the Comware routing switch (or router) that implements portal authentication (Web-Auth) on this guest client's VLAN.

To complete the minimum setup on UAM, you must follow these steps:

3. Create IP groups. You can assign each IP group to *one* portal access device. In this scenario, you have a single guest VLAN, and you simply create an IP group for it.
4. Define a portal access device, which is the Comware routing switch (or router) that implements Web-Auth. (Note that you must define the switch as a portal access device *in addition to* defining it as a UAM access device.)

You need to choose the correct version. Choose 2.0 if this device is the default router for the guest subnet and 3.0 if another router stands between this device and the guest subnet.

You should specify 2000 for the listening port because this is the default port on which Comware switches listen for messages from the UAM portal server, and you cannot change the port. You also set a password for securing communications, which you will match when you set up the switch.

5. After defining the device, you must create a port group for it. Each port group binds an IP group to the portal access device.

Note

You can bind the IP group to particular ports on the access device if you wanted to set up different IP settings for different physical ports on the router or routing switch. For routing switches, you will generally bind all ports (0 to zzzzz) on the portal access device to the IP group.

UAM portal authentication: Set up the Comware switch

UAM provides the easiest way to set up the Comware switch (or router) that implements portal authentication. Because UAM automatically sets many of the settings that must match, you encounter fewer chances for misconfiguration (see Figure 7-29). You must deploy two sets of configuration:

- Deploy a configuration to the portal device that you defined for this switch.
- Deploy an AAA configuration to the access device that you defined for this switch.

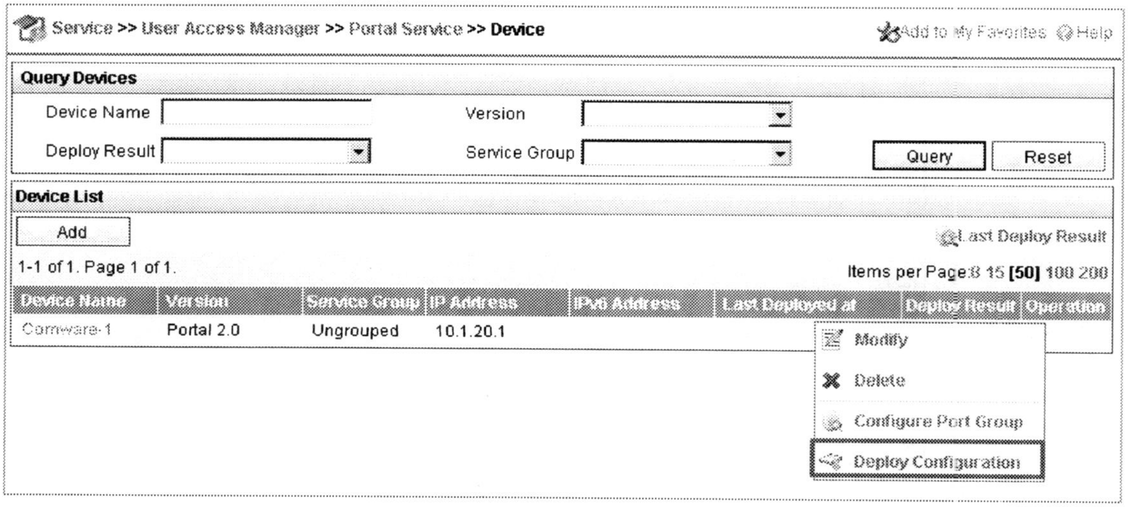

Figure 7-29: Using UAM to configure portal authentication on a Comware switch

When you deploy the settings, UAM prompts you to make various choices. It then formulates the correct CLI commands, which it configures on the switch through Telnet or SSH (depending on the login type specified in IMC).

For your reference, the following sections provide a bit more information about the setup as well as CLI commands for completing the setup on your own.

Defining the portal server

The UAM portal solution's additional features call for close communication between the server and switches implementing Web-Auth. You must define the portal server settings on the Comware switch. These settings not only configure the switch to redirect guests to the correct webpage but also establish parameters for the ongoing communications. Mandatory settings include:

- The portal server's IP address
- The password that the server uses to authenticate its communications
- The URL to which the switch redirects guests

  The CLI does not force you to specify a URL. If you do not, however, the switch redirects the guest to http://<*server IP address*>. Such a redirect does not work for UAM, which hosts the login pages on port 8080 and in the portal folder. You must use this format for the URL:

  http://<*server IP address*>:8080/portal

  Alternatively, you can specify an FQDN, which ensures that the configuration stays up to date even if the portal server's IP address changes:

  http://<*server FQDN*>:8080/portal

  The switch automatically permits DNS messages to pass, so unauthenticated guests can resolve the FQDN.

- You can typically leave the port at the default setting (50100)

  This port defines the port to which the switch *sends* messages to the portal server and does not need to match the listening port that you defined for the portal device in UAM, which is the port on which the switch *receives* messages. UAM listens on port 50100, so you can typically leave this setting.

This CLI command configures these settings:

```
[Comware] portal server <name> ip <IP address>key <password> url <URL>
```

Defining free rules

The UAM portal deployment wizard gives you the option to add free rules, which permit unauthenticated devices to access specified IP addresses. Remember to create such a rule for the DNS server at least.

The free rule can select traffic by one or more of these criteria:

- Destination IP address
- Source:
  - Interface
  - IP address
  - MAC address
  - VLAN

You can also create a free rule in the CLI:

```
[Comware] portal free-rule number { destination { any | ip {IP address mask { prefix length | mask } | any } } | source { any | [ interface interface-type interface-number | ip { IP address mask { prefix length | mask } | any } | mac mac-address | vlan ID ]
```

Enabling Web-Auth on the VLAN interface

After defining the server, you can enable Web-Auth (portal) on the guest VLAN interface. You can choose between three modes. In this scenario, you use direct mode because the switch is the default gateway for the VLAN. This mode corresponds to version 2.0, as defined for the portal device on UAM:

```
[Comware-Vlan-interface<ID>] portal server <name> method direct
```

You would use Layer 3 mode if other network infrastructure devices routed the guest traffic before it reached this switch:

```
[Comware-Vlan-interface<ID>] portal server <name> method layer3
```

You will look at a third mode, **redhcp**, in a moment.

Configuring the RADIUS settings

You configure the RADIUS settings just as you do for a Web-Auth solution that uses local pages. As indicated in Table 7-5 at the beginning of this section, different servers might fulfill various UAM roles. Specify the UAM policy servers' IP addresses in the RADIUS scheme for both the authentication and the accounting server. Also make sure that the NAS ID matches the IP address on which IMC manages the switch. Then apply this scheme for portal (or default) authentication, authorization, and accounting in the appropriate domain.

Important

Accounting plans an important role in the portal solution. The Comware switch must send accounting packets to the server, and if it does not receive a reply, the user cannot get online. You can specify local accounting instead, but in that case, UAM cannot track the user as well and some management functions do not work.

Generally, you can use the default domain—an exception would be if you wanted to authenticate Web-Auth users on different VLANs to different solutions. In that case, remember to specify the desired domain as the portal domain on each VLAN interface that uses portal authentication.

You have examined these CLI commands earlier. For your reference, they include:

```
[Comware]radius scheme <name>
[Comware-radius-<name>] server-type extended
[Comware-radius-<name>] primary authentication <IP address>
[Comware-radius-<name>] key authentication <password>
[Comware-radius-<name>] primary accounting <IP address>
[Comware-radius-<name>] key accounting <password>
[Comware-radius-<name>] nas-ip <IP address>
[Comware-radius-<name>] user-name-format [with-domain | without-domain]
[Comware-radius-<name>] domain <name>
[Comware-isp-<name>] authentication portal radius-scheme <name>
[Comware-isp-<name>] authorization portal radius-scheme <name>
[Comware-isp-<name>] accounting portal radius-scheme <name>
```

Multiple guest VLANs

In this variation, your network features multiple guest VLANs: one for each access layer switch or group of access layer switches. You set up the solution just as you did for the previous scenario with these variations:

- You enable Web-Auth (portal authentication) on multiple guest VLANs.
- You want to define a single portal access device for your core routing switch (IRF virtual device). However, by default, the switch communicates with the portal server using its IP address on the VLAN interface on which it implements Web-Auth. When it implements Web-Auth on multiple VLANs, it uses multiple IP addresses, each of which would require its own portal access device definition. To avoid the complication, you will define the same portal NAS IP address on all of the guest VLAN interfaces. You can then define a single portal access device on UAM that uses that IP address.

- On UAM, you could create multiple IP groups, one for each guest VLAN. This setup enables you to use different portal settings for the different guest VLANs. If you want to apply the same settings, you could create a single IP group that includes the range of IP addresses in all guest VLANs.
- You then bind the IP group or groups to the portal access device as before.

Temporary IP addresses

In this scenario, unauthenticated users receive temporary IP addresses only for the purpose of authenticating. After they authenticate, they receive valid IP addresses for guests. The guests *must* use the iNode client in order for this solution to work.

Set up the solution as indicated in the previous section with these variations:

- On the Comware switch:
  - The Comware routing switch requires two IP addresses on each guest VLAN interface:
    - The primary IP address is the subnet with valid guest IP addresses.
    - The secondary IP address is the subnet for temporary unauthenticated addresses.
  - The Comware switch implements DHCP relay on the guest VLAN interface, and the DHCP server has a scope for both subnets. The scope for the unauthorized guest addresses should have a very short lease (such as 10 seconds). You must set the address check option for DHCP relay.
  - When you implement Web-Auth on the VLAN interface, you set the mode to redhcp.
- On UAM:
  - The UAM portal server needs to know the IP addresses for unauthenticated guests and the addresses for the guests after they authenticate. When you set up the IP group, you specify the *unauthenticated* guest addresses. You then select Re-allocate IP for the Action and specify the postauthentication guest IP addresses for the After-action IP addresses.
  - When you define the port-access device, you must specify **Yes** for **Re-allocate IP**.

Summary of Web-Auth options

You have examined several scenarios for Web-Auth, which are summarized in Figure 7-30.

Switch	Enforcement point	Web server	RADIUS server	DHCP server	Temporary IP addresses?
Comware	Physical port	Internal	Internal or external	External (internal supported but rare)	Requiresmanual set up (short DHCP lease for static VLAN; dynamic VLAN for authorized guests)
	Layer 3 interface	IMC UAM	IMC UAM	External (internal supported but rare)	Optional (iNode client required)
ProVision	Physical port	Internal	External	External + internal for unauthorized	Yes
	Physical port	External	External	External + internal for unauthorized	Yes

Figure 7-30: Summary of Web-Auth options

Comware: Customizing MAC-Auth and Web-Auth connections

In the final section of this chapter, you will learn how MAC-Auth and Web-Auth handle dynamic settings, VLAN assignments, and concurrent clients. You will also learn how to apply multiple authentication methods to the same port.

You learned how to customize 802.1X client connections in several ways. You also learned that the port can support multiple clients at the same time, if required. MAC-Auth and Web-Auth provide many of the same options, but you do need to keep some differences in mind.

Local Web-Auth (portal) and MAC-Auth support an Auth-Fail VLAN, which applies to all clients that do not pass that form of authentication. With MAC-Auth, the switch determines whether the device passes or fails authentication as soon as the device starts sending traffic. With Web-Auth, the device is initially assigned to the port's static VLAN. The Auth-Fail VLAN only applies after the user is redirected to the browser and fails the login.

To use this Auth-Fail VLAN, you must set the interface to hybrid mode and enable MAC-based VLANs—even if only client will connect at a time.

For Web-Auth, you might find that, in practice, it is easier to use free rules and self-service portals to meet the same needs.

You can also customize the RADIUS policies, such as service policies on UAM, used for MAC-Auth or Web-Auth for a variety of use cases. You might create rules for the specific times and locations at which guests are allowed to connect. You can also apply dynamic settings such as dynamic VLAN assignments, ACLs, and rate limits. For example, you might want to set up a rate limit in the guest service policy so that guests cannot consume an unfair amount of bandwidth.

ProVision: Customizing MAC-Auth and Web-Auth connections

ProVision switches can also implement MAC-Auth and Web-Auth to a RADIUS server that applies specific conditions for access or sends specific dynamic settings for individual users and user groups.

However, you should keep in mind some limitations for ports that must support multiple concurrent clients.

ProVision switches do support user-based mode for Web-Auth and MAC-Auth (set the client limit higher than 1). However, they do *not* support MAC-based VLANs on ports that implement these forms of authentication. In other words, the latest client's VLAN assignment applies to all clients. All clients that might connect to the port should have the same assignment; otherwise, behavior will be inconsistent.

Therefore, dynamic VLANs for MAC-Auth and Web-Auth let the switch adapt to different clients that connect *at different times*. Think about why you want to use dynamic VLANs. If you are just trying to place authenticated guests in a different subnet from unauthenticated ones, the temporary Web-Auth addresses already achieve that effect. The dynamic VLANs meet your needs when different types of users or devices authenticate to a port. You do not know which devices will connect to which ports—but only one device connects at time. If multiple devices might connect through the port enforcing MAC-Auth or Web-Auth, ensure that the same VLAN is assigned to all of them.

Comware: Using 802.1X and Web-Auth on the same port

Earlier in this chapter, you saw how unauthenticated-client/guest VLANs can provide limited access for users who cannot perform 802.1X. Sometimes, though, you do not want those users to simply receive access without logging in. You might want to restrict access to authorized guests who know a password, or you might simply want the guests to register their presence in the network.

Of course, the company might have a guest solution within the unauthenticated-client/guest VLAN that provides these services. For companies without such a solution, you can configure an edge port so that it supports either an employee, authenticating with 802.1X, or a guest, authenticating with Web-Auth. (In the more rare circumstances in which multiple clients connect at once, a Comware switch port supports multiple clients authenticating in different ways.)

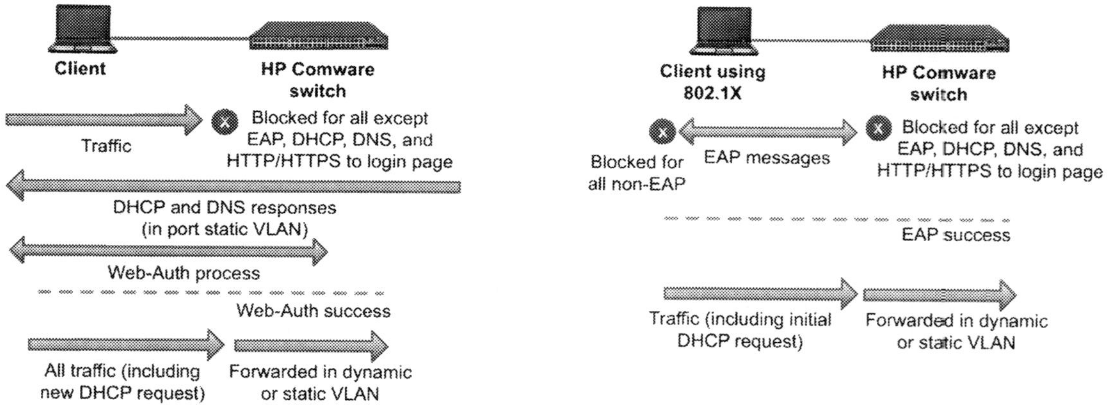

Figure 7-31: Using 802.1X and Web-Auth on the same Comware port

Figure 7-31 shows how a Comware switch port handles authentication in this scenario. As you see, in addition to the EAP messages, the switch allows traffic necessary for Web-Auth to occur. A user can authenticate with either method. Theoretically, the user of a client with a validly configured 802.1X supplicant could open the web browser before the supplicant completes 802.1X. In that case, the user would be redirected to the Web-Auth login page just like a client without a valid 802.1X supplicant. Typically, though, the 802.1X process completes quickly and in the background.

Of course, the user of the client without a properly configured 802.1X supplicant will only succeed to authenticate with Web-Auth.

ProVision: Using 802.1X and Web-Auth on the same port

The scenario and solution for ProVision switches enforcing 802.1X and Web-Auth on the same port is very similar to that for Comware switches (see Figure 7-32). Keep in mind that the ProVision switch supplies the DHCP and DNS services for the unauthenticated client.

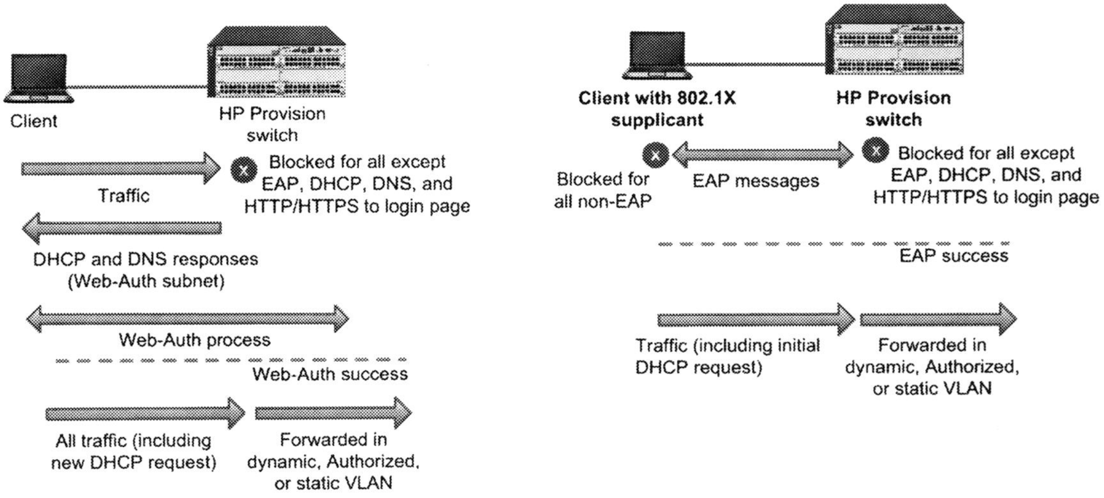

Figure 7-32: Using 802.1X and Web-Auth on the same ProVision port

Guidelines for using 802.1X and Web-Auth on the same port

Keep these guidelines in mind when you enable both 802.1X and Web-Auth (portal authentication) on the same port:

- For Comware switches, the port must operate in MAC-based mode (even if you anticipate only one client connecting at a time). Otherwise, the 802.1X authentication has no effect on the device's status, and the user must pass Web-Auth. For ProVision switches, on the other hand, the port can use a client limit of 1, and the 802.1X and Web-Auth authentication still interoperate properly. (Of course, you can also use a higher client limit and user-based mode if you want.)
- Take steps to ensure that the port adapts to provide the correct level of access to the currently connected client.

  On the ProVision switches, clients without 802.1X supplicants blocking non-EAP messages receive IP addresses in the Web-Auth subnet. After a guest completes Web-Auth authentication, the authenticated guest receives an IP address in the VLAN assigned by the RADIUS server. Or, if the server does send a dynamic assignment, the guest is placed in the untagged VLAN statically configured on the port.

  As long as a client completing 802.1X authentication is blocking non-EAP messages, the client does not receive an IP address until the user authenticates. At that point, it receives an IP address in the dynamic VLAN assigned by the RADIUS server or the port VLAN, if no dynamic VLAN is assigned.

  As a simple setup, set the static untagged VLAN on the port to ID of the VLAN that you want to use for guests. To ensure that the 802.1X users receive different access rights, have the RADIUS server assign 802.1X users to the appropriate VLAN or VLANs.

  On ProVision switches, unauthenticated guests and guests who have completed authentication still receive different IP addresses due to the temporary Web-Auth subnet. On Comware switches, as you recall, you must take further steps if you want this behavior. You might do so if you do not want to place unauthenticated guests in the same subnet as authenticated guests for security and traffic management purposes.

  Also consider whether clients using 802.1X will actually block non-EAP messages. If they do not, using the VLAN for authorized guests as the port's untagged VLAN can cause issues for the 802.1X clients on Comware switches. They receive IP addresses in the static VLAN and their VLAN then changes. In this case, you should reserve the static VLAN for assigning a temporary IP address with a brief lease to an unauthenticated device, which might belong to a guest or an employee. After the employee or guest is authorized and placed in the correct VLAN, the device receives an IP address with a longer lease in that VLAN's subnet.

  For the Comware switches, you can set up multiple VLANs even when multiple devices connect at the same time. For example, a Comware switch port connects to an unmanaged switch in a conference room. You want the switch port to support 802.1X for employees and Web-Auth for guests, *and* you want multiple users to connect to the port at the *same* time. You must make the port a hybrid port and enable MAC-based VLANs. Then you can follow the same guidelines you already discussed. Set the PVID to a VLAN reserved for unauthenticated clients. Create dynamic VLAN assignments on the RADIUS server. On the port, permit untagged traffic in these VLANs.

As a final note, remember that, because ProVision switches support only one untagged VLAN on a port that uses Web-Auth, a guest and an employee could not simultaneously use the port in different VLANs. Therefore, this solution works best when the port supports one client at a time. However, you might be able to safely place employees and guests in the same VLAN if you apply different dynamic ACLs to them.

- Although you can set up a guest/unauthenticated-client VLAN, you generally do not need to. If you simply want to allow employee and guest access on the same port, the 802.1X and Web-Auth solution already does that. These VLANs would add a layer of complexity.

Using 802.1X, Web-Auth, and MAC-Auth on the same port

You can configure a switch port to provide authentication for any of several types of devices:

- Employees authenticating with 802.1X
- Guests authenticating with Web-Auth
- Headless devices (such as printers) authenticating with MAC-Auth

The setup is similar to the previous scenario but adds MAC-Auth to the combination. A port that implements MAC-Auth with another form of authentication immediately uses the device's MAC address to submit an access request, just as it would if MAC-Auth were the only authentication method:

- If MAC-Auth succeeds, the device is authorized for network access with the settings returned by the RADIUS server.
- If MAC-Auth fails, the device still has the option to pass another form of authentication (802.1X, Web-Auth, or either, depending on what is enabled on the port). The user is authorized with the settings returned by the RADIUS server.

Just as in the previous scenario, you should ensure that the port operates in user-/port-based mode.

Make sure that the RADIUS server has the correct accounts and policies for providing MAC-Auth just for the headless devices that require it. You want clients capable of 802.1X to fail MAC-Auth and perform 802.1X authentication instead. You also need to ensure that each type of device and user receives the correct VLAN assignment. For example, you might have the RADIUS server assign MAC-Auth clients to different VLANs from 802.1X users and Web-Auth users.

Summary

In this chapter, you explored appropriate situations for implementing various 802.1X, MAC-Auth, and Web-Auth solutions. You know when and how to set up the basic solutions, and you can also create more advanced solutions that customize ports dynamically, for one or more users, and that combine various forms of authentication.

Learning check

Answer these questions:

1. What are the three components or roles in 802.1X authentication? Give a brief description of each role.
2. A UAM access user requires a service policy. What does the service policy define?
3. What are some of the disadvantages of MAC-Auth as compared to 802.1X?
4. What are the general steps in the Web-Auth process?

Answers

1. What are the three components or roles in 802.1X authentication? Give a brief description of each role.

   - Supplicant—The supplicant requests access to the network and proves that it deserves this access by authenticating—typically in response to a challenge from the far end of the connection (for example, the switch port). The supplicant may also initiate the authentication process on its own behalf (by transmitting an EAP-Start packet).
   - Authenticator—The authenticator controls network access, forcing a supplicant to authenticate before it can send any non-EAP traffic over the connection. The authenticator initiates the authentication process but then relays authentication messages between the supplicant and the authentication server. After the authentication is completed, the authenticator decides how to control the connection.
   - Authentication server—The authentication server makes decisions about whether or not users can access the network. The server also submits its own credentials to the supplicant. In essence, the supplicant and the server authenticate each other, although the authenticator always acts as a proxy in this process.

2. A UAM access user requires a service policy. What does the service policy define?

   The service policy includes the access policies to tell UAM how to authenticate the user and what authorization attributes to apply.

   In more detail, the service policy defines these settings as follows:

   - The service policy's access policy list maps conditions for access requests to an access rule and sometimes to a proprietary attribute policy.
   - If an access request does not match a condition, the default access rule and proprietary attribute policy apply.
   - The access rule defines the type of authentication such as PEAP-MSCHAPv2 or PAP or CHAP. It also defines authorization attributes like user VLAN, rate limits, and ACLs. (It can also bind a user to a specific device and apply other settings not covered in this study guide.)
   - The proprietary attribute policy applies authorization settings that are vendor specific.

3. What are some of the disadvantages of MAC-Auth as compared to 802.1X?

   It is easy for a hacker to discover MAC addresses used in your system whether by eavesdropping or accessing a legitimate device. The hacker can then easily spoof an authorized MAC address. MAC authentication can also be more difficult to manage, requiring directory or domain administrators to create accounts for each MAC address on each device.

4. What are the general steps in the Web-Auth process?

    a. The unauthenticated client connects.
    b. Unauthenticated users receive DHCP and DSN services.
    c. The unauthenticated client is redirected to a login page.
    d. The guest submits credentials in the login page.
    e. The guest is authenticated and authorized for access. The switch allows the authenticated guest traffic.

8 Quality of Service

EXAM OBJECTIVES

In this chapter, you learn to:

- ✓ Configure HP switches to honor the appropriate quality of service (QoS) marks applied by other devices.
- ✓ Create a QoS policy that assigns a specified class of traffic to a priority queue.
- ✓ Select and implement an appropriate strategy for queue scheduling.
- ✓ Implement traffic policing policies that enforce the negotiated committed information rate (CIR), committed burst size (CBS), peak information rate (PIR), and excessive burst size (EBS) for a specified class of traffic.
- ✓ Respond to congestion in advance by applying the appropriate traffic shaping and Weighted Random Early Detection (WRED) policies.
- ✓ Determine the QoS mark that an HP switch will assign to specific outbound traffic and, if necessary, adjust the mark.

ASSUMED KNOWLEDGE

You should understand how switches assign incoming packets to a VLAN, placing the VLAN in the packet's descriptor.

INTRODUCTION

In contemporary enterprise networks, from the branch office to the campus LAN through the network core and into the data center, a complex mixture of network applications share links. HP switches, as key components of HP FlexNetwork solutions, meet the needs of heterogeneous traffic flows.

You can implement QoS mechanisms to improve the experience of users running a variety of network applications—from traditional ones to voice and video applications.

Dealing with congestion

The chapter describes how to implement the many sophisticated QoS features that HP switches support. To implement these features successfully, however, you must first understand their purpose. That is, you should know what problems these QoS features attempt to solve.

In many environments, uplinks are oversubscribed; that is, they provide less bandwidth for outgoing traffic than the total amount of traffic that could be incoming from other devices—often many times less. Usually, endpoints consume a small fraction of their bandwidth, so oversubscription makes sense. However, on a moment-to-moment basis, fluctuating traffic patterns and bursts of traffic can cause congestion. The lower-bandwidth link acts as a bottleneck, and if traffic continues to arrive more quickly than the port can forward it, the port's buffers fill, and it drops traffic.

As you have learned in Chapter 1, the FlexNetwork architecture works to eliminate common sources of congestion. For example, it reduces oversubscription and latency by eliminating the aggregation layer. Nonetheless, even a well-designed network includes oversubscription and the possibility of temporary bursts of congestion, which can wreak havoc on the quality of time-sensitive applications. Network infrastructure devices must implement QoS mechanisms to protect this sensitive traffic and ensure that users have a positive experience running all necessary applications.

QoS mechanisms also help companies to enforce business policies that prioritize certain applications or segments of the network.

TCP and UDP response to congestion

In general, "best-effort" delivery is satisfactory for data traffic. TCP applications are not usually disrupted by packet loss because TCP provides acknowledgment and flow-control mechanisms that ensure dropped packets will be retransmitted. Each segment of a TCP/IP message includes sequence numbers that inform receivers of the proper order and number of packets in the transmission. The receiver in a TCP transmission acknowledges packets by returning messages that include the highest sequence number it has received. TCP applications include a "window" parameter that determines how many bytes can be outstanding and unacknowledged by the receiver at any given moment. This parameter limits the number of packets that can be acknowledged by a single control packet. Delay in delivery of packets or excessive packet loss prompts the sender and receiver to establish a smaller window size. Furthermore, TCP applications back off when network congestion is detected, enabling applications to share bandwidth.

UDP, on the other hand, does not provide acknowledgments or flow control, which enables UDP applications to generate less overhead than TCP applications. Some UDP applications include their own error-checking procedures. However, in most cases, UDP applications are designed to minimize the need for retransmission and use UDP as a transport mechanism precisely because of its lower overhead and network efficiency. UDP also does not provide any backoff procedures.

Because UDP applications do not respond to congestion in any way, they can continue to contribute to congestion, which can also have drastic effects on time-sensitive real-time traffic such as voice and video.

Impact of congestion on real-time traffic

Voice traffic is sensitive to delay, sometimes called latency, which is the amount of time that passes between the sending of a transmission and its arrival at the receiving station. In Figure 8-1, the rectangles represent packets traveling from sender to receiver. In both examples, the packets are sent at a consistent rate.

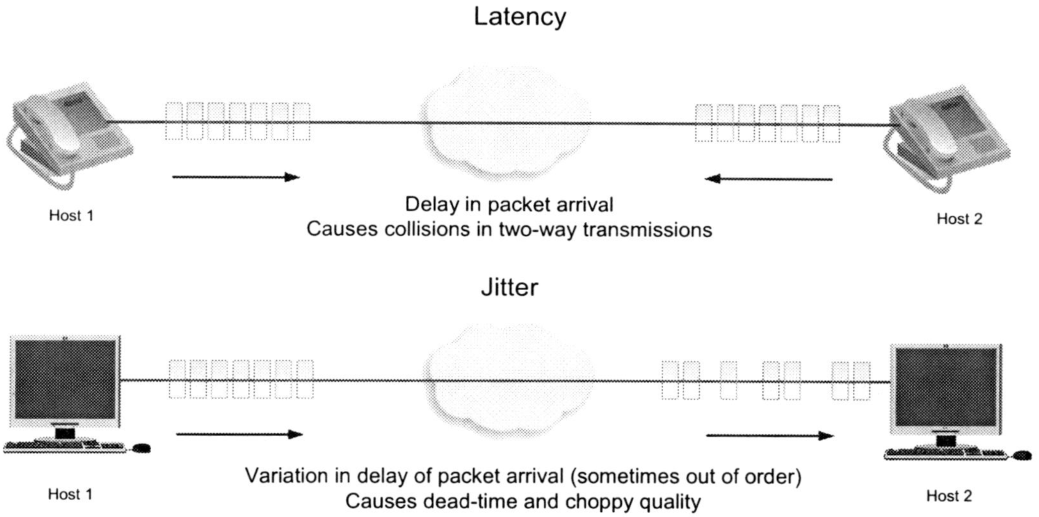

Figure 8-1: Impact of congestion on real-time traffic

In the delay example, latency within the network cloud inevitably causes some elapsed time between transmission and reception. In the case of unidirectional traffic (as video traffic often is), this predictable, measurable period of time is usually tolerable, regardless of its length. However, because multidirectional voice traffic is nearly always interactive, excessive delay can result in conversation collisions. For example, in the diagram, Host 1 begins transmitting in the belief that Host 2 is not transmitting. Because Host 2 is still transmitting, bits from Host 2 begin arriving before Host 1 concludes its transmission. The conversations overlap.

Voice traffic is also sensitive to jitter, which is the variation in intervals between the arrivals of packets or, in other words, the difference between when a packet is expected and when it actually arrives. Video is also sensitive to jitter.

Figure 8-1 shows how Host 1 sends data at a constant rate to Host 2. However, devices within the cloud that represents the network experience varying levels of congestion that result in some packets being delayed. The outcome is a variation in the interval between packet arrival, which results in a choppy voice or video stream.

Jitter can have significant effect on real-time applications even if they require very small data streams. For instance, an IP telephone sends a very small packet every 20–30 milliseconds (ms). In order for the receiver to properly interpret the stream, the packets must arrive at the same rate. If the interval between packets grows to 50 ms, for instance, the transmission will be unsatisfactory for many users.

Supplemental information: Delay over various media

Delay is the interval between the time that a message leaves the sender's mouth and the time that it reaches the receiver's ear. Some delay is inevitable, especially in transmissions over significant distances. While the speed of light in a vacuum is 186,000 miles per second, signals move through fiber or copper cable at approximately 125,000 miles per second. Consequently, a fiber network stretching halfway around the Earth introduces a delay of approximately 70 milliseconds under the best of circumstances. During periods of congestion, the delay can be much longer.

The effects of delay upon application performance depend heavily upon the expectations of users. The International Telecommunications Union (ITU) recommends no more than 150-ms end-to-end delay. Greater delay can cause conversation collisions when each party begins to speak because the line appears quiet. Most users will find this situation to be unacceptable.

However, under some circumstances, users can tolerate higher levels of delay. For instance, the delay over a satellite link can be as high as 500 ms, including 250 ms for the sender's signal to reach the satellite and another 250 ms for the signal to reach the receiver. Frequent users of satellite services, including news organizations, adjust to this delay as a matter of routine.

There is less tolerance for loss when it comes to storage services. Fibre Channel, for example, is inherently a lossless protocol, and as such, priority flow control is one of the requirements for a switch to be able to forward Fiber Channel over Ethernet (FCoE) frames. Contemporary networks must ensure lossless delivery for converged storage traffic.

Note

A discussion of the data center bridging (DCB) technologies that accommodate the sensitivity of storage area network (SAN) traffic to packet loss and latency is outside the scope of this study guide.

Implications and challenges

Think about what you have learned about congestion and its implications. Consider the following questions:

- Which types of applications are most sensitive to congestion, and what is the effect of congestion on the experience of using those applications?
- What could network infrastructure devices do to improve these applications' performance?

You should identify traffic such as:

- VoIP traffic is sensitive to loss, latency, and jitter:
  - If VoIP traffic is queued in a buffer while other traffic is forwarded first, the voice quality will suffer.
  - Similarly, if the VoIP traffic is buffered while other traffic is forwarded first, the latency depends on the congestion at the moment. This behavior can introduce jitter.
  - VoIP often uses UDP, which is more sensitive to packet loss than TCP. If too many packets are dropped during congestion, the voice QoS suffers. (In any case, the latency requirements mean that the source cannot retransmit lost packets without disrupting the voice quality.)
  - In short, you need some way to ensure that switch ports put VoIP traffic at the front of the line rather than buffering it until it is its turn.
- Other real-time applications, including voice conferencing and collaboration applications, have similar needs.
- TCP applications can back off when packets start to be dropped, reducing congestion. However, relying on the applications' backoff mechanisms does not always produce good results. (You will learn more about TCP synchronization in the "Problem with traditional tail drop method" section.)
- When congestion occurs beyond a port's egress buffering capabilities, the port drops all packets. But a network might be carrying some applications that are mission critical and other less important applications. You would want to ensure that the mission-critical applications can use the bandwidth while traffic for less important applications is dropped.
- The network infrastructure should be able to determine which traffic belongs to mission-critical applications. It can then allocate bandwidth to them or prioritize forwarding them.

You could read entire books on the subject of the challenges posed in delivering the best QoS for a heterogeneous mix of network applications; probing all the design principles is beyond the scope of this study guide. However, you should understand some basic implications:

- All traffic does not require the same handling to receive an appropriate level of service. For example, many TCP applications can tolerate a degree of degradation of service because TCP provides for retransmission of packets and otherwise handles congestion well. In particular, applications such as file transfer applications can experience momentary periods of poor service because the user cares only about the end result.
- Time-sensitive traffic requires prioritization over other traffic so that it is transmitted immediately. This prioritization ensures that traffic transmitted at a constant rate by an endpoint such as an IP phone continues to be transmitted at a constant rate throughout the network, minimizing jitter.

- Less important UDP traffic must be prevented from monopolizing a link because the application will not back off on its own.
- In general, an enterprise might wish to enforce rate limits that ensure certain classes of traffic use only their fair share of bandwidth.
- Congestion can cause widespread degradation of service as all packets are suddenly dropped—particularly for non-TCP traffic. Users have a better experience when the network infrastructure responds to congestion gracefully by dropping less important or sensitive traffic.

Queuing and scheduling

You will now learn about how HP switches can prioritize traffic into various groups for transmission or queues. As Figure 8-2 illustrates, queue scheduling enables switches to provide high-priority traffic—particularly voice and video traffic—the priority forwarding that it requires. HP switches support a variety of techniques for queue scheduling, which you will examine in this section.

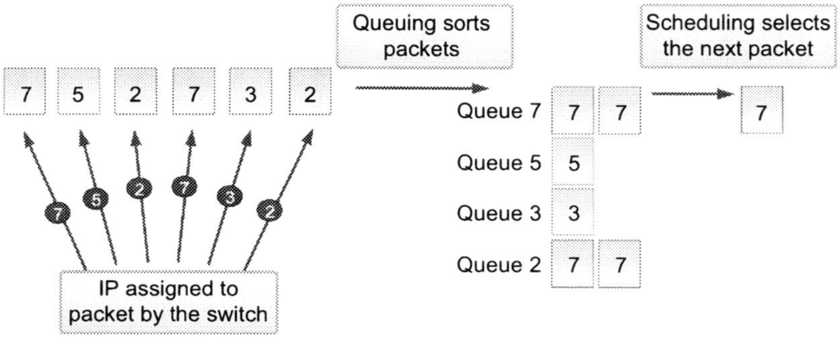

Figure 8-2: Queuing and scheduling on HP switches

As a switch prepares to transmit traffic, it places the traffic in an egress queue for the egress port. (The buffer size determines how much traffic the switch can buffer, and you can find the specifications for your switch in its documentation.)

A switch port that does not implement QoS places all packets in the same egress queue in the order in which they arrive. It follows a simple mechanism, called first in, first out (FIFO), to schedule the packets for transmission. It forwards the first packet that arrived, then the next, and so forth.

As you have learned, however, some packets need priority service. For example, voice packets, transmitted at a constant rate by the endpoint, must be delivered as soon as they arrive to maintain that constant rate (avoiding jitter). To prioritize particular traffic, the switch implements priority queuing, in which it sorts various packets into multiple queues. HP switches support up to eight egress queues per port.

Once a switch has queued traffic in multiple, priority-based queues, the port's scheduling mechanism selects the next packet for forwarding. Scheduling works in tandem with queuing, ensuring that packets in high-priority queues are transmitted more promptly than packets in other queues. HP switches support a variety of scheduling mechanisms, which you will also examine in this chapter.

Overview of priority-based QoS protocols

To communicate the desired priority for traffic between network infrastructure devices, the devices use class of service (CoS) protocols. These protocols define specific values, which request specific priorities for traffic marked with that value. As you will see, HP switches can take these values into account when determining how to queue traffic.

The values reside within frame or packet headers and can be passed on with the traffic from device to device. Ethernet LANs carrying IP traffic typically support two QoS protocols:

- 802.1p, which defines values carried in the Layer 2 802.1Q VLAN tag
- Differentiated Services Code Point (DSCP), which defines values carried in the type of service (ToS) field of an IP header

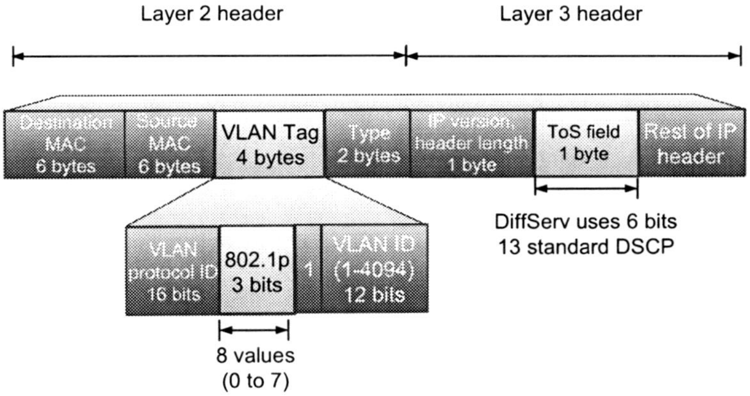

Figure 8-3: Priority-based QoS protocols

Layer 2: 802.1p

The 802.1p standard uses a 3-bit field in the 802.1Q VLAN tag to indicate priority (see Figure 8-3). It defines eight values between 0 and 7. Note that because the VLAN tag carries the priority value, untagged traffic cannot use 802.1p to request a specific priority level. Although Figure 8-3 shows an Ethernet frame that encapsulates an IP packet, 802.1p is a strictly Ethernet protocol; the frame can encapsulate any Layer 3 protocol supported by Ethernet.

Layer 3 protocols

Some QoS protocols function at Layer 3.

ToS and IP precedence bits (deprecated)

As Figure 8-3 illustrates, every IP packet includes an 8-bit field called the ToS field. As originally defined, the ToS field contained two subfields—a 3-bit IP precedence field used for relative prioritization and a 4-bit subfield for the specific ToS desired by an individual packet. The remaining bit, bit 7, is called the must be zero (MBZ) bit and is unused.

The three IP precedence bits in the most significant position were to be used to provide eight levels of precedence or priority. The levels are labeled 0 to 7, with 7 being the highest priority and 0 the lowest. The values 6 and 7 are reserved for "Internet" and "network" uses, respectively. This enables routing updates and other critical traffic to have a higher priority than user-generated traffic. The highest user-defined value is 5, which represents "critical" priority.

The four ToS bits were designed to allow applications to instruct routers to choose routing paths with one of the following four characteristics: minimized delay, maximized throughput, maximized reliability, and minimized cost. Only one of the four bits could be "turned on," that is, set to 1.

Differentiated Services (DiffServ) (Current)

Seeking to improve IP ToS, in 1998 an Internet Engineering Task Force (IETF) working group redefined the use of the ToS header in Request for Comments (RFC) 2474. Under the new definition, the first six bits define 64 Differentiated Services (DiffServ) "code points" (DSCPs). Instead of defining priority relative to each other, the 64 DSCPs are intended to define distinct forwarding behaviors, or per-hop behaviors. Any one of 64 values may appear in a 6-bit field; however, RFC 2597 defined 12 Assured Forwarding (AF) DSCPs. These 12 DSCPs break down into four classes indicated by the first three bits (class 1 has the lowest priority and class 4 the highest). Each class includes three DSCPs, indicated by the second three bits, which request a drop precedence (dp).

RFC 2598 defines an Expedited Forwarding (EF) DSCP that is intended to produce the lowest levels of jitter and delay. Table 8-1 indicates the 13 defined DSCPs. You will see later in this chapter how HP switches map these DSCPs to priority queues.

Administrators are free to use the remaining 51 DSCPs in any way they choose. By default, the undefined DSCPs are mapped to traffic class 0 and thus receive normal priority. However, they can be mapped to other traffic classes as well.

Under the original DiffServ definition, two bits of the 8-bit ToS field were initially allocated as "unused." In 2001, however, RFC 3168 redefined the two unused bits as Congestion Experienced (CE) bits, meaning that routers can use them to signal other devices when they experience congestion. If the CE bits are not in use, a router that experiences significant congestion that causes buffer overflows will drop newer packets rather than place them in its queues. This signals end stations that they should use a smaller TCP window.

Table 8-1: AF and EF DSCPs

PHB	AF class	AF drop precedence	Decimal DSCP
AF11	1 001	Drop last 010	10
AF12	1 001	Drop second 100	12
AF13	1 001	Drop first 110	14
AF21	2 010	Drop last 010	18
AF22	2 010	Drop second 100	20
AF23	2 010	Drop first 110	22
AF31	3 011	Drop last 010	26
AF32	3 011	Drop second 100	28
AF33	3 011	Drop first 110	30
AF41	4 100	Drop last 010	34
AF42	4 100	Drop second 100	36
AF43	4 100	Drop first 110	38
EF	Requests highest priority for lowest jitter and latency		46 (101110)

DSCP compatibility with IP precedence

DiffServ intentionally preserves the three high-order bits for traffic class.

If some applications look for only the three high-order bits, they will recognize the IP precedence value. An application or device that is using the DiffServ interpretation will interpret those same three bits from its own perspective. Applications and devices will arrive at the same relative priority, although the actual values are different.

Methods for prioritizing and queuing traffic

You can follow two strategies for configuring your HP devices to divide traffic into priority queues for transmission:

- Honoring QoS marks set by other devices
- Classifying traffic

Honoring QoS marks

As you have learned, devices can mark traffic with QoS values, which request a specific priority level for that traffic. HP switches can recognize these marks and honor them by assigning the traffic to the corresponding queue; however, you must configure Comware switches to do so.

You should implement this strategy on switch ports that connect to trusted devices—such as servers—that run applications that are capable of selecting and marking their own priority. You should also implement this strategy on switch-to-switch ports (or router-to-switch ports) when the connecting device provides the intelligence, classifying traffic and inserting the correct QoS mark.

Classifying traffic

HP switches can also classify traffic according to a variety of criteria on multiple OSI layers. Then, based on those criteria, they can map the traffic to the desired outbound queue. As you will learn later, they can also mark the traffic with a QoS value, requesting that other devices in the traffic path implement a similar prioritization strategy.

Classify traffic locally when you want this local switch to control precisely how it queues traffic.

Generally, you should configure QoS classification as close to the edge of a domain as possible, enabling the domain to implement consistent controls for traffic across its entire journey through the domain. Ideally, you should deploy intelligent switches, capable of making these decisions, at the server and endpoint access layer. The switches can then classify the traffic correctly from the beginning of its journey. Switches at the edge of a service provider network might also use this strategy to control and classify incoming customer traffic.

Of course, as part of your QoS configuration, you must examine downstream switches and other network infrastructure devices, ensuring that they are prepared to honor the QoS marks.

Mix of strategies

Note that, although for clarity this section will describe how to implement each strategy separately, you can implement both strategies on the same switch. For example, you could configure ports that connect to servers to honor QoS marks set by some applications and also create policies that classify traffic from other applications—or override the honored priorities according to more important business policies as interpreted by QoS policies.

HP Comware: Honoring 802.1p

You will now explore the strategy in which the switch honors QoS marks set by other devices. You will learn how to ensure that HP Comware and HP ProVision switches honor the values used in your environment, beginning with the Comware switches.

To understand the Comware behavior, you must understand several important principles:

- Local precedence (lp)
- QoS trust on a port
- QoS maps

lp

Comware switches use a setting called local precedence (lp) to determine the queue to which packets belong.

You know how a Comware switch assigns incoming packets to a VLAN, placing the VLAN in the packet's descriptor. Similarly, as each frame arrives, the switch assigns the enclosed packet its lp.

Eventually, when the switch prepares to transmit the traffic, it uses the lp to assign the packet to a queue. On a switch with eight queues (0 to 7), an lp of 7 places the packet in queue 7 (highest priority), an lp of 6 places the packet in queue 6 (next highest priority), an lp of 0 places the packet in queue 0 (lowest priority), and so forth. The switches can derive the lp in several ways, which you can customize according to the particular QoS strategy that you are pursuing.

A packet's lp is locally significant. That is, each Comware switch determines a packet's lp on its own and does not communicate this value to other network infrastructure devices. This value essentially enables the switch to track the queue to which it wants to assign a packet according to the rules that you have configured on it.

Note

For convenience, this chapter will sometimes refer simply to assigning packets to a queue. You can infer from such a statement that the Comware switch actually assigns the packet to an lp, which maps to the queue.

QoS trust on a port

On Comware switches, you must explicitly configure switch ports to trust QoS marks on incoming traffic. Otherwise, the port ignores any 802.1p value or DSCP and does *not* use the value to assign traffic to a queue.

Let us assume that your switch port connects to a trusted endpoint or network infrastructure device, and you want to honor the 802.1p priority that it sets on frames. You access the interface view and enter qos trust dot1p.

When you configure the port to trust 802.1p, the port ignores any DSCP included in the IP header. However, it leaves any incoming DSCP intact in the outgoing packet, just as it leaves the incoming 802.1p value intact in outgoing tagged frames. Another device in the path can use either value to prioritize the packet.

Important

As you have learned, 802.1p values are carried in 802.1Q or VLAN tags. Therefore, the port must support tagged traffic in order for frames to provide the 802.1p value. Any untagged frames are assigned the port priority (default, 0).

QoS maps

Once you have allowed the port to trust 802.1p values, the switch maps the 802.1p value in incoming traffic to the lp, which, as you have learned, ultimately places the outgoing packet in the corresponding queue.

Figure 8-4 shows how most Comware switches map 802.1p values to priority queues. Value 7 is reserved for network control traffic. The highest value that an application can request is 6. The lowest values, intended for high-bandwidth but delay-tolerant applications such as file transfers, are 1 and 2. The 0 value, for normal traffic and non-802.1p-capable devices, requests best-effort service. As you see, the 802.1p value 0 does *not* map to the lowest-priority queue, but rather to the third lowest (lp 2).

Comware queue description	802.1p value	Default queue (dot1p-lp map)
Best effort	0 (000)	2
Background	1 (001)	0
Spare	2 (010)	1
Excellent-effort	3 (011)	3
Controlled-load	4 (100)	4
Video	5 (101)	5
Voice	6 (110)	6
Network-related	7 (111)	7

Figure 8-4: Comware: Honor 802.1p

Note

The latest 802.1Q standard recommends a slightly different mapping: Values 6 and 7 are reserved for network control traffic, and the highest value that an application can request is 5. You should check the configuration guide for your precise switch model and see how it maps the values. You should also determine whether the switch lets you change these mappings and whether you need to resolve any discrepancies in the planned priority values or between different switch models.

You can alter the queues associated with specific 802.1p values by configuring the dot1p-lp map:

1. Enter this command to access the map view:

```
[Switch] qos map-table dot1p-lp
```

2. Alter the mapping as you desire:

```
[Switch-maptbl-dot1p-lp] import <802.1p value> export <lp>
```

Note

The Comware switch also maps the 802.1p value to a number of other settings, which you will examine later in this chapter.

HP ProVision: Honoring 802.1p

In general, HP ProVision switches honor 802.1p values in a similar manner to Comware switches. They examine the 802.1p value in a frame's header and use that value to assign the frame to a "traffic class," an internal forwarding priority that determines the queue to which a packet is assigned. This should not be confused with a Comware traffic classifier or a ProVision IPv4/IPv6 class, both of which are discussed later in this chapter. Used in this context, a "traffic class" is somewhat similar to a Comware lp.

However, the ProVision switches differ in several important ways:

- For Comware switches to honor 802.1p values set by other devices, you have to configure that trust on the port. ProVision switches trust by default.
- On Comware switches, you can alter the queues associated with specific 802.1p values by configuring the dot1p-lp map. On ProVision switches, the queues are hard-coded. You can change a frame's 802.1p value but a specific 802.1p value always maps to the same traffic class.

  In other words, there is a one-to-one correspondence between traffic class and Layer 2 802.1p priority. For "trusted" traffic, to which no overriding policies are applied, the value in the 802.1p field of a packet's 802.1Q tag is mapped directly to the traffic class with the same number. In the absence of other prioritization policies, untagged traffic is assigned a value of "0" and forwarded with normal priority.

  The traffic classes map to queues in much the same manner as with the default Comware dot1p-lp map. The 802.1p value 7 maps to the highest-priority queue. Normal priority traffic (value 0) is not mapped to the lowest numbered queue. Instead, there are two priority levels lower than normal.
- You can change the number of queues supported on a port.

  Depending on the switch, ProVision switches support up to eight egress queues, identified as 1–8, per port.

  If you really do need the granularity of eight queues in a ProVision switch that supports them, you can keep the number of queues at the default setting. However, you can also reduce the number of queues. For example, you might reduce the number of queues to four if you only need to differentiate between high, normal, and lower-than-normal traffic. Reducing the number of queues increases the amount of buffering space available to each queue, which might be important for switches with less overall memory for buffering. Figure 8-5 shows how the 802.1p values map to four queues and to two queues.

802.1p value/ traffic class	8 configured queues	4 configured queues	2 configured queues
1 (low)	1	1	1
2 (low)	2	1	1
0 (normal)	3	2	1
3 (normal)	4	2	1
4 (medium)	5	3	2
5 (medium)	6	3	2
6 (high)	7	4	2
7 (high)	8	4	2

Figure 8-5: Honoring 802.1p on ProVision switches

Comware: Honoring DSCP

Sometimes you want a port to honor IP packets' DSCP rather than frames' 802.1p values. For example, the port might not support tagged traffic, so frames cannot carry 802.1p values. In addition, many applications typically use DSCP in preference to 802.1p values.

On Comware switches, you must explicitly configure the port to honor DSCP rather than 802.1p values on a port. Access the interface view and enter qos trust dscp.

The switch then maps the DSCP to an 802.1p value (replacing any 802.1p value that might exist already in the frame). The dscp-dot1p map determines the 802.1p value; Figure 8-6 shows the default settings. From the 802.1p value, using the same dot1p-lp map that you just examined, the switch determines the correct priority queue.

As the switch forwards the traffic in the correct priority queue, it adds the corresponding 802.1p value and the DSCP. Figure 8-6 shows how the outgoing DSCP depends on the dscp-dscp map, which, by default, preserves the incoming mark.

ProVision: Honoring DSCP

On ProVision switches, the QoS type of service setting determines whether the switch honors incoming Layer 3 values. When type of service is disabled, the switch does not refer to a packet's DSCP or ToS value to determine the packet's priority. It only examines the frame's 802.1p value. If the frame does not have such a value, the switch assigns the frame to the normal priority queue

DSCP	Default outgoing 802.1p (dscp-dot1p map)	Default queue (dot1p-lp map)	Default outgoing DSCP (dscp-dscp map)
0 to 7	0	2	Same as incoming
8 to 15	1	0	
16 to 23	2	1	
24 to 31	3	3	
32 to 39	4	4	
40 to 47	5	5	
48 to 55	6	6	
56 to 63	7	7	

Figure 8-6: Honoring DSCP with Comware switches

(traffic class 2). Generally, type of service is disabled by default, but you should check your switch's documentation.

To allow the switch to honor incoming DSCP, set the type of service to DiffServ. The switch then checks each incoming packet's DSCP. If a DSCP map exists for that DSCP, the switch assigns the packet to the 802.1p priority value (and corresponding traffic class) specified in that map. Unlike the Comware switches, the ProVision switches do not automatically map all DSCPs to a queue. Many switches have default mappings for the 13 IETF-defined code points, as shown in Figure 8-7.

With type of service set to DiffServ, the 802.1p value determined by the DSCP map overrides any incoming 802.1p value. However, if the switch receives a packet with a DSCP that has no mapping, it assigns the packet to the queue associated with the 802.1p value in the Ethernet header. If the traffic does not have such a value, the switch assigns the traffic to the normal priority queue.

You can change the default mappings and also add mappings for other DSCPs with the **qos dscp-map <dscp> priority <802.1p value>** command. You can view the maps with the **show qos dscp-map** command. DSCPs without mappings are listed as **no-override** for the 802.1p tag.

You can also set the type of service to IP precedence, which maps incoming packets' 3-bit ToS to an 802.1p value and queue.

Note

Figure 8-7 shows the default mappings, which correspond to RFC recommendations. The default mappings might differ slightly from product family to family. You should refer to the documentation for your product.

PHB	Class	Drop prec.	Decimal value	802.1p/ Traffic class	Mapped queue*
AF Class 1	001	010	10	1	1
	001	100	12	1	1
	001	110	14	2	2
AF Class 2	010	010	18	0	3
	010	100	20	0	3
	010	110	22	3	4
AF Class 3	011	010	26	4	5
	011	100	28	4	5
	011	110	30	5	6
AF Class 4	100	010	34	6	7
	100	100	36	6	7
	100	110	38	7	8
EF	101	110	46	7	8

* When using default setting of 8 queues

Figure 8-7: Honoring DSCP on ProVision switches

Using classification to prioritize traffic

HP switches can select traffic and apply a priority on their own based on a variety of features. First, they can classify all traffic incoming on a physical interface with the same priority. Typically, you would use this strategy when you want to prioritize all traffic from a particular endpoint in the same way. For example, you could prioritize a server's traffic. In a less common use case, a service provider might prioritize all traffic from a particular customer device, depending on a negotiated service-level agreement (SLA).

In addition, HP switches can classify traffic based on a variety of Layer 2 to Layer 4 settings using flexible QoS policies. You would use QoS policies to classify traffic whenever you need to select traffic on a more granular basis. For example, you might prioritize traffic based on application. Or you might place all traffic from a particular VLAN or range of addresses—such as a bank of mission-critical servers—in a high-priority queue.

Note

ProVision switches can classify traffic according to Layer 2 to Layer 4 settings with or without QoS policies. We will take a closer look at this shortly.

Classifying at Layer 2 versus Layer 3

Before you look at the specific methods for classifying traffic, you should understand the two types of priority values that you can apply to traffic: 802.1p priority and DSCP.

Switches and routing switches, including HP switches, generally use a frame's 802.1p value as the ultimate determinant of the frame's priority queue. Therefore, you generally assign this value as you are configuring the switch to classify traffic in the proper queue, as you will see.

However, the switch can only transmit an 802.1p value in tagged traffic. In addition, the priority field in the 802.1Q tag is discarded whenever a switch or router replaces the Layer 2 header. For instance, if a Layer 2 prioritized frame is forwarded through an untagged port, the entire tag, including the 802.1p prioritization value, will be discarded. A WAN router that forwards the packet over a different media also removes the Ethernet header.

For these reasons, when you classify traffic, you might need to assign the traffic a DSCP in addition to an 802.1p value. The DSCP is not modified by the Layer 2 or Layer 3 forwarding processes (unless an infrastructure device is explicitly configured to change the DSCP), allowing the traffic to maintain its priority across the LAN and WAN links.

As you will see, you should specify a DSCP in addition to an 802.1p value and not instead of an 802.1p value. On Comware switches, applying only a DSCP marks traffic for prioritization on the next switch (if it honors the DSCP) *but not on this one*. On a ProVision switch, you might seem to set just a DSCP. However, the DSCP must be mapped to an 802.1p value, and the switch always sets that value in addition to the DSCP.

Note

Although the tag will be destroyed whenever the frame is forwarded at Layer 3 because the forwarding process requires the routing switch or router to create a new Layer 2 header, Q-compliant routers such as the HP switches can preserve the tag. They copy the priority value from the tag of the inbound frame into the tag of the outbound frame before forwarding it. However, the copying only works if the traffic is forwarded as tagged.

Using port priority to classify traffic

Now that you understand how to classify traffic, you will examine a simple way to classify traffic: using the port priority.

Comware

On a Comware switch, to assign all traffic incoming on a port to the same queue, disable QoS trust on the port. Access the interface and enter the undo qos trust command. Then configure the port priority with the qos priority command. As an 802.1p priority, this priority ranges in value from 0 to 7, with 1 and 2 being the lowest values.

The switch uses the dot1p-lp map to assign traffic to the queue (and lp) associated with that priority—regardless of any 802.1p value or DSCP that the incoming traffic might carry. The switch also places the new 802.1p priority value in the tag of the transmitted traffic. However, it leaves an existing DSCP intact, if present.

The Comware switches do not allow you to set a DSCP value for incoming traffic directly on the port. You must use a QoS policy for that purpose (you will learn how to configure such policies in a moment).

ProVision

ProVision switch ports do not have port trust settings. You simply specify the priority that you want to assign to the port by accessing the interface and entering the **qos priority** (for 802.1p) or the **qos dscp** command.

As you have learned, you should apply a DSCP when you want to help downstream devices maintain the priority across the complete traffic path. However, the 802.1p value is required to place the traffic in the correct queue. Therefore, when you apply a DSCP, the switch also applies an 802.1p value in the background using the value specified for that DSCP in the DSCP map. You will receive an error message if you attempt to use the dscp option with a DSCP that does not have a mapping.

On ProVision switches, the port priority takes precedence over incoming 802.1p values but not incoming DSCPs *if* a DSCP map exists.

Figure 8-8 summarizes using port priority on Comware and ProVision switches.

Comware:

- Disable QoS trust on the incoming port.
- Configure the port priority.

Port priority	Outgoing 802.1p value	Outgoing DSCP	lp/queue
1-7	Specified value	Existing (or none)	Set by the dot1p-lp map

ProVision

- Configure the port priority.

Setting	Outgoing 802.1p value	Outgoing DSCP	Traffic class/queue
priority	Specified value	Existing (or none)	Fixed according to 802.1p
dscp*	Priority in DSCP map for specified value	Specified value	Fixed according to 802.1p

*Error if no DSCP map is defined for this value

Figure 8-8: Using port priority to classify traffic on Comware and ProVision switches

Comware: Using port priority with 802.1p

On Comware switches, you can also use a port priority in conjunction with honoring 802.1p. In this case, you both set the port's QoS trust setting to dot1p and configure a port priority (see Figure 8-9). The port then accepts the 802.1p values that are included in tagged frames—including the value 0. On the other hand, the switch assigns the port priority to untagged frames, which cannot carry an 802.1p value.

If you use a port priority when the port trusts DSCP, on the other hand, the port priority does not apply to any IP traffic. This is because all IP traffic has a DSCP even if the DSCP is 0.

This configuration applies to Comware switches. On ProVision switches, you cannot configure the port to trust 802.1p values. It always trusts these values (assuming they are not overridden by a DSCP), but a port priority always overrides the 802.1p value.

Port priority	Outgoing 802.1p value for *incoming* untagged	Outgoing 802.1p value for *incoming* tagged	Outgoing DSCP	lp/queue
1-7	Specified value	Incoming 802.1p value (including 0)	Existing (or none)	Set by the dot1p-lp map

Figure 8-9: Using port priority with 802.1p on Comware switches

ProVision: Classifying traffic without QoS policies

You will now examine more granular methods for classifying traffic. You will first look at classification without QoS policies, which is an option for ProVision switches.

To classify traffic according to any of criteria listed in Figure 8-10, you simply create a global statement that specifies the criterion and the priority that you want to associate with that criterion.

- ProVision switches can classify traffic globally based on:

Precedence (Higher to Lower):

1. Destination UDP/TCP port number
2. Source or destination IP address
3. IP ToS field (precedence bits or DSCP)
4. LAN protocol type (IP, ARP, and others)
5. VLAN ID
6. Ingress switch port
7. Incoming 802.1p value

- When traffic matches multiple criteria, the switch applies the criteria with the highest precedence.

Figure 8-10: Classifying traffic globally based on different precedence on ProVision switches

For example, you can specify priority 4 for TCP port 5060. Now the ProVision switch assigns all traffic that is destined to TCP port 5060 to traffic class 4 and the associated egress queue. (On a switch using eight queues, that queue would be queue 4.)

Because you set up multiple global settings, the criteria might overlap. For example, you might specify 802.1p priority 2 for ingress switch port 1, priority 3 for traffic to and from 10.1.2.10, and priority 4 for TCP port 5060. Figure 8-10 shows which settings take precedence over other settings. In this example, traffic that arrives on switch port 1 from 10.1.2.10 destined for TCP port 5060 is assigned to traffic class 4; the TCP port takes precedence. Traffic that arrives on switch port 1 from 10.1.2.10 destined for TCP port 80 (no configured priority) is assigned to traffic class 3; the IP address takes precedence over the ingress port.

Note that various ProVision switches might support only some of these criteria. Figure 8-10 shows all classifiers that ProVision switch models have in common, plus some that are unique to specific models. For information on a specific product or product family, consult the product documentation.

Just as for the ingress port, you can set the priority for any of these criteria to an 802.1p value or DSCP (see Figure 8-11). Remember that, if you attempt to set a DSCP, that DSCP must be mapped to an 802.1p value already.

Setting	Outgoing 802.1p value in tagged traffic	Outgoing DSCP in IP traffic	Traffic class/queue
priority	Specified value	Existing (or none)	Fixed according to 802.1p
dscp*	Priority in DSCP map for specified value	Specified value	Fixed according to 802.1p

*Error if no DSCP map is defined for this value

Figure 8-11: Classifying traffic without QoS policies on ProVision switches

Comware: QoS policy overview

Comware switches can classify traffic according to much the same criteria as ProVision switches use, but Comware switches use QoS policies. On Comware switches, a QoS policy consists of one or more statements, each of which maps a traffic classifier to a traffic behavior, as shown in Figure 8-12.

Traffic classifiers

Traffic classifiers, which are defined globally and can be reused within multiple policies, select traffic based on a variety of criteria from Layer 2 to Layer 4. To create a classifier, you enter this command:

```
[Switch] traffic classifier <name> [operator {and | or}]
```

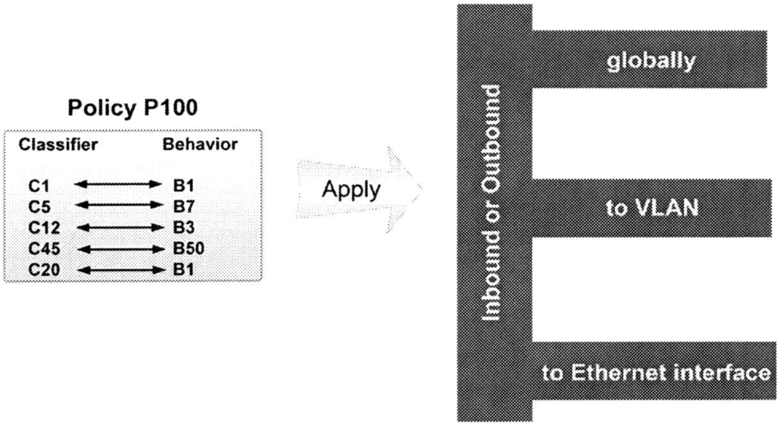

Figure 8-12: QoS policies on Comware: switches

You then specify one or more if-match statements, which select traffic based on a variety of Layer 2 and Layer 3 header fields.

A classifier can include multiple if-match statements. The classifier uses either OR logic or AND logic to determine whether a particular frame or packet matches the classifier. That is, when the classifier uses OR logic, the traffic must match only one of the statements to be selected. In a classifier using AND logic, the traffic must match every statement. (If you do not specify the operator when configuring the classifier, the classifier uses AND by default.)

The "Layer 2," "Layer 3," and "Layer 2 through 7 parameters" sections list parameters against which Comware switches can classify traffic.

Layer 2

Layer 2 parameters include:

- Source MAC address (source-mac <*MAC address*>)
- Destination MAC (destination-mac <*MAC address*>)
- The 802.1p value in the customer VLAN tag of a QinQ frame (customer-dot1p <*802.1p value*>)
- The VLAN ID in the customer VLAN tag of a QinQ frame (customer-vlan <*VLAN ID*>)
- The 802.1p value in the service provider VLAN tag of a QinQ frame (service-dot1p <*802.1p value*>)
- The VLAN ID in the service provider VLAN tag of a QinQ frame (service-vlan <*VLAN ID*>)

Layer 3

Layer 3 parameters include:

- IPv4 or IPv6 protocol (ipv4 or ipv6)

- IP precedence value (ip-precedence <0-7>)
- DSCP (dscp <0-63>)

Layer 2 through 7 parameters

You can also select traffic with IPv4 (and IPv6) access control lists (ACLs) (acl [ipv6] *<number>*). In a moment, you will learn about the wide array of traffic that you can select with an ACL.

Traffic behaviors

Create a traffic behavior by entering this command:

```
[Switch] traffic behavior <name>
```

Within the traffic behavior, you configure an action that implements one of several QoS features:

- Marking traffic with a QoS value and placing it in the corresponding priority queue as dictated by the maps you examined in the "QoS Maps" section (remark [dot1p | dscp | ip-precedence])
- Assigning traffic to a queue but not changing its QoS mark (based on marking the traffic with an lp, remark local-precedence, which is less commonly used)
- Configuring traffic policing for class-based bandwidth management (car)

You will look at an example of marking traffic in a moment. You will look at the traffic policing features later in the chapter.

Traffic behaviors also enable a variety of other features, which are not covered in this chapter:

- Selective QinQ
- VLAN mapping
- Traffic accounting
- Traffic filtering
- Traffic mirroring
- Traffic redirection

Applying QoS policies

You can apply QoS policies in a variety of ways. When you apply the policy, you must also select whether it affects inbound or outbound traffic:

- Globally—The QoS policy applies to all traffic that enters or leaves the switch.
- VLAN interface—The QoS policy applies to all inbound traffic that arrives on this VLAN or to all outbound traffic that is forward on a port in this VLAN (in either case either tagged or untagged).

- Ethernet interface—The QoS policy applies to all traffic that is inbound or outbound on this physical port:

```
[Switch-<interface ID>] qos apply policy <name> [inbound | outbound]
```

```
[Switch] qos apply policy <name> global [inbound | outbound]
```

On some switches, you can apply QoS policies to specific user profiles, which relate to LAN access users who authenticate to the switch locally. However, this study guide does not cover this feature.

In addition, different HP switch models support different methods of applying QoS policies. For example, you might not be able to assign a policy to a VLAN interface. Check the documentation for your switch.

Comware: Using QoS policies to classify traffic in queues

You now have a background in the building blocks of a QoS policy. Next, you will focus on configuring a QoS policy specifically to classify traffic in queues:

1. Determine which traffic you want to prioritize and the factor that this traffic has in common. Then create a traffic classifier that selects this traffic.

   For example, you might want to prioritize all traffic from a specific subnet range, which corresponds to a bank of mission-critical credit card processing servers. You could then create an ACL that selects the source addresses and create a classifier that matches that ACL.

   If you want to classify several different types of traffic, create a classifier for each one.

2. Define a traffic behavior to place the traffic in the correct queue. Use the remark action for this purpose (see Figure 8-13). Typically, you should specify an 802.1p value with this action. This approach both assigns the traffic to a priority queue (according to the dot1p-lp map) and adds the mark to the outbound traffic, communicating the decision to other network infrastructure devices.

   If you want to apply a DSCP as well, add a second action to the behavior. (Usually, a behavior only supports one action, but remarking traffic in various ways is an exception.) For example, whenever you create a behavior that specifies some action be applied to traffic, you might want to add a remark option that sets the DSCP, influencing other devices in the path to apply similar actions. In this way, you might be able to avoid configuring QoS policies on the upstream switches. You might also be able to influence the decisions of devices outside of your administrative control.

   Note, however, that you should generally remark both the 802.1p value and the DSCP not the DSCP *only*. This rule is because remarking only the DSCP does not affect the priority queue.

   As a less common alternative, you can create a traffic behavior that includes the remark local-precedence action and specifies the priority queue to which you want to assign the traffic.

This command places the traffic in the desired priority queue but does not alter the QoS mark, so that decision will not be communicated to other devices.

Of course, if you want to classify traffic in different queues, create a behavior for each queue.

3. Create a QoS policy. Create a statement that maps each classifier to the behavior that specifies the correct priority for traffic selected by that classifier.
4. Apply the QoS policy globally, to a VLAN interface, or physical interface.

Note

An inbound QoS policy that remarks traffic's 802.1p or DSCP value takes precedence over any priority trusted by the port or set by the port priority command.

remark	Outgoing 802.1p value in tagged traffic	Outgoing DSCP in IP traffic	lp/queue
dot1p	Specified value	Existing (or none)	Set by the dot1p-lp map.
dscp	Existing (or none)	Specified value	Set by the dot1p-lp map (and not affected by this action).
Two actions: dot1p and dscp	Specified value	Specified value	Set by the dot1p-lp map.
local-precedence	Existing (or none)	Existing (or none)	Specified value

Figure 8-13: QoS policies for classifying traffic in Comware queues

ProVision: Using QoS policies to classify traffic

Rather than use the global settings to classify traffic on ProVision switches, you might want to take an approach closer to that on the Comware switches. Starting in software release K.14.01, ProVision switches support a classifier-based QoS model that features QoS policies. You can now create and manage QoS policies across a network consisting of HP switches as well as OEM and legacy devices.

Note

This feature in supported by only some ProVision switches. For information on a specific product or product family, consult the product documentation.

Classifier-based QoS is designed to work with existing globally configured, switch-wide QoS settings by allowing you to zoom in on a subset of port or VLAN traffic. Or you can use QoS policies alone to finely select and define the classes of traffic that you want to manage. For example, when you use the global settings, you must place all traffic destined to TCP port 5060 in the same priority queue. With a QoS policy, on the other hand, you can create a class that selects traffic destined to subnet 10.1.2.0/24 and TCP port 5060. You can then handle that traffic differently from traffic destined to TCP port 5060 in other subnets.

A classifier-based QoS policy is also useful when you want to manage different types of traffic in the same way (for example, to prioritize both IP subnet and voice traffic). You can create a series of classes and then assign the same priority to those classes.

Classifier-based QoS also gives you more options for actions—for example, rate limiting and IP precedence marking. You will look at the rate-limiting options later.

To configure classifier-based QoS, you first configure the traffic class and then you specify a priority for the class. This is a bit different from the Comware switches where you create a separate behavior that you then bind to the class in the policy.

Important

Classifier-based policies take precedence over and may override globally configured, switch-wide QoS settings. Carefully plan your QoS strategies in advance, identifying the network traffic that you can globally configure and the traffic on which you want to execute customized, classifier-based QoS actions.

Traffic classes

In classifier-based packet classification, match criteria provide a way to select the packets on which you want to execute QoS actions, such as prioritization.

You configure match criteria by creating a class, which contains one or more match or ignore statements.

Note

A traffic class may be used by any classifier-based software feature, such as QoS or port mirroring, but this chapter focuses on QoS.

To define a traffic class, use this command:

```
[no] class ipv4 <classname>
```

The **no** form of the command removes the existing class.

Note

You can also create IPv6 traffic classes. However, this study guide does not cover IPv6.

After you enter the class command, you enter the class configuration context to specify match criteria. Enter one or more match or ignore commands from the traffic class configuration context to filter traffic and determine the packets on which policy actions will be performed. Use multiple match criteria that search multiple fields in packet headers to select the exact traffic you want to prioritize on a port or VLAN interface.

The list of match criteria includes:

- IP source address
- IP destination address
- LAN protocol type (such as Internet Control Message Protocol [ICMP]or SNMP)
- Layer 3 IP precedence bits
- Layer 3 DSCP code point
- Layer 4 UDP/TCP application port
- VLAN ID

You can look up the precise syntax and options for these match statements in your ProVision switch's *Traffic Management Guide.*

The optional seq-number sequentially orders the match/ignore statements that you enter in a traffic class configuration. Packets are checked by the statements in numerical order.

Configuring QoS policies

A service policy consists of one or more actions that are configured for each class of traffic. The ProVision switches support a variety of service policies, including the ones for policy-based routing (PBR). However, this study guide focuses on the QoS policies. To create a QoS service policy that classifies and prioritizes traffic, follow these steps:

1. Enter the policy qos *policy-name* command from the global configuration context.
2. Enter one or more class action commands from the policy configuration context (see Figure 8-14). This command specifies the actions that you want to execute on packets that match the match criteria in the specified class.

   `no] [seq-number] class ipv4` *`<classname>`* `action` *`<action-name>`*`] [action` *`<action-name>`*`...]` defines the actions to be applied on a preconfigured IPv4 traffic class when a packet matches the match criteria in the class.

The traffic-marking actions valid for a QoS policy include:

- **priority <value>**—Controls the priority queue for traffic leaving the switch and (if traffic exits through a VLAN-tagged port) adds the 802.1p priority value in transmitted traffic

Action	Outgoing 802.1p value in tagged traffic	Outgoing DSCP in frames	Traffic class/queue
priority	Specified value	Existing (or none)	Fixed according to 802.1p
dscp	Priority in DSCP map for specified value	Specified value	Fixed according to 802.1p
Two actions: priority and dscp	Specified value	Specified value	Fixed according to 802.1p

Figure 8-14: QoS policies for classifying traffic in ProVision

- **dscp** ***<value>***—Sets the DiffServ bits in the ToS byte of IPv4 headers (and traffic class byte of IPv6 headers). Either you must specify both a priority and a DSCP or the DSCP must already be mapped to a priority by the DSCP map. Otherwise, you will receive an error message.

Note

You can also configure IP precedence settings; however, this study guide does not focus on IP precedence, which is deprecated.

Finally, you can also configure rate limits, which are discussed in the next section of this chapter.

The configured actions are executed on packets that match a match statement in a class. No policy action is performed on packets that match an ignore statement.

You can configure multiple classes in a policy.

Ensuring the correct mark for outgoing traffic

You have learned about two ways to place traffic in egress priority queues:

- Honoring incoming QoS marks
- Classifying traffic with global settings (ProVision only) or QoS policies

As you have learned about using these strategies to prioritize traffic, you have also learned how the switches apply QoS marks in outgoing traffic:

- As long as the traffic is tagged, the switch includes an 802.1p value in the outgoing frame. This value corresponds to the egress priority queue, which the switch has determined based on a QoS policy, global QoS setting, or an honored incoming priority (and might be 0 if the traffic is in the normal priority queue). Untagged traffic, of course, cannot carry an 802.1p value.

- On ProVision switches, the egress priority queue and the 802.1p value correspond in a fixed way. Refer to Figure 8-5.
- On Comware switches, the egress priority queue and the outgoing 802.1p value correspond according to the dot1p-lp map. For example, a Comware switch port could trust incoming 802.1p values, and you could alter the dot1p-lp map to assign traffic with 802.1p value 5 to lp/queue 6. When the port receives a frame with 802.1p value 5, the switch places the frame in priority queue 6. The outgoing 802.1p value remains 5, the 802.1p value that corresponds to queue 6.

- IP traffic (tagged or untagged) might or might not include a nonzero DSCP. First, the switch applies any DSCP that is set by a QoS policy or global setting. If such a DSCP does not apply, the switch maintains the incoming DSCP, if any.

  An exception occurs for traffic that arrives on Comware switch ports that trust DSCP. The switch transmits such traffic with the outgoing DSCP mapped to the incoming DSCP in the dscp-dscp map. By default, the incoming and outgoing DSCPs are identical, but you can change that behavior.

Review Comware behavior

A Comware switch has the following settings:

```
interface GigabitEthernet1/0/1
 port link-type trunk
 port trunk permit vlan 1 to 3
 qos apply policy B inbound
 qos priority 1
 qos trust dot1p
interface GigabitEthernet1/0/2
 port link-type trunk
 port trunk permit vlan 1 to 3
 qos apply policy B inbound
 qos priority 1
 qos trust dscp
interface GigabitEthernet1/0/3
 port link-type trunk
 port trunk permit vlan 1 to 3
```

```
 qos apply policy B inbound
 qos priority 1
 undo qos trust
acl number 2000
 rule permit source 10.1.3.0 0.0.0.255
traffic classifier 1
 if-match 2000
traffic behavior 2
 remark dot1p 6
 remark dscp 46
qos policy B
 classifier 1 behavior 2
```

The switch is using the default QoS maps.

Consider three frames:

- Frame 1:
    - Includes a 802.1Q tag
    - 802.1p—5
    - DSCP—302
    - Source IP address—10.1.3.5
- Frame 2:
    - Includes a 802.1Q tag
    - 802.1p—5
    - DSCP—32
    - Source IP address—10.1.2.7
- Frame 3:
    - Does not include a 802.1Q tag
    - DSCP—32
    - Source IP address—10.1.1.3

Note

This configuration and the QoS values in the incoming frames are not necessarily typical for a production QoS configuration. Instead, they are designed to ensure that you understand all of the principles covered in the previous section.

Based on this information, answer the following questions (assume that the frame is in the correct VLAN and is accepted):

1. Frame 1 arrives on port gi1/0/1. To which priority queue does the switch assign to the packet within? What QoS marks does the outgoing traffic include (assume that the traffic is transmitted with a tag)?
2. Frame 2 arrives on port gi1/0/1. To which priority queue does the switch assign to the packet within? What QoS marks does the outgoing traffic include (assume that the traffic is transmitted with a tag)?
3. Frame 3 arrives on port gi1/0/1. To which priority queue does the switch assign to the packet within? What QoS marks does the outgoing traffic include (assume that the traffic is transmitted with a tag)?
4. Frame 2 arrives on port gi1/0/2. To which priority queue does the switch assign to the packet within? What QoS marks does the outgoing traffic include (assume that the traffic is transmitted with a tag)?
5. Frame 1 arrives on port gi1/0/3. To which priority queue does the switch assign to the packet within? What QoS marks does the outgoing traffic include (assume that the traffic is transmitted with a tag)?
6. Frame 2 arrives on port gi1/0/3. To which priority queue does the switch assign to the packet within? What QoS marks does the outgoing traffic include (assume that the traffic is transmitted with a tag)?

Answers

1. Frame 1 arrives on port gi1/0/1. To which priority queue does the switch assign to the packet within? What QoS marks does the outgoing traffic include (assume that the traffic is transmitted with a tag)?

   The switch assigns the packet to priority queue 6. The outgoing frame has 802.1p value 6 and DSCP 46.

   The packet is selected by the traffic classifier in the inbound QoS policy (10.1.3.5 matches the permit statement in ACL 2000). The behavior associated with this classifier remarks the packet with 802.1p value 6 and places the traffic in the corresponding queue, which is queue 6 based on the default dot1p-lp map. The policy also remarks the DSCP.

2. Frame 2 arrives on port gi1/0/1. To which priority queue does the switch assign to the packet within? What QoS marks does the outgoing traffic include (assume that the traffic is transmitted with a tag)?

   The switch assigns the packet to priority queue 5. The outgoing traffic has 802.1p value 5 and DSCP 32.

   The port trusts 802.1p values, and the frame has 802.1p value 5. The default dot1p-lp map assigns 802.1p 5 to lp/priority queue 5. When the port trusts 802.1p, it ignores and passes the DSCP. This packet does not match the traffic classifier in the inbound QoS policy, so the policy does not affect it.

3. Frame 3 arrives on port gi1/0/1. To which priority queue does the switch assign to the packet within? What QoS marks does the outgoing traffic include (assume that the traffic is transmitted with a tag)?

   The switch assigns the packet to priority queue 0. The outgoing traffic has 802.1p value 1 and DSCP 32.

   The port trusts 802.1p values, but the frame is untagged. Therefore, the switch applies the port priority of 1. The default dot1p-lp map assigns 802.1p 1 to lp/priority queue 0. When the port trusts 802.1p, it ignores and passes the DSCP. This packet does not match the traffic classifier in the inbound QoS policy, so the policy does not affect it.

4. Frame 2 arrives on port gi1/0/2. To which priority queue does the switch assign to the packet within? What QoS marks does the outgoing traffic include (assume that the traffic is transmitted with a tag)?

   The switch assigns the packet to priority queue 4. The outgoing traffic has 802.1p value 4 and DSCP 32.

   The port trusts DSCP values, and the packet has DSCP 32. The default dscp-dot1p map assigns DSCP 32 to 802.1p value 4. The dot1p-lp map assigns 802.1p 4 to lp/queue 4. The default dscp-dscp map uses the incoming DSCP for the outgoing DSCP. This packet does not match the traffic classifier in the inbound QoS policy, so the policy does not affect it.

5. Frame 1 arrives on port gi1/0/3. To which priority queue does the switch assign to the packet within? What QoS marks does the outgoing traffic include (assume that the traffic is transmitted with a tag)?

   The switch assigns the packet to priority queue 6. The outgoing frame has 802.1p value 6 and DSCP 46.

   The packet is selected by the traffic classifier in the inbound QoS policy (10.1.3.5 matches the permit statement in ACL 2000). The behavior associated with this classifier remarks the packet with 802.1p value 6 and places the traffic in the corresponding queue, which is queue 6 based on the default dot1p-lp map. The policy also remarks the DSCP.

6. Frame 2 arrives on port gi1/0/3. To which priority queue does the switch assign to the packet within? What QoS marks does the outgoing traffic include (assume that the traffic is transmitted with a tag)?

   The switch assigns the packet to priority queue 0. The outgoing frame has 802.1p value 1 and DSCP 32.

   This port does not trust QoS values, so it does not examine either the 802.1p value or the DSCP. It replaces the 802.1p value with the port priority, which is 1. The default dot1p-lp map assigns 802.1p 1 to lp/priority queue 0. When the port does not trust QoS, it ignores and passes the DSCP. If you wanted to clear the DSCP, you would need to configure a QoS policy to do so.

   This packet does not match the traffic classifier in the inbound QoS policy, so the policy does not affect it.

Review ProVision behavior

A ProVision switch has the following settings:

```
interface 1
 service-policy "B" in
interface 2
 service-policy "B" in
 qos priority 1
vlan 1
 name "DEFAULT VLAN"
 untagged 1-24
vlan 2
 name "VLAN2"
 tagged 1-3
vlan 3
 name "VLAN3"
 tagged 1-3
class ipv4 "1"
 10 match ip 10.1.3.0 0.0.0.255 0.0.0.0 255.255.255.255
```

```
policy qos "B"
 10 class ipv4 "1" action priority 6 action dscp 46
qos device-priority 10.1.1.0/24 priority 4
qos device-priority 10.1.3.0/24 priority 3
qos type-of-service diff-services
```

The switch is using the default DSCP maps.

Consider three frames:

- Frame 1:
    - Includes a 802.1Q tag
    - 802.1p—5
    - DSCP—34
    - Source IP address—10.1.3.5
- Frame 2:
    - Includes a 802.1Q tag
    - 802.1p—5
    - DSCP—34
    - Source IP address—10.1.2.7
- Frame 3:
    - Does not include a 802.1Q tag
    - DSCP—34
    - Source IP address—10.1.1.3

Note

This configuration and the QoS values in the incoming frames are not necessarily typical for a production QoS configuration. Instead, they are designed to ensure that you understand all of the principles covered in the previous section.

Based on this information, answer the following questions (assume that the frame is in the correct VLAN and is accepted):

7. Frame 1 arrives on port 1. To which priority queue does the switch assign to the packet within? What QoS marks does the outgoing traffic include (assume that the traffic is transmitted with a tag)?

8. Frame 2 arrives on port 1. To which priority queue does the switch assign to the packet within? What QoS marks does the outgoing traffic include (assume that the traffic is transmitted with a tag)?

9. Frame 3 arrives on port 1. To which priority queue does the switch assign to the packet within? What QoS marks does the outgoing traffic include (assume that the traffic is transmitted with a tag)?

10. Frame 2 arrives on port 2. To which priority queue does the switch assign to the packet within? What QoS marks does the outgoing traffic include (assume that the traffic is transmitted with a tag)?

Answers

7. Frame 1 arrives on port 1. To which priority queue does the switch assign to the packet within? What QoS marks does the outgoing traffic include (assume that the traffic is transmitted with a tag)?

   The switch assigns the packet to priority queue 7 (same as queue 6 on Comware). The outgoing frame has 802.1p value 6 and DSCP 46.

   The packet is selected by the IPv4 class in the inbound QoS policy (10.1.3.5 matches the match statement). The action associated with this class remarks the packet with 802.1p value 6. A second action also remarks the DSCP. The queue associated with the 802.1p value is used.

   Although the packet also matches a global QoS device-priority setting, the QoS policy applied to the interface takes precedence.

8. Frame 2 arrives on port 1. To which priority queue does the switch assign to the packet within? What QoS marks does the outgoing traffic include (assume that the traffic is transmitted with a tag)?

   The switch assigns the packet to priority queue 7 (same as 6 on Comware). The outgoing frame has 802.1p value 6 and DSCP 34.

   This traffic has both an 802.1p value and a DSCP. The ProVision switch honors the DSCP in preference to the 802.1p value when the DSCP has a map and the type of service is set to DiffServ. The default DSCP map does include a mapping for DSCP 34 to 802.1p priority 6. Therefore, the switch replaces the current 802.1p value with the value in the map and places the traffic in the corresponding queue. It keeps the current DSCP.

   This packet does not match the IPv4 class in the inbound QoS policy, so the policy does not affect it. The packet also does not match the global QoS settings.

9. Frame 3 arrives on port 1. To which priority queue does the switch assign to the packet within? What QoS marks does the outgoing traffic include (assume that the traffic is transmitted with a tag)?

The switch assigns the packet to priority queue 4 (same as 3 on Comware). The outgoing frame has 802.1p value 3 and DSCP 34.

This packet does not match the QoS policy applied to the port. However, it does match the global QoS setting for traffic to and from 10.1.3.0/24. Therefore, the switch replaces the incoming 802.1p value to the indicated priority, 3. It keeps the current DSCP. If you wanted to change the DSCP, you would need to set the desired DSCP and ensure that that DSCP has a map to the desired 802.1p value.

10. Frame 2 arrives on port 2. To which priority queue does the switch assign to the packet within? What QoS marks does the outgoing traffic include (assume that the traffic is transmitted with a tag)?

    The switch assigns the packet to priority queue 1 (same as 0 on Comware). The outgoing frame has 802.1p value 1 and DSCP 34.

    The port priority overrides the incoming priority. However, because a DSCP was not specified, the switch maintains the existing DSCP. If you wanted to change the DSCP, you would need to set the desired DSCP and ensure that that DSCP has a map to the desired 802.1p value.

    This packet does not match the IPv4 class in the inbound QoS policy, so the policy does not affect it.

Scheduling methods

You can now assign traffic to queues in a variety of ways and can turn your attention to the method that the switch uses to schedule traffic for transmission. Scheduling actually delivers the prioritization, as it specifies the precise method by which the switch delivers preferential forwarding to higher-priority queues.

Comware switches support a variety of priority queue scheduling methods, including:

- Strict prioritization (SP)
- Weighted round robin (WRR)
- SP and WRR together
- Weighted fair queuing (WFQ)

You can configure the scheduling method for outbound traffic per physical interface, allowing you to select the best method for particular traffic on a granular basis.

Note

The supported queue scheduling methods depend on the switch model. Check the documentation for your switch.

ProVision switches support SP, but they can add guaranteed minimum bandwidth (GMB) on a per-port basis.

SP

SP scheduling always forwards the highest-priority traffic first. That is, the port forwards all packets in queue 7. When that queue empties, the port forwards all packets in queue 6. If that queue does not include any packets, the port forwards all packets in queue 5 and so forth.

Under SP, the port will never forward a packet in a priority queue as long as a higher-priority queue includes even one packet. For example, in Figure 8-15, queue 7, queue 5, queue 3, and queue 1 each include four packets. The port schedules the four packets in queue 7 for transmission and then the four packets in queue 5. Next, the port forwards a packet in queue 3—unless another packet arrives for transmission and is assigned to queue 4 through 7. In that case, the higher-priority packet is forwarded first. If higher-priority traffic continues to arrive, traffic in the lower priority never receives a chance to be forwarded, and traffic in those queues is starved out entirely.

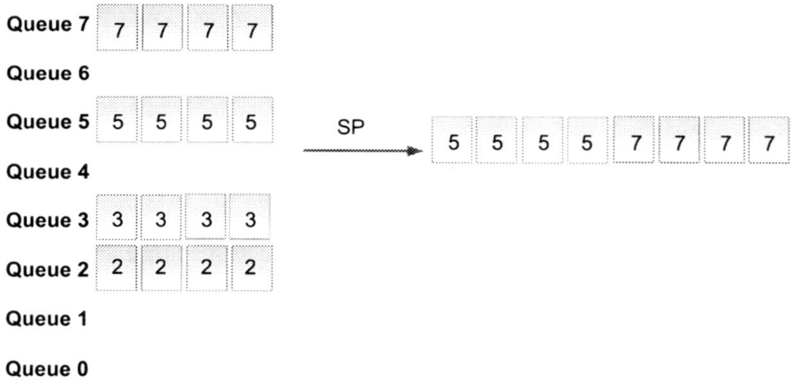

Figure 8-15: SP

When deciding whether to implement SP on a port, you need to consider how much bandwidth the high-priority queues might take. SP works best when the network assigns such a few applications, which take a relatively low amount of bandwidth, the higher-priority queues (network control traffic and VoIP, for example), allowing the majority of traffic in the best-effort queue 2 to be transmitted in between.

If you plan to segment your traffic into many priority queues, each of which might require a significant percentage of the bandwidth, consider implementing one of the other scheduling methods.

For Comware switches, you can enable SP on a port by entering this command:

```
[Switch-<port>] qos sp
```

ProVision switches always use SP.

ProVision: GMB

On ProVision switches, you should use GMB to ensure that high-priority queues do not starve out lower-priority queues. The GMB allocates the specified percentage of the port's service cycle to each queue. In other words, that queue is guaranteed the specified percentage of the port's bandwidth. If, at any moment, the queue does not need the full guaranteed percentage, the unused bandwidth is allocated to other queues according to SP scheduling. If the queue needs more bandwidth than the guaranteed level, it *might* receive the required bandwidth depending on whether any bandwidth remains and whether higher-priority queues require it.

For example, consider a port using the settings shown in Figure 8-16.

```
8212_A# show bandwidth-min output
  Outbound Guaranteed Minimum Bandwidth %
  Port    Priority    Priority    Priority    Priority
  -----   ---------   ---------   ---------   ---------
  A1      10          40          20          30
  A2      10          40          20          30
  A2      10          40          20          30
  A4      10          40          20          30
  ...
```

Figure 8-16: Outbound guaranteed minimum bandwidth on ProVision switches

At the moment, the port needs to use:

- 20% of the cycle to serve traffic in queue 4 (the highest-priority queue)
- 25% of the cycle to serve traffic in queue 3
- 50% of the cycle to serve traffic in queue 2 (the normal priority queue)
- 40% of the cycle to serve traffic in queue 1

As you see, the percentages add up to more than 100%. The port is somewhat congested. The port behaves as follows:

- It forwards all of the traffic in queue 4, which requires 20% of the service cycle, less than the guaranteed 30%.
- It forwards most of the traffic in queue 3, up to the guaranteed 20%. Queue 3 still needs 5% of the service cycle.
- It forwards most of the traffic in queue 2, up to the guaranteed 40%. Queue 2 still needs 10% of the service cycle.
- It forwards some of the traffic in queue 1, up to the guaranteed 10%. Queue 1 still needs 30% of the service cycle.

The port has 10% of the service cycle remaining. It now serves the traffic based on SP:

- It forwards all of the remaining traffic in queue 3, which is the highest-priority queue with remaining traffic.
- It uses the rest of the service cycle to forward some of the remaining traffic in queue 2.
- It buffers the rest of the traffic in queue 2 and queue 1 (and, if necessary, eventually drops it).

Without GMB, queue 1 would have been almost entirely starved out. In this example, the behavior for queue 2 was the same as it would have been without GMB. However, if queue 3 started to consume more bandwidth—for example, 50% rather than 25%—queue 2 would be protected up to 40%.

GMB enables administrators to allocate bandwidth to switch queues on a per-port basis. At default settings, the ProVision switches allocate most bandwidth to normal traffic because most traffic usually has this setting. However, you might want to adjust the GMB for your high-priority queues if you expect that they will require more bandwidth. Remember that unallocated and unused bandwidth is assigned based on SP, so you should allocate the lower-priority queues the minimum amount of bandwidth that you want them to receive in the worst-case scenario.

Comware: WRR

On Comware switches, WRR delivers preferential service to higher-priority queues while still guaranteeing a degree of bandwidth to lower-priority queues.

Round-robin scheduling provides the basis for WRR. Round-robin scheduling forwards one packet from each queue in turn, ensuring that each queue can deliver some of its traffic.

WRR also forwards packets from each queue within a service period; however, it does not provide the same number of forwarding opportunities to each queue. Instead, it provides more opportunities to queues with a higher weight.

WRR can operate in one of two modes: packet-based or byte-based. (Some switch models might support just one of these options; check your switch's documentation.)

The packet-based mode (which is called **weight** in the CLI command) allocates forwarding opportunities by packet. Figure 8-17 illustrates a simple example. Queue 7 has the highest weight, so within one service period, the port forwards four packets from that queue. Queue 5 has the second highest weight, and the port forwards three packets from that queue. It then forwards two packets and one packet from queues 3 and 1, respectively, as these queues have lower weights.

In Figure 8-17, all packets are the same size, so queues receive bandwidth proportional to their weight. However, if the packets are of different sizes, queues with bigger packets receive proportionally more bandwidth. The byte-based mode fixes this issue. WRR uses the weight to allocate *bytes* to each queue. For example, queue 7 has double the weight of queue 5, so the port forwards twice as many bytes in queue 7 as in queue 5.

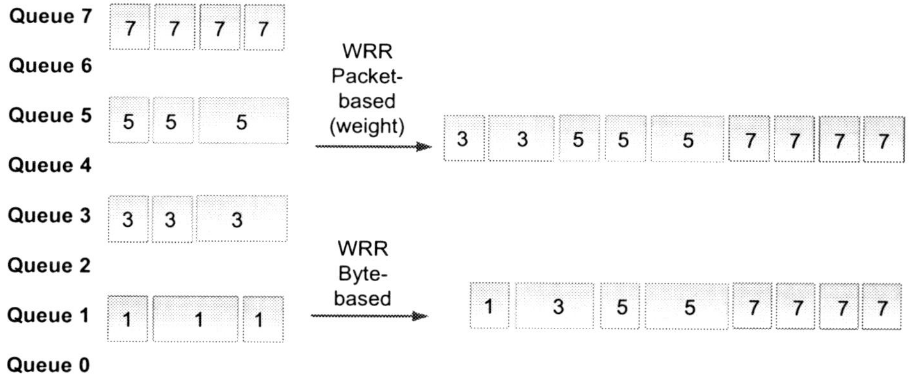

Figure 8-17: Comware: WRR

In either mode, if a queue does not need all of its forwarding opportunities, the scheduler moves on to the next queue.

To configure WRR on a port, enter this command:

```
[Switch-<port>] qos wrr [byte-based | weight]
```

The byte-based option sets the mode to byte-based, and the weight option sets it to packet-based. If you do not specify one of these options, the default one applies (generally byte-based, but check your switch model).

Table 8-2 displays the default weights assigned to each queue. If you are using the byte-based mode, you can use the weight to estimate roughly the percentage of port bandwidth provided to each queue.

Percentage bandwidth = (Queue weight/total weights) * 100

For example, with the default weights, queue 7 receives about 29% of the bandwidth at least (15/52 = 29). If it does not need all of the bandwidth, other queues can use it.

If you are using the packet-based mode, the queue weight determines the proportionate number of packets the queue can forward. For example, queue 7 can forward 15 times as many packets as queue 0 and five times as many packets as queue 2.

You can adjust the weight for any queue with this command:

```
[Switch-<port>] qos wrr <0-7> group 1 <0-15>
```

Table 8-2: WRR weights

Queue	Default WRR weight
0	1
1	2
2	3
3	4
4	5
5	9
6	13
7	15

Comware: SP + WRR

You can combine SP and WRR queuing on the same interface. The port then forwards all traffic in SP queues using the SP scheduling mechanism. When the SP queues are empty, the port forwards traffic in the remaining queues using the WRR mechanism, allocating bandwidth between those queues according to their weight until a packet in an SP queue arrives (see Figure 8-18). At that point, the switch forwards the packet in the SP queue immediately.

Figure 8-18: Comware: SP + WRR

To configure SP and WRR together, set the port scheduling mechanism to WRR with the qos wrr command. Then enter this command to select SP queues:

```
[Switch-<port>] qos wrr <0-7> group sp
```

Finally, configure weights for WRR queues as normal.

Comware: WFQ

The WFQ mechanism is based on fair queuing (FQ). FQ follows the same general principle as round-robin scheduling in that it gives traffic in various queues fair turns to forward traffic. However, FQ attempts to mitigate inherent injustices of round-robin scheduling.

For example, traditional round-robin scheduling only allocates the same amount of bandwidth to each queue only if each queue has the same average packet size. Because traditional round robin forwards one packet per queue, queues with many large packets receive more bandwidth. FQ, on the other hand, provides small packets a statistically better chance to be scheduled for forwarding, so that queues with smaller packets receive just as much bandwidth as queues with larger packets.

FQ also divides traffic into a separate queue for each traffic flow, as defined by source and destination IP address and TCP or UDP port. In other words, a flow roughly corresponds to traffic related to a particular network application. With each flow in its own queue, each flow's behavior affects only that flow.

WFQ applies the same mechanisms as FQ, but assigns different weights to traffic flows based on the flows' priority queue. The weight affects the chances that the next packet scheduled for delivery will be from a flow within a particular queue. WFQ's scheduling algorithms function such that each queue receives bandwidth proportional to its weight according to this formula:

Percentage of assignable bandwidth allocated to the queue = (queue's weight/total weight of all queues) * 100

Table 8-3 provides an example. (The default weight for each queue is 1, meaning that, unless you configure new weights, each queue receives up to 12.5% of the bandwidth.)

Note that, as with other scheduling mechanisms, if a queue is empty, the port skips it and schedules other traffic. In other words, although queue 7 might be allocated up to 30% of a port's bandwidth, if the queue does not require this much bandwidth, packets in other queues can use the bandwidth.

Table 8-3: WFQ weights

Queue	Example WFQ weight	Percentage of bandwidth
0	1	2% (1/50)*100
1	2	4% (2/50)*100
2	3	6% (3/50)*100
3	4	8% (4/50)*100
4	7	14% (7/50)*100
5	8	16% (4/50)*100
6	10	20% (4/50)*100
7	15	30% (4/50)*100

In effect, WFQ provides much the same functionality as WRR in byte-count mode. However, using FQ's mechanisms, it can sometimes provide fairer service and lower latency to small packets in high-priority queues.

Configure an interface to use WFQ with this command:

```
[Switch-<port>] qos wfq
```

Remember to set the weights. Otherwise, WFQ might not provide the expected behavior:

```
[Switch-<port>] qos wfq queue <0-7> <weight>
```

Some switches support two WFQ groups. You can assign some queues to group 1 and some queues to group 2. The port then guarantees an equal amount of bandwidth to group 1 and to group 2. The port allocates the group's bandwidth to queues within the group using WFQ. In this way, you can create two equivalent service sets, each with their own priorities. If you want to use WFQ as normal, simply assign all queues to group 1. For these switches, the command to set the weight for a queue is:

```
[Switch-<port>] qos wfq queue <0-7> group [1 | 2] <weight>
```

Note

On some switches, you can even configure WFQ to use packet count rather than byte count to allocate forwarding opportunities. However, byte count is more typical.

Comware: WFQ minimum guaranteed bandwidth

As you know, WFQ provides each queue a percentage of the "assignable bandwidth." This bandwidth differs slightly from the bandwidth supported by the port. In another difference from WRR, WFQ provides each queue with a minimum guaranteed bandwidth. After each queue receives this minimum level of bandwidth, WFQ allocates the rest of the bandwidth, termed the assignable bandwidth, according to the percentage dictated by the weight.

Table 8-4 provides an example. If one of the queues does not require its full minimum guaranteed bandwidth at the moment, the remaining bandwidth is also allocated to queues that require it, according to the WFQ weight.

The minimum guaranteed bandwidth is a very important feature for WFQ; it allows you to specify precisely the bandwidth that the queue requires. To set the minimum guaranteed bandwidth for a queue, use this command:

```
[Switch-<port>] qos bandwidth queue <0-7> min <kbps>
```

The value for Kbps must be divisible by 8.

Comware: SP + WFQ

On some Comware switches, you can combine SP and WFQ much as you do with WRR.

Set the QoS mode to WFQ:

```
[Switch-<port>] qos wfq
```

Table 8-4: WFQ weights and minimum bandwidth

Queue	Example minimum guaranteed bandwidth (out of 1000 Mbps)	Example WFQ weight	Example percentage of assignable bandwidth	Example bandwidth allocated from assignable bandwidth	Example queue guarantee bandwidth
0	20 Mbps	1	2% (1/50)*100	2 Mbps (2% of 100 Mbps)	22 Mbps
1	30 Mbps	2	4% (2/50)*100	4 Mbps	34 Mbps
2	250 Mbps	3	6% (3/50)*100	6 Mbps	256 Mbps
3	80 Mbps	4	8% (4/50)*100	8 Mbps	88 Mbps
4	80 Mbps	7	14% (7/50)*100	14 Mbps	94 Mbps
5	90 Mbps	8	16% (8/50)*100	16 Mbps	106 Mbps
6	150 Mbps	10	20% (10/50)*100	20 Mbps	170 Mbps
7	200 Mbps	15	30% (15/50)*100	30 Mbps	230 Mbps

Enter this command to select SP queues:

```
[Switch-<port>] qos wfq <0-7> group sp
```

The SP group is sometimes called group 0. Other queues belong to WFQ group 1. (Or, on some switches, as you have learned, you can assign the queues to WFQ group 1 or group 2.) You configure settings for the WFQ queues as normal.

A port has two options for scheduling traffic:

Option 1

In the first option, the port serves the SP queues first. It serves each SP queue in order until the queue is empty. Only then does the port allocate the minimum guaranteed bandwidth to each WFQ queue in order of the queue priority. It is possible that a queue does not receive its minimum guaranteed bandwidth because SP queues have starved them out.

Option 2

In the second option, the queue allocates each WFQ its minimum guaranteed bandwidth *first*. Then it serves each SP queue in order until empty. In this case, you must be careful not to guarantee too much bandwidth to the WFQ queues, which could starve out the SP queues.

In either option, if any bandwidth remains after both serving SP queues and allocating minimum guaranteed bandwidths, WFQ queues share the remaining bandwidth by weight.

Which option a port uses depends on the switch model and, on modular switches, the type of I/O module. You should check your switch documentation.

Port-based rate limiting

You will now examine another QoS feature supported by HP switches: rate limiting.

Both the Comware and ProVision switches enable you to configure rate limits that apply to traffic outbound on a port. As shown in Figure 8-19, the egress rate limit causes the port to drop traffic destined to the connected device down to the configured level. Often, traffic flows more heavily from server to client. Therefore, creating an egress rate limit on an endpoint's port can prevent that endpoint from demanding too much traffic and indirectly from consuming too much of the bandwidth on its switch's uplink.

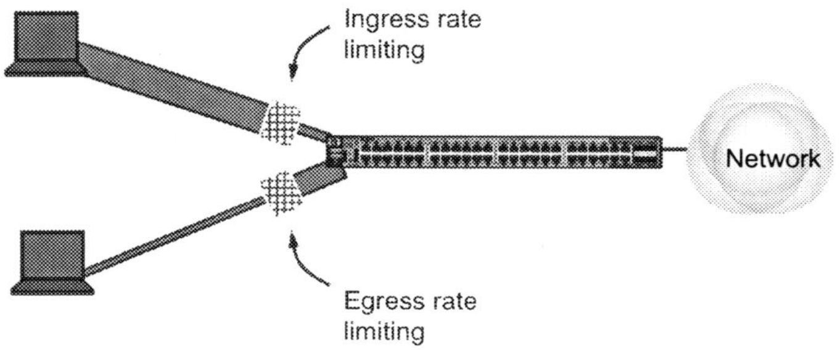

Figure 8-19: Port-based rate limiting

On some switches, you can also configure an inbound rate limit, which drops traffic incoming on the port that exceeds the specified limit. Finally, some switches allow administrators to set a total rate limit that applies to both inbound and outbound traffic. For example, the inbound traffic might not exceed the inbound limit, but, taken with the outbound traffic, it does exceed the total limit. Therefore, the switch port drops the traffic. On switches that support rate limits in both directions, the limits can be configured separately for each direction.

Furthermore, the ProVision switches allow you to apply limits specifically for ICMP traffic, which can be used for attacks such as denial of service (DoS).

Depending on the switch model, you might set a rate limit as a percentage of the total bandwidth or in Kbps. You might also be able to set a committed burst size (CBS), which allows the port to burst traffic briefly past the configured limit.

You should check the documentation for your switch's precise capabilities.

Comware: Traffic policing

On Comware, switches support traffic policing, which regulates the bandwidth consumed by a particular class of traffic. The class of traffic can be defined according to any of the Layer 2 through 7 criteria supported by Comware traffic classifiers. Traffic policing helps to prevent congestion before it occurs by ensuring that each traffic class uses a sensible amount of bandwidth and by scaling back high-bandwidth flows that would otherwise consume too much of the network resources (see Figure 8-20).

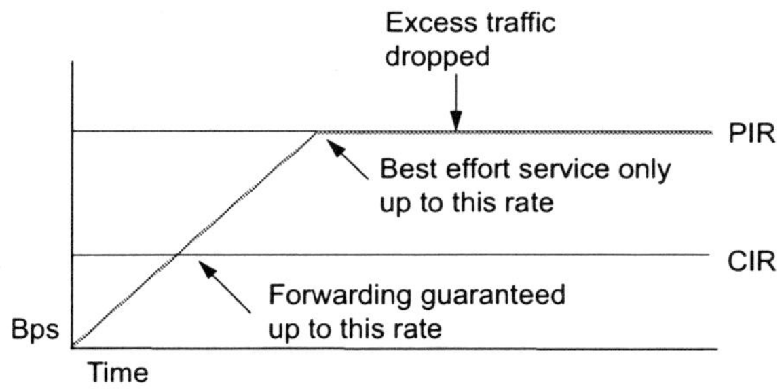

Figure 8-20: Traffic policing on Comware switches

Traffic policing also ensures that particular traffic flows conform with business policies. In a service provider environment, for example, this QoS feature often enforces service-level agreements (SLAs) for allowable bandwidth.

Traffic policing does more than set a simple rate limit, also called a committed access rate (CAR), below which traffic is allowed and above which it is dropped. You can shape the traffic in a variety of ways, ensuring priority service for traffic up to a particular rate, providing best-effort service for traffic up to another rate, and dropping traffic beyond that.

Comware: Evaluating if traffic conforms with the policing policy

Traffic policing uses the concept of a token bucket to determine whether a particular packet conforms to the rate specified by a traffic policing policy. As Figure 8-21 illustrates, the committed bucket (C bucket) contains a certain number of tokens, each of which corresponds to a bit.

When a packet needs to be forwarded, the switch checks whether the token bucket contains enough bits to forward the entire packet. If it does, the switch marks the packet as conforming or, in the terminology used by these HP switches, green. If the bucket does not contain enough tokens, the packet exceeds the policing policy and is marked as red.

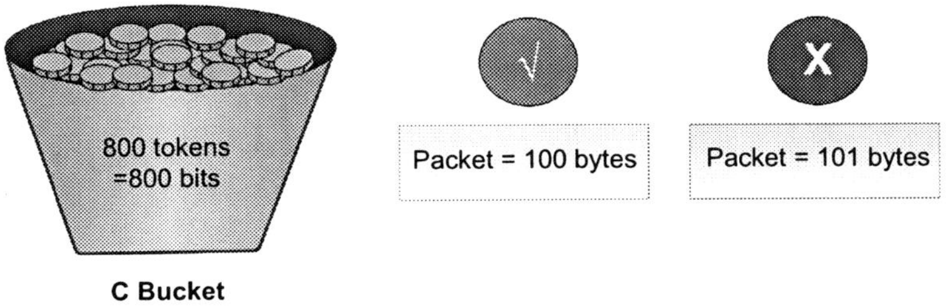

Figure 8-21: Determining if the token bucket has enough bits to forward the packet

Comware: CIR

A traffic policing policy defines how a token bucket is filled, which in turn determines how much bandwidth a class of traffic can consume. First consider the committed information rate (CIR), which determines the rate at which the token bucket is filled. For example, if the CIR is 640 Kbps (.64 Mbps), the switch adds 640,000 tokens to the bucket each second. As packets arrive for forwarding, they deplete the bucket, but the CIR continually refills it. As long as, on average, packets arrive to be forwarded at the CIR or lower, the policy allows them to be forwarded.

Thus, the CIR defines the maximum bandwidth that is guaranteed to the traffic class at all times.

For example, a service provider might offer a subscriber a CIR of 2 Mbps, which guarantees that the subscriber can always send traffic up to 2 Mbps (see Figure 8-22). That is, as long as the subscriber network is sending that much traffic, the service provider cannot allow the rate to fall below 2 Mbps even during congestion. The service provider must configure traffic policing policies for other subscribers to ensure this is the case. Conversely, if the customer's traffic exceeds 2 Mbps, it is noncompliant, and the service provider does not guarantee to forward it.

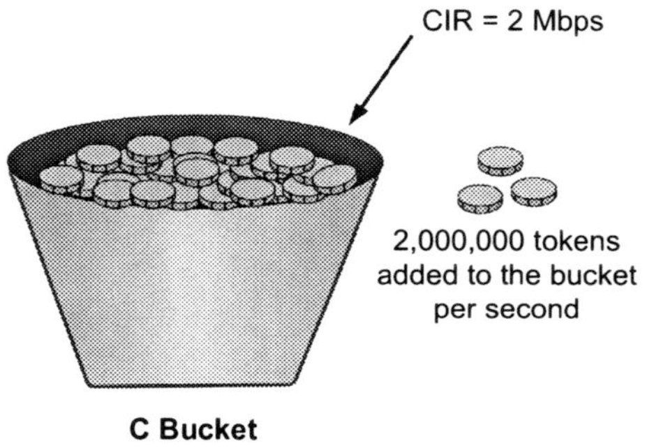

Figure 8-22: CIR on Comware switches

Comware: CBS

If the token bucket had no maximum size, the token bucket would continue to fill indefinitely when the traffic class is not using its full CIR. In essence, the devices and applications affected by the traffic policing policy would be "saving up" their bandwidth. However, the traffic policing policy is not designed to work this way but rather to ensure a guaranteed average bandwidth. Therefore, while the token bucket is continually filling at the CIR, it has a set size beyond which it will not fill—the committed burst size (CBS).

The CBS defines the guaranteed maximum amount of traffic in the class that the switch will transmit at any given moment (regardless of the CIR and past utilization).

The CBS must be greater than the maximum packet size; otherwise, the traffic policing policy would never allow the traffic class to transmit packets of this size. Generally, the CBS should be many times greater than this size to allow the class to burst multiple packets.

For the traffic policing policy to work as expected, it is recommended that you set the CBS at or greater than 1/20 the number of bytes that the CIR adds to the bucket in one second. Because you set the CIR in Kbps and the CBS in bytes, the CBS value should be greater than the CIR value multiplied by 100/16. A CBS smaller than this could prevent normal traffic flows from receiving the CIR because the bucket cannot fill enough to accommodate them. In addition, a traffic class that features bursty traffic—such as TCP applications or a mix of TCP and UDP—requires a greater CBS than a traffic class that tends to send traffic at a constant rate (such as voice or video applications).

In the example shown in Figure 8-23, the CIR is 2 Mbps (2000 Kbps), and the CBS is 16 KB (16,000 bytes or 128,000 bits). The CBS should be at least 12,500 bytes (2000*100/16), so the example CBS meets the recommendation.

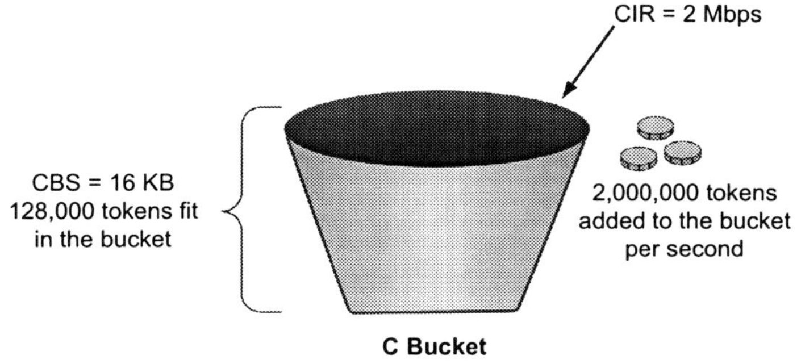

Figure 8-23: Implementing CBS on Comware switches Comware: CBS

Assume that the bucket is full, as it is initially. Five packets in the traffic class arrive, separated by 1 ms. These packets are 100 bytes, 400 bytes, 1500 bytes, 1500 bytes, and 1500 bytes. The first packet depletes the 16,000-byte bucket to 15,900 bytes. Within 1 ms, the CIR adds 250 bytes, but the bucket only fills to 16,000 bytes (the CBS). The second packet depletes the bucket to 15,600 bytes, but the bucket is filled back to 15,850 bytes before the third packet arrives. That packet reduces the bucket to 14,350 bytes, which becomes 14,600 bytes by the time the third packet arrives. The bucket will continue to drain slowly as long as 1500-byte packets arrive at this rate (because one 1500-byte packet per ms = 12 Mbps, far beyond the 2 Mbps CIR). However, as long as the class falls back in utilization, sending fewer packets or smaller packets, it can stay within the CIR.

Comware: Using policies to deliver a PIR

In addition to enforcing a CIR, a traffic policing policy can provide a peak information rate (PIR), which is the maximum amount of bandwidth that the traffic class is permitted when the network is not experiencing congestion.

To provide the PIR, the traffic policing policy evaluates packets against up to two buckets, the C bucket and the excess bucket (E bucket). As Figure 8-24 illustrates, the E bucket fills at the PIR and has a maximum size of the excessive burst size (EBS).

If the C bucket provides enough tokens to forward the packet, the packet is colored green and evaluation ceases. This packet conforms to the CIR. If the C bucket does not provide enough tokens, the packet is evaluated against the E bucket. As long the E bucket provides enough tokens, the packet is colored yellow; it conforms to the PIR. On the other hand, if the E bucket does not provide enough tokens either, the packet is colored red. It exceeds both the CIR and the PIR.

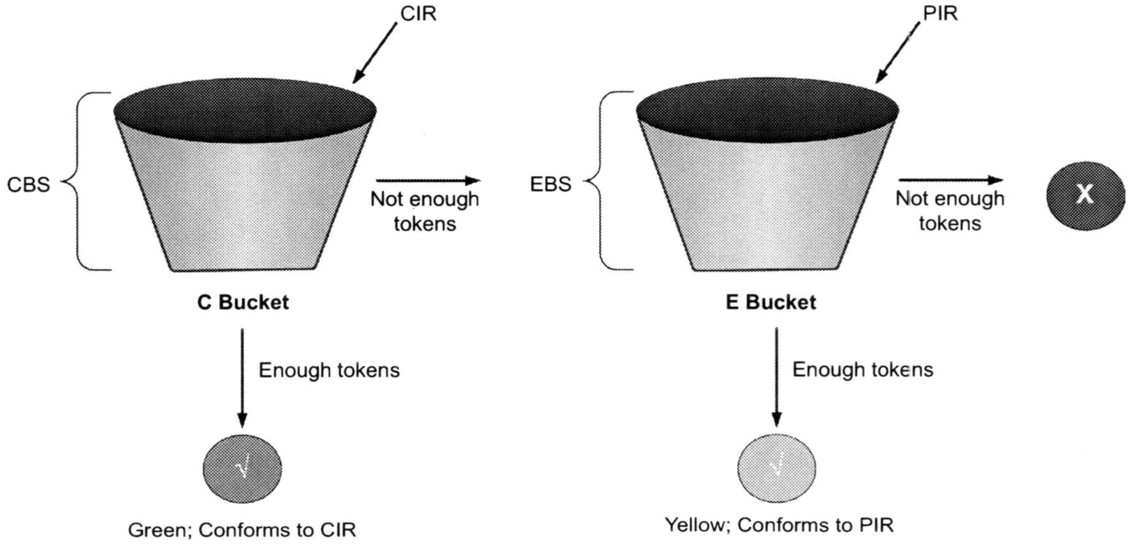

Figure 8-24: Using policies to deliver PIR on Comware switches

For example, the SLA that guarantees a subscriber 2 Mbps for the CIR might further allow the subscriber a PIR of 4 Mbps. If the service provider can accommodate this amount of traffic, it will transmit it. However, if the switch begins to experience congestion, it will drop traffic that exceeds the CIR of 2Mbps. In addition, the switch will never allow the customer network to transmit more than an average of 4 Mbps no matter the network conditions.

Comware: Enforcing the limits

As you have seen, green packets conform with the CIR, yellow packets exceed the CIR but conform with the PIR, and red packets exceed both the CIR and PIR. Or, if the policing policy does not define a PIR, green packets conform with the CIR, and red packets exceed it. The switch must apply actions to packets with specific colors in order to enforce the expected behaviors.

You can set these actions within a traffic policing policy. By default, Comware switches pass green and yellow packets, decrementing the token buckets by the appropriate number of bits. Typically, they discard red packets. (However, as you will see later, you could configure the switch to mark the red packets for later discard if congestion occurs on the egress port).

In order to distinguish between green and yellow packets, you should customize the actions. Within the traffic policing policy, you can configure the switch to pass yellow packets but alter their QoS marks (different switches allow you to change different QoS marks, so you should check the capabilities of the particular switch that you are configuring). For example, you might set a lower IP precedence, DSCP, or 802.1p value for these packets because the PIR only provides best-effort service.

The dp associated with different traffic colors also provides a powerful mechanism for controlling traffic. Comware switches set the dp for green packets to 0, yellow packets to 1, and red packets to 2 (if you set the traffic policing policy to pass red packets). In the next section, you will learn how to configure ports to drop the traffic with a higher dp value when congestion starts to occur—thus, you allow the nonconforming traffic to pass as long as it does interfere with conforming traffic.

Comware: Creating QoS policies to police traffic

Follow the standard procedure to configure a QoS policy that polices traffic:

1. First, determine how you want to define the traffic that the policy controls and create a traffic classifier that selects that traffic.

   You can use any of the traffic classifier types discussed previously, selecting traffic by subnet or TCP port, for example. In a network using QinQ, you might select traffic according to the service provider QoS mark in order to apply traffic policing to traffic defined by a downstream service provider switch. Or you might classify traffic according to the service provider VLAN ID in order to deliver a CIR and PIR to specific customer associated with that ID.

2. Next, create a traffic behavior and define a committed access rate (CAR) action. The parameters for this action enable you to set the CIR in Kbps and the CBS in bytes as well as an optional PIR in Kbps and EBS in bytes. You can also customize the actions applied to green, yellow, and red packets as discussed previously:

   ```
   [Comware] traffic behavior <name>

   [Comware-behavior-<name>] car [name] cir <kbps> [cbs <bytes>] [pir <kbps>] [ebs <bytes>] [green <action>] [yellow <action>] [red <action>]
   ```

3. Create a QoS policy that maps the traffic classifier to the CAR behavior with the appropriate settings:

   ```
   [Comware] qos policy <name>

   [Comware-qos-policy-<name>] classifier <name> behavior <name>
   ```

4. Finally, apply the QoS policy either globally or to an interface. You can apply traffic policing policies to either outbound or inbound traffic, depending on when you want to apply the average rate limits. (However, some switches support applying policies only to outbound traffic).

ProVision: Class-based rate limiting

While ProVision switches do not support traffic policing, you can apply rate limits to a specific traffic class using a QoS policy. However, you can only specify a single rate limit. The concepts of CIR, PIR, various traffic colors, and so on do not apply.

Follow these steps to create the class-based rate-limiting policy:

1. Create a traffic class that selects the traffic to which you want to apply the rate limit:

    a. Create the traffic class:

    ```
    ProVision(config)# class ipv4 <classname>
    ```

    b. Use **match** and **ignore** commands to select traffic for the class. You can match (or ignore) traffic based on IP source and destination address, protocol, TCP or UDP port, and so on. Refer to your switch documentation for the exact syntax.

2. Create a QoS policy. Enter:

    ```
    ProVision(config)# policy qos <policy-name>
    ```

3. Enter one or more class action commands from the policy configuration context. For the action select rate-limit:

    ```
    ProVision(qos-policy)# [seq-number] class ipv4 <classname> action
    rate-limit kbps <0-10000000>
    ```

4. Apply the QoS policy to an Ethernet port or to a VLAN:

    ```
    ProVision(<eth or vlan ID>)#service-policy <policy-name> in
    ```

Comware: Traffic shaping with GTS

Networks provide better service when they gracefully scale traffic back rather than begin to drop it on a large scale. Traffic shaping provides such functionality by adjusting a traffic flow to ensure that it meets the requirements of a downstream device that is enforcing traffic policing. Comware switches support generic traffic shaping (GTS), which sets a rate limit much like a traffic policing policy's CIR/CBS or PIR/EBS. However, rather than drop traffic that exceeds the limit, GTS buffers the traffic as much as possible and then gradually forwards it at the specified limit (see Figure 8-25).

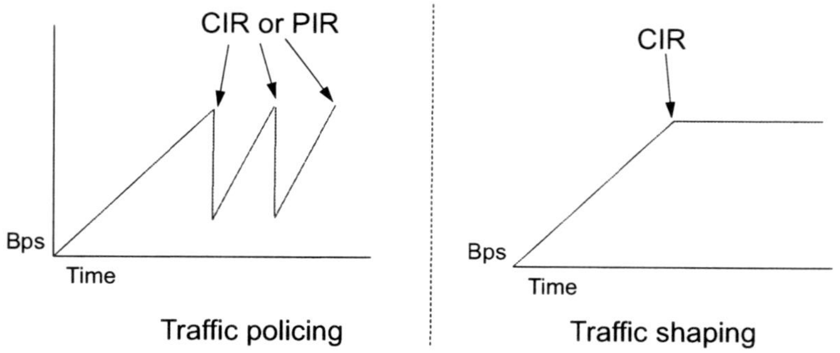

Figure 8-25: Traffic shaping on Comware switches

For example, a subscriber has an SLA that provides a 4-Mbps PIR. Rather than allow the network to forward traffic at a higher rate—causing some traffic to be dropped on the downstream device—the subscriber applies a GTS rate limit of 4 Mbps to the interface that connects to the service provider. Then, if the network begins to exceed the PIR, the subscriber switch automatically adjusts the flow to increase the chances that all traffic can be forwarded.

Configuring traffic shaping on Comware switches

You configure GTS on physical interfaces, and the traffic shaping affects traffic transmitted on that port. The Comware devices permit a degree of granularity by permitting you to set the traffic shaping CIR and CBS per queue. Generally, you should attempt to match the CIR and CBS to the CIR and CBS that a downstream device is expected to apply to traffic within that queue.

Enter this command:

```
[Comware-<interface>] qos gts queue <0-7> cir <CIR in Kbps> [cbs <CBS in bytes>]
```

If you do not specify the CBS, the CBS defaults to a value, which you can check on your specific Comware switch using the display qos gts interface *<interface type interface number>* command.

Problem with traditional tail drop method

You have learned how to configure queue scheduling to prioritize the delivery of mission-critical and time-sensitive traffic, particularly as congestion begins to occur. However, when a link become too congested—that is, its buffers run out of space to queue more traffic—the port must begin to drop packets, as Figure 8-26 illustrates. Without special provisions, the port does not drop packets in an intelligent manner, degrading QoS for applications and even worsening congestion.

The "Comware: Solution—RED" section explains how to reduce congestion by dropping traffic proactively in order to avoid these ill effects. You can use these mechanisms alone or in conjunction with traffic policing.

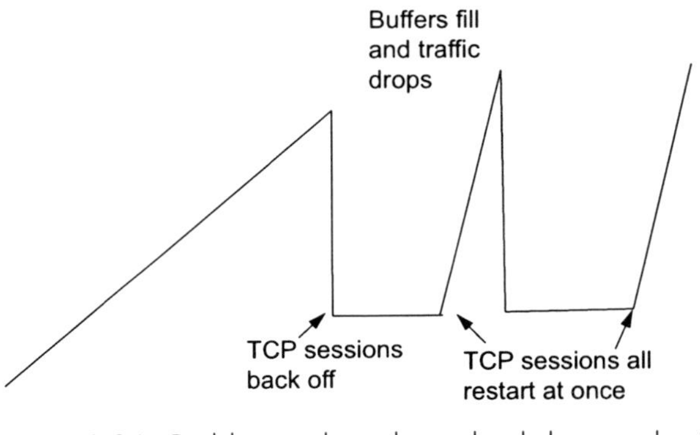

Figure 8-26: Problem with traditional tail drop method

By default, when a port's queues exceed its buffer space, the port begins to drop all traffic, a mechanism called tail drop. Tail drop introduces several problems. First, when the buffer is exceeded, the port drops packets regardless of their priority. Therefore, high-priority traffic might be dropped while low-priority traffic is forwarded simply because the high-priority traffic arrives at a moment of congestion, while the low-priority traffic arrives after the congestion clears.

Second, tail drop tends to cause TCP synchronization. Because the port begins to drop all packets at the same time, multiple TCP sessions are disrupted simultaneously. The TCP endpoints implement their backoff mechanisms, and because TCP specifies a standard backoff time, they restart at the same time. Now, rather than a more natural, and easily managed, traffic pattern in which TCP sessions begin at different times, the TCP sessions all start in a burst. Congestion worsens, causing more TCP sessions to be dropped and increasing the number of synchronized sessions.

Thus, tail drop can lead to a situation in which links are used inefficiently with alternating bursts of underutilization (as tail drop causes many TCP sessions to back off at once) and congestion (as all sessions start up again).

Comware: Solution—RED

Random early drop (RED) avoids these problems by proactively dropping some packets before buffers fill entirely and all packets are dropped. Dropping these packets in advance triggers a few TCP sessions to slow down, relieving congestion on the link as a whole and avoiding more widespread dropping of traffic.

As shown in Figure 8-27, RED defines a lower and upper threshold for the average queue size (defined in bytes). When the average queue size exceeds the lower threshold, RED determines that congestion might soon occur, and it must take action to avoid such congestion. RED begins to randomly drop packets.

To determine when a threshold has been exceeded, RED checks the queue's average length over a period of time rather than its current length. This provision prevents RED from dropping packets due to a transitory burst of traffic rather than a sustained trend toward congestion.

Comware: Discard probability

The rate at which RED drops packets depends on a configurable discard probability, as well as how close the average size is to approaching the upper threshold—at which point all packets are dropped. That is, RED drops fewer packets when the average queue size slightly exceeds the lower threshold and more as the size lengthens. The configured discard probability defines the maximum percentage of packets dropped just before the upper threshold is reached (see Figure 8-28).

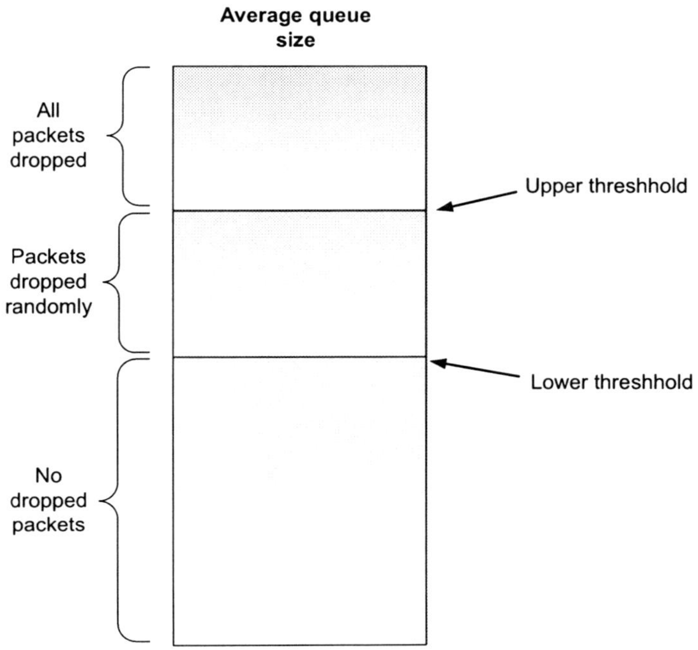

Figure 8-27: RED on Comware switches

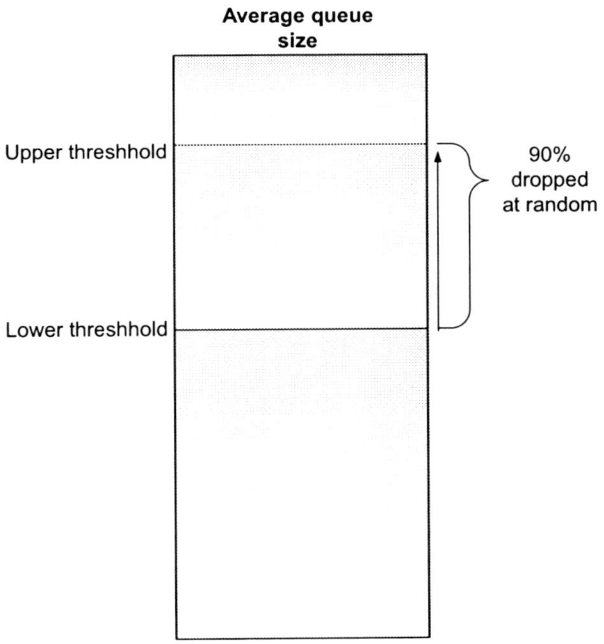

Figure 8-28: Discard probability on Comware switches

Comware: Solution—WRED

RED does introduce one issue: the port drops traffic randomly regardless of its priority. If the switch happens to drop a packet that is associated with an application that does not tolerate packet loss well—such as a UDP voice or video application—RED causes a much greater negative affect than if it had happened to drop a packet associated with a less sensitive or less important application.

Comware switches support a more sophisticated version of RED called weighted RED (WRED), which resolves this issue. WRED functions just like RED except that the port intelligently drops certain traffic in preference to other traffic.

First, you can define different average queue size thresholds for different priority queues. By setting the thresholds lower on lower-priority queues, you trigger the port to begin dropping lower-priority traffic sooner than higher-priority traffic relative to the degree to which the queue is contributing to congestion. Dropping some lower-priority packets reduces congestion, making it less likely that higher-priority traffic must be dropped.

You can also adjust the drop probability per queue. For example as Figure 8-29 shows, you might want to drop a higher percentage of packets in a lower-priority queue than in a higher-priority queue. In some priority queues, you might want to set the percentage to 0 so that no traffic is dropped.

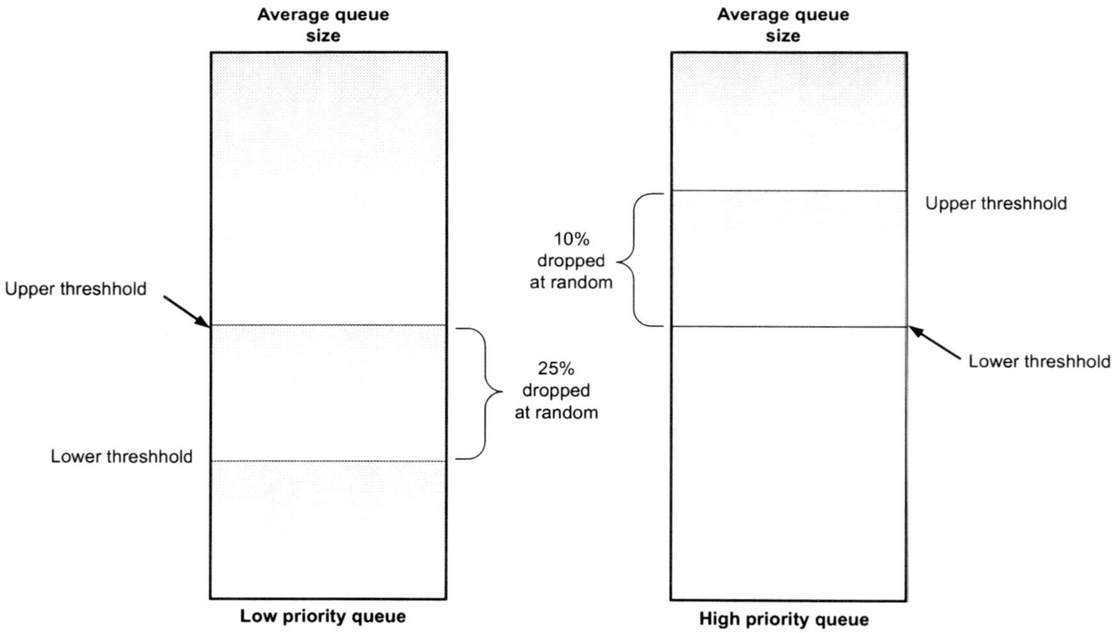

Figure 8-29: WRED on Comware switches

Comware: WRED and dp

You can further customize WRED's thresholds and drop probabilities for different traffic within priority queues based on the traffic's drop precedence (dp). Like lp, Comware switches define a dp, a locally significant setting, for each packet. The dp is intended to select specific traffic within priority queues as more deserving of being dropped than other traffic.

The three dp values, 0, 1, and 2, specify traffic that is increasingly preferable to drop. Traffic with a dp of 0 (the default) should be dropped after traffic with a dp of 1, which should be dropped after traffic with a dp of 2. Therefore, you would set the threshold lower and the drop probability higher for traffic with a dp of 2 than for traffic with a dp of 1 or 0 (see Figure 8-30).

On the Comware switches, you will sometimes see the dp referred to as colors:

- Green—0
- Yellow—1
- Red—2

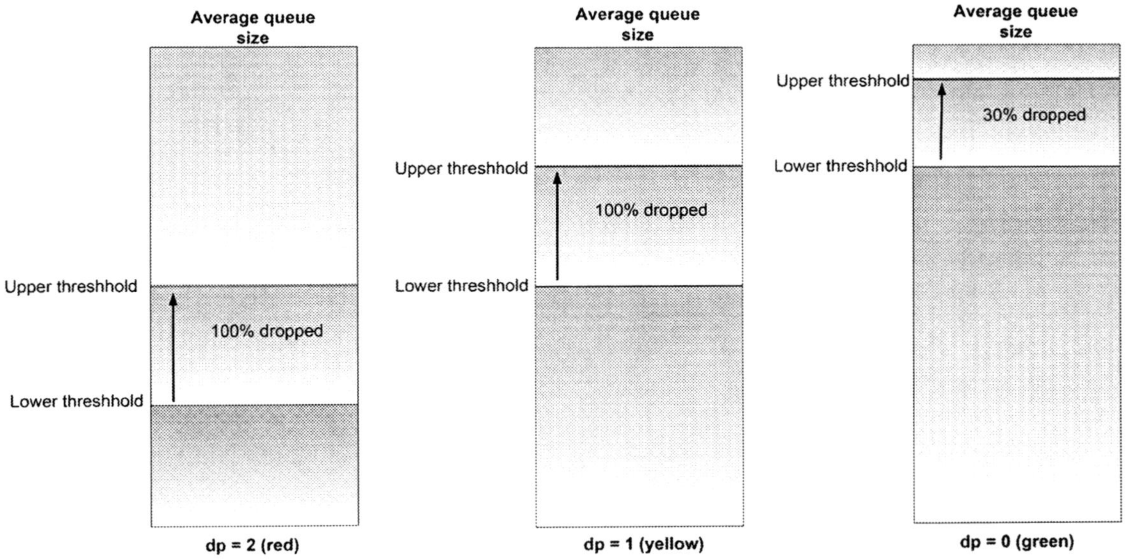

Figure 8-30: WRED and dp on Comware switches

dp based on traffic policing policies

As you have learned, the switch can derive a packet's dp from its conformance with traffic policing policies, which provide rate limiting for specific types of traffic. In other words, traffic that exceeds rate limits is dropped preferentially.

The following sections explain two secondary methods for setting the dp: based on QoS marks on incoming traffic and based on port priority.

dp based on QoS marks

You have already learned how to trust QoS marks on incoming traffic on a port to determine the lp—and corresponding priority queue—for that traffic. Just as each 802.1p value and DSCP maps to an lp, each maps to a dp. By default, the dp is 0 for all values. However, you can adjust the dot1p-dp or dscp-dp map to assign a different dp to traffic with a specific 802.1p value or DSCP.

For example, DSCPs 33 to 39 all map to the same lp by default. You might adjust the dp for some values as recommended by the DSCP AF classes. You can then configure different WRED settings for different dp within the queue associated with all of those DSCPs to reflect the different drop preferences for different DSCPs.

Generally, you should select a port's trust method based on the type of mark that you want to use for sorting traffic into queues. You can then adjust the appropriate map based on that decision.

dp based on port priority

When you do not want to trust QoS marks on a port, all incoming traffic is assigned the port priority, an 802.1p value that maps to an lp/queue. This value also maps to a dp, as specified by the dot1p-dp map, which you can configure.

Note

Untagged traffic arriving on ports that trust 802.1p is also assigned the port priority and corresponding dp.

---

dp based on QoS policy

Finally, you can create a QoS policy that selects a particular traffic class and assigns a dp to that class.

Comware: Configuring WRED

To configure WRED, you must first configure at least one WRED table. This table specifies for each queue and dp:

- Lower threshold (min in Figure 8-31)
- Upper threshold, which is configurable on some switches only (max in the table)
- Drop probability

The table in Figure 8-31 shows the separate settings for the three dp values (0 or G for green, 1 or Y for yellow, and 2 or R for red).

Queue	G min	G max	G %	Y min	Y max	Y %	R min	R max	R %
0	5000	8000	20	4100	5000	100	3100	4000	100
1	6000	8000	30	4000	5000	100	3000	4000	100
2	6000	8000	15	4200	5000	100	3200	4000	100
3	6000	8000	10	4300	5000	100	3300	4000	100
4	7000	8000	10	4400	5000	100	3400	4000	100
5	7000	8000	0	4500	5000	100	3500	4000	100
6	7000	8000	0	4600	5000	100	3600	4000	100
7	7000	8000	0	4800	5000	100	3800	4000	100

Figure 8-31: Comware: Configuring WRED

You can customize these settings as much or as little as you like so that they follow your business policies and SLAs precisely. For example, the dp might reflect your company's contractual agreements. You cannot allow traffic that exceeds its PIR (red packets) to interfere with traffic from another customer that conforms with the PIR (yellow packets). Neither red nor yellow packets should crowd out traffic that conforms with the guaranteed CIR (green). Therefore, you set the limits such that the upper limit for red traffic coincides with the lower limit for yellow traffic of *any* priority. Thus, all excess packets are dropped before any best-effort PIR packets are dropped. A similar concept holds for yellow versus red traffic. You can further adjust the min and max queue sizes such that the switch begins dropping red or yellow traffic with a higher-priority queue a bit later.

Network control traffic, which is placed in priority queue 7, provides for the proper functioning of the network. For example, BPDUs maintain the proper spanning tree topology, and without them, loops can form. Therefore, the WRED table specifies that green traffic in this queue is never dropped. Similarly, the green traffic in queue 6 is never dropped.

Note

The number of settings that you can customize depends on your switch model. Some models support only two dp, for example. Check the documentation for your switches.

After you configure the WRED table, you must apply it to a physical interface in order for it to take effect.

ProVision: Configuring voice VLANs and LLDP-MED

You now know how to protect voice traffic as it travels across the network infrastructure. However, you also need to set up the edge ports that connect to IP phones in a particular way. These initial settings ensure that the phone receives the service that it requires as soon as it connects and for all of its traffic.

It is best practice to assign voice devices such as IP phones to their own VLAN. By doing so, you protect the phones to a degree from excessive broadcasts and other congestion. Of course, the voice VLAN usually needs to share links with other VLANs, so you must implement QoS priorities. The voice VLAN also helps you in this endeavor.

As Figure 8-32 illustrates, LLDP-MED provides the second component of the solution, helping the switch port to autoprovision the connected IP phone.

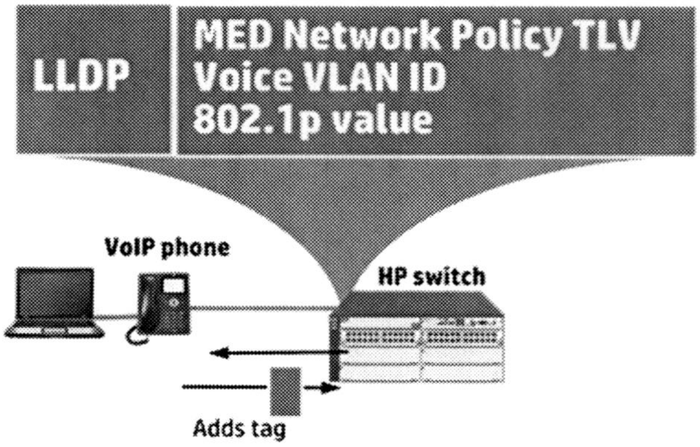

Figure 8-32: Voice VLANs and LLDP-MED on ProVision switches

To create a voice VLAN on a ProVision switch, simply choose the VLAN to which you want to assign IP phones and enter the voice command for that VLAN:

```
ProVision(vlan-<ID>)# voice
```

It is also recommended that you configure an 802.1p value of 5 or above for this VLAN and an optional DSCP (the RFC-defined DSCP typical for voice is expedited forwarding or 46). The previous sections of this chapter explained the importance of these QoS marks, allowing the switch and other devices throughout the network to forward the traffic in a high-priority queue. These settings take another role when you configure the VLAN as a voice VLAN. The switch communicates them to the LLDP-MED-capable VoIP phone so that the phone can apply the priorities itself.

To continue configuring this solution, you assign the voice VLAN to switch ports as normal, depending on whether IP phones send untagged or tagged traffic. You must check the capabilities of the IP phones you are using and the way they are set up. As a general guideline, today's IP phones typically support 802.1Q and send tagged traffic; however, legacy IP phones may support only untagged traffic.

When the phone sends tagged traffic, it needs to learn its VLAN ID so that it can insert the correct tags in the Ethernet header. LLDP-MED provides an option for autoprovisioning the phone with its VLAN ID. ProVision devices support LLDP and the appropriate LLDP-MED extensions by default. When a phone that also supports LLDP-MED boots up and connects, it completes a handshake with the switch port, negotiating the PoE power that the phone needs (if required). The port also sends the

VLAN ID for the voice VLAN that is configured on the port and the priority settings associated with this VLAN. The LLDP-MED Type Length Value (TLV) for this purpose is called the Network Policy TLV. You can double-check that the switch port is transmitting this TLV with this command:

```
ProVision# show lldp config <port ID list>
```

If you need to re-enable LLDP globally or on a port, or if you need to enable the LLDP-MED Network Policy TLV, use these commands:

```
ProVision(config)# lldp run

ProVision(config)# lldp admin-status <port ID list> tx_rx

ProVision(config)# lldp config <port ID list> medTlvEnable network_policy
```

You might also want to use other LLDP-MED TLVs, which you can read about in the configuration guide. (They are also enabled by default.)

There are a few more settings that the phone's switch port should implement when a computer connects through the phone. For example, you should set the port's untagged VLAN to the appropriate user or data VLAN. Also remember the correct setup for implementing 802.1X with multiple users, which you examined in Chapter 7.

Comware: Supporting LLDP-capable VOIP phones

Comware switches also support LLDP-capable VoIP phones (see Figure 8-33). You have several choices for configuring Comware switch ports to support these phones.

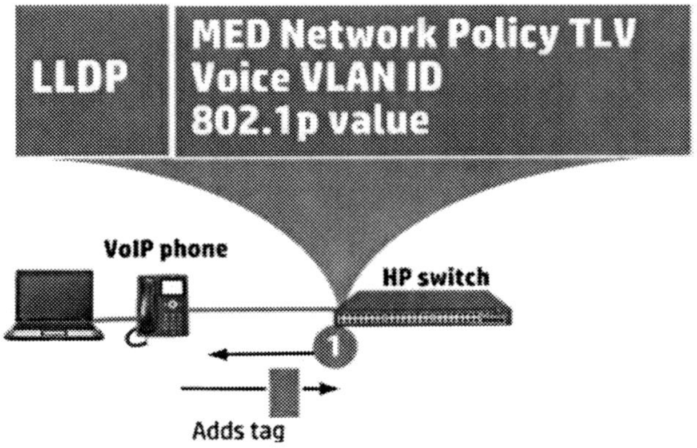

Figure 8-33: Voice VLANs and LLDP-MED on Comware switches

If an LLDP-MED-capable phone and a computer connect to a port, the basic settings are quite similar to those on the ProVision switch:

- LLDP enabled globally and also enabled on the port

  When you enable LLDP globally, it is, by default, enabled on the port.

- LLDP-MED TLVs enabled on the port, in specific the Network Policy TLV

  When LLDP is enabled on a port, by default it supports the LLDP-MED TLVs.

- The port defines the data VLAN as the untagged VLAN and permits tagged traffic in the voice VLAN

  Set the port to trunk mode, configure the data VLAN as the port VLAN ID (PVID), and permit both the data VLAN and the voice VLAN. Alternatively, you can also use a hybrid port, but a hybrid port is not necessary.

- Define the voice VLAN ID advertised by LLDP-MED:

  ```
  [Comware-<interface ID>]lldp voice-vlan <ID>
  ```

- Define the QoS settings advertised by LLDP-MED:

  ```
  [Comware-<interface ID>]voice vlan qos <802.1 value> [dscp]
  ```

Comware switch ports alternatively support a **voice vlan <*ID*> enable** command. However, you do *not* need to use this command when you define the voice VLAN command with the **lldp voice-vlan <*ID*>** command. Entering both commands can lead to unintentionally blocking voice traffic (when voice security mode is enabled and your phones have Organizationally Unique Identifiers (OUIs) not added to the list that security mode references). The LLDP option also provides faster processing and less latency when switches operate in an IRF virtual device.

Remember also to consider the authentication settings, if your network requires security. (Refer to "Chapter 7: Port-Based Authentication.") If you enforce 802.1X, make sure that the VoIP phone supports an 802.1X supplicant. You should also generally use MAC-based mode and authenticate the client and phone separately. The RADIUS server might or might not send a VLAN assignment for the phone. If it does, the new ID overrides the ID defined by the **lldp voice-vlan** command, and LLDP-MED advertises the phone's RADIUS assigned VLAN instead.

Comware: Supporting other VoIP phones

Comware provides an alternative setup for legacy phones that do not support LLDP-MED. You can use the **voice vlan <*ID*> enable** command to define the voice VLAN on the non-LLDP-MED-capable phone's switch port. For example, you might use this command when the port uses Cisco Discovery Protocol (CDP) to send a Cisco phone its correct VLAN ID or when the phone uses Dynamic Host Configuration Protocol (DHCP). For the CDP setup, you enable LLDP as explained in the previous section. However, you also enable CDP compatibility both globally and on the port:

```
[Comware] lldp compliance cdp

[Comware]interface <interface ID>
```

```
[Comware-<interface ID>]lldp compliance admin-status cdp txrx
```

Set up the QoS voice priorities before you enable the voice VLAN:

```
[Comware-<interface ID>]voice qos <802.1 value> [dscp]
```

You must then make several choices about how the voice VLAN becomes permitted on the port.

Automatic mode or manual mode

To understand the proper setup, you must first understand automatic mode and manual mode:

- Automatic mode (default setting when the voice VLAN is enabled)

  In automatic mode, you enable the voice VLAN on the switch, and the switch automatically makes the port a member of this VLAN if it detects a frame that contains a voice packet. The switch then starts an aging timer. If no other voice traffic is received before the timer expires, the switch removes the port from the voice VLAN (until it receives voice traffic again).

  To determine if a frame contains a voice packet, the switch looks for the Organizationally Unique Identifier (OUI) in the first 24 bits of a frame's source MAC address and compares this OUI to its OUI list. The OUI identifies phones from a particular vendor. For example, OUI 0001-e-300-0000 is assigned to Siemens phones, while OUI 0004-0d00-0000 is assigned to Avaya phones. The switch has a default OUI list, but you can add to or change entries in the list as needed.

  You might use automatic mode whenever you are unsure of the precise locations for IP phones. Then the switch adds the tagged voice VLAN to a port only when truly necessary.

- Manual mode

  With manual mode, you must explicitly assign the port connected to an IP phone to the voice VLAN, in addition to enabling the voice VLAN on the port.

  The OUI still affects operations in manual mode. When the switch receives a frame, it examines the first 24 bits of the frame's source MAC address and tries to match it against its OUI list. If the frame matches an OUI, the switch applies the QoS settings associated with the voice VLAN and forwards the frame. However, CDP- and LLDP-capable phones can receive these priorities and implement them on their own.

  You might use the manual mode in environments that connect users' IP phones and workstations to different switch ports.

Note that the OUI list affects how the switch automatically applies the voice VLAN to ports and QoS priorities to frames; it does not affect whether the switch sends the voice VLAN settings using CDP (or LLDP if you chose that setup). Therefore, it is possible for a phone to receive its voice VLAN without your having adding its OUI to the list. However, not adding phones' OUI to the list could cause the switch to block voice traffic, depending on the security mode.

Ensuring voice VLAN traffic can be transmitted on access, trunk, or hybrid ports

Often you configure hybrid ports to support IP phones; however, other types of ports can support these phones as well. The type of port you use and the mode you select (automatic or manual) depend on whether the IP phones send tagged or untagged traffic in the voice VLAN.

IP phones send tagged traffic

If the IP phones send tagged voice traffic, you can configure the port to be a trunk or hybrid port, and you can use either automatic or manual mode. See Table 8-5 for more detailed information about these settings.

If you want to use automatic mode, configure a data VLAN as the port's PVID for either a trunk or a hybrid port. The voice VLAN cannot be the PVID because the traffic in the PVID is untagged. For manual mode, you must additionally configure the trunk port or hybrid port to permit the tagged traffic in the voice VLAN.

Table 8-5: IP phones send tagged voice traffic

Type of port	Automatic or manual mode	Support for tagged voice traffic?	Configuration notes
Access	Automatic or manual	No	
Trunk	Automatic	Yes	Configure the PVID, which cannot be the voice VLAN, and permit the PVID on the port. Enable the voice VLAN on the port
	Manual	Yes	Complete the steps for setting up automatic mode. In addition, permit the voice VLAN on the port and undo automatic voice VLAN mode
Hybrid	Automatic	Yes	Configure the PVID, which cannot be the voice VLAN, and configure the port to permit untagged frames in the PVID. Enable the voice VLAN on the port
	Manual	Yes	Complete the steps for setting up automatic mode. In addition, configure the port to permit tagged frames in the voice VLAN and undo automatic voice VLAN mode

IP phones send untagged traffic

If IP phones send untagged traffic, you can configure the port as an access, trunk, or hybrid port. However, you cannot use automatic mode; you must use manual mode. Table 8-6 lists general configuration guidelines for each type of port.

Table 8-6: IP phones send untagged voice traffic

Type of port	Automatic or manual mode	Support for untagged voice traffic?	Configuration notes
Access	Automatic	No	
	Manual	Yes	Make the voice VLAN the PVID. Enable the voice VLAN and undo automatic mode
Trunk	Automatic	No	
	Manual	Yes	Configure the voice VLAN as the PVID. Permit the PVID on the port. Enable the voice VLAN and undo automatic mode
Hybrid	Automatic	No	
	Manual	Yes	Configure the voice VLAN as the PVID and configure the port to permit untagged frames in the voice VLAN. Enable the voice VLAN and undo automatic mode

Security mode

If you are using the **voice vlan <*ID*> enable** command to define ports' voice VLAN, you need to understand how the voice security mode can affect your solution.

Security mode, which is enabled by default, checks the source MAC address of each frame against this list. This mode is intended to prevent the wrong devices from sending traffic on the voice VLAN:

- An unassuming user accidentally connected a PC (rather than an IP phone) to a port that is configured for receiving untagged traffic in the voice VLAN.
- A malicious user created a large number of voice VLAN packets to consume the voice VLAN bandwidth, affecting normal voice communication.

To prevent this from happening, the switch uses security mode. When the voice VLAN works in security mode, the switch checks all the frames received in the voice VLAN to ensure the source MAC address matches an entry in its OUI list (the same list used by automatic mode; however, security mode uses this list to check all frames, whereas automatic mode just checks the initial frame). If the source MAC address does not match an entry in the OUI list, the switch drops the frame.

Because checking frames consumes system resources, you may want to change the security mode to normal mode, eliminating this OUI check. The switch will then automatically forward frames that are placed in the voice VLAN by their tagged or untagged status. Disabling security mode also ensures that the switch does not block legitimate voice traffic that has an OUI not on the voice list. If you find it difficult to manage the OUI list for your VoIP phones, you might choose to disable security mode.

Note that HP recommends that you do not transmit both data and voice traffic in the voice VLAN. However, if you do, you should also disable security mode so that the switch does not drop the data traffic (because the source address for data traffic does not include the OUI).

Table 8-7 shows how the switch handles tagged and untagged frames when security or normal mode is being used.

Table 8-7: Processing frames in security or normal mode

Security or normal mode	Tagged or untagged	Processing frame
Security mode (default setting)	Untagged frames	If a frame's source MAC address matches an entry in the OUI list, the switch forwards the frame. Otherwise, the switch drops the frame
	Frames with the voice VLAN tag	
	Frames tagged with other VLANs	Switch forwards the frame if the port is configured to permit the tagged VLAN. It drops the frame if the port does not permit the associated VLAN
Normal mode	Untagged frame	The switch does not compare the frame's source MAC address against the OUI
	Frames with the voice VLAN tag	
	Frames tagged with other VLANs	Switch forwards the frame if the port is configured to support the tagged VLAN. It drops the frame if the port does not permit the associated VLAN

To use the normal mode, enter:

```
[Comware] undo voice vlan security enable
```

To return the switch to security mode, enter:

```
[Comware] voice vlan security enable
```

HP IMC QoS Manager

The QoS Manager, an add-on component for HP Intelligent Management Center (IMC), can help you to define QoS settings and policies for your network infrastructure devices. The IMC QoS Manager service comes with several already configured QoS policies, which you can immediately deploy to various devices across your network. You can also customize policies as required for your environment and create policies of your own. In this way, IMC QoS Manager helps you to define complex policies in minutes with fewer chances of misconfiguration. You can then deploy the policies across many different devices almost as quickly.

QoS Manager also manages devices at a smaller scale, letting you configure various settings such as rate limits on particular managed devices.

Using predefined policies

Defining the appropriate criteria for selecting traffic for particular QoS settings presents one of the primary challenges of implementing a QoS solution. Administrators also often struggle to follow best practices for choosing settings for particular types of traffic.

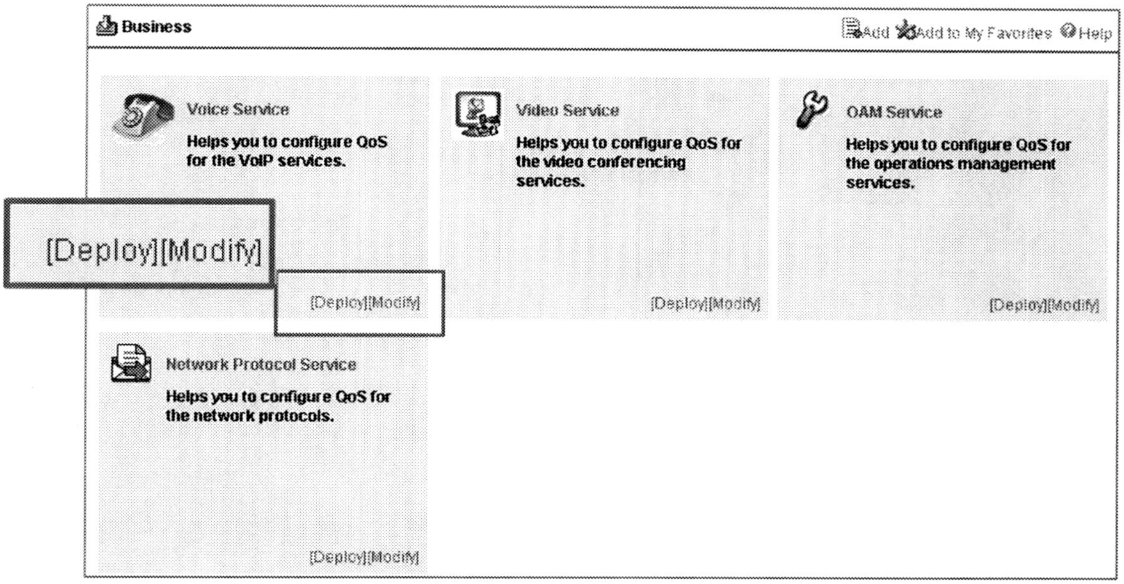

Figure 8-34: IMC QoS Manager

Figure 8-34 shows some of the QoS policies that IMC includes as part of the QoS Manager package. These policies configure best practice QoS settings for various common business scenarios such as voice and video traffic.

You can select a QoS policy for one of these scenarios and deploy the policy to your devices immediately. However, you should typically examine the policy first (click **Modify**) and double-check that its traffic classes select *your* traffic and its behaviors work well with your particular devices. Check behaviors in particular. Many define scheduling mechanisms that are appropriate for router ports, rather than the 802.1p values that are more appropriate for switches.

Creating and deploying policies

You can use the predefined policy to simplify and streamline the creation and deployment of policies in other ways (as Figure 8-35 illustrates).

For example, you can draw on their components as you create your own policies. You can add a policy by navigating to list of QoS devices, choosing devices, and choosing to add a QoS policy to one or more interfaces. You then create and deploy the policy at once. As you deploy the policy, you can select predefined traffic classifiers and choose just the components relevant to you. Then select the behavior and add the appropriate 802.1p value or other setting.

Alternatively, you can create QoS policies in advance. QoS Manager lets you define traffic classes and behaviors in intuitive, easy-to-manipulate pages. You combine these objects in QoS flow policies, following the same logic that you do when dealing with these objects in the CLI.

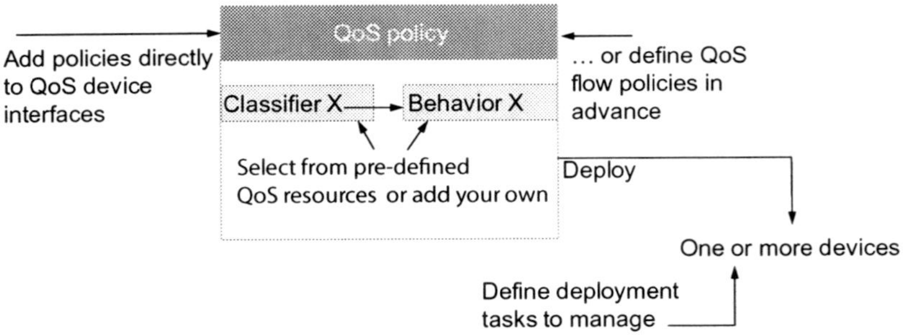

Figure 8-35: Using IMC QoS Manager to create and deploy policies

When you deploy the policy, you must select the device and interface to which the policy applies—or devices and interfaces. QoS Manager eases configuration with batch operations and scheduled deployment tasks.

Monitoring and managing other QoS settings

As you see, you can monitor managed devices as QoS devices, defined as any managed device for which QoS Manager supports at least some features. As Figure 8-36 shows, one useful piece of information that QoS Manager provides is a list of the QoS capabilities that the device supports. You can use this list as you plan policies However, when you deploy a policy, QoS Manager simply skips settings not supported by the targeted device, so you can plan the same policy for devices with different capabilities.

The device's detailed QoS settings reveal the policies, priorities, and rate limits applied to interfaces. You can view some of these settings even for switches that do not support QoS Manager's full policy provisioning settings.

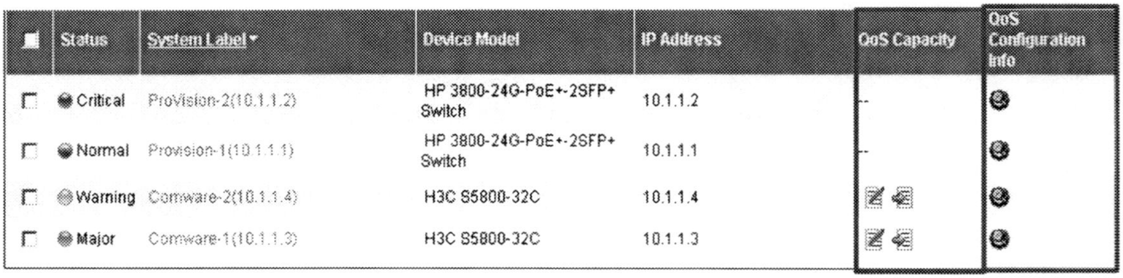

Status	System Label	Device Model	IP Address	QoS Capacity	QoS Configuration Info
Critical	ProVision-2(10.1.1.2)	HP 3800-24G-PoE+-2SFP+ Switch	10.1.1.2	--	
Normal	Provision-1(10.1.1.1)	HP 3800-24G-PoE+-2SFP+ Switch	10.1.1.1	--	
Warning	Comware-2(10.1.1.4)	H3C S5800-32C	10.1.1.4		
Major	Comware-1(10.1.1.3)	H3C S5800-32C	10.1.1.3		

Figure 8-36: Using IMC QoS Manager to monitor and manage QoS settings

Summary

This chapter described the need for switches to handle each traffic flow appropriately, depending on the network application, the traffic's business importance, contractual agreements for furnishing bandwidth, and other, minutely adjustable, criteria. You can now configure HP switches to prioritize incoming traffic and assign it to the appropriate queue for transmission. And in order to implement a consistent QoS strategy throughout the network, you can mark outgoing traffic with the appropriate QoS value.

You can also implement supported traffic policing and traffic shaping policies. Further, on Comware switches, you can apply WRED to help ports handle congestion in a graceful manner that minimizes its impact on the system as a whole as well as on important applications.

You can not only place VoIP phones in the correct VLAN but also use the voice VLAN and LLDP-MED to provision the phone with the correct priority settings.

Finally, you understand how all of these features work together—and how IMC QoS Manager helps you to configure and manage complete solutions.

Learning check

Answer these questions:

1. How do Provision switches honor 802.1p values in incoming frames? Do the switches always honor these values? How do the values map to queues?

2. You have defined a **car** action in a Comware traffic behavior. The action has a committed information rate (CIR) but not a peak information rate (PIR). It is using the default action for red traffic. What does the CIR define for a traffic class mapped to this behavior?

   a. The guaranteed minimum bandwidth

   b. The maximum bandwidth permitted when the network is experiencing congestion

   c. The maximum bandwidth permitted (with some allowance for bursting)

   d. The amount of bandwidth for the class that receives lower-priority transmission

3. What function does Comware weighted RED (WRED) provide? How does it help a port to avoid congestion? How does it protect high-priority traffic from being dropped?

4. A company has LLDP-MED-capable VoIP phones. What are the basic steps for setting up ProVision switches to support these phones?

Answers

1. How do Provision switches honor 802.1p values in incoming frames? Do the switches always honor these values? How do the values map to queues?

   They examine the 802.1p value in a frame's header and use that value to assign the frame to a "traffic class," an internal forwarding priority that determines the queue to which a packet is assigned.

   ProVision switches trust incoming 802.1p values by default, but can change a frame's 802.1p value by configuring a different value on the port or VLAN, by configuring a global setting, or by applying a QoS policy. The relationship between an 802.1p value and a queue is hard-coded; a specific 802.1p value always maps to the same traffic class.

   The 802.1p value 7 maps to the highest-priority queue. Normal priority traffic (value 0) is not mapped to the lowest numbered queue. Instead, there are two priority levels lower than normal.

   Depending on the switch, ProVision switches support up to eight egress queues, identified as 1–8, per port.

2. You have defined a **car** action in a Comware traffic behavior. The action has a committed information rate (CIR) but not a peak information rate (PIR). It is using the default action for red traffic. What does the CIR define for a traffic class mapped to this behavior?

   a. The guaranteed minimum bandwidth

   b. The maximum bandwidth permitted when the network is experiencing congestion

   c. The maximum bandwidth permitted (with some allowance for bursting)

   d. The amount of bandwidth for the class that receives lower-priority transmission

3. What function does Comware weighted RED (WRED) provide? How does it help a port to avoid congestion? How does it protect high-priority traffic from being dropped?

   If congestion causes a port drop all traffic at once, TCP sessions can synchronize, backing off and bursting together. This behavior makes congestion worse. WRED randomly drops more and more packets as a port handles more traffic but before congestion becomes severe. This behavior allows some TCP sources to back off.

   At the same time, WRED protects higher-priority traffic from being dropped. You can configure different WRED settings for different priority queues so that high-priority traffic is not dropped.

4. A company has LLDP-MED-capable VoIP phones. What are the basic steps for setting up ProVision switches to support these phones?

   First, make sure that LLDP is enabled on the switch and the switch interfaces. Also make sure that the appropriate LLDP-MED TLVs are enabled. (They should be by default.)

Identify the VLAN for IP phones and enter the voice command for that VLAN. It is also recommended that you configure an 802.1p value of 5 or above for this VLAN and an optional DSCP (the RFC-defined DSCP typical for voice is expedited forwarding or 46). The switch communicates these settings to the LLDP-MED-capable VoIP phone so that the phone can apply the priorities itself.

Assign the voice VLAN to switch ports as normal, depending on whether IP phones send untagged or tagged traffic. You must check the capabilities and configuration of your IP phones.

If a computer connects through the phone, you would typically set the voice VLAN as tagged on the port and the data VLAN as untagged.

9 Advanced OSPF

EXAM OBJECTIVES

In this chapter, you learn to:

- ✓ Deploy HP products in single-area and multiarea Open Shortest Path First (OSPF) systems
- ✓ Use area definitions and summaries to create efficient and scalable multiple-area designs.
- ✓ Advertise routes to external networks in a variety of OSPF environments.
- ✓ Promote fast, effective convergence during a variety of failover situations.
- ✓ Use virtual links as required to establish nondirect connections to the backbone.
- ✓ Implement OSFP authentication.

ASSUMED KNOWLEDGE

Before completing this chapter, you should understand basic OSPF operations. You should also be familiar with single-area OSPF systems.

INTRODUCTION

This chapter explains how to implement OSPF routing in multiple-area designs. It presents several advanced scenarios and then describes how to configure OSPF to meet the specific need outlined in the scenarios.

Reviewing basic OSPF concepts

Before delving into multiple-area OSPF designs, you will briefly review basic OSPF concepts.

OSPF, as an Interior Gateway Protocol (IGP), functions between routing devices within the same domain, or autonomous system (AS), which is generally defined as a group of networked devices under the control of the same entity. OSPF enables the routing devices to exchange information with each other about the IP subnets within the AS to discover routes between them.

Typically, an OSPF router automatically discovers its OSPF neighbors on each OSPF interface (which is a Layer 3 interface with an IP address on which OSPF is enabled). The router then establishes adjacency with certain neighbors:

- The single neighbor on a point-to-point interface
- The designated router (DR) and backup designated router (BDR) of a broadcast or nonbroadcast multiaccess (NBMA) network

  An Ethernet network is an example of a broadcast network.

An OSPF router generates a router link-state advertisement (LSA), or type 1 LSA, with information about:

- Its own OSPF networks or links
- The cost of using those links

The DR on a broadcast or NBMA network also generates a network, or type 2, LSA with information about all the routers on that network.

The cost of an LSA generally relates to the bandwidth. Higher-bandwidth links have a lower cost. However, administrators can set the cost as they choose. The router labels itself as the advertising router for these LSAs, using a consistent router ID which is typically a loopback interface address.

To achieve full adjacency, OSPF neighbors exchange LSAs until their link-state databases (LSDBs) match. In this way, information about the complete topology spreads from neighbor to neighbor throughout the network. Although every router on a broadcast link does not necessarily achieve full adjacency, all routers achieve full adjacency with the DR and BDR, ensuring they all have identical LSDBs.

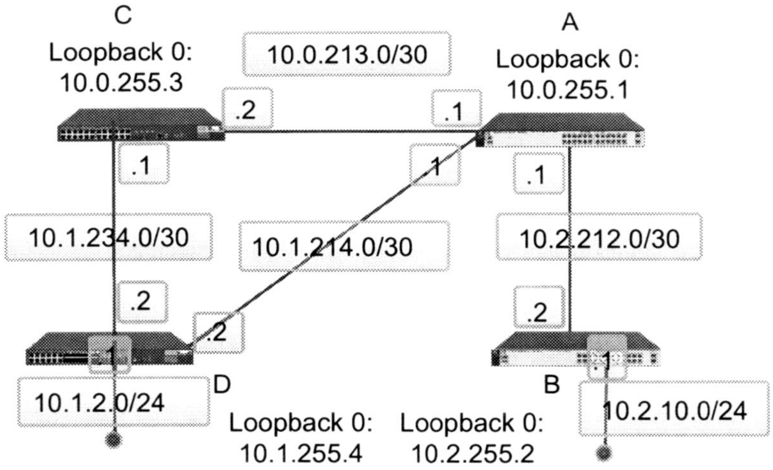

Figure 9-1: OSPF single area

For example, the routers in Figure 9-1 have learned that:

- Router A (10.0.255.1) connects to:
  - 10.0.213.0/30 (same network as router C, 10.0.255.3)

  - 10.1.214.0/30 (same network as router D, 10.1.255.4)
  - 10.2.212.0/30 (same network as router B, 10.2.255.2)
- Router B (10.2.255.2) connects to:
  - 10.2.10.0/24
  - 10.2.212.0/30 (same network as router A, 10.0.255.1)
- Router C (10.0.255.3) connects to:
  - 10.0.213.0/30 (same network as router A, 10.0.255.1)
  - 10.1.234.0/30 (same network as router D, 10.1.255.4)
- Router D (10.1.255.4) connects to:
  - 10.1.2.0/24
  - 10.1.214.0/30 (same network as router A, 10.0.255.1)
  - 10.1.234.0/30 (same network as router C, 10.0.255.3)

The designated router (DR) in each broadcast network has also sent out network LSAs for its networks, indicating which routers connect to it.

The routers have enough information to create a topology (sometimes called a directed graph), which is similar to the one in Figure 9-1. That is, they know which routers connect to other routers by the links that they share, and they know which routers offer gateways to which stub networks, which are networks advertised by only that router.

Each router then calculates the best route to each network—each advertised link—based on this topology. Specifically, they run the Djikstra algorithm, which determines the lowest-cost path between the router and the network.

If information changes—for example, one of the OSPF router's links fails—the router with the new information sends a link-state update (LSU) with the new or withdrawn LSA and waits for an acknowledgment. Neighbors flood this information among each other until it propagates throughout the system. The DR in a broadcast network is responsible for forwarding the information on the Ethernet network.

Table 9-1 provides information about LSA types. You already learned about type 1 and 2 LSAs. You will learn about LSA types 3 to 7 in the rest of this chapter.

Table 9-1 also lists extended or opaque LSAs, which enable features such as OSPF graceful restart (GR) and traffic engineering. Routers only advertise opaque LSAs to neighbors that indicate that they are opaque capable during adjacency establishment. You will learn about graceful restart in the "OSPF Graceful Restart Process" section. Other opaque LSAs are beyond the scope of this study guide. As indicated in Table 9-1, some LSAs have special functions in OSPFv3, which is the OSPF version for IPv6 routing. OSPFv3 is also not covered in this study guide.

Table 9-1: LSA types

Type	Name	Information advertised	Advertised by	Scope
1	Router	The OSPF links on a routing device (transit and stub networks)	Each OSPF routing device	Area
2	Network	The routers on broadcast or NBMA networks	DR for the network	Area
3	Summary	Routes to networks (or aggregated ranges of networks) in another area (copied from or into the backbone area)	ABR	Area
4	ASBR	Route to an ASBR	ABR	All nonstub areas
5	External (or ASE)	Routes imported into OSFP from another routing protocol or method	ASBR	All nonstub areas
7	NSSA external	Routes imported into OSFP from another routing protocol or method	ASBR	NSSA
8	Link-local (OSPFv3 only)	List of link-local IPv6 addresses on a link	Each OSPFv3 routing device	Link
9	Opaque link-local (OSPFv2) Intra-area prefix (OSPFv3 only)	OSPFv2, extensions (such as graceful restart requests) OSPFv3, the network prefixes associated with stub and transit networks in type 1 and 2 LSAs	Each OSPF routing device (which is capable)	Link (OSPFv2) Area (OSPFv3)
10	Opaque area-local	Extensions (such as traffic engineering)	Each OSPF routing device (which is capable)	Area
11	Opaque AS	Extensions	ASBR (which is capable)	All nonstub areas

OSPF multiarea overview

OSPF allows you to divide autonomous systems into multiple areas. In this chapter, you will not only learn how to establish multiple areas but also *why* you would want to do so. You will explore the differences between intra-area OSPF routes and interarea OSPF routes, as well as the LSAs that help to propagate routing information across the area boundaries.

Figure 9-2 displays an OSPF network, which consists of two segments: segment 1 and segment 2. These segments might be a campus LAN and a data center, a campus LAN and a large branch office, buildings in a large three-tier campus topology, or any other segments within a larger network. Segment 1 and segment 2 clouds each represent many routing switches, which ultimately connect to two core routing switches. These core routing switches connect through a network backbone, which might also include more routers and routing switches. Figure 9-2 shows only two segments for simplicity, but a large enterprise network might have several such segments.

Every OSPF router in an area must maintain LSAs from every other router in the area in its LSDB. gate across the system. As you recall, type 1 LSAs are router LSAs. Routers use type 1 LSAs

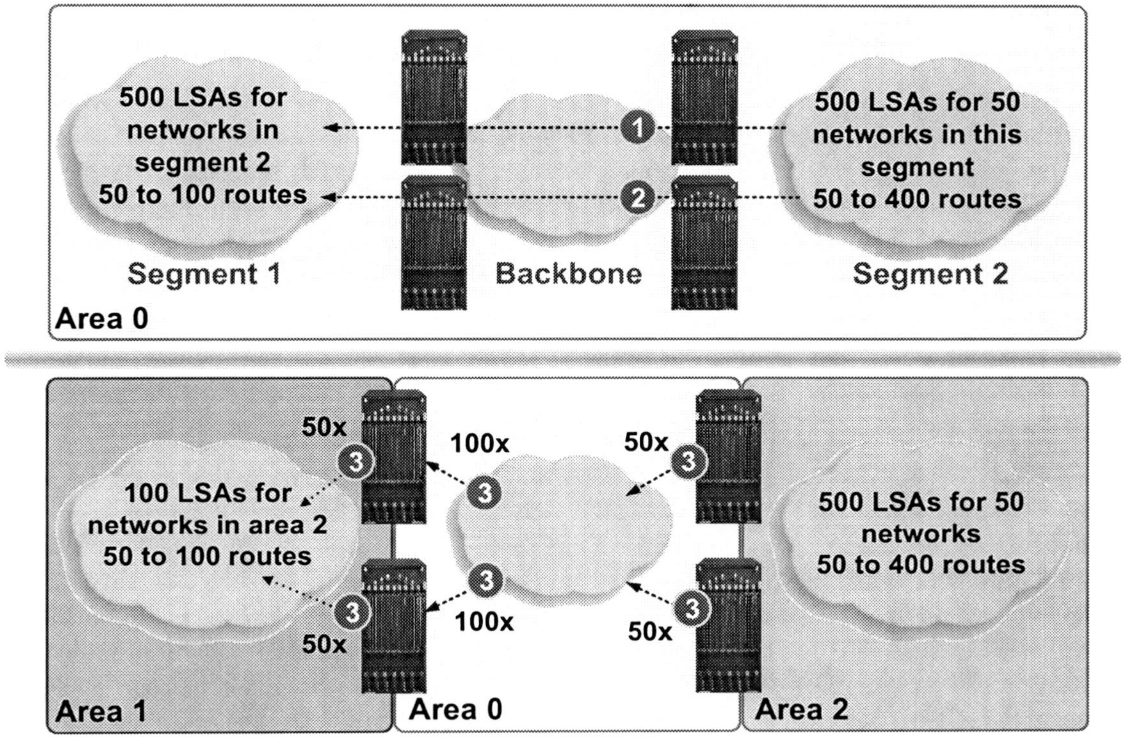

Figure 9-2: Multiple OSPF areas

to advertise that they are on the network and list links and the related metrics to other routers in the area. Type 2 LSAs are network LSAs. OSPF designated routers (DRs) use type 2 LSAs to provide information about all the routers on a network segment. (Table 9-2 lists LSA types and their use.)

Table 9-2: LSA types

LSA type	Description
Type 1	Router LSA
Type 2	Network LSA
Type 3	Summary LSA
Type 4	ASBR summary LSA
Type 5	External LSA
Type 6	Group membership LSA
Type 7	Routers in a not-so-stubby area (NSSA)
Type 8	Link-local-only LSA for OSPFv3
Type 9	Link-local-only "opaque" LSA in OSPFv2 and the intra-area prefix LSA in OSPFv3
Type 10	Area-local "opaque" LSA
Type 11	AS "opaque" LSA

In addition, Figure 9-2 only mentions LSAs associated with segment 2. Each OSPF routing device would also maintain LSAs for routers and networks in segment 1 and the backbone.

As a domain grows, the LSDB becomes excessively large and difficult to process. The shortest-path algorithm also becomes complicated, and the OSPF routing switch must dedicate valuable processing resources to running it. Often, this resource consumption comes with little value. As shown in Figure 9-2, networks often feature aggregation such that one segment connects to another through a few core devices. With only two exit points from segment 1 toward segment 2, the routing switches in segment 1 have no reason to maintain so much granular information about networks in segment 2.

Modern routers and routing switches often do have the power to handle large LSDBs and many calculations. However, there is no reason to burden the devices, network administrators, and support staff with the large, complex LDSBs and routing tables when most of the routes will use the same one or two next hops.

In addition, suppose that two routing switches in segment 2 provide a connection to the subnet A. One of those switches has a flapping interface. It loses its connection to subnet A and sends an LSU, withdrawing that subnet from its list of links. Every routing switch in the network backbone and segment 1 receives this LSU, adjusts the LSDB, and reruns the Djikstra algorithm on the topology. However, these routers still reach subnet A through the same core routing switches. When the flapping interface comes back up, the switch sends another LSU, which propagates throughout the entire network, causing every OSPF router to recalculate shortest paths, again without any significant change.

OSPF areas segment the domain into more stable and manageable pieces. Each OSPF routing device maintains the LSDB and runs the shortest-path algorithm only for its own areas. OSPF routers called area border routers (ABRs) forward information necessary for routing between areas. As you see in Figure 9-2, the ABRs act as bottlenecks for LSAs, forwarding only one LSA per network into other areas.

In the simplest configuration, a multiarea design decreases LSDB size dramatically and increases stability—routers do not have to rerun the shortest-path algorithm as often—but does not greatly affect the number of routes. OSPF, however, provides options for further optimizing the LSDB and routing table sizes. As you will see, you can aggregate advertisements of networks in between areas, simplifying the routing table. You can also control routes at this boundary, filtering them if you so desire. These benefits still apply even for powerful routers that could support very large areas.

ABRs

You now understand that each OSPF area defines an isolated region, and type 1 and type 2 LSAs and shortest-path topologies do not extend beyond this area. However, some routing information must flow between the areas. Area border routers (ABRs) are responsible for forwarding this information.

ABRs have interfaces in at least two areas, one of which is area 0. Area 0 is the backbone area, and OSPF design requires all areas to connect to this backbone (you will explore an exception later in this chapter).

As a member of multiple areas, the ABR maintains multiple LSDBs, one for each area. In Figure 9-3, ABR-A receives type 1 and type 2 LSAs from router C and router D, which reside entirely within area 1 and are called internal routers. ABR-A, router C, and router D have identical area 1 LSDBs.

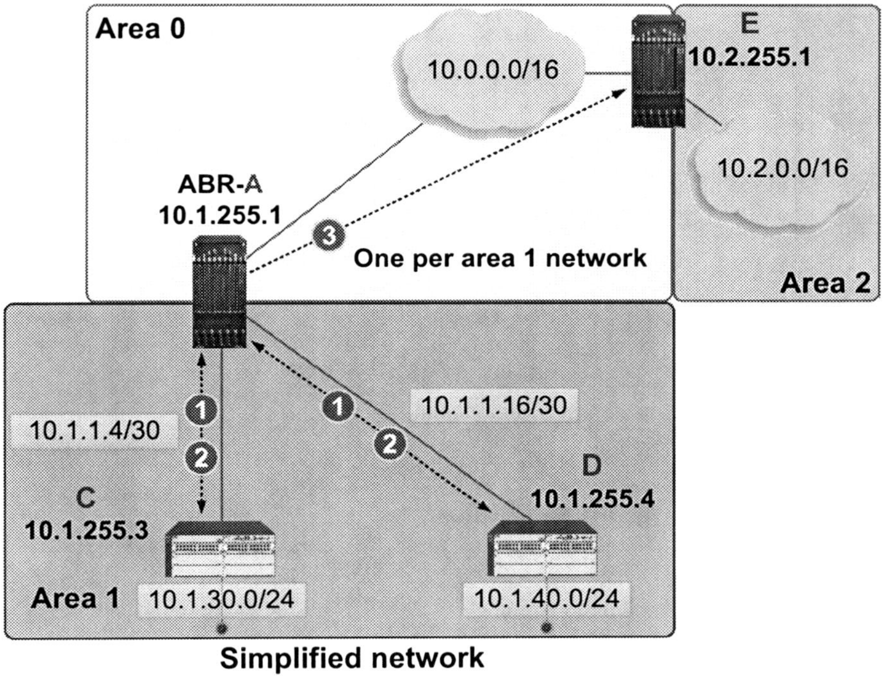

Figure 9-3: ABRs

The ABR follows the same process in area 0, exchanging type 1 and type 2 LSAs and synchronizing its area 0 LSDB with routers such as ABR-E.

In each area, the ABR also creates type 3, or summary, LSAs which advertise routes that it knows in *other* areas. Figure 9-3 shows how the ABR performs this function for area 0. The ABR creates a type 3, or summary, LSA for each network in *area 1*. It places these type 3 LSAs in its LSDB for *area 0*. The ABR can then advertise its routes to area 1 networks to other routers in area 0.

For example, ABR-A has an interarea route to 10.1.30.0/24. It learned about this network from router C's type 1 LSA, and its area 1 LSDB contains that LSA. ABR-A generates a type 3 LSA for 10.1.30.0/24 and places the type 3 LSA in its area 0 LSDB. Other area 0 routers, such as ABR-E, receive ABR-A's type 3 LSA and add it to their area 0 LSDB. They can then calculate an interarea route to 10.1.30.0/24.

In addition to the network address, the type 3 LSA includes a metric, which enables routers to select the best interarea route when multiple ABRs advertise a type 3 LSA with the same network ID. (You will learn more about the interarea route calculation process in a moment.)

Figure 9-4 shows how ABR-A also generates type 3 LSAs to advertise into nonbackbone areas.

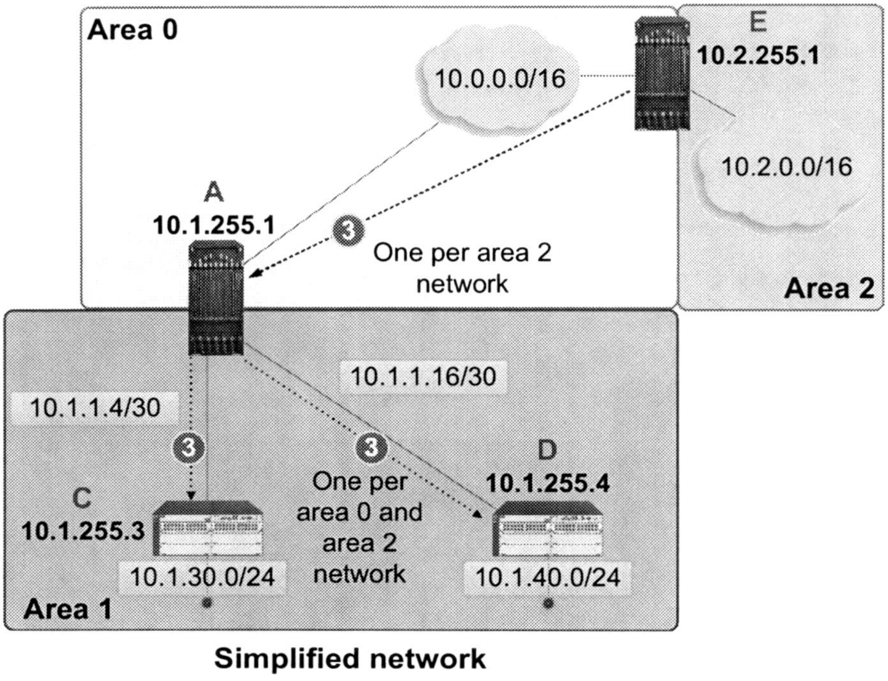

Figure 9-4: ABRs

First, just as ABR-A advertised in area 0 a type 3 LSA for each area 1 network, the ABR advertises in area 1 a type 3 LSA for each area 0 network.

The ABR also advertises a type 3 LSA for each *interarea* route that it has learned from type 3 LSAs advertised by other ABRs in area 0. For example, ABR-E has generated type 3 LSAs for each network in area 2 and advertised them in area 0. Using these LSAs, ABR-A has calculated an interarea route to each area 2 network. Now ABR-A generates type 3 LSAs for these routes and advertises them in area 1, using the metric in the routing table. (It does *not* advertise these LSAs in area 0 because the interarea routes' next hops are in area 0. The original ABR, which, of course, always resides in area 0, has already advertised these LSAs in area 0.)

If ABR-A connected to another area, such as area 3, it would place type 3 LSAs for each area 3 intra-area route in the area 0 LSDB and the area 1 LSDB. In other words, for each area, an ABR creates type 3 LSAs with routes to *other* areas to which the ABR connects. Remember, though, that the interarea routes always pass through area 0 even if the same ABR supports multiple nonbackbone areas. That is, you cannot configure an OSPF router with area 1 and area 3 and expect the router to advertise area 3 networks in area 1 and area 1 networks in area 3. The router must have interfaces in area1, area 3, *and* area 0.

Figure 9-4 shows how internal routers C and D in area 1 receive the type 3 LSAs for area 0 and area 2. Note that, although area 1 routers learn routes to area 2, the type 3 LSAs themselves do not cross area boundaries. Instead ABR-E, connected to area 2 and area 0, generates type 3 LSAs for area 2 in area 0. It is the advertising router for these LSAs. ABR-A receives these LSAs in area 0 and generates *new* type 3 LSAs for area 1, in which it is the advertising router.

Calculating the best interarea routes

You have seen how ABRs generate type 3 LSAs for routes to networks in their nonbackbone areas and how they exchange these LSAs through area 0. You have also seen how the ABRs advertise all their interarea routes into nonbackbone areas, again using type 3 LSAs.

Next, you will examine a bit more closely how routers use the type 3 LSAs to generate interarea routes (see Figure 9-5).

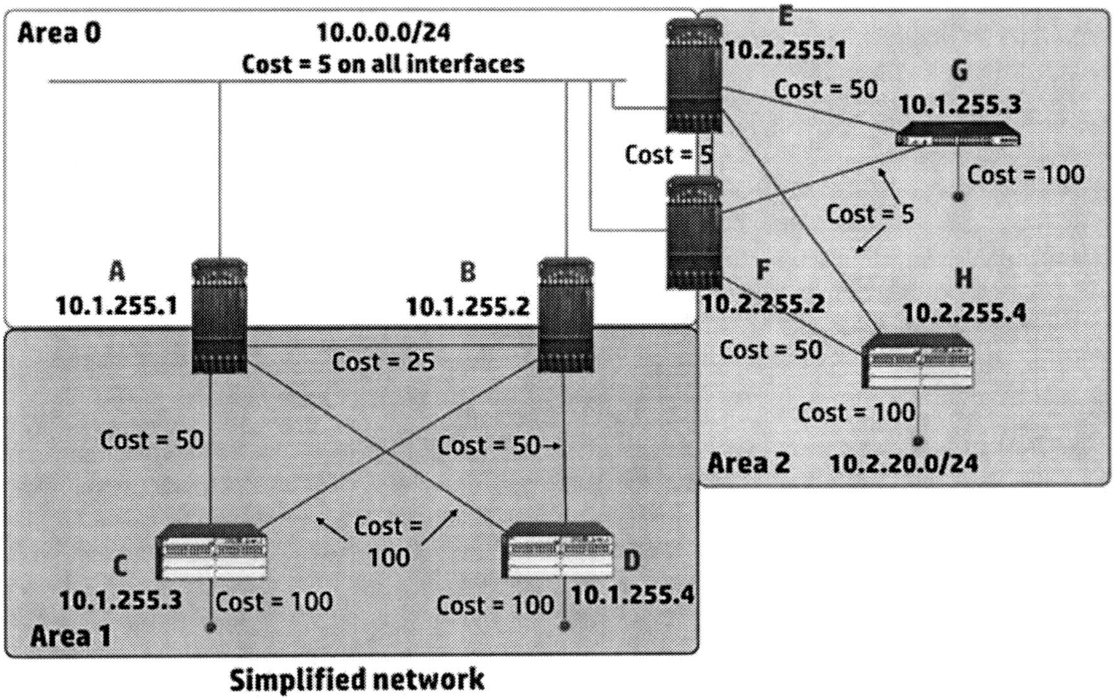

Figure 9-5: Calculating the best interarea routes

A type 3 LSA advertises a *route* to the network. The routers that receive these advertisements do *not* create a topology of the networks in other areas. Instead, they treat the ABR as a gateway to all of the networks which the ABR advertises in type 3 LSAs. That is, those networks are leaves hanging off the ABR in the shortest-path tree.

Each router determines the shortest path to the ABR or ABRs advertising type 3 LSAs to it. It calculates this path as normal for interarea routes by applying Djikstra's algorithm to the area in which the router connects to the ABR. The next hop for reaching the ABR becomes the next hop for all interarea routes advertised by that ABR in type 3 LSAs.

If a router receives multiple type 3 LSAs for the same network, as it does when the area includes multiple ABRs, it learns multiple interarea routes for that network. It uses the metric to choose between the routes. The metric for each interarea route consists of:

- Metric advertised in the type 3 LSA, which the ABR copies from its metric for the route in question
- Metric for reaching the advertising ABR (from area shortest-path topology)

Note that, because each ABR advertises the cost of its own route to the network, cost increments across the complete system. For example, ABR-E advertises 10.2.20.0/24 as a type 3 LSA with metric 105 in area 0, and ABR-F advertises this route with metric 110. ABR-A adds 5 to each of these metrics because that is its cost for reaching each of the routers. It selects ABR-E as offering the best route and adds an interarea route with metric 110 to its OSPF table. When ABR-A generates an LSA for 10.2.20.0/24 in area 1, it advertises metric 110. Router C adds its metric for reaching ABR-A to 110, assigning the interarea route to 10.2.20.0/24 a metric of 160.

Interarea routing in OSPF combines the best aspects link-state and distance-vector routing. Router C, for example, uses link-state routing to determine its shortest path to its two ABRs. It can then use those same two paths for all routes to other areas without considering the link states of networks in those areas. Even though router C only knows a small part of the topology, the traffic that it routes toward other areas still receives the benefits of link-state routing across the complete path. For example, router C sends traffic to 10.2.20.0/24 on its shortest path to ABR-A. ABR-A, in turn, routes the traffic using the shortest path to ABR-E, which it has calculated from its detailed area 0 topology. Finally, ABR-E routes the traffic using the best path calculated by the detailed area 2 topology. In each area, routers maintain just the local segment of the complete path for interarea traffic.

Also consider how this system enhances stability. When an OSPF router receives an update for a type 3 LSA, it does not run the complete shortest-path algorithm. Instead, it adjusts the networks behind each ABR, as indicated by the update, and adds a new interarea route, removes an interarea route, or selects a different interarea route (due to a cost change), as appropriate. The router *does not*, on the other hand, need to recalculate the path to the ABR, which is the only segment of the path with which it is concerned.

Need for route aggregation between areas

You will now explore several strategies for optimizing a multiarea configuration.

First, you should minimize the LSAs advertised between areas. In the example network shown in Figure 9-6, both area 1 and area 2 have only two exits toward the network backbone and other areas. Internal area 1 routers and backbone routers might benefit from some level of detail in the information that they receive about area 2. For example, each ABR's type 3 LSAs reflect the metrics of reaching the network in question, enabling routers in other areas to select the best ABR for reaching each network in area 2.

However, with only two entry points into area 2, routers (and, in particular, internal area 1 routers) are probably receiving too granular an amount of information about every subnet. This information clutters the routers' LSDB and routing tables.

In addition, small changes in the area 2 topology might send a ripple through the backbone and area 1. For example, suppose that network 10.2.20.0/24 in area 2 becomes unavailable. The two area 2 ABRs withdraw their type 3 LSA for that specific network in area 0, and then the two area 1 ABRs do the same in area 1. However, the area 1 routers have no other path to this network. It might be simpler not to inform routers beyond area 2 of the small change. If they continue to forward traffic for this network, the area 2 ABRs simply drop it.

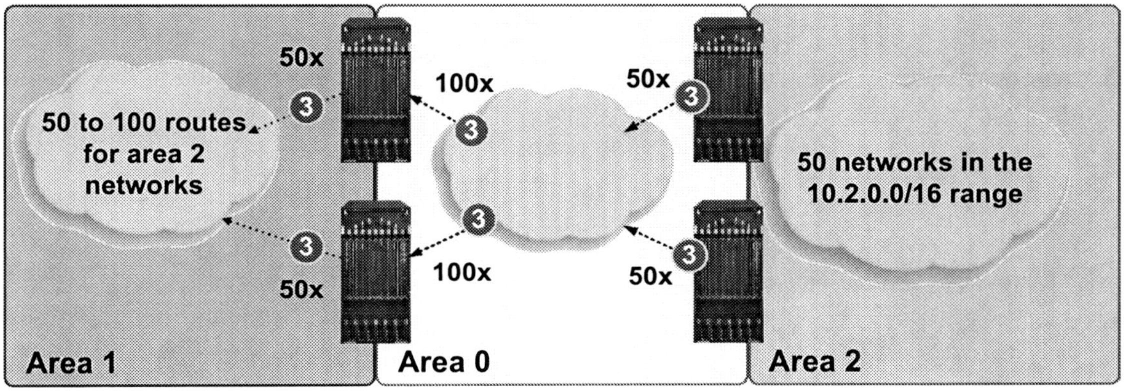

Figure 9-6: Route aggregation between areas

You need to aggregate the type 3 LSAs to hide the unnecessary information and to smooth over small changes, thereby gaining the true benefits of a multiple-area OSPF configuration.

Note

Although this example focuses on aggregating advertisements for area 2 networks, you would aggregate summaries of area 1 or area 0 networks for similar reasons.

Aggregating summaries

You can configure an ABR to aggregate its summary routes. Rather than generate a type 3 LSA for every network in an area, the ABR generates a few type 3 LSAs, each of which includes many networks in its range. In other words, the type 3 LSA advertises a network with a shorter prefix length such as a/16 subnet that includes many/24 subnets.

You configure the area summaries on the ABR that connects to the area for which you are aggregating the summaries. For example, in the network illustrated in Figure 9-7, you configure area summaries for area 2 on the two ABRs on the right.

These ABRs then advertise the summaries to other backbone routers and ABRs in area 0. If you create area summaries for area 1 on the ABRs shown on the left, those summaries aggregate area 1 networks. The area 1 ABRs advertise them into area 0 and other ABRs readvertise them into area 2.

Figure 9-7 illustrates how area summaries drastically reduce the number of LSAs associated with area 2. In this example, a single summary includes all of the networks in the area.

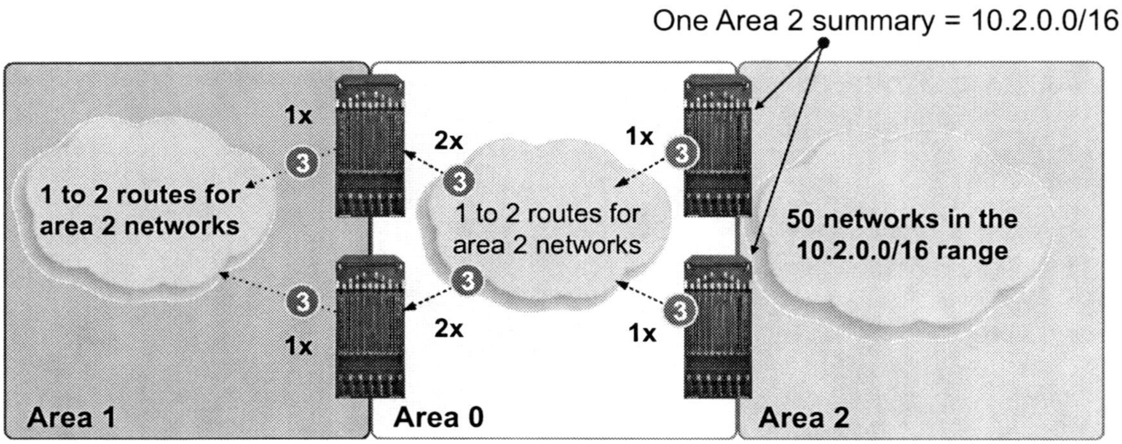

Figure 9-7: Aggregating summaries

You enter the HP Comware command for creating this area summary in the area view of the OSPF process:

```
[ComwareABR-ospf-1-area-0.0.0.2] abr-summary 10.2.0.0 16
```

You enter the HP ProVision command for creating this area summary in the OSPF configuration mode context:

```
ProVisionABR(ospf)# area 2 range 10.2.0.0/16
```

When you enter one of these commands, the ABR automatically suppresses the generation of individual type 3 LSAs for subnets within the specified network. However, it *continues* to advertise individual type 3 LSAs for networks that do not fall into the range. This is why it is recommended to assign an area contiguous IP addresses within a binary block. Then you can easily aggregate the routes.

Note also that the ABR only advertises the summary if it is receiving advertisements for at least one network within that range.

The ABR actually adds an OSPF route to the aggregated subnet to its routing table. The forwarding interface for this route is null (or black hole), which means the ABR drops traffic that matches this route. Traffic matches the more specific intra-area routes first, ensuring that the ABR can still route traffic correctly. The null route lets the ABR drop any traffic that arrives for an address in the aggregated range that does not actually exist in the area.

An ABR sets the metric for a type 3 LSA by adding the cost of the advertising interface to the cost of the interarea route. When you configure an area summary, the ABR adds the cost of the advertising interface to the cost of the interarea route with the highest cost in the aggregated range.

In an environment with multiple ABRs for the same area, you might want to alter this behavior. For example, one ABR might have a higher cost for a couple of networks but a lower cost for most networks overall. You can set the metric for the area summary route manually, giving you more control. The manual metric might also stabilize the system by ensuring that the cost never changes.

The Comware command is:

```
[Comware-ospf-<ID>-area-<ID>] abr-summary <network address> <prefix length>
cost <metric>
```

The ProVision command is:

```
ProVision(ospf)# area <ID> range <network address></prefix length> type
summary cost <metric>
```

Return to the example. The ABRs for area 1 now receive *one* type 3 LSA with the aggregated summary from *each* of the two ABRs that connect to area 2. Each area 1 ABR selects the best LSA for the summary range and creates an interarea route to 10.2.0.0/16 through the best next hop.

Each area 1 ABR then generates an area 1 type 3 LSA to advertise this interarea route to routers in area 1. The internal area 1 routers now have two LSAs for all area 2 networks, and they learn routes to only one aggregated network (up to two routes if the ABRs advertise equal metrics and the internal routers support Equal-Cost Multipath [ECMP]).

Usually, you want to create as few area summaries as possible. Sometimes, though, you need to create more summaries. For example, areas might have grown organically and without advance planning, creating noncontiguous ranges or ranges that do not fit neatly within a Classless Inter-Domain Routing (CIDR) subnet. For example, area 2 includes networks in the 10.2.0.0-10.2.255.255 range as well as the 10.3.20.0-10.3.23.255 range. Area 3 includes networks in the 10.3.0.0-10.3.19.255 range and the 10.3.30.0-10.3.39.255 range. You cannot simply use 10.2.0.0/15 for the area 2 summary and 10.3.0.0/16 for the area 3 summary due to the noncontiguous and overlapping ranges. Instead, you would need to use several area summaries for area 2 and several summaries for area 3.

You could also use multiple-area summaries when an area has multiple ABRs, and you want to prefer one ABR for one segment of the area but another ABR for another segment. This approach is more complicated, however. You should take care in configuring matching summaries on each ABR (although the costs would not match). Otherwise, you can accidentally introduce routing loops.

Area 1 routers' routing tables

After you configure area summaries on an area's ABR, you can see the summaries in the routing tables for routers *not* connected to that area.

The sections below show examples of Comware and ProVision routing tables with the area 2 summaries.

HP Comware

The HP Comware example shows the route to 10.2.0.0/16 for an ABR connected to area 1. Other routes have been removed for clarity. As you see, the cost for the route has incremented across area 0. The area 1 ABR has selected a route through the ABR closer to it.

Note also that the area for this route is area 0 because the ABR receives the type 3 LSA from the area 2 ABR in area 0:

```
<ABRArea1> display ip routing

Destination/Mask    Proto    Pre    Cost    NextHop     Interface

10.2.0.0/16         OSPF     10     400     10.0.0.2    Vlan1000

<ABRrea1> display ospf routing

Destination         Cost     Type   NextHop  AdvRouter   Area

10.2.0.0/16         400      Inter  10.0.0.2 10.0.255.2  0.0.0.0

  Total Nets: 551

  Intra Area: 50 Inter Area: 1 ASE: 500 NSSA: 0
```

HP ProVision

The HP ProVision example shows the route to 10.2.0.0/16 for an internal router in area 1. Other routes have been removed for clarity. As you see, the internal router has learned the route through the ABR. The route's metric reflects the cost of reaching the ABR and of the ABR.

```
InternalArea1# show ip route

                                    IP Route Entries

Destination    Gateway       VLAN   Type   Sub-Type      Metric   Dist.
-------------  ----------    ----   -----  ---------     -------  -----

  10.2.0.0/16  10.1.1.5      104    ospf   InterArea     500      110

  10.2.0.0/16  10.1.1.13     112    ospf   InterArea     500      110
```

Area 1 routers' LSDB

The sections below include examples of type 3 LSAs associated with the aggregated summary route.

Comware

The Comware example shows the LSAs on the ABR connected to area 1. As you see in the area 0 database, the ABR has received a summary LSA for 10.2.0.0/16 from both ABRs connected to area 2. Both LSAs have the same metric because, in this example, both area 2 ABRs, from 10.0.255.2, have a lower metric.

In the area 1 database, you see the summary LSA for 10.2.0.0/16 advertised by this ABR into area 1. The area 1 ABR has added the cost of reaching the closer area 2 ABR to the metric. This database also includes the summary LSA for 10.2.0.0/16 being advertised by the other area 1 ABR (10.0.255.3).

(Only type 3 LSAs associated with 10.2.0.0/16 are shown for clarity.)

```
<Internal>display ospf lsdb summary

OSPF Process 1 with Router ID 10.0.255.1

                 Area: 0.0.0.0

                 Link State Database

Type        : Sum-Net

LS ID           : 10.2.0.0

Adv Rtr         : 10.0.255.2

LS Age          : 930

Len             : 28

Options         : DC E

Seq#            : 80000009

Checksum        : 0x3bb4

Net Mask        : 255.255.0.0

Tos 0   Metric: 300


Type            : Sum-Net

LS ID           : 10.2.0.0

Adv Rtr         : 10.0.255.4

LS Age          : 1099

Len             : 28

Options         : E

Seq#            : 80000009

Checksum        : 0x11fc
```

```
Net Mask       : 255.255.0.0
Tos 0 Metric: 300
                         Area: 0.0.0.1
                    Link State Database
Type           : Sum-Net
LS ID          : 10.2.0.0
Adv Rtr        : 10.0.255.3
LS Age         : 1227
Len            : 28
Options        : DC E
Seq#           : 8000000a
Checksum       : 0x1f6a
Net Mask       : 255.255.0.0
Tos 0 Metric: 400

Type           : Sum-Net
LS ID          : 10.2.0.0
Adv Rtr        : 10.0.255.1
LS Age         : 1320
Len            : 28
Options        : E
Seq#           : 8000000a
Checksum       : 0xd9e
Net Mask       : 255.255.0.0
Tos 0 Metric: 400
```

ProVision

Below you see the summary LSAs for the aggregated area 2 routes on an internal router in area 1. Note that the LSAs are associated with area *1* (not 2 or 0) because the area 1 ABRs generated new summary LSAs for this area. The advertising routers are the area 1 ABRs (but the original aggregation was configured on the area 2 ABRs):

```
InternalArea1# show ip ospf link-state type summary

OSPF Link State Database for Area 0.0.0.1

                              Advertising
LSA Type     Link State ID    Router ID     Age    Sequence #    Checksum
---------    --------------   -----------   ----   -----------   --------

Summary      10.2.0.0         10.0.255.1    888    0x8000000a    0x00000d9e

Summary      10.2.0.0         10.0.255.3    794    0x8000000a    0x00001f6a

InternalArea1# show ip ospf link-state detail type summary

 OSPF Link State Database for Area 0.0.0.1

LSA Age                            : 1235

LSA Type                           : 0x3 (Summary)

Advertising Router                 : 10.0.255.1

Link State ID                      : 10.2.0.0

LSA Sequence                       : 0x80000007

LSA Checksum                       : 0x139b

LSA Option Bits                    : E=1 MC=0 N/P=0 EA=0 DC=0

LSA Metric                         : 400

Network Mask                       : 255.255.0.0

LSA Age                            : 1392

LSA Type                           : 0x3 (Summary)

Advertising Router                 : 10.0.255.3

Link State ID                      : 10.2.0.0

LSA Sequence                       : 0x80000015

LSA Checksum                       : 0x975

LSA Option Bits                    : E=1 MC=0 N/P=0 EA=0 DC=1

LSA Metric                         : 400

Network Mask                       : 255.255.0.0
```

Filtering interarea routes

Remember that failing to configure an area summary for a specific range makes the routing switch continue to advertise the individual routes. If you want to *stop* advertisement of certain routes to certain networks entirely, you must configure a nonadvertised area summary on the ABR.

For example, a company has a strict security policy for one of the sites that make up OSPF area 2. The policies allow only local access to high-security subnets 10.2.128.0/17; devices in other areas can reach subnets 10.2.0.0/17. Of course, a firewall can filter out prohibited traffic, but in this case, you want to prevent any routing of traffic from other areas to the restricted networks.

Therefore, you configure two area summaries on the area 2 ABRs, one for 10.2.0.0/17 and one for 10.2.128.0/17. ABRs advertise the first but not the second (see Figure 9-8).

Note that the filtering occurs at the area 2 boundary; you cannot apply the filtering to some areas and not others. For example, you *cannot* use this configuration to block interarea routes to the area 2 subnets in area 1 but allow them in area 0.

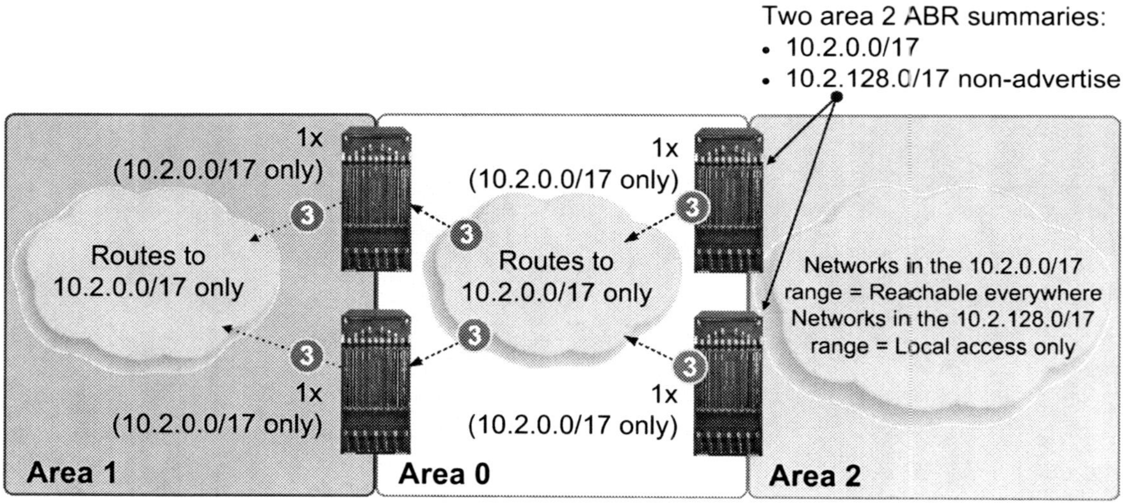

Figure 9-8: Filtering interarea routes

The Comware commands for configuring this example are:

```
[ComwareABR-ospf-1-area-0.0.0.2] abr-summary 10.2.0.0 17

[ComwareABR-ospf-1-area-0.0.0.2] abr-summary 10.2.128.0 17 non-advertise
```

The ProVision commands for configuring this example are:

```
ProVisionABR(ospf)# area 2 range 10.2.0.0 17

ProVisionABR(ospf)# area 2 range 10.2.128.0 17 non-advertise
```

To enable the advertisement again, configure the same summary but specify the **advertise** option explicitly:

```
[ComwareABR-ospf-1-area-0.0.0.2] abr-summary 10.2.128.0 17 advertise

ProVisionABR(ospf)# area 2 range 10.2.128.0 17 advertise
```

Simply removing the non-advertised summary would enable the advertisement of individual routes without aggregation.

Need for redistributing external routes into OSPF

Sometimes, the OSPF AS needs to learn routes to networks outside the OSPF AS. As you have learned, OSPF is a link-state protocol. OSPF routers advertise network interfaces that run OSPF, and ABRs advertise interarea routes. But sometimes an OSPF router has other routes in its IP routing table (a static route, for example), and you want the router to advertise those routes to its OSPF neighbors.

Options for routes redistributed into OSPF	Example use case
Static	The OSPF router needs to advertise its default static route toan ISP router.
Another routing protocol supported by the device	During a migration from RIP to OSPF, OSPF segments need to learn about RIP segments.
Directly connected	Youdo not want to configure connected stub networks as passive OSPF interfaces.
Another OSPF process (Comware only)	Two organizations that use OSPF are merging. For now, you want to maintain separate OSPF AS with some exchange of information.

Figure 9-9: Redistributing external routes into OSPF

Examples include:

- A company uses OSPF internally. One or two of the company's routers connect to an ISP. They have static default routes to this ISP and need to communicate the default route to the rest of the OSPF AS.

Note

Comware switches do not redistribute default routes into OSPF even when you redistribute the default route's type with the **import-route** command. You use a different OSPF command for this function: **default-route-advertise**.

- You are upgcsrading a network from Routing Information Protocol (RIP) to OSPF. During the upgrade, some segments of the network will continue to run RIP. The OSPF AS needs to know about the networks in the RIP segment, just as the devices running RIP need to know about the networks in the OSPF segment.
- Often, you make stub networks (networks with just one OSPF router) passive OSPF interfaces. However, you might *not* to enable OSPF on the stub network interfaces at all. However, you still need OSPF to advertise routes to these networks.

You can make any of these types of routes available to OSPF by redistributing (or importing) the routes into OSPF. In detail, you can redistribute the routes listed in Figure 9-9.

- Static routes
- Any routes discovered by another routing protocol supported on the routing switch, which might include:
  - RIP
  - Intermediate System to Intermediate System (IS-IS) (some Comware switches)
  - Border Gateway Protocol (BGP) (some Comware and ProVision switches)
- Connected routes on which OSPF is not enabled

You redistribute routes by type (static, specific routing protocol, or directly connected). By default, the OSPF process receives all routes discovered by the router through the specified mechanism. However, you can filter the redistributed routes using route maps.

External routes and ASBRs

The router on which you redistribute the routes is called an Autonomous System Border Router (ASBR) because it communicates routing information between the OSPF AS and other routing processes. A redistributed route becomes an *external* OSPF route, which is sometimes referred to as an AS external (ASE) route (see Figure 9-10). (All of the routes calculated from type 1, type 2, and type 3 LSAs, including both intra-area and interarea routes, are *internal* OSPF routes.)

ASBRs advertise external routes in type 5 LSAs. The LSA includes the network address for the redistributed route, the router ID of the advertising ASBR, and the metric, which you assign when you redistribute the route into OSPF.

OSPF routers propagate the type 5 LSA in LSUs throughout the entire AS, making type 5 LSAs different from type 1, type 2, and type 3 LSAs. (Remember that although type 3 LSAs advertise interarea routes, each area has its own set of type 3 LSAs.)

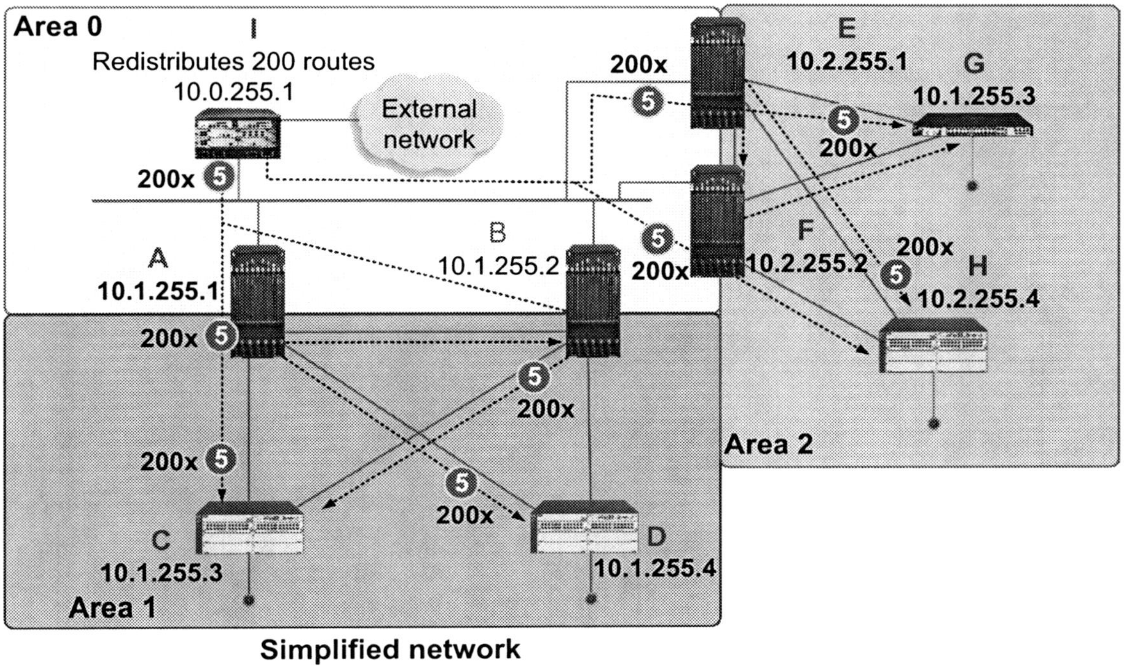

Figure 9-10: External routes and ASBRs

To calculate an external route from the type 5 LSA, a router refers to the LSA's advertising ASBR. The router has calculated a shortest path to each ASBR. The forwarding interface and next hop for that path become the forwarding interface and next hop for all external routes advertised by that ASBR.

In other words, routers do *not* add the external networks to the directed graph for the shortest-path algorithm. Instead, they add them to a list of networks for which the advertising ASBR is the gateway. Therefore, changes to the particular external networks advertised by an ASBR do not affect the area topology.

Example OSPF external routes in routing tables

The sections below provide examples of external OSPF routes. The routes belong to a router in area 0 that has learned them from an ASBR.

All routes except the external OSPF ones have been removed for clarity. Of course, in an actual enterprise environment, the table would probably include more external routes, as well.

Examples of type 5 LSAs are also provided. Note that:

- These LSAs are not associated with an area in the LSDB; they extend across the AS.
- The LS ID and the network mask indicate the external network in the route.
- The metric is the metric advertised by the ASBR.

Comware external routes

Here, you see external routes on a Comware routing switch. As you see, these routes are considered external OSPF routes and have a higher administrative distance (150) than internal OSPF routes:

```
[Area1ABR]display ip routing

Routing Tables: Public

        Destinations : 204          Routes : 208

Destination/Mask     Proto     Pre   Cost        NextHop      Interface

172.16.0.0/16        O_ASE     150   1           10.0.0.2     Vlan1000
```

Comware type 5 LSA

Here, you see the command for viewing a type 5 LSA on a Comware switch and example command output:

```
 [Area1ABR]display ospf lsdb ase

      OSPF Process 1 with Router ID 10.0.255.1

                 Link State Database


Type          : External

LS ID         : 172.16.0.0

Adv Rtr       : 10.0.255.2

LS Age        : 100

Len           : 36

Options       : DC E

Seq#          : 80000001

Checksum      : 0x1fc9

Net Mask      : 255.255.0.0

TOS 0 Metric: 1

E Type        : 2

Forwarding Address : 0.0.0.0

Tag           : 0

Type          : External
```

```
LS ID        : 172.16.0.0
Adv Rtr      : 10.0.255.4
LS Age       : 76
Len          : 36
Options      : E
Seq#         : 80000001
Checksum     : 0x7fe
Net Mask     : 255.255.0.0
TOS 0        Metric: 1
E Type       : 2
Forwarding Address : 0.0.0.0
Tag          : 1
```

ProVision external route

The command output below shows an external route on a ProVision routing switch. Note that ProVision switches use the standard OSPF administrative distance for both internal and external OSPF routes:

```
Area1ABR# show ip route

                        IP Route Entries

Destination      Gateway   VLAN  Type  Sub-Type    Metric     Dist.
------------     --------  ----- ----  -------     -------    ------
172.16.0.0/16    10.0.0.6  1004  ospf  External2   1          110
```

ProVision type 5 LSA

You must use a Comware **display** command, which is provided in the ProVision CLI for software version K15.08 and above, to view the external LSAs:

```
Area1ABR# display ospf lsdb ase

      OSPF Process 1 with Router ID 10.0.255.1

                 Link State Database

Type         : External
LS ID        : 172.16.0.0
Adv Rtr      : 10.0.255.2
```

```
LS Age        : 217
Len           : n/a
Options       :
Seq#          : 0x80000004
Checksum      : 0x7369
Net Mask      : n/a
TOS 0  Metric:  10
E Type        : 1
Forwarding Address : 0.0.0.0
Tag           : 0
```

Summary routes to ASBRs

When a router receives a type 5 LSA from an ASBR in the same area, it can use the area's topology to calculate the shortest path to that ASBR, the ASBR route. The next hop in this route becomes the next hop for all external networks for which that ASBR is the gateway.

Routers in different areas from the ASBR also receive the type 5 LSA, but they might need help determining how to reach the ASBR. The type 5 LSA indicates the advertising ASBR by router ID. You have learned how to make router IDs routable. However, because OSPF does not *require* routable router IDs, it must provide another mechanism for informing routers in other areas how to reach the ASBR.

Also, even if OSPF did require routable router IDs, an OSPF router in another area might not want to use its typical interarea routes to reach the ASBR. The ABR with the lowest cost for the broader area range might not provide the best path to the ASBR within the area. For example, in Figure 9-11, ABR-A might advertise a lower cost for the 10.0.0.0/16 range. However, ABR-B might have a lower cost for reaching ASBR-I, specifically.

OSPF defines the type 4 summary LSA so that the routers in other areas can learn the best ABR through which to reach particular ASBRs. For each area, an ABR generates one type 4 LSA for each ASBR that it can reach in *another* area. As the metric for this LSA, the ABR specifies the cost for its shortest path to the ASBR. For example, in Figure 9-11, ABR-A would specify a metric of 5 for its type 4 LSA to router I.

Internal routers that receive the type 4 LSAs use them to select the best path to each ASBR in other areas.

In summary, every router in the OSPF AS calculates an ASBR route:

- If the router is in the same area as the ASBR, it knows the correct forwarding interface and next hop from the area's shortest-path topology. Equal-Cost Multipath (ECMP) routing might apply.

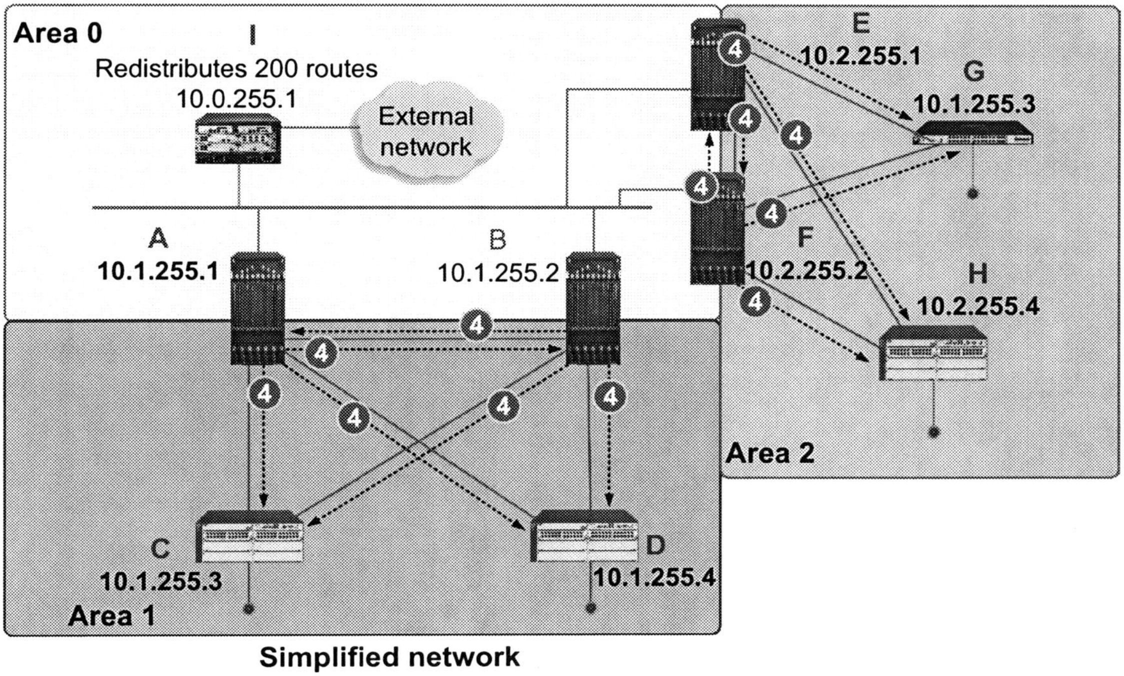

Figure 9-11: Summary routes to ASBRs

- If the router is in a different area from the ASBR, it calculates a metric for the ASBR route by incrementing the metric in the type 4 LSA with the metric for reaching the advertising ABR. If multiple ABRs advertise a type 4 LSA to the same ASBR, the router chooses the ASBR route with the lower *total* cost.

  For example, assume that ABR-A has cost 10 on the interface that connects to the ASBR, while ABR-A has cost 5 on this interface. However, router C has cost 50 on the interface that connects to ABR-A and cost 100 on the interface that connects to ABR-B. It selects a path to the ASBR through ABR-A (total metric 60 rather than 105).

- The forwarding interface and next hop for the ASBR route are taken from the shortest path to the advertising ABR, which the router knows from the area's shortest-path topology.

- A router might have multiple equal-cost paths to an ASBR. Multiple ABRs might offer the router an equal-cost path to the ASBR, and the router has a different next hop for each ABR. Or the router might have multiple equal-cost paths to one ABR.

The router now has a lowest-cost ASBR route. It uses the next hop in that router as the next hop for all external routes advertised by that ASBR.

Example OSPF external routes (ASBR in another area)

The sections below provide examples of an external OSPF route learned by an internal area 1 router from the ASBR in area 0. As you see, the router has *one* external route for which ABR-A is the next hop even though the router has *two* interarea routes for the range that includes the ASBR's router ID.

As you have learned, this situation occurs because the router selects the lowest-cost path to the ASBR, using costs advertised in type 4 LSAs.

You can see those type 4 LSAs in the sections below. You can also see the type 5 LSA in the router's LSDB. Note that the type 5 LSA matches the one in area 0, and the advertising router remains the ASBR that redistributed the route. Type 5 LSAs extend across the AS.

Comware routing tables

```
<InternalArea1> display ip routing

Destination/Mask     Proto   Pre   Metric        NextHop        Interface

10.2.0.0/16          OSPF    10    500           10.1.1.5       Vlan104

                     OSPF    10    500           10.1.1.13      Vlan112

172.16.0.0/16        O_ASE   150   1             10.1.1.5       Vlan104
```

```
<InternalArea1> display ospf routing

Routing for ASEs

  Destination        Metric    Type    Tag       NextHop       AdvRouter

10.2.0.0/16          500       Inter   10.1.1.5  10.0.255.1    0.0.0.0

                     500       Inter   10.1.1.13 10.0.255.3    0.0.0.0

172.16.100.0/30      1         Type2   1         10.1.1.5      10.0.255.1

Total Nets: 551

  Intra Area: 50 Inter Area: 1 ASE: 500 NSSA: 0
```

Comware type 4 LSA

```
<InternalArea1>display ospf lsdb asbr

         OSPF Process 1 with Router ID 10.1.255.3

                         Area: 0.0.0.1

                 Link State Database


    Type        : Sum-Asbr

    LS ID       : 10.0.255.2

    Adv Rtr     : 10.0.255.3

    LS Age      : 165
```

```
    Len         : 28
    Options     : DC E
    Seq#        : 80000001
    Checksum    : 0x506c
    Tos 0 Metric: 100

    Type        : Sum-Asbr
    LS ID       : 10.0.255.2
    Adv Rtr     : 10.0.255.1
    LS Age      : 159
    Len         : 28
    Options     : E
    Seq#        : 80000001
    Checksum    : 0x2a50
    Tos 0 Metric: 200
```

Comware Type 5 LSA

```
<InternalArea1> display ospf lsdb ase
OSPF Process with Router ID 10.1.255.3
                 Link State Database

    Type        : External
    LS ID       : 172.16.0.0
    Adv Rtr     : 10.0.255.2
    LS Age      : 565
    Len         : n/a
    Options     :
    Seq#        : 0x80000009
    Checksum    : 0x696e
    Net Mask    : n/a
```

```
    TOS 0 Metric: 10
    E Type     : 1
    Forwarding Address : 0.0.0.0
    Tag        : 0
```

ProVision routing tables

```
InternalArea1# show ip route
                      IP Route Entries

Destination     Gateway     VLAN Type     Sub-Type   Metric   Dist.
--------------- ----------- ---- -------- ---------- -------- -----
10.0.0.0/16     10.1.1.5    104  ospf     InterArea  300      110
10.0.0.0/16     10.1.1.13   112  ospf     InterArea  300      110
172.16.0.0/16   10.1.1.5    104  ospf     External2  1        110
```

ProVision type 4 LSA

```
InternalArea1# show ip ospf link-state type as-summary
  OSPF Link State Database for Area 0.0.0.1
                                 Advertising
LSA Type   Link StateID  RouterID      Age  Sequence # Checksum
---------- ------------- ------------  ---- ---------- ----------
AsbSummary 10.0.255.4    10.0.255.1    144  0x80000001 0x00002a50
AsbSummary 10.0.255.4    10.0.255.3    149  0x80000001 0x0000506c

InternalArea1# show ip ospf link-state detail type as-summary
OSPF Link State Database for Area 0.0.0.1

LSA Age                      : 496
LSA Type                     : 0x4 (AS Boundary)
Advertising Router           : 10.0.255.1
Link State ID                : 10.0.255.2
LSA Sequence                 : 0x80000006
LSA Checksum                 : 0x4893
LSA Option Bits              : E=1 MC=0 N/P=0 EA=0 DC=0
```

```
LSA Metric                          : 100
Network Mask                        : 255.255.255.255

LSA Age                             : 1601
LSA Type                            : 0x4 (AS Boundary)
Advertising Router                  : 10.0.255.3
Link State ID                       : 10.0.255.2
LSA Sequence                        : 0x80000007
LSA Checksum                        : 0x4410
LSA Option Bits                     : E=1 MC=0 N/P=0 EA=0 DC=1
LSA Metric                          : 200
Network Mask                        : 255.255.255.255
```

ProVision type 5 LSA

```
InternalArea1# display ospf lsdb ase
OSPF Process with Router ID 10.1.255.3
                 Link State Database

    Type      : External
    LS ID     : 172.16.0.0
    Adv Rtr   : 10.0.255.2
    LS Age    : 565
    Len       : n/a
    Options   :
    Seq#      : 0x80000009
    Checksum  : 0x696e
    Net Mask  : n/a
    TOS 0 Metric : 10
    E Type    : 1
    Forwarding Address  : 0.0.0.0
    Tag       : 0
```

Choosing between external routes to the same network: Type 2 metric

OSPF routers might calculate external routes to the same network from type 5 LSAs advertised by different ASBRs. They use the LSA's metric to choose which external route to install in the routing table.

Example 1

Figure 9-12 outlines the selection process when OSPF is choosing between identical external OSPF routes. Usually, a network advertised in a type 5 LSA should not be associated with an internal OSPF route, but a misconfiguration could cause this situation to occur. In that case, the OSPF always prefers internal (intra-area and then interarea) routes.

And, of course, the router uses administrative distance to choose between the OSPF route and routes learned through other means.

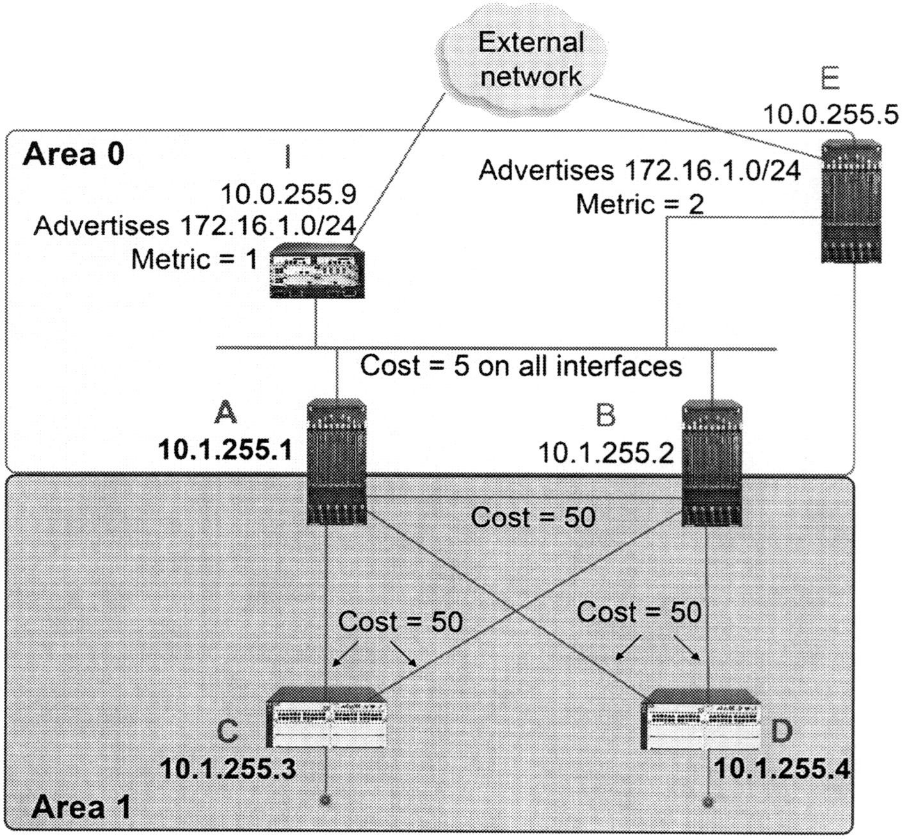

Figure 9-12: Choosing between external routes to the same network using type 2 metric

Type 5 LSAs introduce type 1 and type 2 metrics. When routers examine a type 2 metric, which is the default, they consider only the metric advertised by the ASBR. OSPF follows this high-level process to choose the best external route:

1. The router uses the type 5 LSA to calculate the external route, setting the correct forwarding interface and next hop for the path to the advertising ASBR.
2. The router sets the external route's metric equal to the metric advertised in the type 5 LSA. The advertising ASBR added this metric when it redistributed the route into OSPF.
3. When the OSPF routing table already includes an external (type 2) OSPF route to this network, the router selects the route with the lower metric.
4. If both routes have the same metric, the router examines the metric for the forwarding path to the ASBR. It selects the route associated with the ASBR to which it has a lower-cost path.
5. If forwarding paths to the ASBRs have equal costs and the router supports ECMP, the router installs multiple routes in the OSPF routing table. (Otherwise, it uses another mechanism such as the router ID to break the tie.)

When all ASBRs use type 2 metrics, an OSPF router will always select the external route through the ASBR that advertises the lowest metric for it, *even if the router is closer to another ASBR that advertises the same route*. However, if multiple ASBRs advertise the route with the same metric, other routers do take the cost of reaching the ASBR into account.

The reasoning behind this behavior is that the cost of reaching the ASBR through the OSPF AS is often negligible compared to the cost of reaching the external network. In addition, the ASBR might not be assigning metrics to the external routes that correspond to the metrics used in the OSPF AS. In other words, a metric of 10 for the external route might not mean the same thing that a 10 metric on an OSPF link means, in terms of how strongly the company prefers using a particular path.

Review questions

Consider what you have learned by answering the following questions:

1. Which ASBR does router C, in area 1, select to reach the external route 172.16.1.0/24?
2. What is the next hop in router C's external route to 172.16.1.0/24? (This router supports ECMP.)

Answers to review questions

1. Which ASBR does router C, in area 1, select to reach the external route 172.16.1.0/24?

   Router C selects router I as the ASBR for 172.16.1.0/24 because this router advertises the low metric.

2. Which router is the next hop in router C's external route to 172.16.1.0/24? (The router supports ECMP.)

   Router C has the same cost to reach the ASBR through both ABRs, router A and router B. Therefore, it learns two routes to 172.16.1.0/24, one through each ABR.

Example 2

Figure 9-13 shows a slightly different topology in which the ABRs have different costs to the ASBRs.

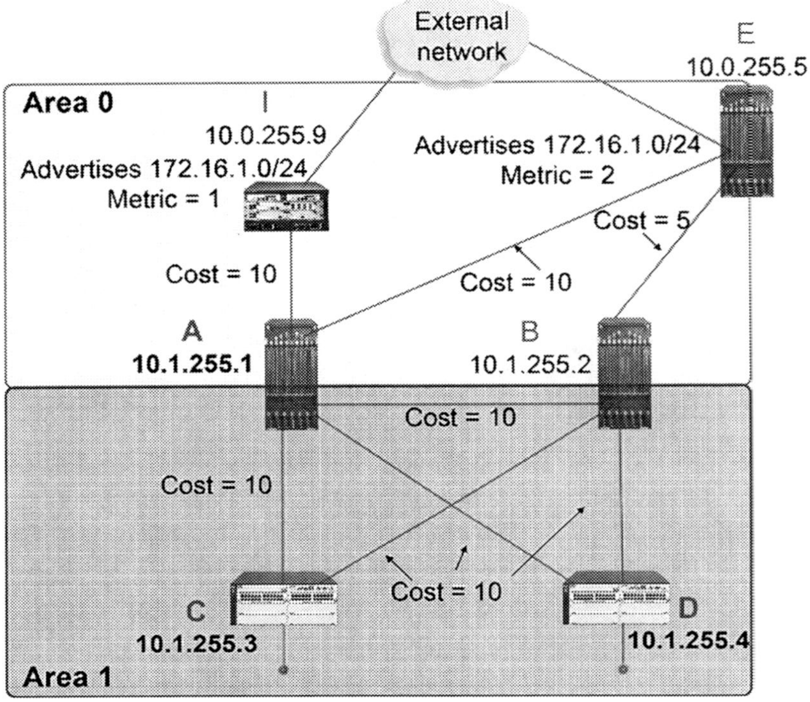

Figure 9-13: Another example of choosing between external routes to the same network

Review questions

Answer the following questions:

1. Which ASBR does router C, in area 1, select to reach the external route 172.16.1.0/24? Is this the same ASBR as it selected in the topology shown on the previous example?
2. What is the next hop in router C's external route to 172.16.1.0/24? (This router supports ECMP.) Is this the same next hop used for this route in the topology shown on the previous example?

Answers to review questions

1. Which ASBR does router C, in area 1, select to reach the external route 172.16.1.0/24? Is this the same ASBR as it selected in the topology shown on the previous example?

   Router C still selects ASBR-I because this ASBR advertises the lowest cost for the external route. With type 2 metrics, the router always selects the ASBR that advertises the lowest cost regardless of the cost of reaching that ASBR.

2. What is the next hop in router C's external route to 172.16.1.0/24? Is this the same next hop used for this route in the topology shown on the previous example?

   Router C learns just one route to 172.16.1.0/24 because it has a single lowest-cost path to the ASBR through ABR-A. This is different from the previous topology because ABR-A offers a lower-cost path to ASBR-I in this topology.

Choosing between external routes to the same network: Type 1 metric

Sometimes, the cost of reaching the ASBR is not negligible. In that case, you might want routers to take the cost of reaching the ASBR into account when selecting the ASBR for a particular external network (as opposed to the path to the already selected ASBR). Use a type 1 metric, which signals routers to add the cost of reaching the ASBR to the metric advertised in the type 5 LSA (see Figure 9-14). You should also ensure that ASBRs redistribute the external routes with metrics that make sense when compared to costs used within the OSPF process.

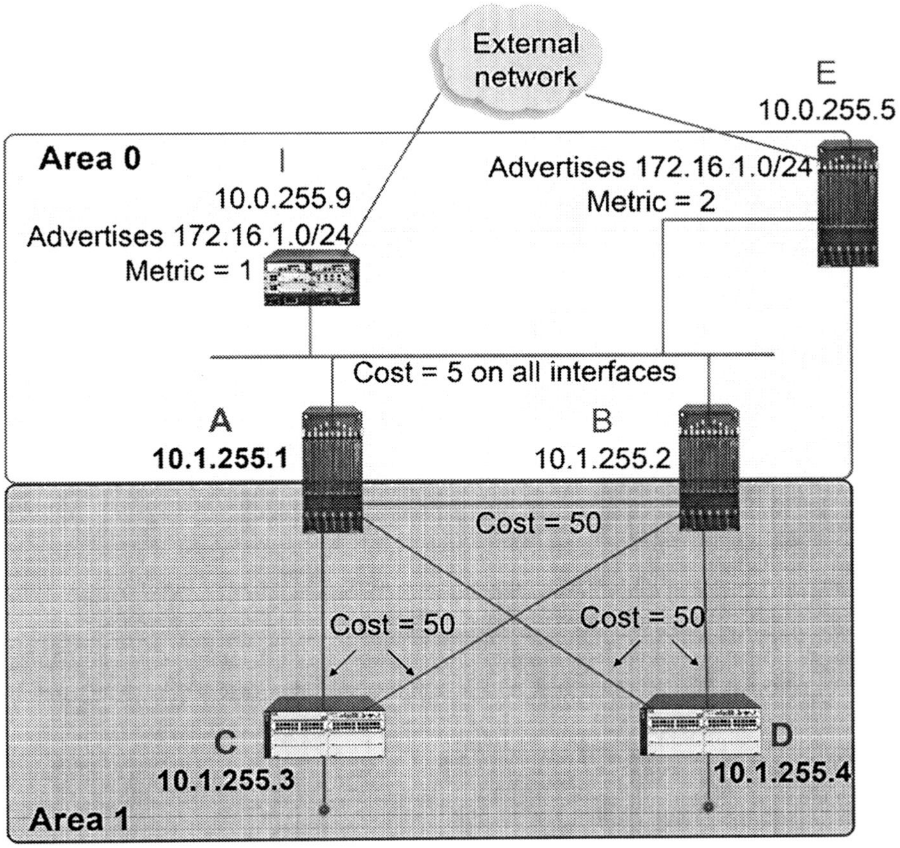

Figure 9-14: Choosing between external routes to the same network using type 1 metric

In more detail, an OSPF router chooses between type 1 external routes to the same network as follows:

1. The router uses the type 5 LSA to calculate the external route, setting the correct forwarding interface and next hop for the path to the advertising ASBR.
2. The router sets the external route's metric equal to:

   The metric advertised in the type 5 LSA + the metric for the forwarding path to the advertising ASBR
3. When the OSPF routing table already includes an external (type 2) OSPF route to this network, the router selects the route with the lower metric.
4. If both routes have the same metric, the router installs multiple routes in the OSPF routing table. (Otherwise, it uses a mechanism such as a router ID to break the tie.)

Review questions

1. Which ASBR does router C, in area 1, select to reach the external route 172.16.1.0/24? Is this choice different from the choice when ASBRs use type 2 metrics?
2. What is the next hop in router C's external route to 172.16.1.0/24? (This router supports ECMP.)

Answers to review questions

1. Which ASBR does router C, in area 1, select to reach the external route 172.16.1.0/24? Is this choice different from the choice when ASBRs use type 2 metrics?

   The total metric for the LSA received from ASBR-I is 56:

   1 (external metric) + 55 (metric of lowest-cost path to ASBR-I)

   The total metric for the LSA received from ASBR E is 57:

   2 (external metric) + 55 (metric of lowest-cost path to ASBR E)

   Therefore, router C selects ASBR-I. In this topology, router C's choice is the same with type 1 and type 2 metrics because router C has the same cost for reaching both ASBRs.
2. What is the next hop in router C's external route to 172.16.1.0/24? (This router supports ECMP.)

   As before, router C learns two external routes, one through each ABR, because both paths to the ASBR have the same cost.

Additional considerations

Now consider the effects of this topology when the ASBRs advertise type 1 LSA (see Figure 9-15).

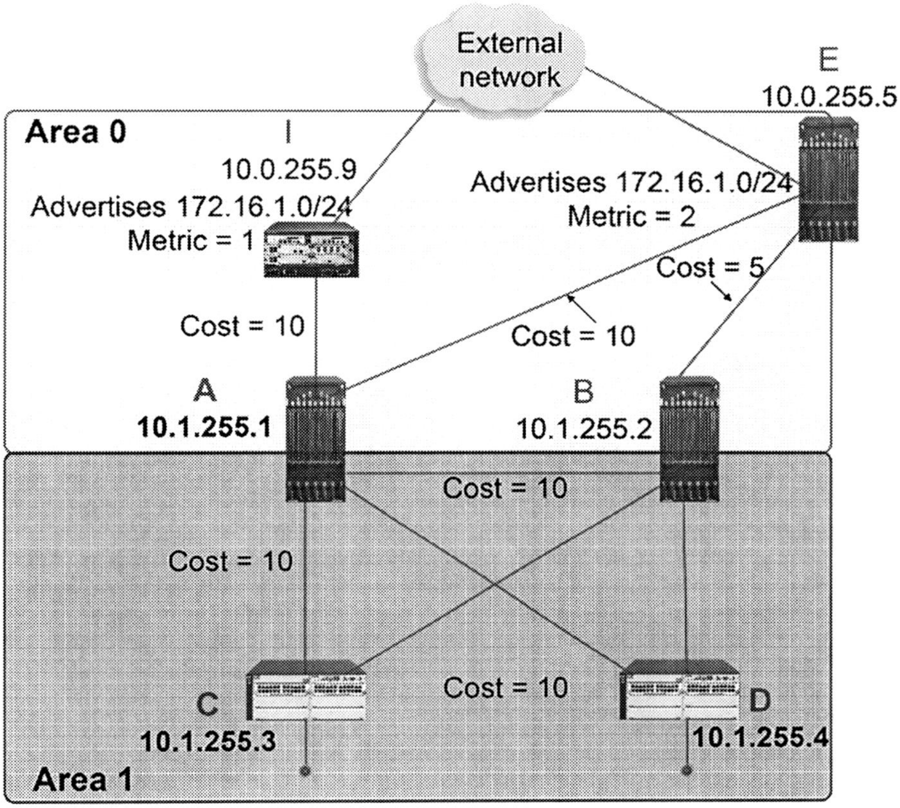

Figure 9-15: Choosing between external routes to the same network using type 1 metric

Review questions

1. Which ASBR does router C, in area 1, select to reach the external route 172.16.1.0/24? Is this choice different than the ASBRs use type 2 metrics?
2. What is the next hop in router C's external route to 172.16.1.0/24? (This router supports ECMP.)

Answers to review questions

1. Which ASBR does router C, in area 1, select to reach the external route 172.16.1.0/24? Is this choice different from the choice when ASBRs use type 2 metrics?

   The total metric for the LSA received from ASBR-I is 21:

   1 (external metric) + 20 (metric of lowest-cost path to ASBR-I)

   The total metric for the LSA received from ASBR E is 17:

2 (external metric) + 15 (metric of lowest-cost path to ASBR E)

Therefore, router C selects ASBR E. In this topology, router C makes a different choice when the ASBR uses type 1 metrics versus type 2 metrics because the cost of its path to ASBR E is lower enough to counteract the higher metric advertised by ASBR E.

2. What is the next hop in router C's external route to 172.16.1.0/24? (This router supports ECMP.)

   In this case, ABR-B offers router C a lowest-cost path to ASBR E. ABR-B is the next hop in this path, so router C learns one external route, in which ABR-B is the next hop.

Comware: Aggregating external routes

In some networks, external routes, carried in type 5 LSAs, can make up the bulk of the LSAs. These LSAs propagate throughout the entire system regardless of area. For the same reasons that you might want to minimize type 3 LSAs and interarea routes, you might want to minimize type 5 LSAs and external OSPF routes.

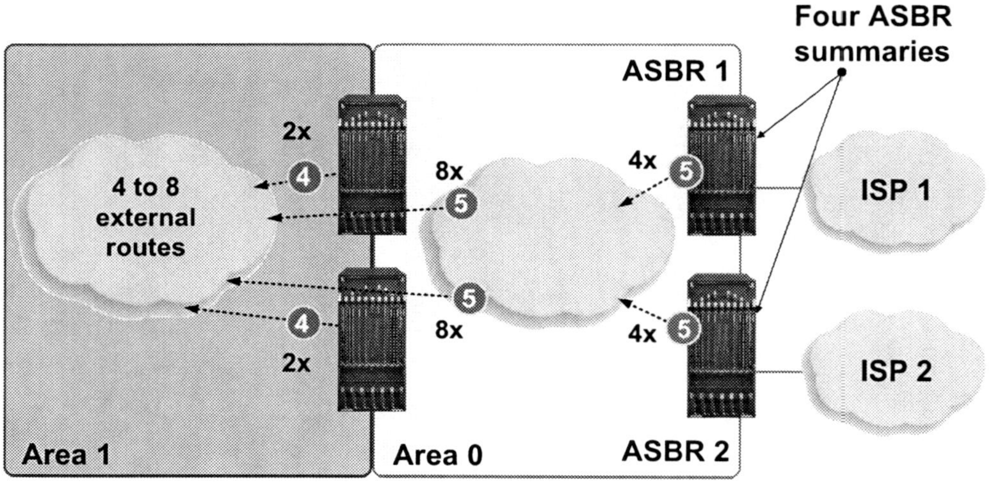

Figure 9-16: Aggregating external routes on Comware switches

You can aggregate external routes on Comware ASBRs. Simply create an ASBR summary that includes the networks in individual redistributed routes (see Figure 9-16).

ASBR summaries function much like ABR summaries but for external routes. The ASBR advertises the summary in a type 5 LSA, as long as it has one external route to advertise in that range. The summary also suppresses all individual type 5 LSAs in the range.

This Comware command aggregates external routes:

```
[Switch-ospf-<ID>] asbr-summary <network address> <mask | prefix length>
[not-advertise] [cost <metric>]
```

Need for eliminating excessive external routes

Aggregating external routes with ASBR summaries can help to streamline routes and advertisements. However, in some cases, the number of external routes remains high. In a nonbackbone area with just one or two exit/entry points, you might want to eliminate the external routes entirely.

Remember the primary benefit of individual type 4 and type 5 LSAs. The type 5 LSAs help routers choose the best ASBR for reaching a particular external network. If that ASBR is in another area, type 4 LSAs further help the routers choose the best ABR for reaching that ASBR. (Depending on the external metric type, the type 4 LSA might also help the routers select the best ASBR). You should examine each nonbackbone area and determine whether routers in the area truly need to choose between paths for particular external networks (see Figure 9-17).

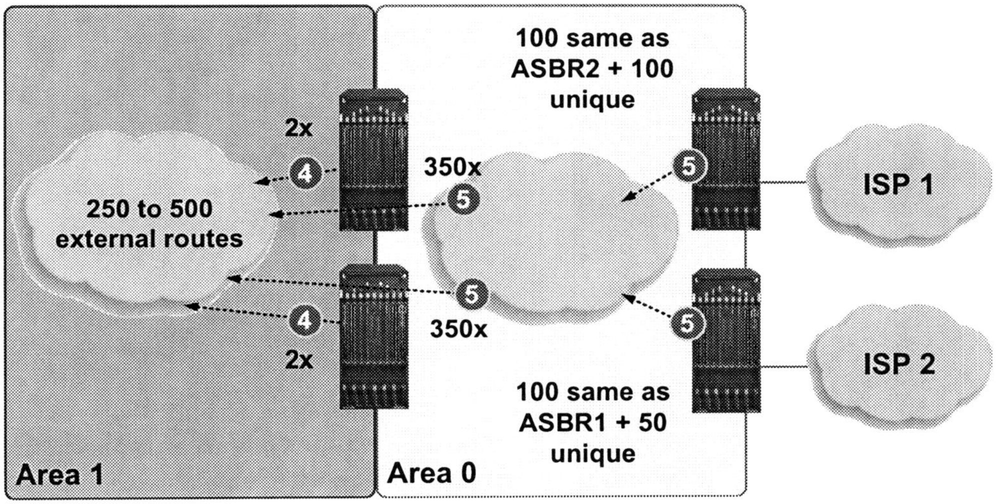

Figure 9-17: Eliminating excessive external routes

First, consider whether the area includes any ASBRs. If it does not, then all external traffic must pass through the area's ABRs. The number of ABRs in the area restricts the number of paths for external routes. If the area has only one ABR, the area has only one path for external traffic. A default route through the ABR provides the same benefit as individual external route advertisements much more efficiently.

If the area has multiple ABRs, internal routers *might* benefit from choices for external paths. As you have learned, these choices extend not just to choosing the best ASBR but also the best ABR for reaching a particular ASBR. You need to assess, however, whether the internal routers truly receive a benefit from this level of detail and choice. In many cases, the ABRs offer roughly equivalent paths to the external networks, so the internal routers can receive the same benefit from a default route advertised by both.

Using stub areas to eliminate external routes

To eliminate external routes and advertisements from an area, define a stub area, which is illustrated in Figure 9-18. ABRs for stub areas do not forward type 4 or type 5 LSAs into those areas, and internal routers in those areas do not generate or accept them. You must define the stub area setting on each OSPF routing device within the stub area as well as the ABR.

Otherwise, the devices cannot become neighbors and establish adjacency.

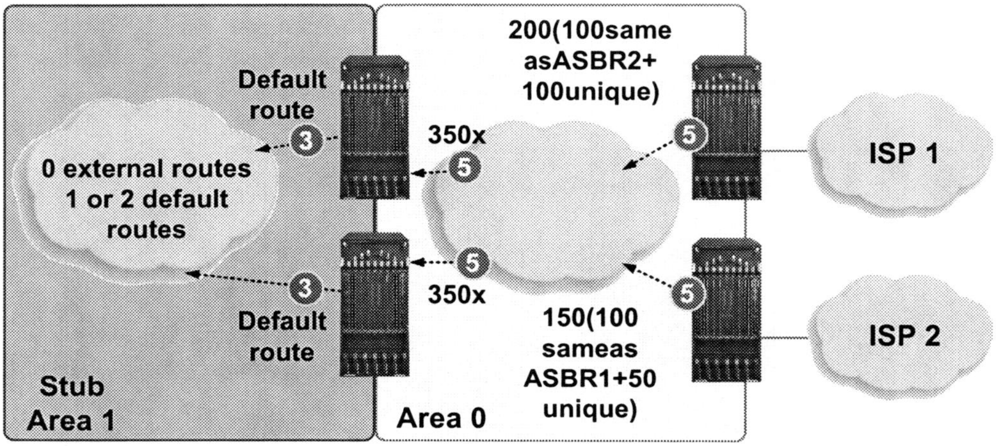

Figure 9-18: Using stub areas to eliminate external routes

When you activate this option on the ABR, it automatically generates a default route, which it injects as a type 3 LSA into the stub area. As usual, the ABR assigns a metric to the default route, and you can manipulate metrics to set up one ABR as a preferred exit for external traffic.

Note that, although type 3 LSAs for interarea routes are not shown in this picture, the stub area acts like a normal area in this respect. It receives summary LSAs for other areas through ABRs; the summary LSAs include individual or aggregated routes as configured on the ABRs.

The Comware command for configuring a stub area is:

```
[Comware-ospf-<ID>-area-<ID>] stub
```

To set the metric for the default route, enter this command on a Comware ABR:

```
[Comware-ospf-<ID>-area-<ID>] default-metric <metric>
```

The ProVision command for configuring a stub area and the default metric is:

```
ProVision(ospf)# area <ID> stub <metric>
```

Note that, on ProVision products, you must set the metric when you configure the area as a stub on internal routers as well as on ABRs. However, only ABRs advertise the default route.

Stub internal area 1 router global and OSPF routing table

With area 1 defined as a stub, the internal router receives a single default route from each ABR instead of the external routes (it selects the lowest-metric default route). The default route is considered an interarea OSPF route, and the router still has two other interarea routes. (Although the output is cut off before these routes, the totals indicate five interarea routes.)

The sections below provide examples.

Comware

```
<InternalArea1> display ip routing

Destination/Mask  Proto  Pre  Metric     NextHop        Interface

0.0.0.0/0         OSPF   10   11         10.1.1.13      Vlan5

<InternalArea1> display ospf routing

Destination     Metric   Type       NextHop      AdvRouter     Area

0.0.0.0/0       11       Inter      10.1.1.13    10.1.255.2    0.0.0.1

Total Nets: 55

Intra Area: 50 Inter Area: 5 ASE: 0 NSSA: 0
```

ProVision

```
InternalArea1# show ip route

                       IP Route Entries

Destination    Gateway    VLAN Type    Sub-Type    Metric    Dist.
------------   ---------- ---- ------- ----------  --------  -----
0.0.0.0/16     10.1.1.13  5    ospf    InterArea   11        110
```

Need for eliminating all interarea and external routes

In the use case shown in Figure 9-19, area 1 remains a stub area. All traffic destined to it terminates there. In addition, the area has only one exit for all nonlocal traffic, including traffic destined to the backbone, to other OSPF areas, and to the ISP. In such an environment, the internal area 1 routers do not gain a clear benefit from multiple interarea routes. (In fact, in some cases, even areas with two ABRs do not gain a clear benefit from these routes.)

You can configure the OSPF system to suppress all interarea routes in this area in addition to all external routes.

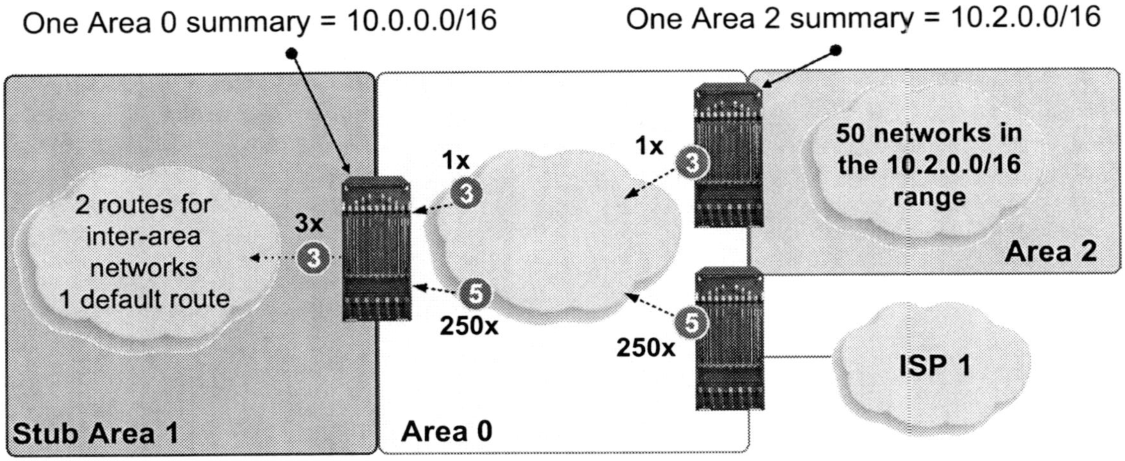

Figure 9-19: Eliminating all interarea and external routes

Stub area 1 routing tables

The tables that follow show interarea routes on an internal router in stub area 1, including a default route. The tables also show several intra-area routes, but in the real world, the table would show more intra-area routes.

Comware routing tables

```
<InternalArea1> display ip routing

Destination/Mask     Proto  Pre  Metric     NextHop      Interface

0.0.0.0/0            OSPF   10   101        10.1.1.5     Vlan104

0.0.0.0/0            OSPF   10   101        10.1.1.13    Vlan112

<InternalArea1> display ospf routing

Destination    Metric   Type    NextHop     AdvRouter     Area

0.0.0.0/0      101      Inter   10.1.1.5    10.1.255.1    0.0.0.1

0.0.0.0/0      101      Inter   10.1.1.13   10.1.255.3    0.0.0.1

10.0.0.0/16    301      Inter   10.1.1.5    10.1.255.1    0.0.0.1

10.0.0.0/16    301      Inter   10.1.1.13   10.1.255.3    0.0.0.1

10.2.0.0/16    500      Inter   10.1.1.5    10.1.255.1    0.0.0.1

10.2.0.0/16    500      Inter   10.1.1.13   10.1.255.1    0.0.0.1

  Total Nets: 56

  Intra Area: 50 Inter Area: 6 ASE: 0 NSSA: 0
```

ProVision routing tables

```
InternalArea1(ospf)# show ip route

                              IP Route Entries

Destination     Gateway      VLAN Type      Sub-Type    Metric   Dist.
-------------   ----------   ---- -------   ----------  -------- -----
0.0.0.0/0       10.1.1.5     104  ospf      InterArea   101      110
0.0.0.0/0       10.1.1.13    112  ospf      InterArea   101      110
10.0.0.0/16     10.1.1.5     104  ospf      InterArea   301      110
10.0.0.0/16     10.1.1.13    112  ospf      InterArea   301      110
10.2.0.0/16     10.1.1.5     104  ospf      InterArea   500      110
10.2.0.0/16     10.1.1.13    112  ospf      InterArea   500      110
127.0.0.0/8     reject            static                0        0
127.0.0.1/32    lo0               connected             1        0
```

Using totally stub areas

You define a totally stub area by configuring the ABR or ABRs connected to that area to suppress summaries (see Figure 9-20). Only Comware switches support true totally stub areas. The ProVision switches have a similar command, but its effect is a bit different.

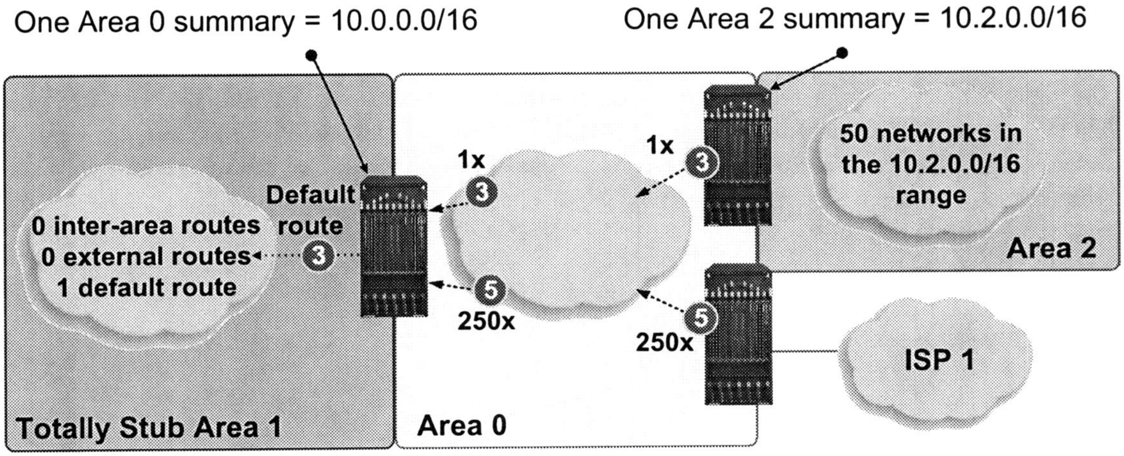

Figure 9-20: Totally stub areas

Comware

You enter this command on all ABRs that connect to the area:

```
[Comware-ospf-<ID>-area-<ID>] stub no-summary
```

Then ABR then suppresses all external route advertisements (LSAs 4 and 5) into this area. It also suppresses inbound type 3 LSAs, which advertise networks in other areas. It does, however, inject a default route into the totally stub area using a type 3 LSA. In addition, the ABR continues to advertise outbound summary LSAs for the totally stub area into the area 0 backbone, and other ABRs can readvertise those summaries in other (non-totally stub) areas.

You do not need to configure any additional options on routers internal to the totally stub area (just enter the typical command for configuring the area as a stub area).

ProVision

You enter this command on all ABRs that connect to the area:

```
ProVision(ospf)# area <ID> stub no-summary <metric>
```

This command suppresses all external route advertisements as normal for a stub area. It also prevents the ABR from generating nonaggregated summary routes for this area. For example, the ABR in Figure 9-20 would not generate area 1 summary LSAs for the routes that the ABR knows to area 0 networks.

However, if you aggregate the area 0 networks with an **area range** command, the ABR would advertise that summary.

Internal area 1 router global and OSPF routing table

The tables below show how the internal router in totally stub area 1 now has only one interarea route (the default route) and no external routes. (It still maintains all intra-area routes, but these are not shown below.) Both the Comware and the ProVision routing tables reflect the state for a totally stub area with a Comware ABR.

Comware

```
<InternalArea1> display ip routing
Destination/Mask  Proto  Pre   Metric  NextHop    Interface
0.0.0.0/0         OSPF   10    11      10.1.1.13  Vlan112
<InternalArea1> display ospf routing
```

```
Destination     Metric  Type        NextHop     AdvRouter    Area
0.0.0.0/0       11      Inter       10.1.1.13   10.1.255.2   0.0.0.1
  Total Nets: 51
  Intra Area: 50 Inter Area: 1 ASE: 0 NSSA: 0
```

ProVision

```
InternalArea1# show ip route

                          IP Route Entries

Destination     Gateway      VLAN Type     Sub-Type    Metric      Dist.
--------------- ------------ ---- -------- ----------- ----------- -----
0.0.0.0/0       10.1.1.13    112  ospf     InterArea   11          110
```

Need for advertising local external routes in a stub area

A stub (or totally stub) area might have its own local connection to an external network (see Figure 9-21). The area would still benefit from a suppression of external routes from other sources. You need to define the area such that the internal OSPF routers receive the local external routes but not external routes from other areas.

Remember that any redistributed route is defined as an external route. Therefore, you would encounter a similar issue if you decided to redistribute connected routes (interfaces) into OSPF.

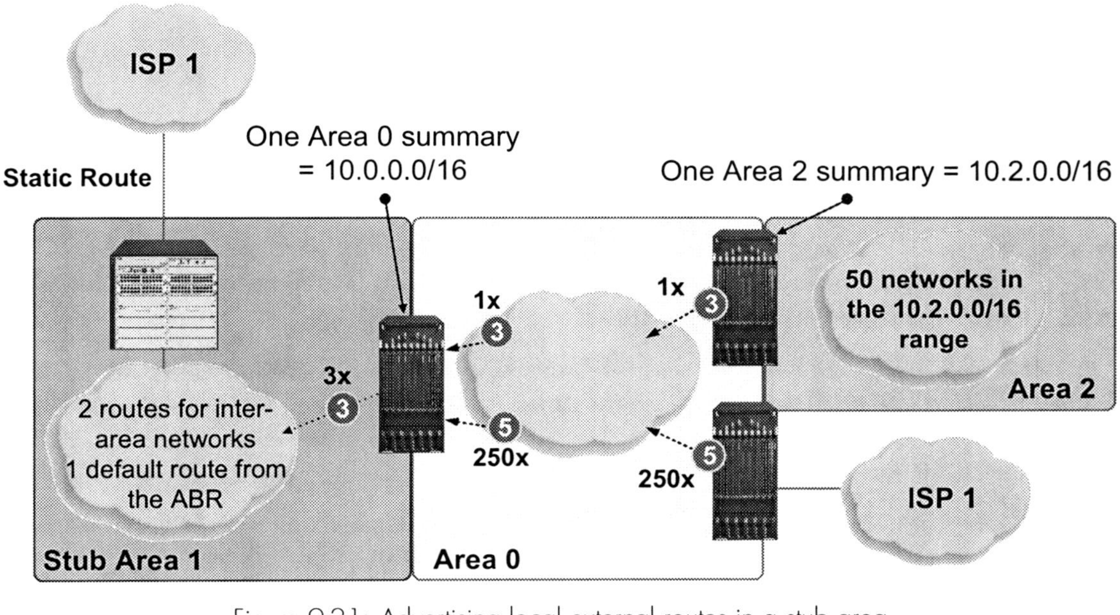

Figure 9-21: Advertising local external routes in a stub area

Using NSSAs

A not-so-stubby area (NSSA) furnishes the solution for this problem (see Figure 9-22). You must define this area type on every OSPF router in the area.

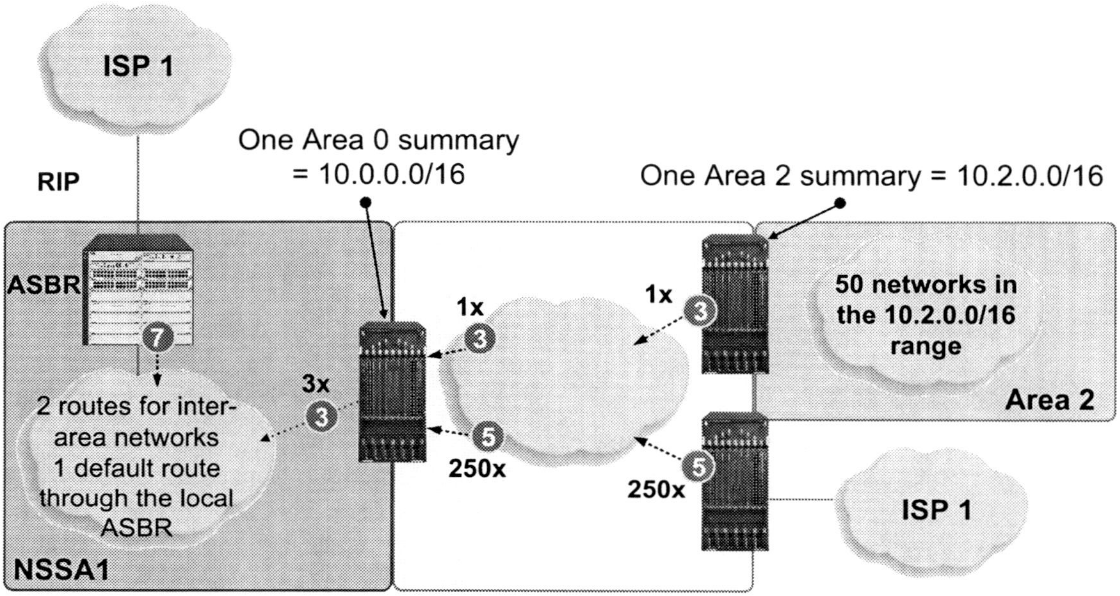

Figure 9-22: Using NSSAs

The Comware command is:

```
[Comware-ospf-<ID>-area-<ID>] nssa [no-summary]
```

The ProVision command is:

```
ProVision(ospf)# area <ID> nssa <metric> [no-summary]
```

An NSSA acts like a stub area or totally stub area (depending on whether the ABR uses the **no-summary** option). However, the area permits external routes that are advertised in OSPF by a local ASBR. These external routes are carried in type 7 LSAs. (The ABR injects the type 7 LSAs into other areas as type 5 LSAs that it appears to originate.)

Of course, you must configure the local ASBR to redistribute routes, advertise a default route, or both, as you have learned in the section on configuring ASBRs and external routes. Like other ASBRs, the NSSA ASBR assigns a metric (type 1 or type 2) to the external routes that it advertises in OSPF, which you can adjust to influence route selection.

The ABR can also advertise a default route into the NSSA (this route uses a type 7 LSA rather than a type 3 one). You must enter this command on the ABR:

```
[Switch-ospf-<ID>-area-<ID>] nssa default-route-advertise
```

You could, for example, have the local ASBR advertise a default route with a lower metric. The ABR could advertise a default route with a higher metric:

```
[ASBR-ospf-<ID>] default-route-advertise [always] metric 10
[ABR-ospf-<ID>-area-<ID>] nssa default-route-advertise
[ABR-ospf-<ID>-area-<ID>] default-metric 20
```

The internal routers will then use the default route through the local external connection in most cases. However, if this connection fails, they can use the connection through the OSPF backbone.

A ProVision ABR advertises a default route in stub areas and NSSAs by default.

OSPF failover and convergence

You will now explore how OSPF systems respond to failures of various types (see Figure 9-23). You will also learn how features such as bidirectional forwarding (BFD) and OSPF graceful restart can speed convergence.

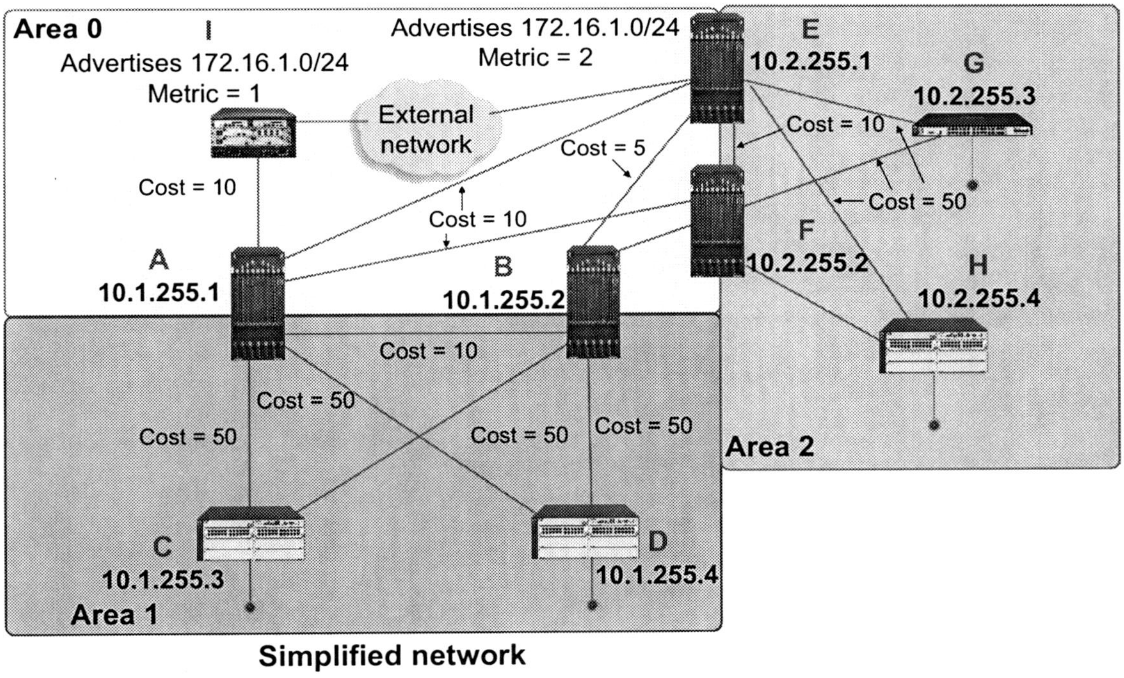

Figure 9-23: OSPF failover and convergence

A number of factors can affect whether or not connectivity is disrupted and for how long. For example:

- How quickly neighbors on a failed link detect lost connectivity

- How quickly routers can generate and transmit LSUs in response to the problem and how quickly their neighbors can process LSUs
- How many routers have to rerun the shortest-path algorithm and how much of the algorithm they have to run
- How far LSUs withdrawing LSAs need to propagate (the more routers that need to receive and generate updates, the longer it takes for the system to converge)

Set aside the first factor for a moment. (You will look at ways to speed lost neighbor detection as the next topic.) For the next factor, the default timers often provide sufficiently speedy LSU generation and propagation. When very fast convergence is required, you can adjust OSPF timers such as the LSA generation interval. This topic is beyond the scope of this study guide.

For the other questions, consider how what you have learned about multiarea designs helps to stabilize the system and enhance convergence.

Area borders acts as boundaries for changes, and often LSUs generated by a change within the area will not prompt LSUs for other areas. In addition, only changes within the OSPF area cause a router to rerun the full shortest-path algorithm. Updates to summary, summary-ASBR, and external LSAs require just a partial rerun. The router simply adds a new network to the list of networks reachable behind the ABR, removes a network from the list, or adjusts the cost for reaching the network. The router does not need to recalculate and run the shortest-path algorithm because it cares only about the segment of the interarea path between itself and the proper ABR for reaching the interarea network. Those paths have not changed.

For example, assume router G loses its link to 10.2.20.0/24. Routers in area 2, including ABR-E, learn this information and begin routing traffic to 10.2.20.0/24 through router H. However, ABR-E still has a route to this network, so it does not withdraw its type 3 LSA for the subnet. It might need to update the cost, but usually it will not even need to do that because it is aggregating the area 2 summaries anyway. As far as routers in area 0 and area 1 are concerned, nothing has changed.

As you see, changes and any attendant disruptions are isolated to area 2.

Detecting failed neighbor relationships

In a well-designed OSPF AS with proper area boundaries, the time to detect failed neighbor relationships often determines how seamlessly OSPF routes fail over.

You will now examine a situation in which an OSPF router has lost its connection to the next hop in one of its routes. Either a link has failed or the neighbor as a whole has failed. To remove the invalid route and begin using an alternate, the OSPF router must detect the failure of the next hop neighbor in the primary route. Often the detection of the problem makes up the bulk of the time required for routes to fail over.

When two neighbors have a point-to-point connection on a direct link, the status of the interface correlates rather well to the status of the neighbor. If the neighbor fails, the interface with the direct connection goes down. The router can immediately remove any routes through that interface.

Similarly, when a VLAN interface is dedicated to a single physical link, the failure of that link brings the VLAN interface down. The routing switch can then remove any routes through that interface.

In either case, the time to detect the link layer problem affects the time for detecting the failed neighbor relationship. This time varies depending on the media and might be acceptable or unacceptable (see Figure 9-24).

Type of link to neighbor	Time for removing neighbor relationship depends on:	Typical times:
Point-to-point direct connection	Time for detecting a Layer 2 issue and bringing the interface down	Under a second
Ethernet, dedicated VLAN on direct links	Time for detecting Layer 2 issue and bringing the VLAN interface down	Under a second
Ethernet with indirect links (Other broadcast or NBMA such as Frame Relay)	Interface OSPF dead timer	30 seconds (default) 3 seconds (minimum)

Figure 9-24: Detecting failed neighbor relationships

Worse, sometimes the status of an OSPF interface and the status of an OSPF neighbor do not directly correlate. Examples include:

- The neighbors connect on a broadcast network such as a VLAN, but the VLAN has multiple links, and possibly multiple neighbors, associated with it.
- The neighbors connect on a NBMA network such as a frame relay network.
- The neighbors do not have a direct Layer 2 connection. For example, a nonrouting switch intervenes between two OSPF neighbors.
- The neighbors have a direct Layer 2 connection, but problems occur at Layer 3. For example, a misconfiguration has prevented one of the neighbors from implementing OSPF on the network.

In those cases, the OSPF router must rely on the OSPF timers to determine the neighbor status. That is, if the dead time expires without a hello from the neighbor, the router considers the neighbor down and removes routes using that neighbor as a next hop from the route table. The default OSPF dead timer is 30 seconds, which is far too long for contemporary networks.

You can adjust the hello and dead timers, but the lowest standard hello timer is 1 second, making the lowest dead timer 3 seconds, still too long for many contemporary networks.

Comware: Using BFD to speed detection

BFD provides a standard, reliable mechanism for detecting failed neighbors in any of the cases you examined previously.

As you have learned with HP Intelligent Resilient Framework (IRF), BFD tests the connectivity between two IP addresses with heartbeat messages, which are transmitted very rapidly (the default Tx rate is 400 ms). BFD reports when connectivity is lost. The router (or routing switch) can then use that information to take the appropriate actions, depending on the functions to which you have tied BFD.

In this case, you enable OSPF BFD on an OSPF interface, and BFD assumes responsibility for testing connectivity to the neighbor. (The interface still sends out OSPF hellos, but BFD will generally detect any problems first.)

In more detail, the interface on which you enable OSPF BFD begins to establish BFD sessions.

As you see in Figure 9-25, a BFD session is point to point. In this example, the 10.0.0.0/24 network has three OSPF routers, each of which implements OSPF BFD. Each router establishes two sessions for six total BFD sessions on the network.

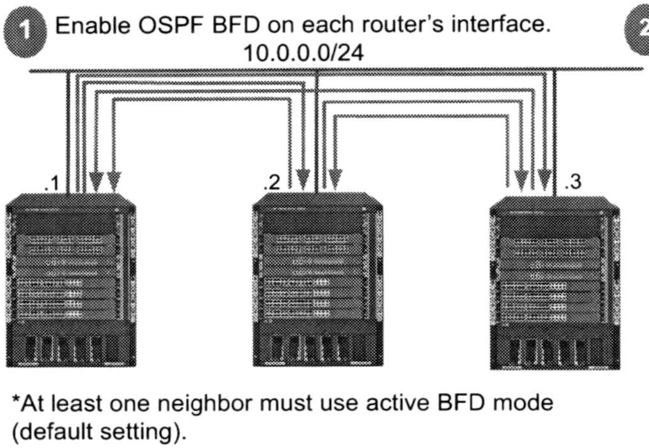

2 Routers automatically* establish a BFD session with each OSPF neighbor.

Router	Neighbor	Act Tx	Detect
10.0.0.1	10.0.0.2	Larger of: • Min Tx time • Nbr Min Rx time	Detect multiplier x Nbr Act Tx time
	10.0.0.3		
10.0.0.2	10.0.0.1		
	10.0.0.2		
10.0.0.3	10.0.0.1		
	10.0.0.2		

Figure 9-25: Using BFD to speed detection on Comware switches

If BFD uses active mode, which is the default mode, the interface automatically attempts to establish a BFD session with each OSPF neighbor on that interface. It transmits a unicast BFD control packets on UDP port 3784 to each neighbor (the port is configurable). A passive-mode Comware router does not initiate BFD sessions, but responds with its own control packets when it receives a neighbor's. Both neighbors can act in active mode, but both cannot act in passive mode or the session will never establish.

To establish a BFD session, the neighbors simply determine that both support BFD and negotiate timers. Each neighbor in the session has its own active Tx timer, which determines how often it transmits BFD control packets. Both neighbors in a BFD session continue to transmit control packets at a set interval, keeping the session up.

The transmit intervals derive from the Min Tx and Min Rx timers, which are configurable intervals on each OPSF interface. An interface's active Tx timer in a particular BFD session is the larger of two values: its Min Tx timer (the Tx timer that the interface wants to use) and its neighbor's Min Rx timer (the smallest interval at which the neighbor can receive control packets). Each neighbor in the session also has its own detect time, which determines how long the interface can go without receiving a control packet from the neighbor before taking the session down. The detect timer derives from the detect multiplier configured on the interface multiplied by the neighbor's active Tx timer (which in turn is the larger of the neighbor's Min Tx timer or this interface's Min Rx timer).

Therefore, the Min Tx timer that you set on an interface can determine how quickly a neighbor can detect a failed link to this interface, and the Min Rx timer that you set can determine how quickly this interface can detect a failed link to a neighbor. However, the neighbor values will also affect these settings, if they are larger. The detect multiplier also affects how quickly this interface detects failed neighbors; the default is 5, and you can lower it to 4 or 3 if you have a reliable link that loses little traffic.

The active Tx timer and detect time on the two neighbors in a BFD session do not have to match. However, they often do.

As long as all links are up, each OSPF BFD interface receives its neighbors' BFD control packets at the expected intervals. Here, you see an example in which router C's link on the OSPF BFD interface has failed, so its BFD control packets no longer reach routers A and B. After the detect time for its BFD session with router C, router A's BFD process reports a problem to OSPF. The OSPF process then removes the neighbor relationship with 10.0.0.3 and, of course, also removes any routes that use that neighbor as a next hop. The router can immediately start using any backup routes. A similar process occurs on router B.

For example, in the network shown in Figure 9-26, both routers B and C connect to 10.0.1.0/24, but router C offers a lower cost. Therefore, initially, routers A and B were both using 10.0.0.3 as the next hop to 10.0.1.0/24. When router C's link on 10.0.0.0/24 fails, router C still has its low-cost link to 10.0.1.0/24, so it does not send an LSU removing that LSA. However, BFD lets routers A and B rapidly remove their neighbor relationship to 10.0.0.3 and all routes through that next hop. Therefore, router A immediately begins using the route to 10.0.1.0/24 through 10.0.0.2, and router B begins using its directly connected link.

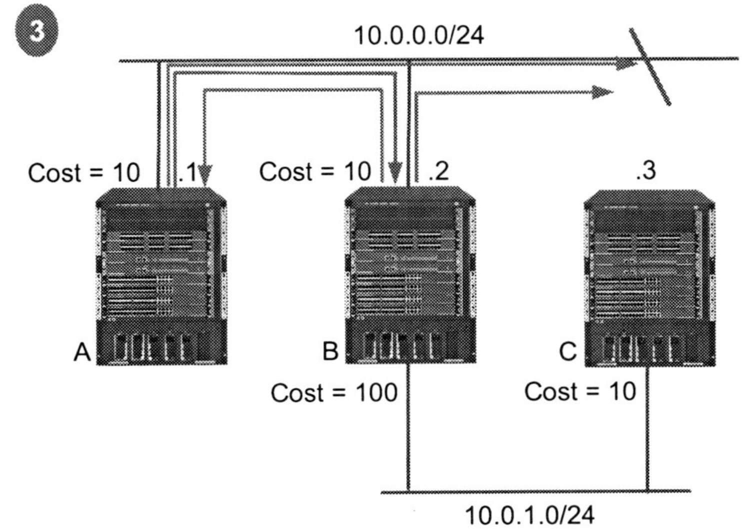

Figure 9-26: Using BFD to speed detection on Comware switches (cont.)

Supplemental information: Echo mode

The default mode for BFD is control mode, which you have examined to this point. You can alternatively configure OSPF BFD echo mode on an interface. This mode is quite similar to control mode. However, rather than both sides of the session sending control packets at the active transmit interval, one side of the session sends BFD echo packets and the other side sends responses.

You only need to *enable* echo-mode OSPF BFD on one of the routers on the interface. However, the neighbor must also support this feature for it to work. Echo-mode monitors both sides of a session just as control mode does and echo-mode uses the same UDP port. Rather than configure Min Tx and Min Rx timers, you configure a Min echo Rx timer. The largest timer on the link determines the rate at which the active neighbor sends the echo packets.

Note that you can set a global BFD echo source for the packets. Choose an IP address that is not used on any actual interface; otherwise, an interface can be flooded by Internet Control Message Protocol (ICMP) redirects.

ProVision: Adjusting timers to speed failed neighbor detection

When possible, you should implement BFD with OSPF. However, the ProVision switches do not support BFD as of the publication of this study guide. You can speed convergence by following these guidelines:

- Even on Ethernet links, simulate point-to-point connections using a VLAN dedicated to each link.

  This guideline helps to ensure that neighbor connectivity problems are often detected at Layer 2 as quickly as possible.

- Set the hello timer to the minimum value (1 second).

  This guideline ensures that even when problems are not detected at Layer 2, the OSPF process detects them within 3 seconds.

Comware: Need for OSPF graceful restart with HP IRF

Next consider situations in which an OSPF system includes routers that also implement other protocols for high availability. You have explored several ways to ensure redundancy of the default router (or routing switching):

- Intelligent Resilient Framework (IRF) (preferred when possible)
- Virtual Router Redundancy Protocol (VRRP)

However, simply ensuring that the default gateway functions are maintained does not ensure that failure of the IRF or VRRP master does not disrupt the routing system as a whole. When you use IRF, OSPF graceful restart ensures nonstop routing as the virtual switch fails over from the master to subordinate. (You will examine strategies for speeding convergence in systems that use VRRP later in this chapter.)

Figure 9-27 illustrates why graceful restart is necessary.

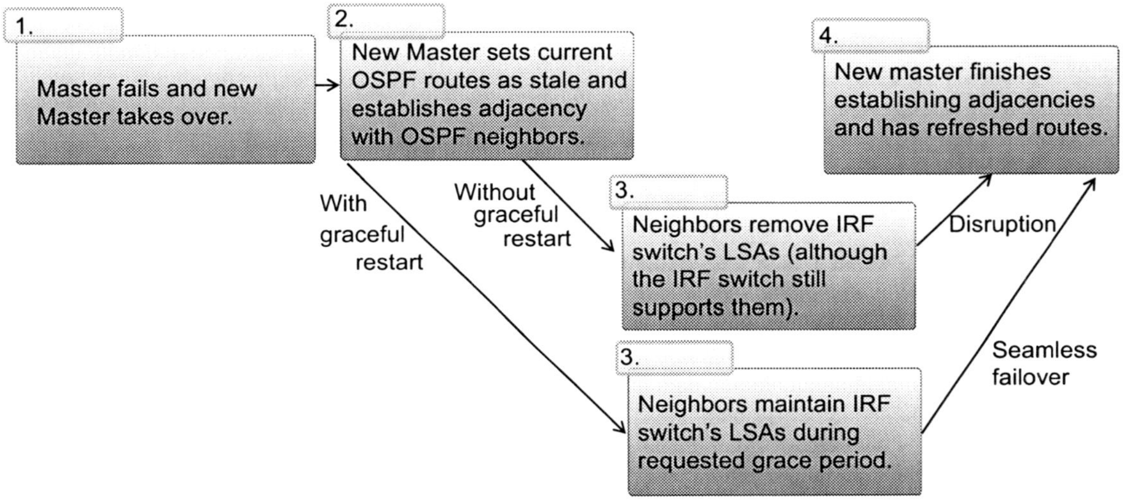

Figure 9-27: OSPF graceful restart with HP IRF

The master of the IRF virtual switch maintains routing tables and routing protocol information in its control plane. (In a chassis switch, the master's main processing unit (MPU) maintains this information.) The master proxies this information to other members' control planes, but if the master fails, the routing processes must restart on the new master.

The active IRF members have routes proxied to the forwarding information databases (FIBs) in their control planes (or the control planes of their LPUs, in the case of chassis-based switches). They can continue to route traffic with these stale routes until the new master builds the routes in its own control plane—and, when the IRF switch is using static routes or Routing Information Protocol (RIP) routes, they do seamlessly.

However, standard OSPF does not allow OSPF neighbors to entirely resynchronize the LSDBs without tearing down the old adjacency relationship. Therefore, when the new master sends hellos to OSPF neighbors and tries to reestablish adjacency, the IRF virtual switch's neighbors remove the IRF switch's LSAs from their LSDB and notify other OSPF routers to do the same. Routes reconverge away from IRF switch—even though that switch is perfectly capable of routing the traffic—causing unnecessary disruption to the traffic flow and possible interruption to service.

OSPF graceful restart, on the other hand, enables the IRF switch to ask its neighbors to maintain the neighbor and/or adjacency relationship undisturbed while the new master restarts the OSPF process. Thus, traffic continues toward its destination undisturbed.

Comware switches support Internet Engineering Task Force (IETF) standard OSPF graceful restart and a nonstandard version.

Need for graceful restart with redundant management modules

An OSPF router or routing switch that has redundant management modules must also implement OSPF graceful restart. The failover of the primary management module to the standby module is analogous to the failover of the IRF master to the subordinate member. Although the OSPF router or routing switch can use its stale routes to forward traffic, it must also send OSPF graceful restart requests to inform its neighbors not to drop its LSAs while the new management module rebuilds the adjacencies (see Figure 9-28).

Note that OSPF graceful restart also helps to maintain routing system stability in *any* circumstance in which an OSPF router must restart its OSPF processes, including when a network administrator restarts the process for maintenance purposes (such as applying new route maps or configuring a new router ID).

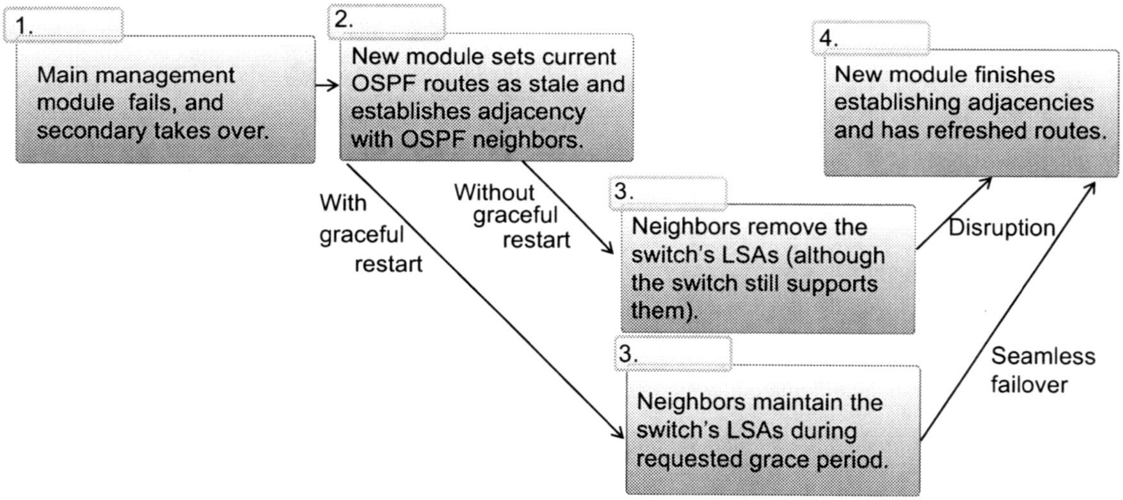

Figure 9-28: Need for graceful restart with redundant management modules

OSPF graceful restart process

Now that you understand when and why graceful restart is required, you can examine the IETF standard graceful restart process (see Figure 9-29).

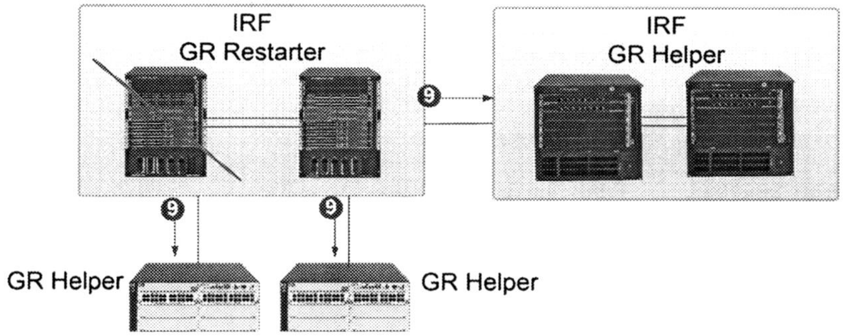

Figure 9-29: OSPF graceful restart process

1. The switch configured to support graceful restart, or the GR Restarter, determines that it needs to restart its OSPF process.

   As you have learned, this often occurs when a new IRF member takes over the master role or when a redundant management module takes over.

2. The GR Restarter sends a type 9 opaque LSA, or grace LSA, to its neighbors, which requests that the neighbors transition to graceful helper mode. The LSA includes:

- A grace period timer, which specifies how long the neighbors stay in graceful helper mode
- A graceful restart reason:
    - 0—Unknown
    - 1—Software restart
    - 2—Software upgrade
    - 3—Failover from an active to a standby MPU (or, for IRF, failover from Master to a subordinate)
- The GR Restarter's IP address

3. The neighbors, which must be configured as IETF standard GR Helpers, place the GR Restarter in graceful restart mode. In this mode, they behave as if they have full adjacency with the GR Restarter no matter what is actually occurring. Therefore, the GR Helpers continue routing traffic to the GR Restarter and maintain LSAs from it, trusting that the GR Restarter can truly continue to route traffic with its stale routes.

4. During this time, the GR Restarter is establishing neighbor relationships and adjacencies with the GR Helpers. As always during this process, the switch obtains LSAs advertised by OSPF routing devices throughout the area. Some of these LSAs were advertised by itself (or the previous IRF master); the Restarter labels those LSAs as stale.

5. After the GR Restarter has established adjacency with all of its neighbors, it sends a grace LSA, requesting that the Helpers leave helper mode (the grace period time is set to 0).
6. The GR Helpers leave helper mode. However, because they have achieved adjacency with the GR Restarter, they continue to route traffic to it without interruption.

The GR Restarter has continued routing traffic with stale routes. However, it must ensure the truth of what it asked the GR Helpers to trust: it knows the correct routes, and the topology that it advertised before the restart remains accurate. During and after the process of establishing adjacency, the GR Restarter recalculates routes, changes stale routes to active routes, and determines whether any updates are necessary:

- The GR Restarter calculates routes from the LSAs that it received from OSPF neighbors. The new routes replace the stale routes. If any stale routes remain, they are deleted as no longer valid. The GR Restarter now knows that it is routing traffic with the most up-to-date information.
- The GR Restarter generates its own LSAs and compares them to its own stale LSAs, which it received from the Helpers. Unless its connections have changed, the GR Restarter's LSAs will match its stale LSAs. The GR Restarter removes the stale designation from those LSAs and takes no further action (its neighbors already know the LSAs).
- If the GR Restarter has generated any new LSAs, it advertises them to neighbors as normal in an LSU message. Similarly, if any stale LSAs remain (no matching current LSA), the GR Restarter deletes those LSAs and withdraws them in an LSU message.

The GR Restarter and the GR Helpers now know the proper LSAs, network topology, and routes. Throughout the process, all of the OSPF routing switches continued routing traffic with stale routes, only removing the routes if they were demonstrated to be truly invalid. Because the stale routes typically prove to be valid, the routing switches typically transition seamlessly from routing with stale routes to routing with active routes.

Note

The OSPF graceful restart standard dictates that GR Helpers must leave helper mode if the topology changes in a way that would affect the GR Restarter. That is, the GR Helper generates or receives an LSA that it would normally send to the GR Restarter. This provision prevents routing loops.

Supplemental information: Comware nonstandard graceful restart

Comware switches also support a nonstandard implementation of graceful restart. The process for a nonstandard graceful restart resembles the IETF standard one in concept, but the Restarter and Helpers use special extensions to OSPF messages negotiate the process. You should typically implement the standard form of graceful restart, but the nonstandard process is provided for your reference:

1. The GR Restarter determines that it needs to restart its OSPF.
2. The GR Restarter sends an OSPF hello to its adjacent neighbors. This hello includes an L bit, indicating that it includes a link-local signaling (LLS) data block. The LLS data block uses two bits to communicates the Restarter's request to restart OSPF without disrupting the relationship:
   - LSDB resynchronization (LR) bit, which indicates that the switch is capable of out-of-band (OOB) resynchronization of the LSDB.
   - In other words, the Restarter can operate briefly while disconnected from the OSPF relationship, and it can resynchronize LSAs without the Helper needing to remove the LSAs that the Restarter formerly advertised.
   - Restart signal (RS) bit, which indicates that the switch actually wants to disconnect and restart the OSPF process.
3. The neighbor must indicate that it is capable of OOB operation as well. If it is *and* if it is configured as a GR Helper, the neighbor returns a hello with the L bit set and an LLS data block with the LR bit sit.
4. The GR Restarter then initiates the restart and begins to resynchronize its LSDB. It uses the typical OSPF messages to do so (but the messages include a special R bit indicating that they are part of a graceful restart). As for IETF standard graceful restart, the Restarter receives LSAs from other routing switches and also its own (or a former master's), which it labels as stale.
5. After the GR Restarter establishes adjacency with its neighbors, the GR Helpers, it updates its routes, generates LSAs, and removes any stale routes or LSAs.

Implementing OSPF graceful restart

Now that you understand why you need to enable OSPF graceful restart, doing so is simple. Configure IRF virtual switches, as well as routing devices that must support nonstop routing during management module failover, as GR Restarters. As a best practice, you could also enable the restarter function on other routing devices that support it, protecting the network in case an administrator accidentally restarts the OSPF process in a live environment.

All neighbors of the GR Restarters must act as GR Helpers. As a best practice, you should generally ensure that all OSPF routing devices in the AS support this function.

Comware configuration process

To configure a Comware routing switch or router as a GR Restarter, follow these steps:

1. Enable support for the OSPF opaque type 9 LSAs:

```
[ComwareRestarter-ospf-<ID>] opaque-capability enable
```

2. Enable graceful restart and set the method to IETF standard:

```
[ComwareRestarter-ospf-<ID>] graceful-restart ietf
```

3. Set the grace period time that the GR Restarter requests from GR Helpers. The default is 120 seconds, and the time must extend long enough for the switch to achieve complete adjacency with its neighbors. (You should give the Restarter a slight buffer.)

```
[ComwareRestarter-ospf-<ID>] graceful-restart interval <seconds>
```

The GR Restarter automatically implements the GR Helper functions as well.

To configure a Comware routing switch or router as a GR Helper, follow these steps:

1. Enable support for the OSPF opaque type 9 LSAs:

```
[ComwareHelper-ospf-<ID>] opaque-capability enable
```

2. By default, the GR Helper then helps all neighbors, and you should generally maintain this behavior. However, you can optionally restrict the neighbors that the GR Helper will help:

```
[ComwareHelper-ospf-<ID>] graceful-restart help {<acl ID> | prefix <prefix list name>}
```

Note

As mentioned before, the Comware devices also support a nonstandard version of graceful restart. To configure this version, instead of enabling opaque LSAs, enable link-local signaling and out-of-band synchronization on both GR Restarters and GR Helpers. Enable graceful restart in nonstandard mode on GR Restarters. Set the interval as for standard mode.

ProVision configuration process

As of the publication of this study guide, the only ProVision switches that can act as OSPF GR Restarters are HP 8200 switches with redundant MPUs. To enable the graceful restart functions, follow these steps:

1. Enable nonstop routing for OSPF:

```
ProVisionRestarter(ospf)# nonstop
```

2. Optionally, adjust the restart interval (default 120 seconds). The time should exceed the time required for the new MPU to achieve adjacency with its neighbors:

```
ProVisionRestarter(ospf)# restart interval <1-800>
```

3. Increase the redundancy rapid switchover time so that it exceeds the restart interval.

   After this timer expires, the switch removes any stale routes—that is, routes not recalculated from the reconstructed LSDB. The restart interval dictates the amount of time that the switch promises its neighbors that can continue to route traffic while it achieves adjacency and reconstructs the LSDB. Therefore, the rapid switchover time must exceed the restart interval. (For example, exceed the restart interval by 3 seconds.)

The default rapid switchover time of 90 seconds does *not* exceed the default restart interval of 120 seconds, so you must complete this step even if you did not adjust the restart interval:

```
Switch(config)# redundancy rapid-switchover <0-2147483647>
```

All ProVision routing switches that support OSPF automatically act as GR Helpers.

Implementing OSPF and VRRP

You should use IRF to provide redundancy for the default router role whenever possible. IRF provides redundancy while allowing the IRF virtual switch to act as a single entity in the routing protocol, making implementing IRF with OSPF almost as easy as implementing OSPF without redundancy. However, sometimes you have to implement OSPF and VRRP. For example, you might have redundant HP ProVision 8200 zl Series switches at the distribution layer or core of your network. You need to design the solution carefully to ensure that the OSPF and VRRP functions will work together.

When a router implements VRRP and OSPF on the same interface, it uses its actual IP address to run OSPF. However, the protocol interoperations can cause complexities. To minimize these complexities, you should avoid running OSPF and VRRP on the same interfaces.

First, connect VRRP routers to upstream OSPF routers on their own networks and do *not* run VRRP on these networks (see Figure 9-30). This design helps an upstream router to immediately determine when it loses connectivity to a VRRP router.

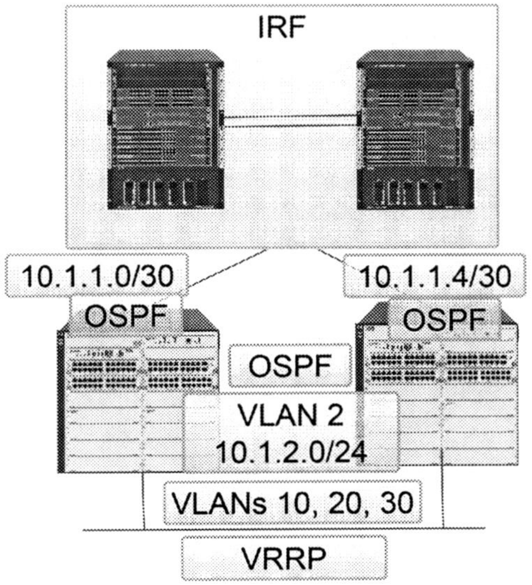

Figure 9-30: Implementing OSPF and VRRP

Second, do not establish OSPF adjacencies on the interfaces that run VRRP. Because VRRP provides redundancy for default gateway services, VRRP runs on directly connected networks, which are generally stub networks. You can redistribute the directly connected networks as external networks. (Make sure to set up the area as a normal area or an NSSA.) You could alternatively configure the interfaces as passive OSPF interfaces.

One exception applies: You should create one OSPF adjacency between the VRRP routers. These routers should exchange routes with each other. For example, one of the routers might lose its connection to the network core. However, it is still the VRRP master for one or more networks. It needs a backup path to the network core, and its adjacency with the other VRRP router provides this path. (An alternative strategy is using a VRRP track to reduce the VRRP priority if the switch loses its upstream connection, but this study guide does not cover tracks.)

Third, choose the VRRP design in which none of the VRRP routers own the virtual IP address. Otherwise, when the VRRP owner fails and comes back up, it immediately takes over routing on the interfaces for which it owns the VRRP address. However, it has not yet established adjacencies with upstream OSPF routers, disrupting connectivity on those VLANs. By configuring a virtual IP address that none of the VRRP routers own, you can prevent this problem. Either disable preempt mode or set a preempt-mode delay time that is longer than the time required to establish adjacency with all neighbors and calculate the OSPF routes (generally, 2 or 3 minutes).

Issues with expanding areas and adding redundant links to the backbone

Well-designed OSPF systems should ensure that you can always connect new areas directly to the backbone. However, sometimes the company expands in unexpected ways, and you cannot meet this requirement.

For example, in the scenario shown in Figure 9-31, the company needs to expand to new sites (10.1.128.0/17). Routers in area 1 are available for connecting to the new routers. However, area 1 has grown so large that you would prefer to establish the new sites in area 3. However, the rules of OSPF prevent you from creating the new area because this area would need to connect directly to routers in area 0, not area 1.

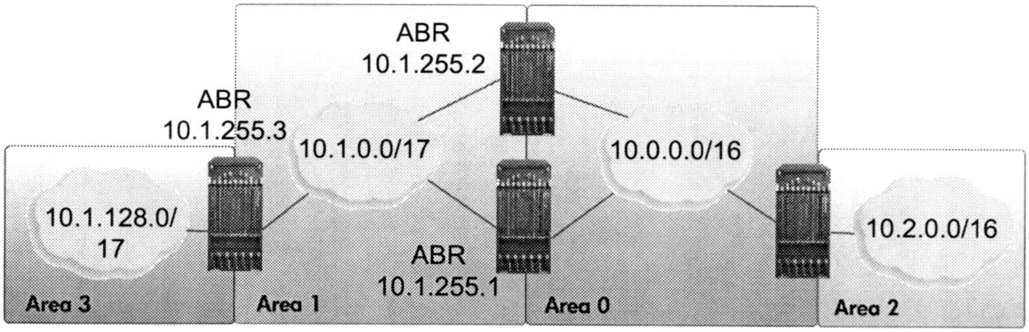

Figure 9-31: Issues when areas expand

Similar issues can arise when you need to add a redundant link between an area and the backbone area 0. As you have learned, you can create link aggregations between internal area routers and the ABR, which might be one router or, for even better redundancy, two physical routing switches in a virtual IRF switch. You can also connect internal area routers to two ABRs in area 0. You should choose one of these solutions whenever possible.

In this case, however, a second connection between a particular site and an ABR at the backbone is not available. You must establish the link through a transit site.

Resolving the issue with virtual links

You can use a virtual link to resolve both of these issues.

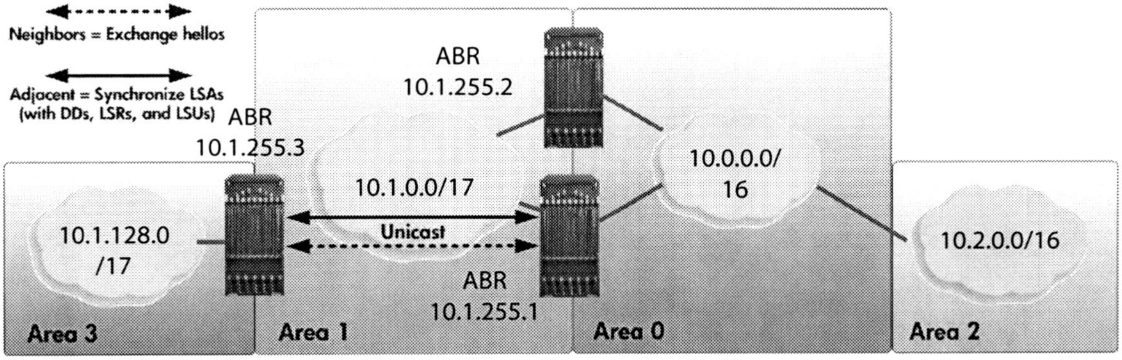

Figure 9-32: Resolving the issue with virtual links

You establish the virtual link across a transit area (area 1, in the example shown in Figure 9-32). A routing device on the border between the nonbackbone connected area and the transit area uses the virtual link to connect to an ABR between area 0 and the transit area. A virtual link acts like a shared network interface, allowing the two OSPF routing devices to exchange hellos and establish adjacency. Logically, the virtual link acts as point-to-point OSPF network on which both ABRs have an interface.

In the example shown in Figure 9-32, you would also configure a virtual link between ABR 10.1.255.3 and ABR 10.1.255.2 for redundancy.

Note that you should not generally rely on virtual links as part of your design because they introduce complexity and negate some of the benefits of multiarea designs by leaking backbone routing information into nonbackbone areas. Instead, use virtual links as a patch when you have no other way to create the necessary link to the backbone at the current time.

Virtual link configuration process

The sections below provide more detailed steps for Comware and for ProVision products. Before completing the following steps, you should configure the basic OSPF settings on all of the routers involved in the solution.

Comware configuration

1. Make sure that the transit area is not defined as a stub area or NSSA.
2. Configure the virtual link on the ABR that does not connect to area 0.
   a. This ABR should have its basic OSPF settings defined:
      - Router ID
      - The area without the direct connection to the backbone (area 3 in the example)
      - The transit area (area 1 in the example)
      - OSPF enabled on the appropriate networks in each area
      - Any other settings required for the OSPF solution
   b. In the transit area (area 1 in the example), configure a virtual link. Specify the router ID of the device at the other end of the link.

```
[NonbackboneABR-ospf-1-area-0.0.0.1] vlink-peer 10.1.255.1
```

3. Configure the virtual link on the ABR that connects to area 0.
   a. Again, this ABR should have its basic OSPF settings defined:
      - Router ID
      - Area 0
      - The transit area (area 1 in the example)
      - OSPF enabled on the appropriate networks in each area
      - Any other settings required for the OSPF solution
   b. In the transit area, configure a virtual link, specifying the router ID of the device at the other end of the link.

```
[BackboneABR-ospf-1-area-0.0.0.1] vlink-peer 10.1.255.3
```

Note

Note that the OSPF routing devices must know how to reach each other's router IDs. It is generally best practice to run OSPF on the network that includes the router ID (including a loopback interface).

4. You must check consistency for the typical settings required for two OSPF routers on a link to achieve adjacency. These settings include the OSPF timers. (The virtual link does not have a network associated with it, so you do not have to check consistency for it.)

ProVision configuration

1. Make sure that the transit area is not defined as a stub area or NSSA.
2. Configure the virtual link on the ABR that does not connect to area 0.
    a. This ABR should have its basic OSPF settings defined:
        - Router ID
        - The area without the direct connection to the backbone (area 3 in the example)
        - The transit area (area 1 in the example)
        - OSPF enabled on the appropriate interfaces in each area
        - Any other settings required for the OSPF solution (such as aggregated area summaries)

    b. In the transit area (area 1 in the example), configure a virtual link. Specify the router ID of the device at the other end of the link.

```
NonbackboneABR(ospf)# area 1 virtual-link 10.1.255.1
```

3. Configure the virtual link on the ABR that connects to area 0.
    a. Again, this ABR should have its basic OSPF settings defined:
        - Router ID
        - Area 0
        - The transit area (area 1 in the example)
        - OSPF enabled on the appropriate networks in each area
        - Any other settings required for the OSPF solution

    b. In the transit area, configure a virtual link, specifying the router ID of the device at the other end of the link.

```
BackboneABR(ospf)# area 1 virtual-link 10.1.255.3
```

Note

Note that the OSPF routing devices must know how to reach each other's router IDs. It is generally best practice to run OSPF on the network that includes the router ID (including a loopback interface).

4. You must check consistency for the typical settings required for two OSPF routers on a link to achieve adjacency. These settings include the OSPF timers. (The virtual link does not have a network associated with it, so you do not have to check consistency for its type.)

OSPF authentication

You can implement authentication for OSPF processes on particular router or routing switch interfaces. When an OSPF interface implements authentication, it includes authentication data in all OSPF packets that it transmits. It also checks this data in all OSPF packets that it receives and rejects any packets that fail the check. In other words, only OSPF routers configured with the same authentication keys can establish neighbor relationships.

A packet's authentication field specifies the type (as the AuthID), so receiving routers know how they are expected to authenticate the packet. They only accept packets that are using the same authentication type configured on their interface. For each OSPF area, you can choose from several authentication options:

- **Null**—The default authentication type, in which OSPF interfaces do not enforce authentication. The OSPF packets have an authentication field, but the field does not include any data. An OSPF interface implementing null authentication does not examine the authentication field but process all valid OSPF packets.
- **Simple**—The OSPF interface places the authentication key in every packet's OSPF header. Other OSPF routers, also implementing this type, examine this field. They only accept OSPF packets for which the field matches their own key.

  Because packets include the key in plaintext, anyone who can collect an OSPF packet transmitted on the network can learn the key. You should use this form of authentication only if you primarily aim to eliminate misconfigurations and routers inadvertently connecting on the wrong networks. You should not use it to secure your OSPF system from unauthorized routers and hackers.
- **Cryptographic (MD5)**—The OSPF interface hashes the OSPF packet data and the key. It then places the hash at the end of the packet. The authentication field includes the key ID, a length for the hash, and a cryptographic sequence number; however, it does not include the key itself, protecting it from eavesdroppers.

  A router receiving the OSPF packet hashes the packet with its secret key, identified by the key ID. If the hash matches, the router knows that the transmitting router has the correct secret key for this key ID, so it accepts the packet.

  The cryptographic sequence number provides a degree of protection against replay attacks. An OSPF interface only accepts packets with a sequence number equal to or higher than the number for the most recent accepted OSPF packet from this neighbor. The sequence number might be a counter, but is usually based on the system clock. In either case, the number is always increasing.

  If a hacker obtains a legitimate OSPF packet and attempts to retransmit it later, the packet's cryptographic sequence number is no longer valid, and the other routers on the network reject it. A hacker could replay a packet in the window before the legitimate neighbor sends a packet with a higher sequence, but this window is quite small.

Cryptographic authentication provides authenticity, ensuring that only OSPF interfaces configured with the same key can communicate. It does not provide privacy: the contents of the packet are not encrypted.

All HP routers and routing switches support MD5 for the cryptographic hashing algorithm.

Comware simple authentication configuration

Enter these commands to set up simple authentication on a Comware router or routing switch:

1. Access all OSPF areas on which you want to enforce this form of authentication. Enable simple authentication:

   ```
   [Comware-ospf-area-<ID>]authentication-mode simple
   ```

2. Access each interface in this area. Set the password, which can have up to eight characters and must match the password on any OSPF neighbors on this interface. Set the **cipher** option to have the simple password encrypted in the configuration file; use the **plaintext** option to show the password in the file:

   ```
   [Comware-<interface ID>]ospf authentication-mode simple [cipher | plaintext] <password>
   ```

For example, you might enter these commands:

```
[Comware]ospf 1
[Comware-ospf-1]area 0
[Comware-ospf-1-area-0.0.0.0]authentication-mode simple
[Comware-ospf-1-area-0.0.0.0]interface vlan 100
[Comware-Vlan-Interface100]ospf authentication-mode simple cipher key100
[Comware-Vlan-Interface100]interface vlan 101
[Comware-Vlan-Interface101]ospf authentication-mode simple cipher key101
```

Comware MD5 authentication configuration

Enter these commands to set up MD5 authentication on a Comware router or routing switch:

1. Access all OSPF areas on which you want to enforce this form of authentication. Enable MD5 authentication:

   ```
   [Comware-ospf-<ID>-area-<ID>]authentication-mode md5
   ```

2. Access each interface in this area. Set the key ID and the password, which must match those values on any OSPF neighbors on this interface. Set the **cipher** option to have the password encrypted in the configuration file; use the plaintext option to show the password in the file:

```
[Comware-<interface ID>]ospf authentication-mode md5 <key ID> [cipher
| plaintext] <password>
```

For example, you might enter these commands:

```
[Comware]ospf 1
[Comware-ospf-1]area 0
[Comware-ospf-1-area-0.0.0.0]authentication-mode md5
[Comware-ospf-1-area-0.0.0.0]interface vlan 100
[Comware-Vlan-Interface100]ospf authentication-mode md5 1 cipher key100
[Comware-Vlan-Interface100]interface vlan 101
[Comware-Vlan-Interface101]ospf authentication-mode md5 1 cipher key101
```

ProVision simple authentication configuration

To set up simple authentication on a ProVision routing switch, access each interface on which you want to enforce this form of authentication. Enable authentication and set the password with this command:

```
ProVision(<vlanID>)# ip ospf [IP address] authentication-key <password>
```

The password can have up to eight characters and must match the password on any OSPF neighbors on this interface.

You only need to set the IP address if the VLAN supports multiple IP addresses, and you want to specify different settings for each.

For example, you might enter these commands:

```
ProVision(config)# vlan 100
ProVision(vlan100)# ip ospf authentication-key simple key100
ProVision(vlan100)# vlan 101
ProVision(vlan101)# ip ospf authentication-key simple key101
```

ProVision MD5 authentication configuration

Enter these commands to set up MD5 authentication on a ProVision routing switch:

1. Configure one or more MD5 key chains from the global configuration mode context:

   ```
   ProVision(config)# key-chain <chain name> key <key ID> key-secret <password>
   ```

2. Access each interface on which you want to enable MD5 authentication. Enable this form of authentication and select the key chain:

   ```
   ProVision(vlanID)# ip ospf md5-auth-key-chain <chain name>
   ```

For example, you might enter these commands:

```
ProVision(config)# key-chain chain100 key 1 key-secret key100
ProVision(config)# key-chain chain101 key 1 key-secret key101
ProVision(config)# vlan 100
ProVision(vlan100)# ip ospf md5-auth-key-chain chain100
ProVision(vlan100)# vlan 101
ProVision(vlan101)# ip ospf md5-auth-key-chain chain101
```

Monitoring OSPF with HP IMC

Throughout this study guide, you have learned how HP Intelligent Management Center (IMC) can help you manage and monitor your network. You can use IMC to quickly gather OSPF information about switches (see Figure 9-33). IMC provides OSPF information such as:

- Router ID
- Whether or not the switch is the ABR or ASBR
- Information about OSPF neighbors
- Link-state database (LDSB) information
- Virtual link

IMC also allows you to put this OSPF information into the context of the entire network by viewing the Layer 3 topology.

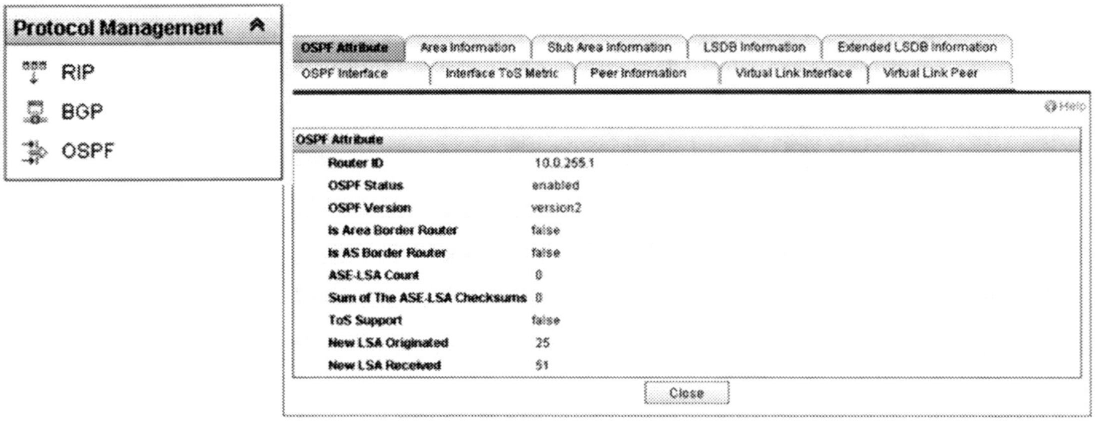

Figure 9-33: Monitoring OSPF with IMC

Summary

In this chapter, you have learned how to implement more complex OSPF systems. You know why you might need to divide an OSPF AS into multiple areas, how to establish the areas, and how the area borders affect LSA propagation. You understand how OSPF routers use metrics to select inter-area routes, and you can aggregate and filter routes between areas.

You also know when and how to redistribute external routes into OSPF. You understand how routers calculate external routes using the best path to the ASBR, and external type 1 and type 2 metrics help them choose between identical external routes. You can also aggregate external networks with ASBR summaries, as well as eliminate the external routes from stub areas entirely.

You know which types of areas benefit from being defined as stub areas and from being defined as totally stub areas. You can also set up an NSSA to allow local external routes within a stub area.

You also explored strategies and technologies for enhancing OSPF convergence. In particular, you learned when and how to use BFD to speed the detection of a failed neighbor. You can also use OSPF graceful restart to ensure seamless failover of an OSPF process on an IRF virtual device or router device with redundant management modules.

You examined several situations when a virtual link might be necessary, and you practiced creating such a link.

Finally, you learned how to implement OSPF authentication to minimize misconfigurations as well as to provide some protection against hackers.

Learning check

Answer these questions:

1. What are advantages of designing an OSPF AS to use multiple areas?

2. How does an OSPF router choose the best external route when it has received multiple type 5 LSAs for the same network? The LSAs are using type 2 metrics.

3. An OSPF router has an Ethernet link to another OSPF router; however, the link is indirect. How can you speed up the time that it takes for a router to detect a failed neighbor?

4. What function does cryptographic (MD5) OSPF authentication serve? How does this type of authentication work?

Answers

1. What are advantages of designing an OSPF AS to use multiple areas?

   OSPF areas segment the domain into more stable and manageable pieces. Each OSPF routing device maintains the LSDB and runs the shortest-path algorithm only for its own areas. The ABRs act as bottlenecks for LSAs, forwarding only one LSA per network into other areas. A multiarea design decreases LSDB size and increases stability—routers do not have to rerun the shortest-path algorithm as often. It also lets you aggregate advertisements of networks in between areas, simplifying the routing table. You can also control routes at this boundary, filtering them if you want. These benefits still apply even for powerful routers that could support very large areas.

2. How does an OSPF router choose the best external route when it has received multiple type 5 LSAs for the same network? The LSAs are using type 2 metrics.

   It chooses the route with the lower metric, which is copied directly from the LSA (no costs are added from OSPF links). If multiple LSAs have the same metric, the router chooses the route with the lowest-cost path to the ASBR. If multiple paths have the same cost, the router can use ECMP, or it can use the advertising ASBRs' router IDs to break the tie. In more detail:

   1) The router uses the type 5 LSA to calculate the external route, setting the correct forwarding interface and next hop for the path to the advertising ASBR.
   2) The router sets the external route's metric equal to the metric advertised in the type 5 LSA. The advertising ASBR added this metric when it redistributed the route into OSPF.
   3) When the OSPF routing table already includes an external (type 2) OSPF route to this network, the router selects the route with the lower metric.
   4) If both routes have the same metric, the router examines the metric for the forwarding path to the ASBR. It selects the route associated with the ASBR to which it has a lower-cost path.
   5) If forwarding paths to the ASBRs have equal costs and the router supports ECMP, the router installs multiple routes in the OSPF routing table. (Otherwise, it uses another mechanism such as the router ID to break the tie.)

3. An OSPF router has an Ethernet link to another OSPF router; however, the link is indirect. How can you speed up the time that it takes for a router to detect a failed neighbor?

   The interface's OSPF dead time determines how long the router takes to detect a failed neighbor. You can reduce this time by reducing the hello time (must match on all routers on the interface). On Comware switches, you can set up OSPF BFD to detect failed neighbors in ms. BFD tests the connectivity to an IP address and pulls down the OSPF adjacencies if it detects a problem.

4. What function does cryptographic (MD5) OSPF authentication serve? How does this type of authentication work?

   Cryptographic authentication provides authenticity, ensuring that only OSPF interfaces configured with the same key can communicate. It does not provide privacy: the contents of the packet are not encrypted.

   The OSPF interface hashes the OSPF packet data and the key. It then places the hash at the end of the packet. The authentication field includes the key ID, a length for the hash, and a cryptographic sequence number; however, it does not include the key itself, protecting it from eavesdroppers.

   A router receiving the OSPF packet hashes the packet with its secret key, identified by the key ID. If the hash matches, the router knows that the transmitting router has the correct secret key for this key ID, so it accepts the packet.

10 eBGP

EXAM OBJECTIVES

In this chapter, you learn to:

- ✓ Establish and monitor exterior Border Gateway Protocol (eBGP) sessions between your routers and Internet Service Provider (ISP) routers.
- ✓ Advertise an IP block to multiple ISP routers.
- ✓ Filter BGP routes as required for a dual-homed ISP connection.
- ✓ Configure a BGP router to advertise a default route in Open Shortest Path First (OSPF) or to redistribute and aggregate BGP routes, as appropriate.

ASSUMED KNOWLEDGE

Before studying this chapter, you should have a basic understanding of interior routing protocols such as Open Shortest Path First (OSPF).

INTRODUCTION

This chapter introduces you to BGP, a powerful routing protocol designed for ISPs and very large enterprise networks. This chapter does not teach you about all the aspects of the protocol or all the ways in which you can manipulate BGP route advertisements and route selection. Rather, this chapter is intended to train you to configure eBGP for a specific use case.

BGP use case: Multihomed ISP connections

You will now examine the BGP use case that is the focus of this chapter. For your reference, this section also introduces other use cases for BGP that are beyond the scope of this study guide. Finally, you will begin to learn ways in which BGP differs in design from routing protocols such as Router Information Protocol (RIP) and Open Shortest Path First (OSPF)—an exploration that you will continue throughout the chapter.

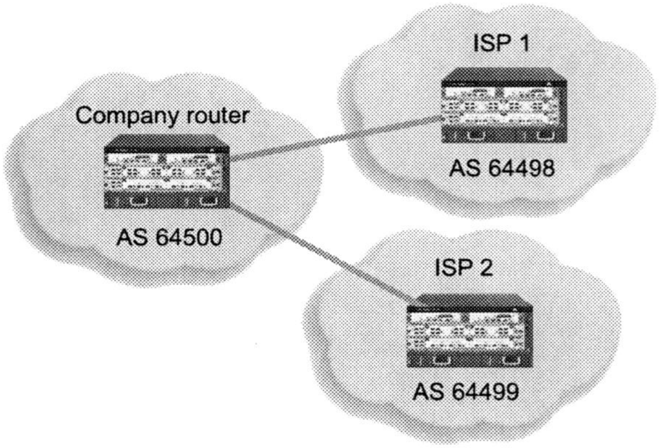

Figure 10-1: Multihomed ISP connections

Figure 10-1 illustrates the primary use case that you will examine in this chapter: a company needs to communicate with multiple ISP routers. You must understand when this scenario and similar scenarios call for BGP, as well as when it does not.

You do not need to implement BGP simply to connect to an ISP. You only need to do so when the company specifically needs to communicate routing information with the ISP. Typically, a company with a single connection to an ISP does not require BGP:

- The ISP handles announcing the block of public IP addresses assigned to the company.
- The company's routers use a default route for reaching all public IP addresses.

When the company has multiple connections to the same ISP, it might benefit from communicating routing information with the ISP:

- The company router can discover more granular information about which connection offers the best route to particular external networks.
- In some special cases, the company might need to advertise different IP blocks over different connections.

Figure 10-1 shows a case in which BGP might become even more useful: the company router is multihomed to different ISPs:

- Again, the company router can discover which connection offers the best connection for particular Internet traffic.
- The company router might need to announce the company's public IP address block to the second ISP. That ISP can then propagate this advertisement throughout the Internet, enabling external devices to reach the company over the best path, which might include either ISP.

  Alternatively, *both* ISPs might be able to handle advertising the block of addresses. The company would need to discuss the proper solution with both ISPs. If the ISPs handle announcing the company's addresses, the company might not need to implement BGP at all. You should assess whether the benefit of selecting the particular connection for particular external routes outweighs the complexities of implementing BGP.

Note

For better redundancy, you might establish the multihomed connections on multiple routers. This solution requires the company routers to implement interior BGP (iBGP), which is beyond the scope of this study guide.

BGP use cases: ISPs and large enterprises

BGP is designed specifically to communicate routes between organizations controlled by different entities and scales to hundreds of thousands of routes. For these reasons, ISPs use BGP to communicate routes with other ISPs, and BGP underlies all Internet routing.

For the same reasons, large enterprises that have outgrown a routing protocol such as OSPF (or IS-IS) can implement BGP. The enterprise can segment itself into multiple OSPF Autonomous Systems (ASes) and use BGP to communicate routes between these ASes in a scalable manner.

Multiple Protocol Label Switching (MPLS) is used by service providers and sometimes in data centers to establish transparent paths across complex routing systems. An MPLS virtual private network (VPN) can connect networks to each other across the MPLS domain at Layer 2 or Layer 3. The VPN can be established with BGP.

Implementing a complete BGP solution in use cases such as these is beyond the scope of this study guide.

BGP overview

With this brief introduction to BGP scenarios, you can turn your attention to the protocol itself and how it is designed specifically for these scenarios.

As you have seen, BGP is an external gateway protocol (EGP) used to connect different autonomous systems (ASes), between a company and an ISP, multiple ISPs, or even segments within a large private network.

Note

Do not confuse the fact that BGP is an EGP with a deprecated protocol, also called EGP.

Figure 10-2 lists some ways in which BGP, as an EGP, differs from the IGPs such as OSPF.

BGP tends to offer network administrators tighter control over neighbor relationships, route advertisements, and routes accepted. For example, when you enable OSPF on an interface, the OSPF router automatically attempts to establish neighbor relationships. However, you must manually configure BGP peer relationships.

As you will learn, you must also explicitly inject routes into BGP. You can also define a variety of attributes that affect route selection on your routers and other BGP routers. In these ways, each AS can enforce its own organizational policies.

OSPF	BGP
Designed as an IGP (thousands of routes)	Designed as an EGP between ASes (hundreds of thousands of routes)
Discovers neighbors automatically on networks on which it is enabled	Establishes connections to manually configured neighbors in TCP session
Sends messages directly at the IP layer	Sends messages within TCP sessions
Advertises link states for networks on which it is enabled	Advertises routes manually injected into a routing table

Figure 10-2: Comparing OSPF and BGP

BGP AS

The AS forms the fundamental building block of BGP, and each organization such as a company or ISP generally has its own AS, identified by a unique number (see Table 10-1). Originally, BGP supported only 16-bit AS numbers, ranging between 0 and 65535.

Table 10-1: BGP AS 16-bit numbers

Purpose	AS numbers
Reserved	0 and 65535
Public IANA-assigned	1-64495
Documentation and examples	64496-64511
Private	64512-65534

Because 16-bit numbers allow for only 65535 unique numbers, the supply will soon be depleted. Request for Comments (RFC) 4893 establishes 32-bit AS numbers, increasing the range of AS numbers to 4 billion.

When a company determines that it needs to run BGP, it must obtain a valid AS from the proper regional authority.

BGP operates in one of two forms depending on the ASes to which the communicating peers belong:

- Peers in different ASes communicate with eBGP.
- Peers in the same AS communicate with iBGP.

The same router might use eBGP with one peer and iBGP with another.

It is important that you understand that iBGP is not the only routing protocol that operates within the AS (see Figure 10-3). The AS also runs an interior gateway protocol (IGP) such as OSPF, RIP, or IS-IS, which is localized to the AS and handles routing the local traffic. Typically, just a few routers in the AS—the routers that connect to other ASes and perhaps some core routers—run BGP in addition to the IGP.

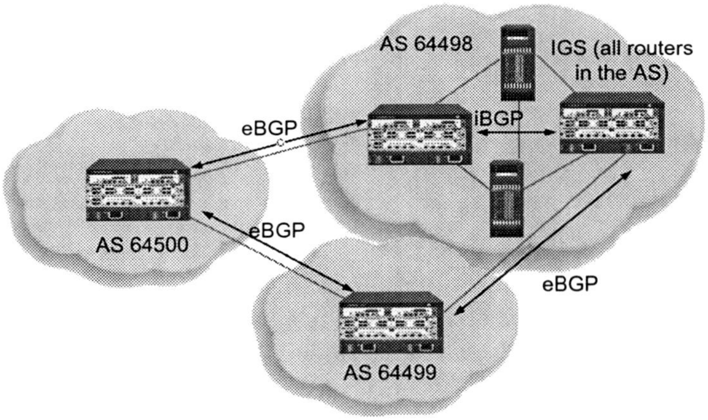

Figure 10-3: BGP AS

As you see, the primary use case for this chapter features only eBGP sessions between a company router and ISP routers. Therefore, this chapter focuses on eBGP.

Note

In some cases, the ISP might place a company within its own AS. In this case, however, the company router would use iBGP to communicate with the ISP routers. The company would also need a different AS to connect to another ISP. This study guide does not cover these use cases.

BGP TCP sessions

BGP peers establish a TCP session on port 179, as shown in Figure 10-4. They then open a BGP connection within the session. This connection carries all messages and routing updates between the peers, and the TCP session provides a degree of reliability for the communications. The BGP connection also features keepalives, which helps the peers determine whether the peer is reachable.

By default, the eBGP peers must establish the session on a direct Layer 3 connection. In other words, they have IP addresses on the same subnet. In most cases, you should keep this behavior.

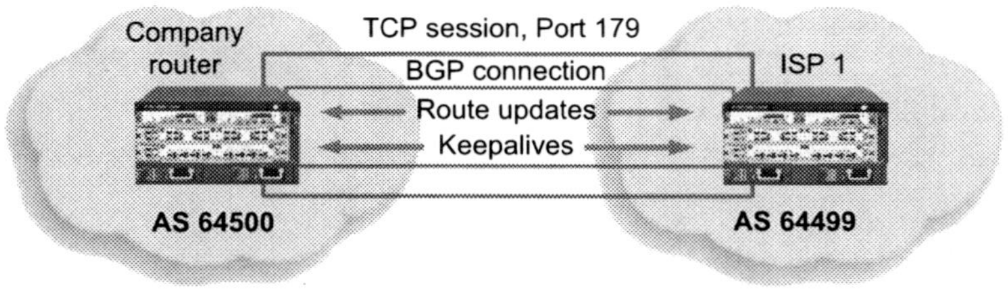

Figure 10-4: BGP TCP sessions

Non-directly connected peers

However, sometimes you need to configure a session between peers on different networks.

For example, your router has a redundant Layer 3 connection to the same router. Each connection has its own IP address, but you want to establish a single connection to the peer with a stable IP address. Therefore, you configure the BGP peer's loopback interface IP address as its IP address. Similarly, you set the local peer's IP address to its loopback interface. You can then set the peer's loopback IP address as the peer IP address.

Although the loopback IP addresses are on directly connected routers and are technically only one hop apart, they are not on the same link. You have two options. You can disable the directly connected subnet check or set eBGP multihop to 2, which has the same effect.

You must also ensure that the routers have routes to each other's loopback interfaces. Create static routes to the peer's loopback interface; the next hop for each route is the peer's IP address on one of the connected interfaces. Depending on whether you want to load-balance traffic over the redundant links to the ISP router, set the same or different administrative distances on the routes.

Similarly, ensure that the peer creates routes to your router's loopback interface.

Another use case for eBGP multihop is when you must reach the ISP router that runs BGP through another router or routers. To set up this solution, you must set the eBGP multihop to the correct number (2 if one router intervenes, three if two routers intervene, and so on). Your router also requires a route to the ISP router's IP address, of course.

BGP connection states

To establish the BGP connection, the peers first open the TCP session on port 179. They then send BGP open messages with a list of parameters for the session and, assuming the parameters match, finally send the first keepalive of the session.

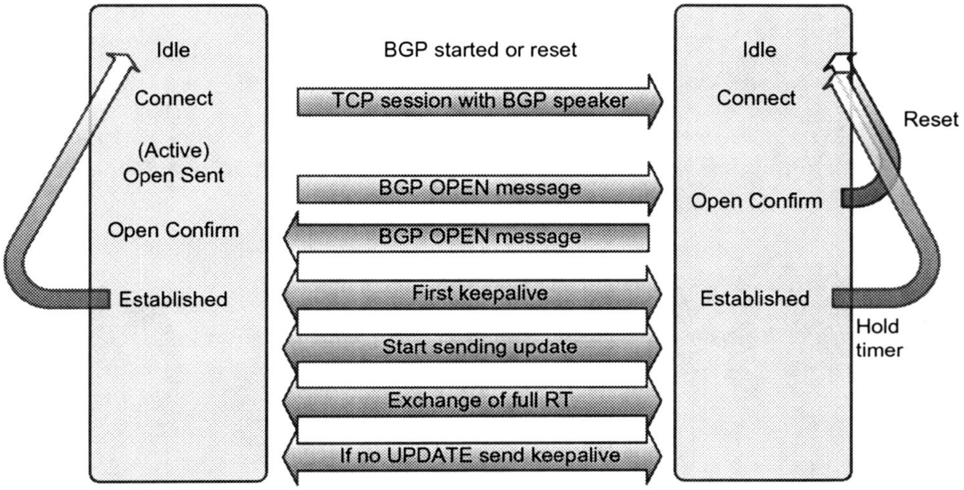

Figure 10-5: BGP connection states

Using Figure 10-5, you can trace the states through which a BGP connection might pass:

- The BGP connection begins in the idle state and returns to this state when the session establishment times out without success or an error occurs.
- The connection typically transitions immediately to the connect state, in which the peer attempts to contact the peer at TCP port 179 and waits for a response.

  If the router is set to passive TCP mode for this peer, the session transitions to active mode, rather than to connect mode. In this mode, the router waits for the peer to initiate the session.

  The connection also moves to the active state if the router does not receive a response to its TCP session initiation. In this state, the router waits for the peer's response and also, after a timer expires, might try to initiate the connection again.
- After the BGP peers have established the TCP session, the router sends an open message, either immediately or after a configurable delay. The connection also transitions to the OpenSent state or directly to the OpenConfirm state depending on whether the router has already received the peer's open message.
- In the OpenSent state, the router waits for the peer's open message. Once it receives that message, it verifies that necessary parameters match (for example, the peer's AS matches the remote AS

configured on your router). Assuming that the parameters match, the router sends a keepalive and transitions to the OpenConfirm state.

If the parameters do not match, the router sends an error notification and reverts to the idle state.

- In the OpenConfirm state, the router waits for the peer's keepalive message and also sends its own keepalive. It also listens for error notifications such as one indicating that the open message parameters do not match.
- After receiving the peer's keepalive, the connection transitions to the established state.

  The connection remains in this state during normal operation. The peers continue to send keepalives. If a BGP router fails to receive a number of keepalives, it ends the connection. Note that the keepalive period is quite long (5 minutes, by default).

Supplemental information

The following sections describe the states in more detail for your reference.

Idle

The BGP routing process is initially in the idle state. If the BGP routing process receives a start event (either an administrator manually starts it or the system starts it automatically), the process initializes the BGP resources, starts the ConnectRetry timer, tries to set up a transport connection to another BGP peer, and listens for a connection from a BGP peer. It then changes its state to connect.

If at any time the BGP routing process is reset, then the peer is reset, and the BGP routing process returns to the idle state.

Note
If the BGP routing process is set to start in passive TCP mode, it listens for connections but does not initiate them and changes its state to active.

Connect

During the connect state, the BGP routing process waits for the TCP connection to another BGP peer to be completed. If the connection is established successfully, the BGP routing process:

- Clears the ConnectRetry timer
- Sends an open message to its peer, either immediately or after a delay, depending on the configuration
- Either:
  - Transitions to the OpenSent state—if it sends the open message first
  - Transitions to the OpenConfirm state (but first sends its open message)—if it receives the peer's valid open message first

If the connection fails to be established, the BGP router resets to the idle state. However, if the process was waiting to send a delayed open message and the connection then fails, the router moves to the active state.

Active

In the active state, the BGP routing process continues to listen for a TCP session with a BGP peer.

- Failures
    - If the ConnectRetry timer expires before the connection is established successfully, the BGP router will restart the timer. It will then move to the connect state and try to initiate a TCP connection to a BGP peer again as well as continue to listen for other connections. If the second attempt is unsuccessful, the BGP routing process will return to the idle state.
    - If the BGP router receives a connection request from an unknown IP address, it will reject the request, restart the ConnectRetry timer, and continue to listen for connections from other BGP peers.
- Success
    - If the connection is established successfully, the BGP routing process takes the same steps that it takes at a successful connection in the connect state. The process sends an open message and transitions to the OpenSent or OpenConfirm state (depending on whether it first received an open message from the peer).

OpenSent

The BGP routing process listens for an open message from its BGP peer. When it receives the message, it determines whether its BGP parameters match those of the BGP peer. If the parameters fail to match, the BGP routing process sends a notification to its BGP peer. The local router's BGP routing process transitions to the idle state.

If the parameters match, the BGP routing process sends a keepalive message and transitions to the OpenConfirm state.

OpenConfirm

The BGP routing process listens for a keepalive from the BGP peer. If its keepalive timer expires first, it sends its own keepalive and continues to wait for the peer's. The process continues to wait for the duration of the HoldTime, which started running when the process sent its open message. (The HoldTime is quite long, typically 4 minutes.) If the HoldTime expires before the process receives a keepalive, the routing process resets the connection and returns to the idle state.

However, if the BGP routing process does receive a keepalive message, it transitions to the established state.

On the other hand, if it receives a notification message (such as one that indicates mismatched parameters), the BGP routing process transitions to the idle state. If an error or configuration change occurs, the BGP routing process sends a notification message with the finite-state machine (FSM) error code and then transitions to the idle state.

Established

The BGP routing process has established a peer and will now exchange update messages with that peer. It will restart the hold timer when it receives an update or keepalive message.

If the BGP routing process receives an error notification, however, it will move to the idle state.

MD5 authentication

The ISP might require your router to authenticate. You simply add a password when you define the peer. The BGP peers use this password to hash the TCP payload of packets that carry the BGP messages, appending the hash to the packet (see Figure 10-6). If the peers' passwords do not match, the TCP session for the BGP connection fails to establish.

Because the peers add the MD5 hash to each packet sent on the TCP session, MD5 authentication provides data integrity for all BGP messages and for underlying TCP session. In other words, each peer knows that the proper peer sent the message and that no one has altered the message. However, MD5 does not provide protection against eavesdropping.

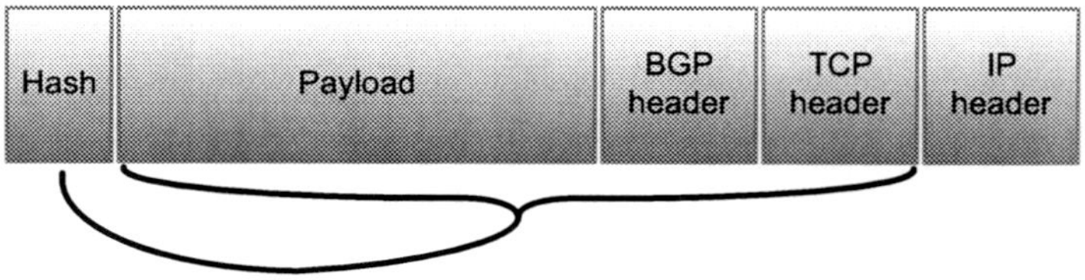

Figure 10-6: MD5 authentication

Configuring eBGP connections

To configure the eBGP connection, you must first set the AS. For both HP Comware and HP ProVision switches, you specify the AS when you first access the BGP configuration view or context. You continue to specify that AS whenever you access that view or context.

You then define the peer AS and specify the peer's IP address.

Remember to set the multihop option if the peer IP address is not on the same network as one of the local router's IP addresses. Of course, your router also requires a route to the peer IP address.

Also remember to specify the password if you need to use MD5 authentication.

After you configure these basic settings, a Comware router (or routing switch) begins to establish a BGP connection with that peer, as you learned earlier. On a ProVision switch, you must first enable BGP manually. In either case, because the router's AS and the peer's AS differ, the peers automatically interact using eBGP.

BGP routing table

With BGP, you must take explicit steps to advertise routes to a peer. This section explains which routes eBGP peers advertise to each other and how you can control those advertisements.

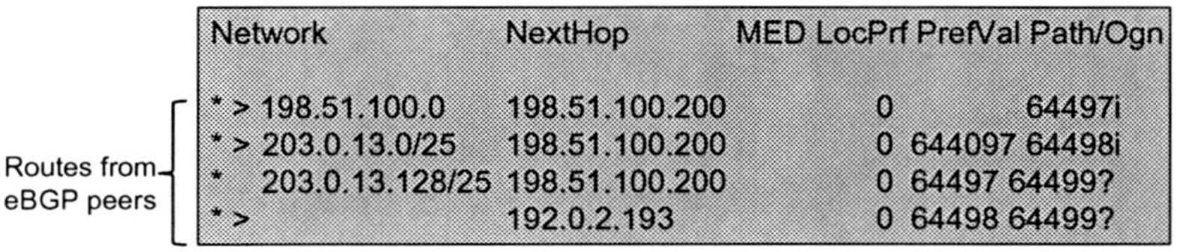

```
Network            NextHop          MED LocPrf PrefVal Path/Ogn

* > 198.51.100.0     198.51.100.200          0       64497i
* > 203.0.13.0/25    198.51.100.200          0 644097 64498i
*   203.0.13.128/25  198.51.100.200          0 64497 64499?
* >                  192.0.2.193             0 64498 64499?
```


Figure 10-7: Example BGP routing table

After establishing an eBGP session, the peers exchange their BGP routing tables. More specifically, they advertise their best route to each network in the table to the peer. The best route is indicated by the > character, as shown in Figure 10-7. The * indicates that the route is valid—that is, the router has a route to the next hop.

The BGP table includes routes that the BGP router has received from its BGP peers. If you want the BGP router to advertise any other routes, such as routes to networks within its own AS, however, you must add those routes explicitly. For example, your BGP router needs to advertise the company's IP block to the ISP. You must add a route to that block to the BGP routing table.

You add routes from the BGP router's IP routing table, either by injecting the routes or redistributing them. The next sections explain these two approaches in more detail.

Injecting routes into the BGP routing table

You can inject any route in the IP routing table into the BGP routing table by entering the network command in the BGP process. You must specify a network address and prefix length that matches the route in the routing table exactly. In the example shown in Figure 10-8, you enter **network 192.0.2.192 26**, and the network is added to the BGP routing table.

You should always view the BGP routing table to check the routes that your router will advertise to its peers. Figure 10-8 shows an example BGP routing table. When you inject the route into the BGP routing table, the BGP routing process adds a number of BGP-specific attributes. The origin of the route is i, indicating that the BGP router has learned this route from one of its IGPs. The next hop is 0.0.0.0, indicating that this router advertises the route.

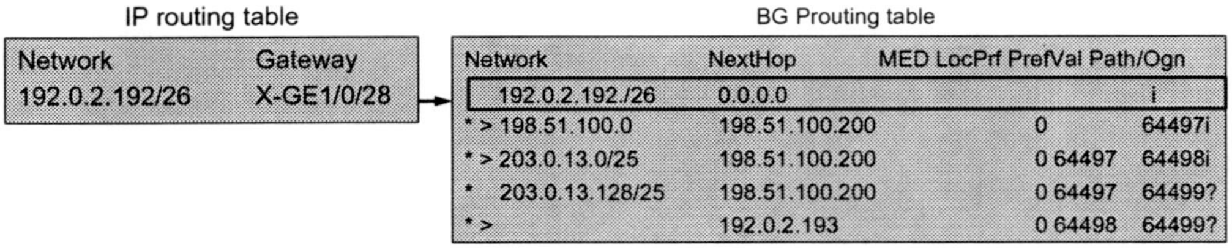

Figure 10-8: Injecting routes into the BGP routing table

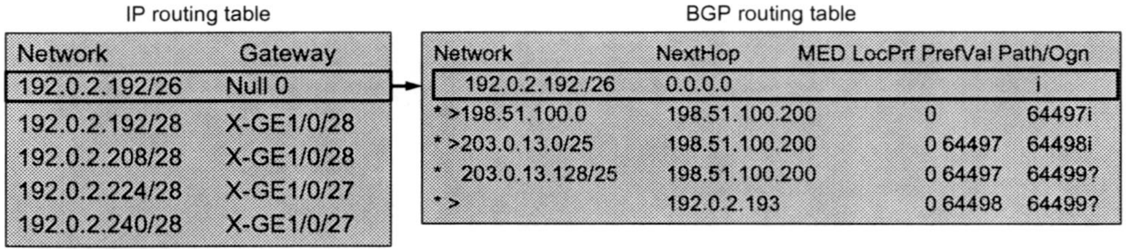

Figure 10-9: Injecting routes into the BGP routing table

Sometimes you want to inject a single route that aggregates several routes known by your router. For example, your IP block consists of 192.0.2.192/26, and your router knows routes to 192.0.2.192/28, 192.0.2.208/28, 192.0.2.224/28, and 192.0.2.240/28 (see Figure 10-9). However, you want to advertise the complete block as one route to 192.0.2.192/26. When the aggregated route does not exist in the routing table, add the route through the null interface. Specifying the null interface also ensures that the route always remains in the routing table.

You can then specify the aggregated network in the BGP process. The network matches the null route, so it is added to the BGP table.

Redistributing a route into the BGP routing table

Instead of injecting routes with the network command, you can redistribute routes of a specific type:

- Static routes
- Any routes discovered by another routing protocol supported on the routing switch, which might include:
    - OSPF (the typical choice)
    - RIP
    - IS-IS (some Comware switches)
- Directly connected routes

Of course, the BGP router can only redistribute active routes in the global IP routing table.

Redistributing routes enables you to import many routes into the BGP routing table with a single command. However, you do not have the same level of control provide by the network command, which lets you choose the precise routes added.

For example, the router might know OSPF routes to networks using private IP addresses as well as to the networks using the public IP addresses. Redistributing OSPF routes into BGP adds all of the OSPF routes indiscriminately. Or you might create a static route through the null interface to the company's complete public IP block, so you redistribute static routes. The BGP routing table also receives any static routes defined on the router for other reasons.

Therefore, if you choose to redistribute routes, you should usually apply a route map to the redistribute command. Configure the route map to filter out all routes except the route or routes to the company's public IP networks.

As always, check the BGP routing table to verify that your router will advertise the correct routes to its peers. As you see, the origin for redistributed routes is ?, indicating an unknown origin. Keep in mind that peers prefer routes with an i origin, so the choice to redistribute rather inject the routes might affect whether peers choose to route traffic to this router.

Receiving routes from the eBGP peer

The eBGP peer also advertises its entire BGP routing table to your BGP router. Your router follows this process when it receives the routes:

- The router adds the routes to its BGP routing table.
- The route specifies the eBGP peer as the advertiser. By default, other BGP attributes remain intact.
- If the BGP router has received routes to identical networks from multiple peers, it must select the best route. You will look at the specific selection process in a moment.
- The router proposes the best routes to the global IP routing table.
- The router advertises the best routes to other eBGP peers (see Figure 10-10).

  The routes received from eBGP peers are already part of the BGP routing table. You do not need to take additional steps to configure the BGP router to announce these routes to other eBGP peers.

  Of course, a router never advertises a route that it received from a peer back to that peer. But it does advertise the route to other peers even if those peers have advertised an identical, but less preferred, route to it.

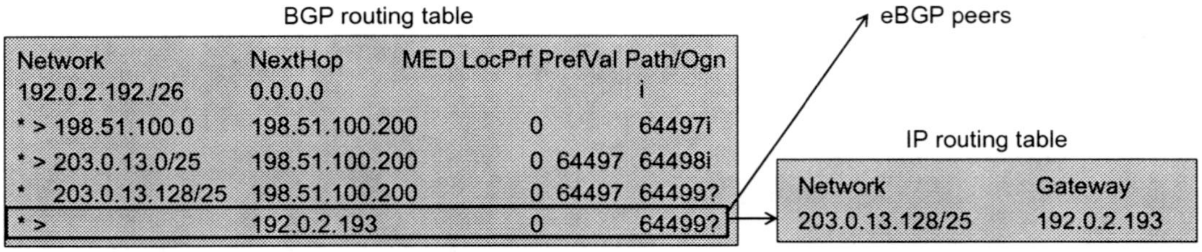

Figure 10-10: Receiving routes from the eBGP peer

Selecting the best BGP routes

This chapter focuses on multihoming ISP scenarios, in which the BGP router connects to at least two ISP routers. Therefore, the router will often receive routes to identical networks (same network address *and* prefix length) from multiple peers. The router uses BGP attributes to choose the best route to each network.

Figure 10-11 shows BGP attributes and the values that are preferred for each attribute. Figure 10-11 lists attributes in the order by which the router examines them when selecting the best route. That is, the router selects the route with the better preferred value first. Only if those values are equal does the router move on to examine the next attribute, local preference. (Your router can set the preferred value for specific routes, and that value has meaning only on the router itself. When it sets the local preference, on the other hand, it communicates that value to iBGP peers also.)

Attribute or criteria	Preferred values
Preferred value (local significance)	Highest
Local preference	Highest
Atomic aggregate	Non-aggregated
AS path	Shortest path (fewestIDs in the list)
Origin	i > e > ?
MED	Lowest

Attribute or criteria	Preferred values
eBGP versus iBGP	eBGP > confederation > iBGP
Metric to next hop	Lowest
Cluster list	Shortest
Originator ID	Lowest
Advertising router ID	Lowest
Advertising router IP address	Lowest

NOTE: You can manipulate attributes to affect route selection, but this topic is covered in other training.

Figure 10-11: Selecting best BGP routes

Without special manipulation on your part or that of your ISPs, your router will tend to prefer routes based on the AS path. It will accept the route from the ISP that reaches the networks in question through fewer ISPs. Presumably, the shorter AS path indicates a shorter latency; however, this is not necessarily the case.

If the ISP routers advertise routes with the same number of AS IDs in the AS path, the origin might come into play. Routes that were originally injected into BGP are preferred to routes learned from

EGP (deprecated) and redistributed routes. However, often the ISPs will have learned the route from a similar ultimate source, so the origin is identical.

The MED would take effect only if the ISP routers have explicitly set this attribute. They might do this if they needed to distinguish between multiple paths into your AS (in a scenario in which the multiple connections are to the same ISP). Generally, a router only compares MED values between routes advertised from the same AS. However, in some cases, you can request different ISPs to send routes with different MED values. You must specify the **compare-different-as-med** command.

Therefore, if the AS path is the same length, your router will probably accept the route from the router to which it has a lower cost connection. After that point, the choice becomes nearly arbitrary with the originator ID or the advertising router ID determining it.

Similarly, assuming no manipulation of attributes, routers throughout the Internet will tend to route traffic to your company through the ISP to which they are closer.

BGP allows administrators to control several of these attributes, granting both the ISP and you the ability to fine-tune the route selection. However, such manipulation is beyond the scope of this study guide.

Ongoing updates

After the eBGP peers have exchanged their BGP routing tables, they only exchange updates. For example, a BGP router sends an update if configuration changes affect routing policies or if a topology change has occurred.

Routes remain in the BGP routing table unless an update occurs. A router can also request a route refresh from a peer (refresh bgp command on Comware), which might be necessary so that the router can apply new policies to routes that it has already received. The next section describes such policies.

Filtering received and advertised routes with route maps

You often require additional control over both the routes that your BGP router advertises to specific peers and the routes that it accepts from these peers. You can use route maps, which are a bit like access control lists (ACLs) for routes, to filter both advertised and accepted routes (see Figure 10-12).

Before examining specific use cases for route maps, you should know a bit about the components of a route map.

Each route map consists of one or more permit or deny statements (Comware devices call the statements "nodes," and ProVision devices, "sequences.") The statement includes match criteria that select routes to be either permitted or denied. A permit sequence might also include a set statement, which alters the selected route.

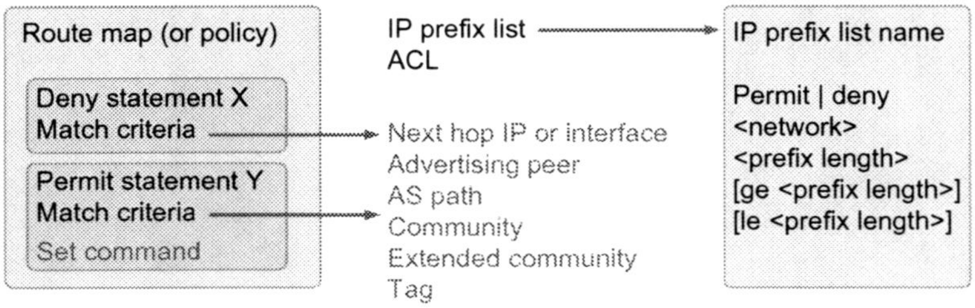

Figure 10-12: Filtering received and advertised routes with route maps

Match criteria

The route map statement can permit or deny routes according to a variety of criteria including:

- Prefix list, which are like ACLs for network addresses and prefix lengths
- ACL
- Next hop IP address or interface
- Metric
- AS list

This study guide focuses primarily on prefix lists. A prefix list includes statements that permit or deny routes according to its network address and prefix length. To match a prefix list statement, the route network address must match the specified bits in the network address indicated by the prefix length. For example, if the prefix list specifies 10.0.0.0/8, the network address must begin with 10, while if the prefix specifies 10.0.0.0/16, the network address must begin with 10.0.

In the prefix list statement, you can set the prefix length precisely, in which case both the network address and the prefix length must match exactly. Alternatively, you can set a range of prefix lengths. For example, the prefix list has statement 203.0.13.0/24 ge 24. The statement matches any route that begins with 203.0.13 and has a prefix length of 24 through 32, including (but not limited to):

- 203.0.13.0/24
- 203.0.13.128/25
- 203.0.13.192/30

Conversely, a prefix list statement might specify 203.0.13.0/24 le 26. This statement matches any route that begins with 203.0.13 and has a prefix length of 0 through 26, including (but not limited to):

- 203.0.13.0/24
- 203.0.13.128/25
- 203.0.13.64/26

Remember that you reference the prefix list in a route map statement. This statement itself specifies permit or deny:

- A permit prefix list statement permits routes that match the prefix list (match permit statements in the list).
- A deny prefix list statement denies routes that match the prefix list (match permit statements in the list).

Set

A permit statement might also include a set command, which alters the attributes for a selected route. For example, you could set a local preference or a MED for specific routes.

You will now examine some reasons for implementing route maps in multihomed ISP scenarios.

Filtering advertised routes (outbound route maps)

As you have learned, by default, your router advertises its entire BGP routing table to every eBGP peer (except any routes received from that peer itself). This behavior means that your router will advertise all the routes that it receives from ISP 1 to ISP 2. Essentially, your router is presenting itself as a transit AS, and one of your ISPs might select your router as the best next hop for reaching the other ISP. Your router will quickly become overburdened by external Internet traffic.

The ISP routers themselves might filter out all advertisements except those to your proper public IP block. However, you should not rely on your ISPs to protect you. As a best practice for the use cases covered in this chapter, always filter outbound advertisements.

Follow these steps:

1. Create a prefix list. This list selects the routes that you have injected (or redistributed) into the BGP routing table. The list implicitly denies all other routes.

   You could enter multiple statements, each of which matches a network address and prefix length exactly. Or you could enter one statement that matches any route within your entire IP block: specify the aggregated network address and prefix length and then add the **ge** option, specifying the prefix length for the aggregated network. For example, if you want to select any route within the 192.0.2.0/24 range, specify 192.0.2.0/24 ge 24.

2. Create a route map and add a sequence that permits the prefix list.

Note

You could also add actions that set attributes, but attribute manipulation is beyond the scope of the study guide.

3. Apply the route map as an outbound filter on each eBGP peer.

Filtering received routes (inbound route maps)

You can also filter the routes that your BGP router accepts from a specific eBGP peer.

You simply create a route map that specifies the routes that you want to accept. You can take a white-list approach—the map permits specific routes and denies all others—or a blacklist approach—the map denies specific routes and permits all others. You then apply the route map as an inbound filter to the eBGP peer.

For example, ISPs' BGP routing tables can be very large, and you can use inbound route maps to reduce the number of routes that your router has to handle.

You must, of course, ensure that your router can reach any Internet address, but you can achieve that purpose by configuring static default routes through the ISP routers. Learning specific BGP routes provides a benefit in load-balancing traffic, presumably over the connection that is specifically better for that destination. However, you might decide that the company does not care about load balancing in this manner; instead, you value a simple routing table. For this use case, follow these steps:

1. Create a static default route to each ISP router.

   You can either use Equal-Cost Multipath (ECMP) (same administrative distance in each route) or an active and standby route (higher administrative distance on the standby route).

2. Create a prefix list that matches all routes (0.0.0.0/0 ge 0).
3. Create a route map.
4. Configure the route map to deny the prefix list.
5. Apply the route map as an inbound filter on all eBGP peers.

Note

You could also ask the ISP to filter routes at its end.

You could also take a hybrid approach, in which your router filters out many of the routes but does accept some.

For example, you might have your router accept routes to large blocks of IP addresses (routes with small prefix lengths) but filter more specific routes. The routes with short prefix lengths represent a high value, in that a single route provides a route to many destinations. The next hop might not be the best next hop for every destination, but it has a better chance of being so than the static default route.

You might also have a particular destination for which you want to ensure optimal routing, so you permit routes to that destination network:

1. Create a static default route to each ISP router.

   You can either use ECMP (same administrative distance in each route) or an active and standby route (higher administrative distance on the standby route).

2. Create a prefix list that permit routes with any IP address and a prefix length that is shorter than your desired limit. For example, you might choose prefix lengths 12 or shorter.
3. If you have a special destination network for which you want to permit routes, you can add a statement that permits that network.
4. The prefix list denies all other routes implicitly.
5. Create a route map that permits the prefix list.
6. Apply the route map as an inbound filter on all eBGP peers.

If you wanted to apply different filtering rules to different eBGP peers, you would create multiple route maps and prefix lists. You could then permit and deny different combinations of prefix lists in each route map.

Note

You could also add actions to the route policy that set attributes for the incoming routes, but attribute manipulation is beyond the scope of the study guide.

Redistributing routes into OSPF

You need to consider how much of the routing information that your BGP router receives from the ISP routers available to the company IGP, typically OSPF. In the use cases for this chapter, your BGP router offers the only connection to the ISPs. All external traffic must pass through this router, so the BGP router simply needs to act as an Autonomous System Border Router (ASBR) and advertise a default route in OSPF.

If the BGP router is a Comware router (or routing switch), you can enter this command:

```
[Comware-ospf]default-route-advertise [always]
```

Note that you enter the command from the OSPF view, which configures this router to act as an ASBR, advertising the default route as an external route. When you enter the command from the area view, as you learned in the previous chapter, the router, acting as an area border router (ABR), advertises the default route to routers in a stub area.

The **always** option determines whether the router advertises the default route only when it has a default route in its routing table (no **always** option) or whether it advertises the default route regardless of the routes in its table (**always** option).

If the BGP router is a ProVision routing switch, you must configure a static default route, if you have not done so already. You can then redistribute static routes:

```
ProVision(ospf)# redistribute static
```

You might need to apply a route map to the command if the ProVision switch has other static routes that you do not want to advertise in OSPF.

Although you can redistribute BGP routes into OSPF, it is not typically recommended. The large BGP routing table might overwhelm the OSPF processes on the local router and on other routers throughout the AS. The main reason to redistribute BGP routes would be to allow OSPF routers to choose among multiple ASBRs, selecting the ASBR that offers the best metric for a particular destination. If you do decide to redistribute OSPF into BGP, make sure to configure ASBR summaries that aggregate routes into manageable sizes.

Configuring eBGP for a dual-home ISP solution

You can now put together everything that you have learned in this chapter to configure a router to connect to two ISP routers.

The company wants to use both ISP connections. However, you will not attempt to manipulate routes to load-balance either incoming or outgoing traffic evenly. Instead, the router will accept the best routes as advertised by the ISP and rely on Internet routers to select the best routes for return traffic.

The company AS runs OSPF, and the router running BGP will announce a default route in OSPF.

You must complete the following steps to set up the company router:

1. Ensure that the router is properly configured to run OSPF.
2. Ensure that the Layer 3 interfaces that connect to the ISP routers are configured and connected. If the ISP wants you to establish a session with a non-directly connected peer IP address, create a route to that address.
3. Configure a static IP route to the company's complete public IP network through the null interface.

This step is not necessary if the router already has a route to the network with the correct prefix length.

4. Create a prefix list that permits only the routes that you have advertised in BGP.
5. Create a route map that permits only that prefix list.
6. Access the BGP configuration view or context, setting the AS number that the company has obtained from the proper authority.
7. Define the first BGP peer, specifying it by AS and IP address.
8. If necessary, configure eBGP multihop and authentication.
9. Apply the route map as an outbound filter to this peer.
10. Repeat the previous two steps to define the second BGP peer.
11. Inject the company's network into the BGP routing table.
12. Create two static default routes, each of which uses one of the ISP routers as the next hop.

    These routes might not be necessary because the router will receive more specific routes from the two ISP routers. However, they ensure that even if the ISP routers do not advertise some networks, this router can route all external traffic.

    In this scenario, the company wants to use both ISP connections, so you leave the default administrative distance for both routes.

Note

On Comware routers, the default administrative distance (preference) for static routes is higher than for OSPF routes. Usually, because your router is not in a stub area, it will not receive an OSPF default route. However, you might set the administrative distance to 1 to make doubly sure that your router does not select the wrong default route.

---

13. Advertise a default route in OSPF.

    If you are configuring a Comware router or routing switch, enter the **default-route-advertise** command for advertising an ASBR default route.

If you are configuring a ProVision routing switch, redistribute the default static routes. However, first check the routing switch's router's configuration for other static routes. You might need to filter those routes out with a route map.

Monitoring BGP with HP IMC

Just as HP Intelligent Management Center (IMC) allows you to monitor OSPF, it provides insight into your BGP network (see Figure 10-13). You follow the same steps to view information about BGP.

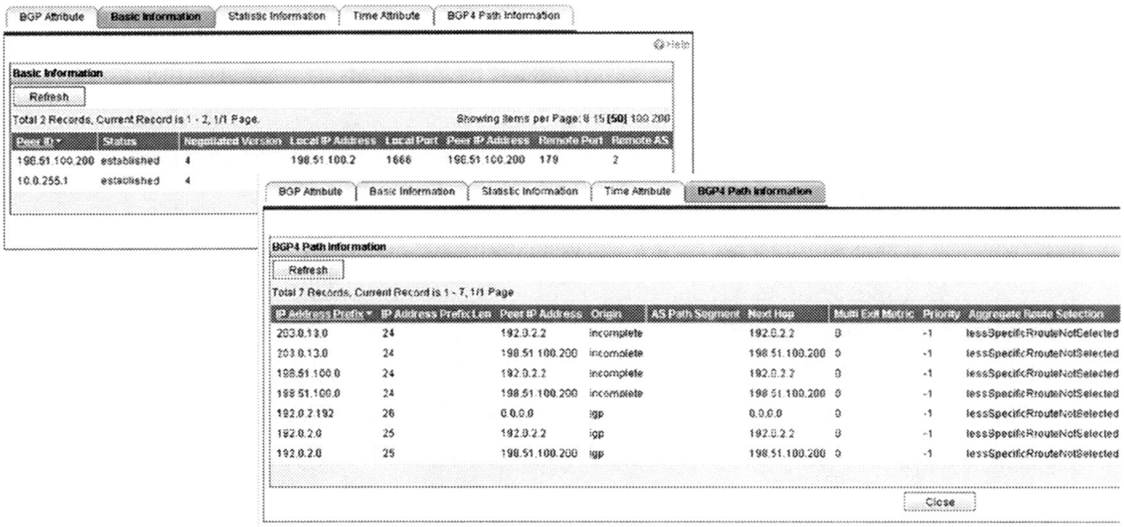

Figure 10-13: Monitoring BGP with IMC

1. Navigate to **Resource** » **Device View**.
2. Click the name of one of the switches.
3. Look at the links provided in the right bar. Scroll down until you see **Protocol Management**.
4. Click **BGP**.

IMC allows you to view information about the BGP attributes as well as the basic information shown in Figure 10-13 (peer ID, negotiated BGP version, peer IP address, and remote AS number). The BGP4 Path information provides additional information about connected networks.

Summary

This chapter introduced you to BGP, focusing on how eBGP peers use the protocol to exchange routing information between different ASes in a stable, highly manageable manner.

You can establish and monitor eBGP connections between a company router and an ISP router. You can also add routes from your router's global IP routing table to the BGP routing table so that your router advertises the correct networks. Although you have not become an expert in manipulating BGP attributes, you do understand basically how attributes affect route selection.

Specifically, you have learned how to configure a BGP solution for a company router that connects to two ISP routers. Not only can you set up BGP correctly for this scenario, you can also implement the proper route maps and OSPF configuration for ensuring that the solution as a whole functions correctly.

Learning check

Answer the following questions:

1. What are two features of the Border Gateway Protocol (BGP)?
    a. Designed as an interior gateway protocol (IGP)
    b. Designed as an external gateway protocol (EGP) between autonomous systems
    c. Sends messages within TCP sessions
    d. Discovers neighbors automatically on networks on which it is enabled
    e. Advertises link states for networks

2. Discuss strategies for advertising routes in BGP.

3. What is one use case for applying an outbound route map to a BGP peer?

Answers

1. What are two features of the Border Gateway Protocol (BGP)?

    a. Designed as an interior gateway protocol (IGP)

    b. Designed as an external gateway protocol (EGP) between autonomous systems

    c. Sends messages within TCP sessions

    d. Discovers neighbors automatically on networks on which it is enabled

    e. Advertises link states for networks

2. Discuss strategies for advertising routes in BGP.

    You can inject routes from the IP routing table using the network command. This approach lets you choose the precise routes that you want to advertise. The exact route, including network address and prefix length, must exist in the routing table, so if you want to aggregate advertisements, you must add a null route to the aggregated network.

    You can also redistribute routes from the IP routing table. This approach gives you less control over which routes are advertised (although you can add a route map to filter).

3. What is one use case for applying an outbound route map to a BGP peer?

    Many use cases exist. This chapter discussed applying a route map to filter out routes that you do not want this router to advertise. For example, a private company BGP router should generally advertise only its own routes, not routes that it has received from its ISPs. So you can apply a route map that selects only these routes and filters others.

    Another use case is adding attributes to specific routes to affect other routers' route selection. For example, you can prepend AS numbers to the AS path in some routes that you advertise to one peer. The routes then appear less desirable to that peer and other routers that receive the route from it. It becomes less likely for traffic destined to these networks to use the link to that peer.

11 IP Multicast

EXAM OBJECTIVES

In this chapter, you learn to:

✓ Route multicast traffic using Protocol Independent Multicast-Dense Mode (PIM-DM) or Protocol Independent Multicast-Sparse Mode (PIM-SM).

✓ Select and configure rendezvous points (RPs) based on particular environmental needs such as redundancy and efficient operation.

✓ Minimize unnecessary multicast flooding.

✓ Apply advanced controls such as source-specific multicasting (SSM) and administrative scopes to a PIM-SM deployment.

ASSUMED KNOWLEDGE

Before studying this chapter, you should understand how a switch forwards unicast traffic and how a Layer 3 switch or a router decides how to route traffic.

INTRODUCTION

In this chapter, you will learn how to implement basic multicast routing solutions that enable large enterprises to deliver a variety of multicast services throughout a complex environment. You will then learn how to optimize multicast solutions by eliminating redundant or unnecessary multicast flooding. Finally, you will learn how to control multicasts precisely, confining streams to specific areas or selecting specific sources.

Multicast overview

You are familiar with unicast traffic, which is directed to a unique destination device, and with broadcast traffic, which every device in a subnet/VLAN (broadcast domain) receives and processes. IPv4 defines a third type of traffic, multicast traffic, which flows from one source to multiple destinations (see Figure 11-1). That is, the multicast source includes the multicast address as the destination IP address for the packet. Any source that is listening for that address processes the packet. Because multiple receivers can listen for the same address, switches might need to flood the packet on multiple interfaces. Similarly, routing switches might need to replicate the packet for transmission on multiple VLANs or other Layer 3 interfaces. (The remainder of this chapter explains how the network infrastructure devices manage this process.)

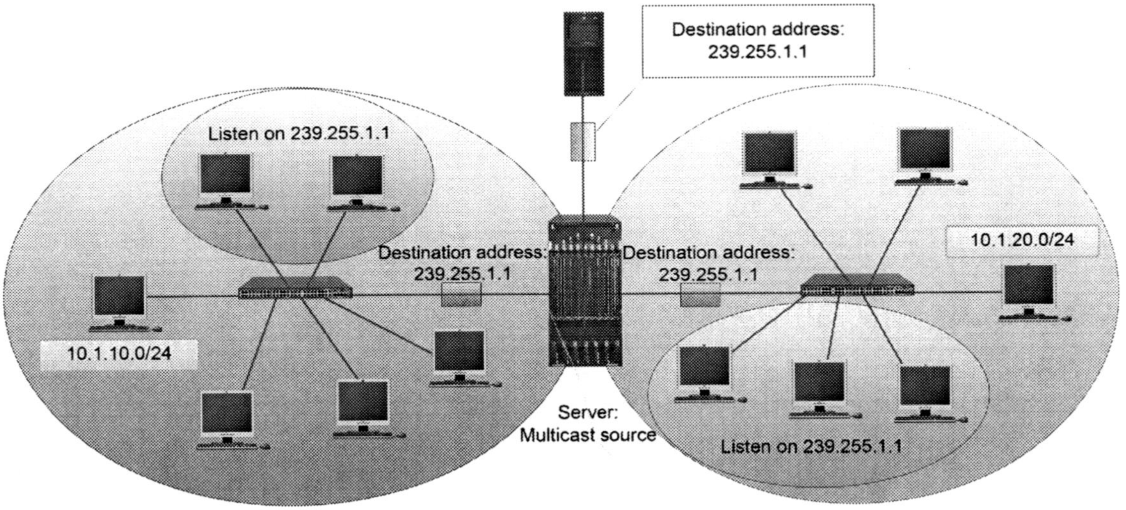

Figure 11-1: Multicast overview

Because multiple IPv4 nodes listen for the same multicast address, this address is often referred to as a *group*. Multicast addresses fall within the 224.0.0.0/4 range (224.0.0.0 to 239.255.255.255). Internet Assigned Numbers Authority (IANA) has assigned specific multicast addresses to some protocols. For example, Open Shortest Path First (OSPF) routers communicate with each other on multicast address 224.0.0.5. As shown in Table 11-1, IANA has assigned other addresses to specific organizations. Some applications, such as applications that use Session Announcement Protocol (SAP), can dynamically select multicast addresses from a certain range. Finally, individual organizations can define administratively scoped multicast addresses in the 239.0.0.0/8 range as they want; these multicasts should not leave the domain.

Thus, depending on the type of application and address that it uses, the multicast source and multicast receivers might be statically configured to stream and listen for the multicast address, or they might negotiate the address dynamically. Determining the multicast address falls to the application; the network infrastructure must simply ensure that the traffic reaches its destination.

Table 11-1: Multicast addresses

Multicast address range	Purpose
224.0.0.0 to 224.0.0.255	Local Network Control Block IANA assigns these addresses to specific protocols (such as OSPF) for communications within a subnet. These multicasts must never be suppressed within a subnet and never be routed outside the subnet.
224.0.1.0 to 224.0.1.255	Internetwork Control Block IANA assigns these addresses to specific protocols for communications globally. These multicasts must never be suppressed and may be routed.

Continued

Table 11-1: Continued

Multicast address range	Purpose
224.0.2.0 to 224.0.255.255	AD-HOC Block I The AD-HOC Block I, II, and II addresses were intended for multicasts that could be routed throughout the Internet but did not fit in the Internetwork Control Block. They are not typically used.
224.1.0.0 to 224.1.255.255	Reserved
224.2.0.0 to 224.2.255.255	SDP/SAP Block These addresses are dynamically selected by applications that use Session Announcement Protocol (SAP) to do so.
224.3.0.0 to 224.4.255.255	AD-HOC Block II
224.5.0.0 to 224.251.255.255	Reserved
224.252.0.0 to 224.255.255.255	DIS Transient Groups
225.0.0.0 to 231.255.255.255	Reserved
232.0.0.0 to 232.255.255.255	Source-Specific Multicast Block These addresses are used by multicast streams that originate from a specific source.
233.252.0.0 to 233.255.255.255	AD-HOC Block III
Scoped addresses	
233.0.0.0 to 233.251.255.255	GLOP Block These addresses are assigned to organizations based on their autonomous system (AS) number.
233.252.0.0 to 233.255.255.255	AD-HOC Block III
234.0.0.0 to 234.255.255.255	Unicast-Prefix-based IPv4 Multicast Addresses These addresses are automatically allocated to organizations based on their public IP address.
235.0.0.0 to 238.255.255.255	Reserved
239.0.0.0 239.255.255.255	Administratively Scoped These addresses are intended for organizations to use locally and should not be routed outside of the domain.
239.192.0.0 to 239.195.255.255 (and 239.0.0.0 to 239.63.255.255 239.64.0.0. to 239.127.255.255 239.128.0.0 to 239.191.255.255) 239.255.0.0 to 239.255.255.255 (and 239.254.0.0 to 239.254.255.255 239.253.0.0 to 239.253.255.255)	Organization-Local Scope Local Scope

Multicast uses

Latest-generation network applications often feature more collaborative components, making them well suited for multicast traffic. Applications that often rely on multicasting include:

- Unified communications and collaboration applications such as:
    - IP conference calling
    - Videoconferencing
    - Telepresence
    - Video streaming applications
    - Virtual classroom applications
    - Stock ticker tape streaming applications

IT administrators might also assign servers or endpoints to multicast groups so that they can deliver patches, software updates, and other controls to a targeted group.

Multicasting presents two primary challenges to network administrators:

- Users and devices that require specific multicast traffic often connect in a variety of locations. Network administrators might need to route the traffic across multiple VLANs and even campus LAN sites.
- Multicast traffic for sophisticated applications, particularly multimedia ones, can consume a great deal of bandwidth, particularly when it is copied for distribution to multiple locations. In fact, even copying multicasts can consume processing power on network infrastructure devices. To maintain a high-performing network environment, administrators must implement multicast technologies properly to minimize unnecessary flooding and duplication.

Minimizing multicast flooding within subnets

You will first learn how switches use Internet Group Management Protocol (IGMP) to optimize forwarding of multicasts within VLANs.

By default, in an Ethernet network, switches flood multicasts on all ports in a VLAN, or broadcast domain (see Figure 11-2). Thus, although multicasts should target only the specific devices that require them, they can act like broadcasts, creating congestion for an entire subnet. Contemporary unified communications and collaboration applications, such as videoconferencing applications, often involve high-bandwidth streams of multicast traffic. If each group must receive the traffic of each other group within the VLAN, overall performance quickly suffers.

In addition, broadcasting multicast packets can introduce security issues because every endpoint in a subnet receives traffic that is actually meant for a particular group. For example, even if most users would not know how to eavesdrop on the multicast traffic from a neighboring meeting room's

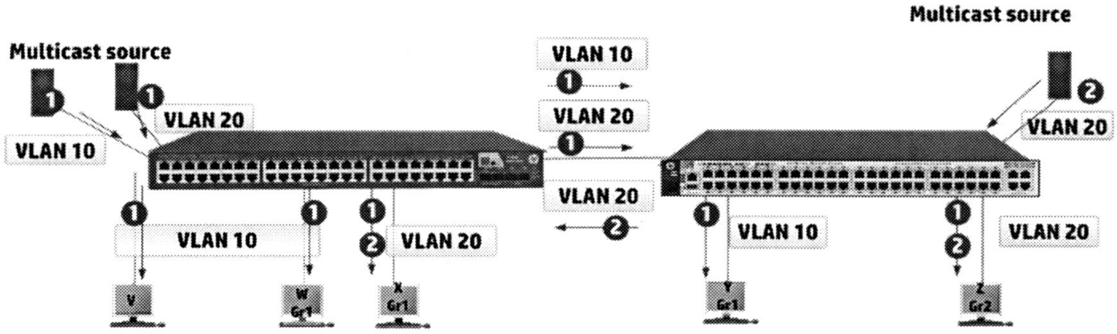

Figure 11-2: Multicast flooding in a network without IGMP

videoconference that shows up unnecessarily on their ports, some users might—which could violate privacy regulations. Finally, service providers that provide for-pay services must limit multicast traffic to the properly registered receivers so that they can track who actually uses the services.

In summary, to deploy multicasting applications effectively, you must enable the switches to suppress specific multicasts on ports that do not require them.

IGMP provides this function.

IGMP queries

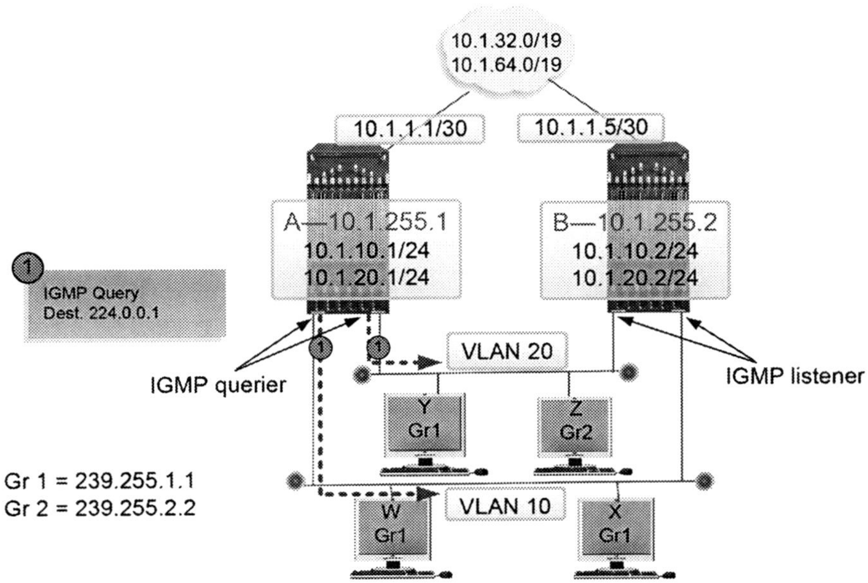

Figure 11-3: IGMP queries

Understanding IGMP queriers and IGMP hosts

IGMP enables switches to discover which hosts within a VLAN require specific multicasts and, therefore, which interfaces must forward the traffic. An IGMP querier periodically sends IGMP queries to multicast address 224.0.0.1. All IGMP-capable devices listen on this address. The next section illustrates how they respond with the groups for which they want to receive traffic.

Generally, the default gateway for a VLAN should act as the IGMP querier. Just as you should usually implement a redundant default gateway (using, for example, Virtual Router Redundancy Protocol [VRRP]), you should usually implement a backup IGMP querier. Simply enable the querier function on all devices that are routing in the VLAN. Even if you do not have redundancy for routing within the VLAN, you can still create a backup IGMP querier. Just add an IP address on the VLAN to one of the nonrouting switches and enable IGMP and IGMP querier.

Enabling IGMP on HP Comware switches

To configure IGMP on Comware switches, you must first enable multicast routing:

```
[Comware] multicast routing-enable
```

You can then enable IGMP on a VLAN interface basis:

```
[Comware] interface vlan <ID>

[Comware-Vlan-interface<ID>] igmp enable
```

In Figure 11-3, for example, both core routing switches implement IGMP on VLANs 10 and 20. IGMP is required on these VLANs because they support endpoints—potential multicast receivers or, in multicast terminology, IGMP hosts.

Note

Some chapters on the high-end Comware switches allow you to set their Ethernet ports to routed mode. Other chapters support routing mode for other types of physical interfaces. In this case, you can enable IGMP on the routing-mode physical port itself. However, this chapter will primarily explain how to configure IGMP on VLAN interfaces.

Comware switches support IGMP v2 by default, but they also support IGMPv3.

Enabling IGMP on HP ProVision switches

On ProVision switches, you enable IGMP on a per VLAN basis. If multiple VLANs are not configured on the network, you simply enable IGMP on VLAN 1.

Enter this command to enable IGMP on a VLAN:

```
ProVision(config)# vlan <ID> ip igmp
```

When the switch has an IP address for a VLAN, it can become an IGMP querier in IGMP-enabled VLANs.

Once IGMP is enabled, the IGMP querier function is enabled by default. You can disable and re-enable this function with this command:

```
ProVision(config)# vlan <ID> ip igmp querier
```

Like Comware switches, ProVision switches support IGMP v2 by default.

Note that if you disable IGMP on a VLAN and then later re-enable IGMP on that VLAN, the switch restores the last-saved IGMP configuration for that VLAN.

Electing the IGMP querier

It is a good idea to enable multiple switches with IP addresses on a subnet to act as queriers. You then have backup for the querier functions.

During normal operation, the switches elect a single IGMP querier for each VLAN. IGMPv1 uses the multicast routing protocol to select a querier, while IGMPv2 selects the device with the lower IP address. In the example shown in Figure 11-3, the routing switches implement IGMPv2, so routing switch A is elected the IGMP querier.

The other routing switch is an IGMP listener. As you will see, it continues to use IGMP to learn the multicast traffic that endpoints desire in the interface just like the querier. It simply refrains from sending queries itself.

Note that IGMP operates separately from Virtual Router Redundancy Protocol (VRRP). That is, if routing switch A and routing switch B implement VRRP and IGMP in VLANs 10 and 20, either the VRRP backup or the VRRP master could be elected as the IGMP querier regardless of its VRRP role. The switches use their actual IP addresses for the election (the VRRP master does not use its virtual address).

IGMP reports and group memberships

Any host that wants to receive multicasts sends an IGMP join report that lists the multicast groups for which it wants to receive traffic (see Figure 11-4).

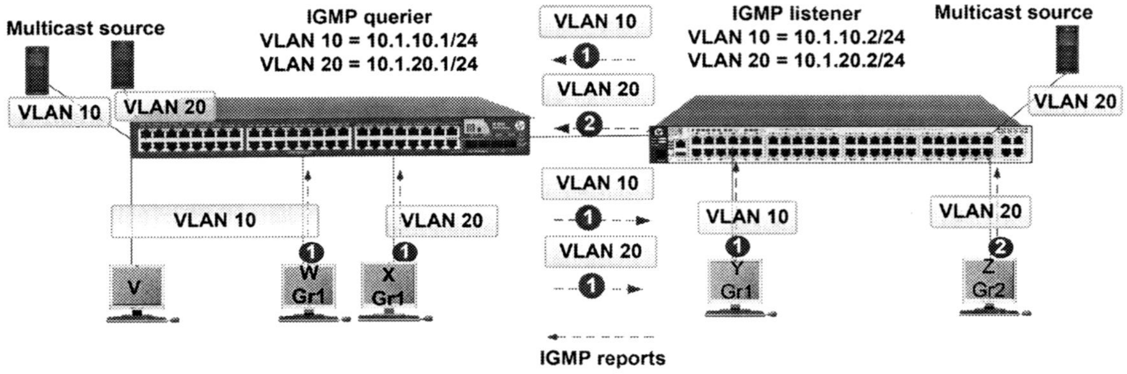

Figure 11-4: IGMP reports

Any switch that implements IGMP (whether or not it is the active querier) uses this information to construct multicast forwarding tables. When the switch receives the first report for a multicast address on a VLAN interface, it adds a table for that address to the VLAN. This table filters traffic for multicasts destined to that address. Rather than forward these multicasts on all ports, the switch forwards multicasts only on ports listed in the table. These are ports on which the switch has heard IGMP joins for that address.

In Figure 11-5, you see an example of the table that the IGMP querier has constructed based on the joins shown in previous section. The querier has created a forwarding table for group 239.255.1.1 in VLAN 10 and VLAN 20. The querier has created a forwarding table for group 239.255.2.2 in VLAN 20 but not in VLAN 10 because no host in VLAN 10 has joined this group.

VLAN 10

Group	Ports
239.255.1.1	To_W, To_Switch

VLAN 20

Group	Ports
239.255.1.1	To_X
239.255.2.2	To_Switch

Figure 11-5: Multicast forwarding table built by the IGMP querier based on the IGMP reports

Although not shown in Figure 11-5, similar tables also exist on the IGMP listeners which have constructed them from the joins that it has heard.

The table controls how the switches forward the multicasts. Now each host only receives multicasts for groups for which it has registered (see Figure 11-6).

The group membership entries do not persist indefinitely. Instead, the switch also sets a timer for the group membership, which it resets every time that its query produces another report. If the timer expires, the switch removes the group membership from the interface.

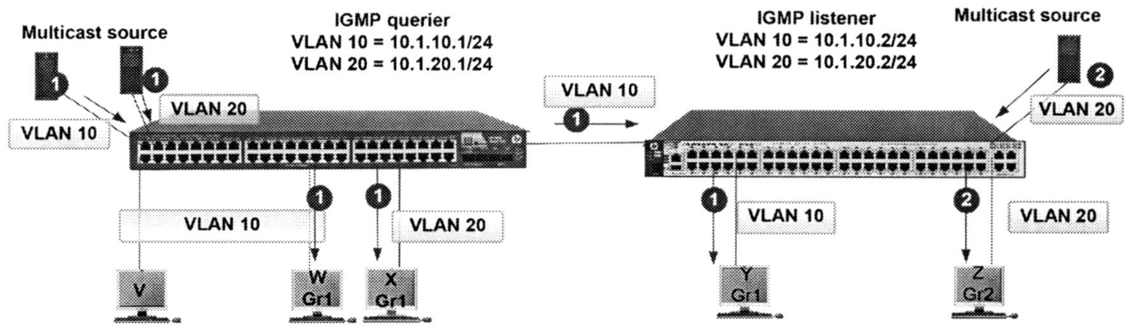

Figure 11-6: Multicast forwarding in a network using IGMP

However, IGMPv2 defines a faster way for an interface to leave a group. When a receiver no longer requires multicast traffic in a particular group, it sends an IGMP group message. The IGMP querier instantly sends a group-specific last member query to determine whether other receivers still require the traffic.

Note also that multicast receivers do not need to wait for an IGMP query to send a report. They can send a report for a group as soon as they join the group.

HP ProVision operating modes: Data-driven and non-data-driven

HP ProVision switches operate in one of two modes for IGMP:

- **Non-data-driven**—With non-data-driven IGMP, the switch filters multicasts only in a VLAN if it has a table for that multicast group—in other words, if at least one host has joined the group for known multicast groups. If the switch happens to receive multicasts for an unknown multicast group, it floods rather than filters the multicasts. The switch might receive unknown multicasts in cases such as these:
  - The switch has removed a multicast group filtering table because the last member left the group. However, a multicast router forwarding the traffic has not yet received the leave, so it continues to forward the multicasts.
  - A multicast source is streaming, and another switch in the path between the source and this switch does not support IGMP.
- **Data-driven (or smart)**—With data-driven IGMP, the switch filters all multicasts for which it does not have a multicast table. If no host has joined the group, but the switch receives multicasts, the switch drops them.

IGMP at Layer 2 (IGMP) snooping

A subnet might include switches operating at Layer 2 with no IP addresses on that subnet. These switches do not act as IGMP queriers or backup queriers, but they still should use endpoints' IGMP join requests to create efficient multicast forwarding tables.

You need to enable IGMP snooping on these switches.

The IGMP snooping switch eavesdrops on IGMP traffic, using what it learns to determine which ports require specific multicast traffic. It stores this information in a table, creating one table for each VLAN interface on which IGMP snooping is enabled (see Figure 11-7 and Figure 11-8).

To create the table, the switch follows these basic rules, which allow it to discover necessary information about group memberships without disturbing IGMP communications between multicast receivers and their default routers:

1. The switch tracks the ports that receive IGMP general queries.

Comware switches set these ports as dynamic router ports in the table for the VLAN on which the query arrived. Or if the port is already defined as a dynamic router port, the switches reset the dynamic router port aging timer (105 seconds, by default).

ProVision switches call these ports querier ports.

The switch needs to identify router or querier ports so that it knows on which ports to forward IGMP reports.

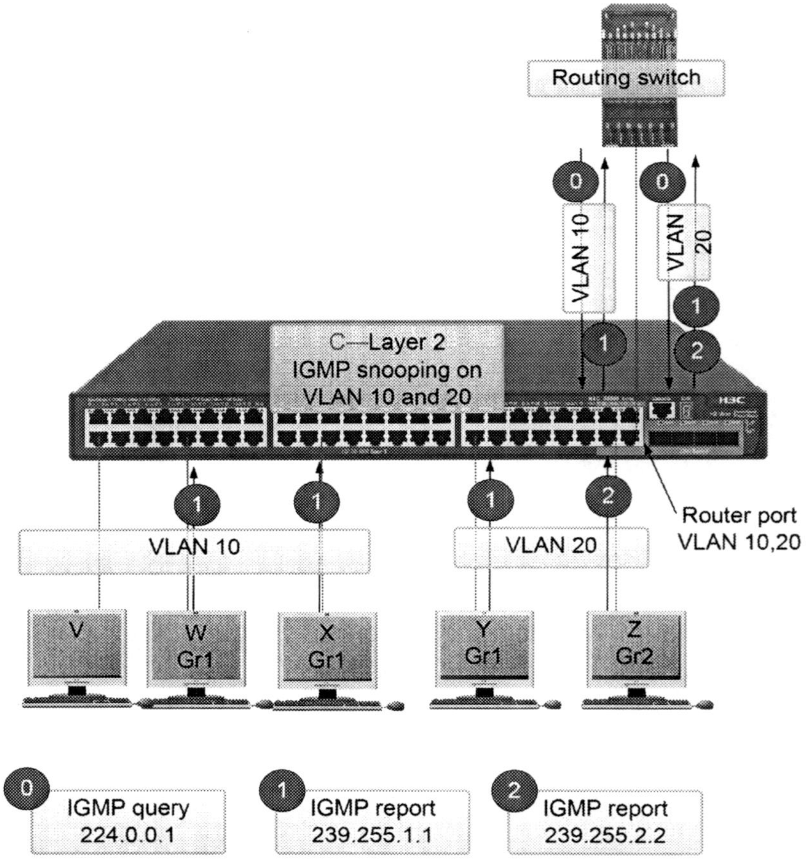

Figure 11-7: IGMP queries and reports received by a switch using IGMP snooping

2. When the switch receives a multicast receiver's IGMP report, it maps the incoming port to the group in the report in the appropriate VLAN IGMP snooping table (see Figure 11-8). Or if the port is already mapped to that group, the switch resets the dynamic port member aging timer for that port and group.

Group	Ports
VLAN 10	
239.255.1.1	To_W, To_X
VLAN 20	
239.255.1.1	To_Y
239.255.2.2	To_Z

Figure 11-8: Multicast forwarding table built by an IGMP snooper based on IGMP reports

When hosts hear IGMP reports from other hosts in the same group, they suppress their own reports. This behavior was intended to minimize floods of reports and still gives the multicast routing switch all the information that it needs: at least one receiver exists on the VLAN. However, the IGMP snooping switch requires more granular information about the receivers on each port. Therefore, the switch only forwards IGMP reports on router or querier ports. Other hosts, not hearing the report, send their own, and the switch can learn their group memberships.

Note

The IGMP querier implements a similar mechanism on its own physical ports within VLANs on which IGMP is enabled.

The snooping switch minimizes floods of reports to the IGMP querier by forwarding a report only when it creates the initial entry for a multicast group in its IGMP snooping table (rather than every time it adds a new port to the entry).

3. Inactive group memberships time out. However, multicast receivers speed up the process by sending messages to leave a group. The IGMP snooping switch eavesdrops on the leave message and removes the incoming port from the group in question, removing the mapping for the group entirely if the port member list has become empty.

Note, however, that ProVision switches that use non-data-driven IGMP disable this function by default. They need to maintain the multicast table for a period after the last host has left the group so that they can continue to filter multicasts during the period that it takes for the multicasts to stop arriving.

In this way, the IGMP snooping switch builds a table such as the one in Figure 11-8. When the switch receives multicast packets for a particular group on a router port, it forwards them only on the ports mapped to the group (see Figure 11-9).

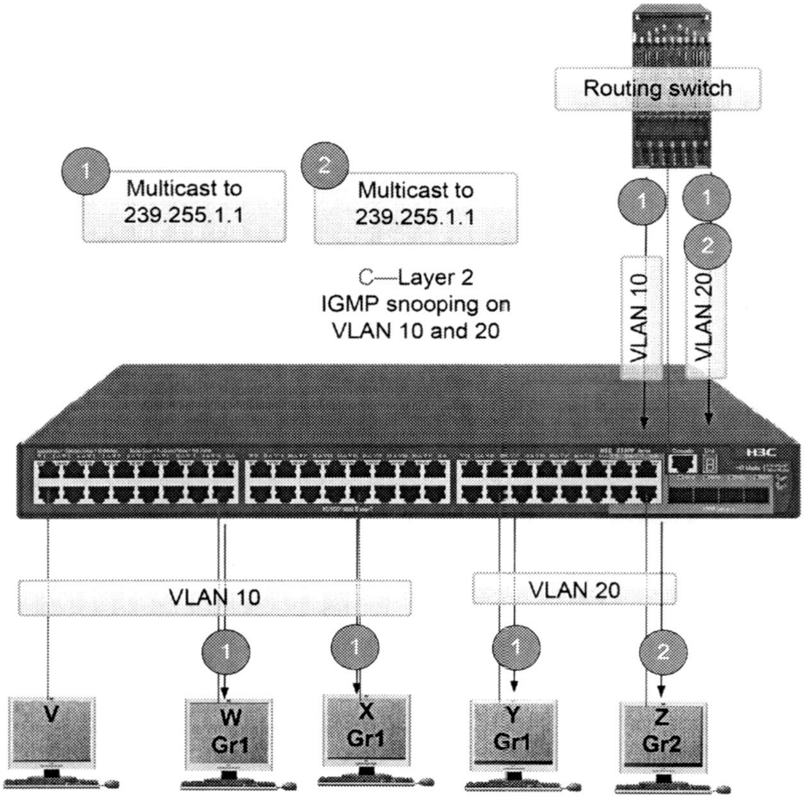

Figure 11-9: Multicast forwarding in a network that uses IGMP and IGMP snooping

Configuring IGMP snooping on HP Comware switches

Complete these steps to configure IGMP snooping on HP Comware switches:

1. Enable IGMP snooping globally:

```
[Switch] igmp-snooping
```

2. Enable IGMP snooping on VLANs:

```
[Switch-Vlan-interface<ID>] igmp-snooping enable
```

You can enable IGMP snooping on individual ports instead, but typically, you should enable it on the entire VLAN.

Example IGMP table

If you enter the **display igmp-snooping group** command on the Comware switch in Figure 11-9, you would see the following output:

Note
In the output that follows, the ! indicates comments, which explain the output.

```
Total 3 IP Group(s).
Total 3 IP Source(s).
Total 3 MAC Group(s).

Port flags: D-Dynamic port, S-Static port, C-Copy port
Subvlan flags: R-Real VLAN, C-Copy VLAN
Vlan(id):10.
  Total 1 IP Group(s).
  Total 1 IP Source(s).
  Total 1 MAC Group(s).
  Router port(s):total 1 port.
        TE1/0/50                        (D)
```

!Port TE1/0/50 has been detected as a dynamic router port (connects to the routing switch) because it has received an IGMP query:

```
  IP group(s):the following ip group(s) match to one mac group.
    IP group address:239.255.1.1
      (0.0.0.0, 239.255.1.1):
        Host port(s):total 2 port.
          GE1/0/10                      (D)
          GE1/0/20                      (D)
```

!The GE1/0/10 and GE1/0/20 ports connect to W and X, which have sent reports in group 239.255.1.1:

```
  MAC group(s):
    MAC group address:0100-5e7f-0101
        Host port(s):total 2 ports.
          GE1/0/10
          GE1/0/20
```

```
 Vlan(id):20.
   Total 2 IP Group(s).
   Total 2 IP Source(s).
   Total 2 MAC Group(s).
   Router port(s):total 1 port.
         TE1/0/50                    (D)
```

!The switch has also detected that port TE1/0/50 is a router port in VLAN 20:

```
 IP group(s):the following ip group(s) match to one mac group.
   IP group address:239.255.1.1
      (0.0.0.0, 239.225.1.1):
       Host port(s):total 1 port.
         GE1/0/40                    (D)
```

!In VLAN 20, the switch has detected one member of group 239.255.1.1 on port GE1/0/40:

```
 MAC group(s):
   MAC group address:0100-5e7f-0101
       Host port(s):total 1 ports.
         GE1/0/40
IP group(s):the following ip group(s) match to one mac group.
   IP group address:239.255.2.2
      (0.0.0.0, 239.225.2.2):
       Host port(s):total 1 port.
         GE1/0/45                    (D)
!This port connects to Z, which has sent reports in this group:
 MAC group(s):
   MAC group address:0100-5e7f-0202
       Host port(s):total 1 ports.
         GE1/0/45
```

Configuring IGMP snooping on ProVision switches

Remember that a ProVision switch's VLAN interface controls both Layer 2 and Layer 3 functions. You can enable IGMP whether or not the VLAN has an IP address. When the VLAN does not have an IP address, entering the **vlan <ID> ip igmp** command enables IGMP snooping.

The switch listens for IGMP join requests on a VLAN. The switch learns the ports on which the requests for a particular request were received. It adds those ports to the table of ports for forwarding multicast traffic in that group. It filters multicasts for that group on other ports.

The switch also listens for IGMP queries and notes the ports on which these messages are received as querier ports. It forwards IGMP reports on the querier ports but not others.

Comware: IGMP snooping and MSTP

In a network that features redundancy, you must consider how IGMP snooping and MSTP interact. Figure 11-10 shows the situation for VLAN 10, which is mapped to an MSTP instance in which the

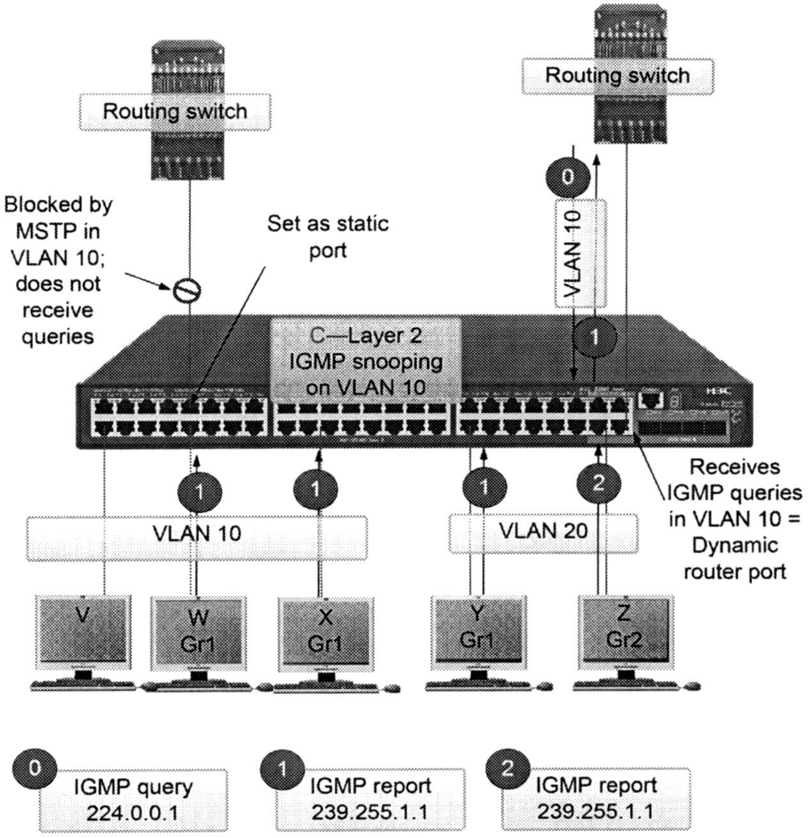

Figure 11-10: IGMP snooping and MSTP

routing switch on the left is root. MSTP typically blocks the link to the other routing switch, which is the secondary MSTP root in this instance.

If the edge switch loses its link to the primary root, MSTP brings the link back up. However, switch C does not realize that this port is the router port until the querier sends a report, which can interfere with a multicast stream. Therefore, you should generally set redundant uplinks as router ports statically.

The opposite situation would occur in VLAN 20, for which the routing switch on the right is root. To ensure smooth operation, you might want to set all uplinks as static router ports in all VLANs that implement IGMP snooping.

Routing multicast traffic between subnets

IGMP helps routing devices implement multicast routing solutions. As you know, unicast traffic can only reach destinations within the same broadcast domain, such as a VLAN, unless the network includes a unicast routing solution. Similarly, multicast traffic is confined to a broadcast domain unless the network has a multicast routing solution. You have seen how unicast routes direct traffic for a specific destination to a next hop. Due to the nature of multicasting, multicast routes function a bit differently.

Because a destination multicast address defines a group of devices, traffic destined to a single destination might need to be forwarded along multiple paths. Routers and routing switches need to track multiple interfaces on which to forward traffic.

These forwarding paths often change because destinations that require multicasts frequently change. For example, a user might close the application that receives a multicast video stream, causing the endpoint to leave the multicast group. Routers and routing switches require a way to discover which destination devices require which multicast traffic.

Note

Many of the technologies discussed in this chapter can be implemented by routers or routing switches. Because this study guide focuses on switches, it will refer to "routing switches" which you should generally take to imply routers as well.

IGMP and Protocol Independent Multicast (PIM) in either Dense Mode (DM) or Sparse Mode (SM) address these needs. Together, they enable multicast routing (see Figure 11-11).

IGMP operates between multicast receivers and their default gateways. This protocol enables the receivers to inform the routing switches for which multicast groups they want to receive traffic. The routing switches can then forward the multicasts in the correct subnets (typically, for routing switches, subnets on VLAN interfaces) and suppress the multicasts in other subnets.

PIM operates between multicast routers or routing switches, enabling the devices to construct trees for routing multicast traffic from the source to the receivers.

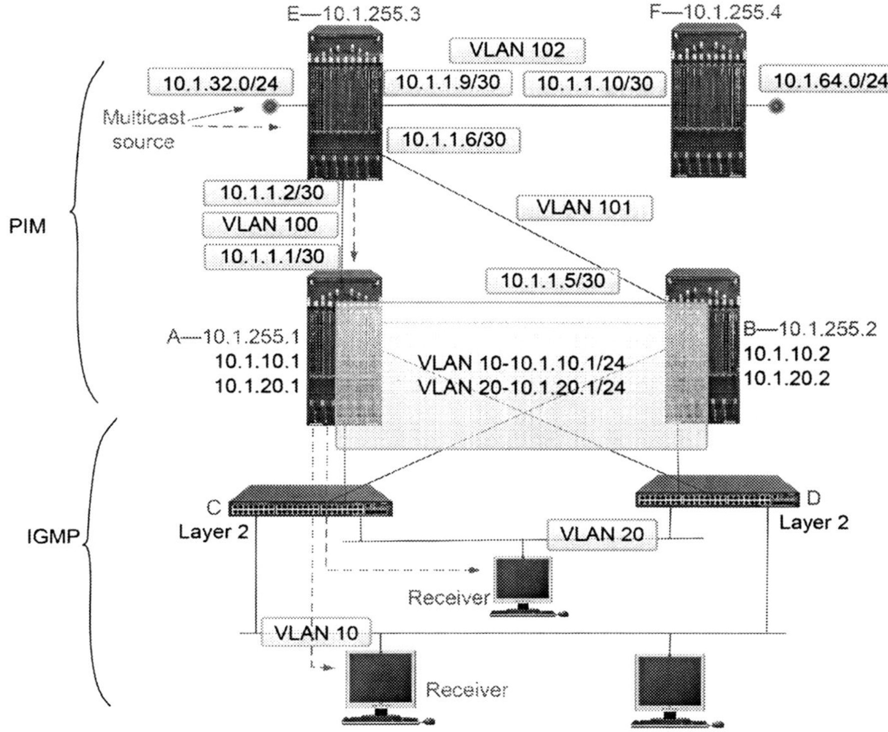

Figure 11-11: Routing multicast traffic between subnets

PIM

You will now learn about Protocol Independent Multicast (PIM), which allows switches to determine which interfaces should receive multicast traffic. You will be introduced to two types of PIM: PIM Dense Mode (PIM-DM) and PIM Sparse Mode (PIM-SM), and you will then learn more about PIM-DM.

You now understand how routing switches use IGMP to determine in which interfaces to forward multicast traffic. In order to receive the traffic from the subnet in which it originates, a routing switch must participate in a multicast routing protocol. Unicast traffic journeys toward a single, unique destination, so unicast routes require only a forwarding interface/next hop. Many endpoints throughout many subnets might need to receive multicasts; therefore, the forwarding path takes the shape of a tree with many leaves, all of which require the traffic.

Reflecting that tree shape, the multicast routing table entry features multiple downstream interfaces rather than a single destination. The routing switch forwards multicasts on each downstream interface, which are the interfaces that IGMP has indicated require the traffic. Figure 11-13 shows a multicast table on a Comware switch. The table displays information about the example network shown in Figure 11-12.

Unlike unicast routes, the source, in addition to the destination, is significant for multicast routes. The downstream routing switch (nearer the receivers) needs to indicate to the upstream switch that it requires the multicasts, so it must know the path one hop backward as well as one hop forward. It must also know this upstream hop so that it can ensure that it accepts multicasts on the correct interface; otherwise, loops could occur.

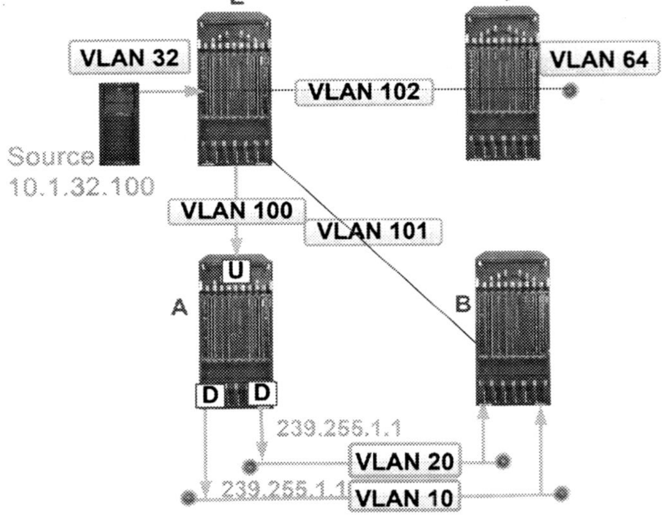

Figure 11-12: Multicast routing tree

In an (S, G) entry, which denotes a multicast entry from a specific active source (S) to a specific multicast address (G), the upstream interface faces the source. The routing switch refers to its unicast routing table to determine this interface; the forwarding interface for the route to the source being the upstream interface. The routing switch can learn the unicast route in any manner, which is why PIM is called protocol independent.

Some downstream interfaces are discovered by IGMP, in which case they forward multicast traffic into the ultimate destination subnets. Other downstream interfaces deliver multicast traffic to PIM neighbors that require it.

```
<A> display multicast routing-table
00001. (10.1.32.100, 239.255.1.1)
  Uptime: 00:00:28
  Upstream Interface:
    Vlan-interface100
  List of 2 downstream interfaces
    1: Vlan-interface10
    2: Vlan-interface20
```

Table component	Discovered by
Upstream interface	Forwarding interface for unicast route to source
Downstream interfaces	Interfaces with this IGMP group

Figure 11-13: Multicast routing table

PIM-DM versus PIM-SM

The routing switches can construct the multicast routing tree in one of two modes:

- PIM Dense Mode (DM)—PIM-DM-enabled routing interfaces flood multicasts to all neighbors until the neighbors indicate that they do not need them by sending a prune. In addition, the upstream neighbor sets a timer for the pruned interface. When the timer expires, the interface reverts to the default behavior—forwarding multicasts until explicitly requested not to.

  This mode works best in environments with high bandwidth and high tolerance for congestion such as a high-speed Ethernet core.

- PIM Sparse Mode (SM)—PIM-SM-enabled routing interfaces must specifically indicate that they want to receive multicast traffic in a particular group by sending a join for the tree. The interface must then periodically refresh its join. If its upstream neighbor does not receive joins, it reverts to its default behavior: not forwarding traffic.

  When you examine this protocol in more detail, you will see how the routing switches can become part of a multicast forwarding tree even before a source starts streaming. Because PIM-SM establishes efficient forwarding paths, it is well suited for a lower-bandwidth environment such as one built on WAN links.

HP switches can run either PIM-DM or PIM-SM but not both.

PIM-DM

To configure PIM-DM, you enable it on all Layer 3 interfaces between the multicast sources and the multicast receivers:

- Interfaces on which multicast sources reside
- Interfaces on which multicast receivers reside (IGMP-enabled interfaces)
- Interfaces between all routers and routing switches that connect sources and receivers

On Comware switches, you enable PIM-DM on physical Layer 3 interfaces or on VLAN interfaces. On ProVision switches, you enable PIM-DM on VLAN interfaces.

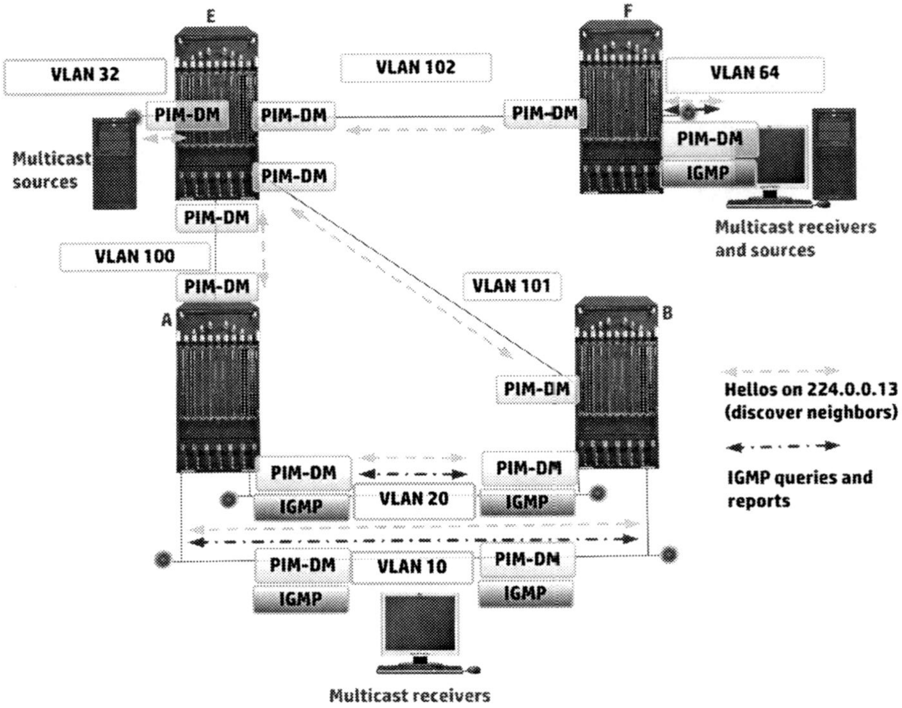

Figure 11-14: Interfaces that run PIM-DM and IGMP

Figure 11-14 shows an example network that includes Comware switches and indicates the interfaces on which you would enable PIM-DM as well as the interfaces on which you must enable IGMP.

On Comware switches, the command for configuring PIM-DM on a VLAN interface is:

```
[Switch-Vlan-interface<ID>] pim dm
```

On ProVision switches, you must first enable both multicast routing and PIM-DM globally. You can then enable PIM-DM on the VLAN interface, as shown in the following:

```
Switch(config)# ip multicast-routing
Switch(config)# router pim enable
Switch(config)# vlan <ID> ip pim-dense
```

PIM-DM interfaces automatically discover neighbors on the same subnet by transmitting and listening for hellos. PIM-DM routers communicate on multicast address 224.0.0.13.

Figure 11-15 illustrates a more complex system. Note that the intervening routing switch G must support PIM.

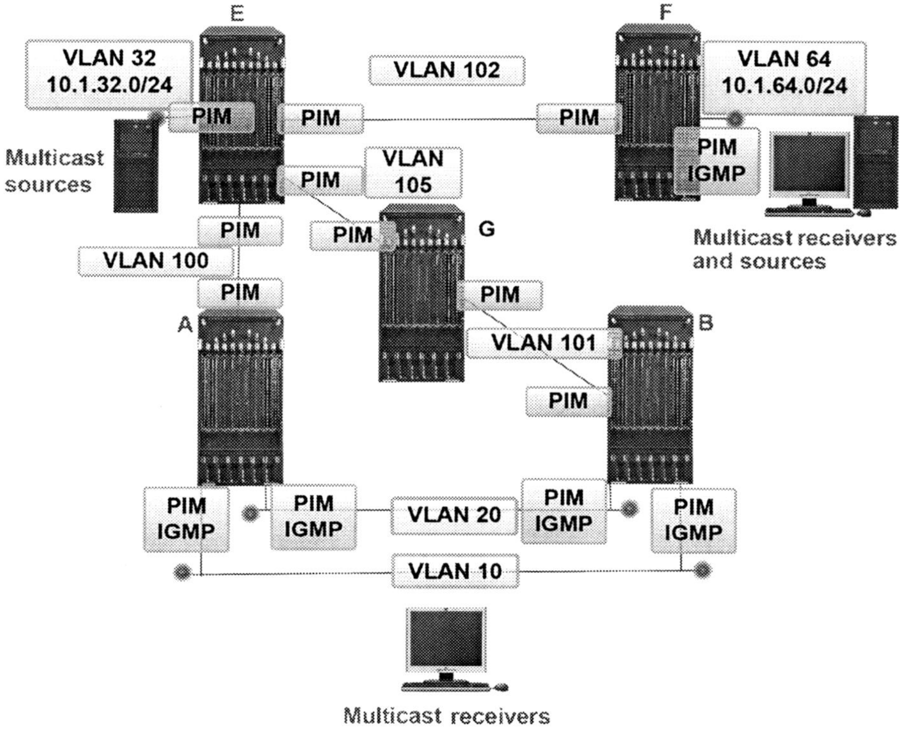

Figure 11-15: PIM and IGMP in a larger system

PIM-DM (*, G) entries

As soon as PIM-DM is enabled on a switch VLAN interface, the switch creates a (*, G) entry for each group membership discovered by IGMP, as shown in Figure 11-16. This entry specifies the downstream interfaces from which a tree can be formed when multicasts begin to arrive from an upstream interface.

The (*, G) entry essentially acts as a placeholder, listing the downstream interfaces that IGMP has discovered require multicasts in this group. The framework of the tree that delivers these multicasts is not filled in. The upstream interface list is NULL (the routing switch does not know where multicasts will arrive). For the same reason, the downstream interface list does not specify interfaces that connect only to PIM neighbors (the switch does not know which interface will be the upstream one).

However, because PIM-DM operates on the assumption that each PIM-DM neighbor sends multicast traffic to all PIM-DM neighbors until requested otherwise, without taking any action, this

```
<A> display igmp group
...
Vlan-interface10 (10.1.10.1) :
  Total 1 IGMP Groups reported
  Group Address  Last Reporter   Uptime    Expires
  239.255.1.1    10.1.10.100     00:06:30  00:01:52

Vlan-interface20 (10.1.20.1) :
  Total 2 IGMP Groups reported
  Group Address  Last Reporter   Uptime    Expires
  239.255.1.1    10.1.20.64      00:06:30  00:01:52
```

```
<A> display pim routing-table
VPN-Instance: public net
 Total 1 (*, G) entries; 0 (S, G)en

 (*, 239.255.1.1)
     Protocol: pim-dm, Flag: WC
     UpTime: 00:24:59
     Upstream interface: NULL
         Upstream neighbor: NULL
         RPF prime neighbor: NULL
     Downstream interface(s) infor
     Total number of downstreams: 2
       1: Vlan-interface10
            Protocol: igmp, UpTime: 00:2:59, Expires: never
       2: Vlan-interface20
            Protocol: igmp, UpTime: 00:3:01, Expires: never
```

Figure 11-16: PIM-DM (*, G) entry created based on IGMP membership

PIM-DM routing switch is a de facto member of that yet-to-be established forwarding tree. It is ready to route multicast traffic into the necessary user VLANs as soon as this traffic arrives.

The (*, G) entry remains in the PIM routing table for as long as an IGMP membership for the group exists on at least one interface. It signals the switch to accept multicasts on behalf of the endpoints for which it routes traffic whether or not PIM neighbors require the traffic.

Note that the (*, G) entry exists only in the PIM routing table; it is not an active route forwarding multicasts, so it is not copied in the multicast routing table

Comware switches

On Comware switches, you can enter the **display pim routing-table** to view (*, G) entries and **display igmp group** to view information about downstream interfaces discovered through IGMP.

ProVision switches

To view this information on ProVision switches, enter **show ip pim pending** command. To view information about IGMP groups, enter **show ip igmp group**.

PIM-DM SPT construction: Flooding

Figure 11-17 illustrates the PIM-DM Source Path Tree (SPT), which is formed when a multicast source initially begins to stream traffic to a multicast group.

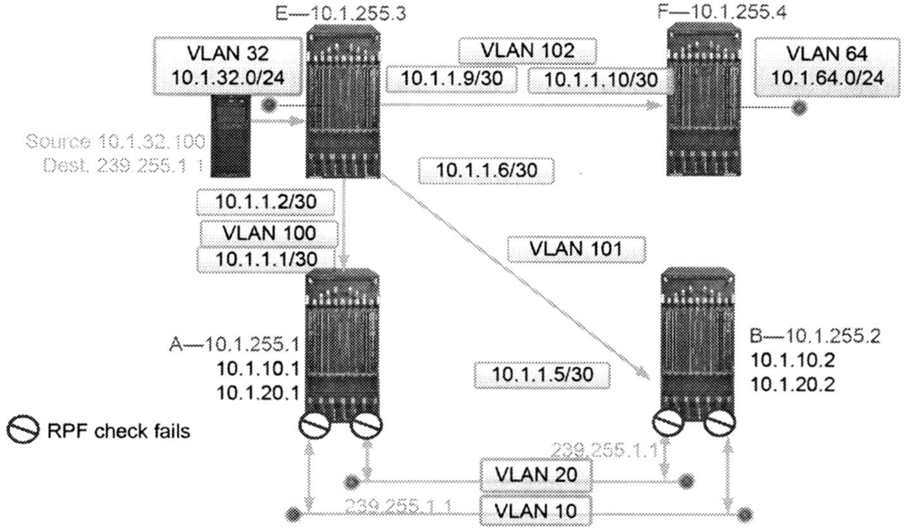

Figure 11-17: PIM-DM SPT construction: flooding

1. The multicast source's default router sends the multicasts to all PIM neighbors. In this example, E sends multicasts to F, A, and B.
2. The PIM neighbors run a reverse path forwarding (RPF) check on the traffic, which verifies that the upstream interface on which the traffic arrives is the forwarding interface for the unicast route to the source.
3. If the multicast passes the RPF check, the routing switch accepts it and forwards it to all of its own neighbors (except the one from which it received the multicast). The routing switch also forwards the traffic on downstream interfaces (interfaces with IGMP memberships in the multicast group).

As you see, both A and B send multicasts on VLANs 10 and 20. However, the routing switches do not accept these multicasts because they do not pass the RPF check.

PIM-DM SPT (S, G) entries

Consider the situation that you just examined as a topology sketch as you would see it by entering **display** commands on Comware switches.

When the switch receives a multicast, it creates an (S, G) entry, which includes:

- The source for the incoming multicasts.
- The upstream interface on which the multicast arrived.

- The ACT flag, indicating that the switch has received multicast traffic for this group.

  The switch also runs an RPF check. By setting the RPF flag, the switch avoids running the RPF check every time a multicast arrives, which would consume a great deal of processing power. Instead, the switch accepts traffic on the upstream interface listed in the entry with the RPF bit.

- The downstream interfaces, which initially include every interface on which PIM is enabled.

  The switch maintains the downstream interfaces discovered by IGMP for as long as the IGMP membership remains on those interfaces. The switch might maintain the interfaces added by PIM alone, or it might remove (prune) them—as you will see in a moment.

If the (S, G) entry has downstream interfaces or if the source is connected directly to the switch (LOC), the routing switch adds the (S, G) entry to its active multicast routing table.

Figure 11-18 illustrates routing switch A's (S, G) entry for the multicast traffic in this example. Routing switch B would have the same entry, except that its upstream VLAN would be VLAN 11.

```
<A> display pim routing-table
VPN-Instance: public net
 Total 1 (*, G) entries; 1 (S, G) entry
. . .
  (10.1.32.100, 239.255.1.1)
      Protocol: pim-dm, Flag: ACT
      UpTime: 00:00:29
      Upstream interface: Vlan-interface100
          Upstream neighbor: 10.1.1.2
          RPF prime neighbor: 10.1.1.2
          1: Vlan-interface10
              Protocol: - , UpTime: 00:04:32, Expires: 00:01:25
          2: Vlan-interface20
              Protocol: - , UpTime: 00:04:34, Expires: 00:01:2
```

Figure 11-18: PIM-DM SPT (S, G) entries

PIM-DM SPT construction: Asserts

At this point, the PIM-DM SPT is not operating efficiently. First, examine VLANs 10 and 20 in Figure 11-19. Active receivers for the multicast traffic do reside in these VLANs; however, both routing switch A and routing switch B are forwarding traffic in the VLAN. This inefficient state of affairs can also cause problems for hosts receiving duplicate copies of the multicasts.

You have already seen that routing switch A and B reject the multicasts from each other because they do not pass the RPF check. However, the switches take further action. Whenever a PIM-enabled interface receives a multicast on one of its own downstream interfaces for that group, it realizes that another PIM routing interface is active on the subnet and sends an assert.

The PIM routing switches examine each other's asserts to elect a multicast forwarder for the interface. The assert includes the (S, G) entry under dispute and information about the PIM routing switch's unicast path to the source. The routing switch with the more favorable path becomes the multicast forwarder as determined first by administrative distance (preference) and then, if the administrative distance ties, by metric or cost. If the metric ties as well, the routing interface with the higher IP address is elected.

In this example shown in Figure 11-19, both switches have OSPF routes (preference 10) to the multicast source. However, the cost of A's route is 11, and the cost of B's route is 20, so A is elected multicast forwarder. This process occurs on both VLAN 10 and VLAN 20. B removes these VLANs from the list of downstream interfaces in its (S, G) entry.

B also sends a prune to A on the interfaces on which it lost the assert. A continues to forward multicasts in these VLANs because multicast receivers connect to them. However, the prune would be important if this VLAN did not include receivers; A would need to know that B has a better path to the SPT root (which is the reason for the assert occurring in the first place).

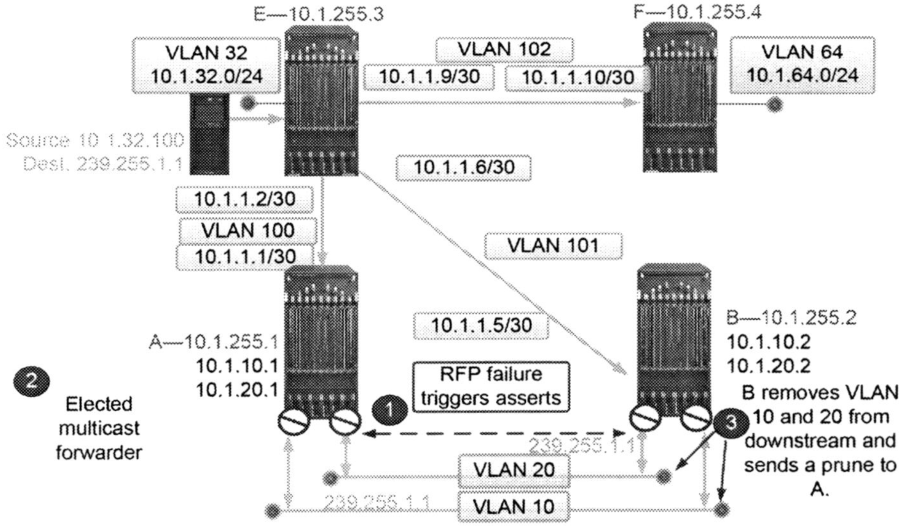

Figure 11-19: PIM-DM SPT construction: asserts

PIM-DM SPT construction: Pruning

When a PIM-DM routing switch has an (S, G) entry with no downstream interfaces, it prunes itself from the tree by sending a prune message on the upstream interface. The upstream neighbor removes the interface on which it receives the message from its (S, G) entry.

In the example shown in Figure 11-20, routing switch F never had a (*, G) entry for this multicast tree because it received no IGMP reports for the group. Therefore, it prunes itself from the tree. Routing switch B has removed all of its downstream interfaces due to losing the assert election. Therefore, it sends a prune message as well.

Both routing switch B and F, however, maintain the (S, G) entry, which they can use to receive multicasts later, if necessary.

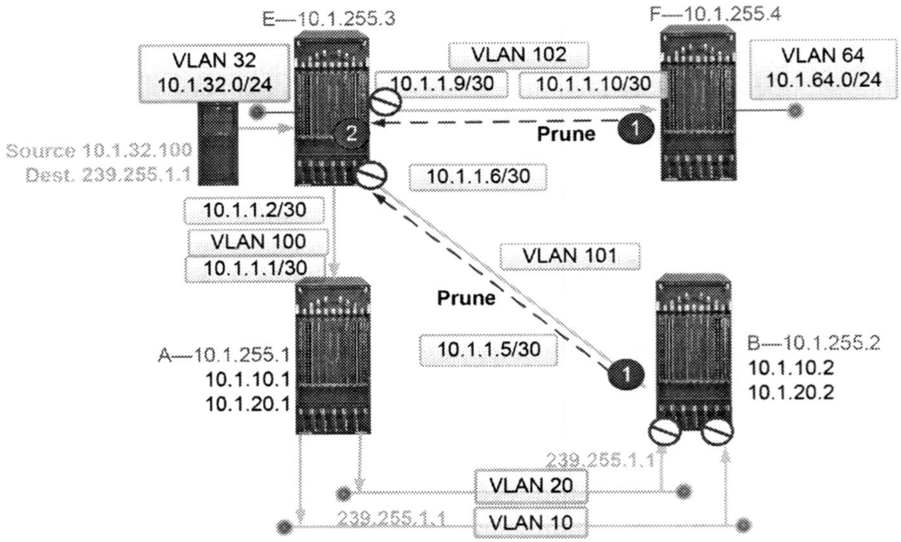

Figure 11-20: PIM-DM SPT construction: pruning

PIM-DM SPT construction: G rafts

Figure 11-21 illustrates the multicast forwarding tree for this source and multicast group after pruning. Routing switch E is forwarding multicast traffic to routing switch A only, which forwards the traffic to receivers in VLANs 10 and 20.

Because group memberships change all the time, the SPT can change as well. First, PIM-DM operates on preferring forwarding unnecessary multicast traffic over failing to forward necessary traffic. Therefore, prune messages expire, and the routing switch that had received the prune message begins forwarding multicast traffic again.

In addition, group members leave groups. For example, if all of the multicast receivers in VLAN 10 and VLAN 20 left the group, then routing switch A would send a prune message to routing switch E, stopping the flow of traffic.

Figure 11-21 illustrates another change in the situation: a multicast receiver for this group becomes active in VLAN 64. (Perhaps a user starts a multicast video streaming application.) The host sends an IGMP report to routing switch F. Eventually, the prune message will expire, and F will receive traffic again. However, PIM-DM provides a faster method of pulling the multicast traffic down. F's (S, G) entry has not yet expired, although it has no downstream interfaces, so F sends a graft message on the upstream interface requesting to be added to the tree.

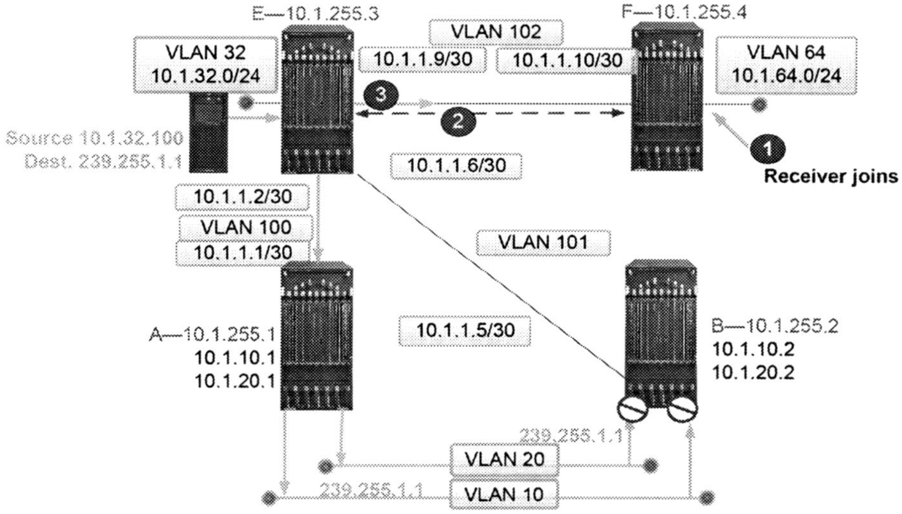

Figure 11-21: PIM-DM SPT construction: grafts

Minimizing reflooding with state refreshes

As mentioned earlier, prunes periodically time out, causing PIM-DM-enabled interfaces to begin again to flood multicasts to neighbors that had sent prune messages. If the neighbor still does not require the traffic, it must send another prune message. These bursts of unnecessary traffic can contribute to congestion.

As Figure 11-22 shows, the routing switch connected to the source can send state refresh messages along the complete multicast routing tree. The state refresh message indicates information about the state of the interface that transmits the message, whether pruned or forwarding. It propagates throughout the system, triggering all PIM-DM routing switches to send joins and prunes only to correct errors. They can then, at the same time, refresh their prune states with confidence without needing to flood multicast traffic.

In more detail, receiving a state refresh message triggers the PIM-DM interface to take these actions:

1. Drop the message unless it has arrived from the RPF upstream neighbor.
2. For messages from the proper upstream neighbor:
   a. Check the indicated state of the upstream interface and determine whether it matches this routing switch's requirements:

- If the indicated state is pruned and this routing switch has no downstream interfaces for the (S, G) entry, the indicated state is correct. Similarly, if the indicated state is forwarding and this routing switch does have downstream interfaces on which it must forward traffic, the state is correct.

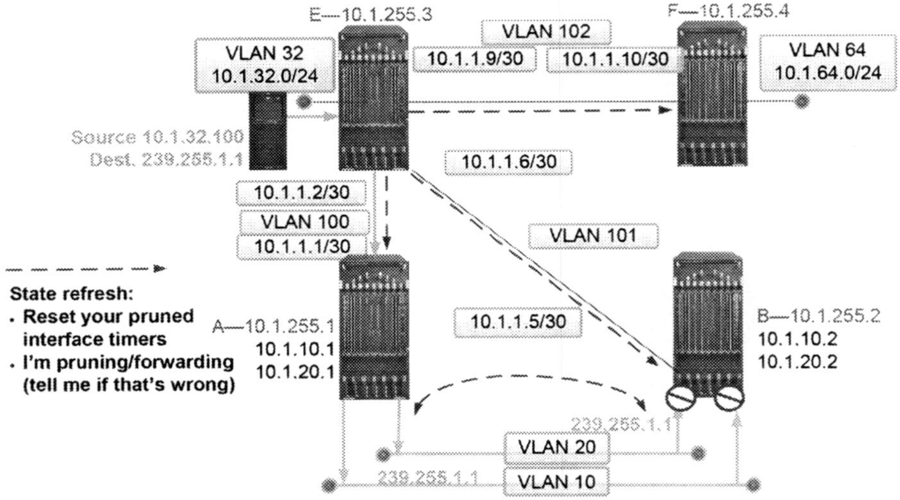

Figure 11-22: Minimizing reflooding with state refreshes

- If the indicated state is pruned, and this switch requires the multicasts, the switch sends an (S, G) join upstream to alter the state. Similarly, if the indicated state is forwarding, but this switch does not have an active entry forwarding multicast traffic downstream, the switch sends an (S, G) prune upstream.

b. Reset all of its pruned interface timers.

   The switch can safely reset all of the timers because its downstream neighbors will correct any mistakes on receipt of the state refresh message, just as this routing switch did.

c. Forward the state refresh message (with one less value in the time to live [TTL]) on all interfaces with PIM-DM neighbors except the interface on which it received the message.

Comware

On Comware switches, you must enable state refresh messages manually on broadcast and multi-access network interfaces such as VLAN interfaces. Enter this command:

```
[Switch-Vlan-interface<ID>] pim state-refresh-capable
```

However, when you enable this feature, you must adjust the time to live (TTL) for state refresh packets. Otherwise, the packets can cycle through a small network for a long time, adding to congestion that they should be eliminating.

In this simplified example network, the state refresh messages only require a TTL value of 2: the furthest PIM routing switches are only two hops apart. In real-world network, the TTL will probably be higher but should always match the maximum number of hops between PIM routing devices.

Remember to count hops from even those devices that you do not expect to connect directly to sources; multicast traffic can turn up in unexpected places. To account for topology changes, you should typically add a few hops to your estimate.

To set the state refresh TTL, enter this command:

```
[Switch-pim] state-refresh-ttl <TTL>
```

ProVision

ProVision switches automatically support state refresh messages, but only the routing switch connected directly to the unicast source initiates state refresh messages. All other PIM routers in the network only propagate these state refresh messages.

You can configure the interval in seconds (10 to 300) between successive state refresh messages. The default setting is 60 seconds:

```
Switch(pim)# router pim state-refresh [10-300]
```

PIM-SM

You will now examine PIM-SM, which involves a little more setup than PIM-DM because PIM-SM routing switches cannot assume that they are part of the initial multicast routing tree for all traffic. Instead, PIM-SM routing switches must explicitly join that tree, and you must configure them to do so.

Configuring PIM-SM is similar to configuring PIM-DM in some ways. For example, you must enable PIM-SM on every interface between multicast sources and receivers, including:

- Interfaces on which multicast sources reside
- Interfaces on which multicast receivers reside (IGMP-enabled interfaces)
- Interfaces between all routers and routing switches that connect sources and receivers

On Comware switches, you enable PIM-DM on physical Layer 3 interfaces or on VLAN interfaces. On ProVision switches, you enable PIM-DM on VLAN interfaces.

As with PIM-DM, PIM-SM-enabled interfaces exchange hellos on 224.0.0.13 (see Figure 11-23). In addition to establishing neighbor relationships, PIM interfaces in broadcast or multiaccess networks elect a designated router (DR). The DR is responsible for forwarding multicasts and is selected according to priority (higher priority is preferred) and by address. The same best practice as for MSTP and VRRP apply: load-balance DR responsibilities across VLANs by setting a higher priority on one routing switch in one VLAN and another routing switch in the other.

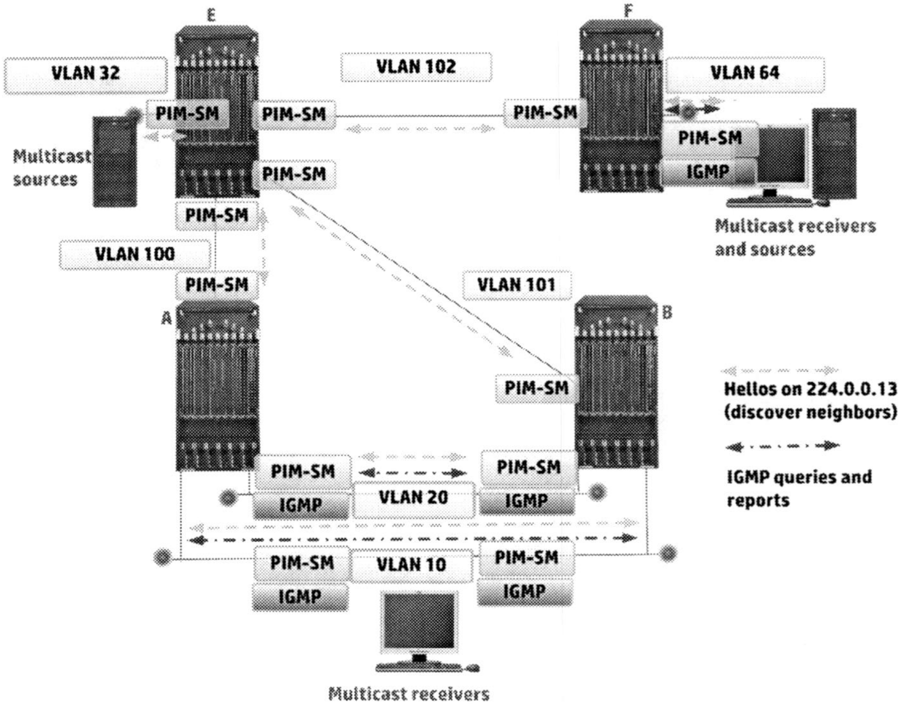

Figure 11-23: PIM-SM

Note

PIM-SM uses the same multicast address as PIM-DM, 234.0.0.13, which helps to explain why you cannot implement PIM-SM and PIM-DM in the same domain. PIM-enabled routing switches operating in one mode would expect certain behavior from neighbors operating in another, leading to failed establishment of the multicast forwarding tree.

RPT

PIM-SM routers must explicitly join multicast forwarding trees for traffic that receivers on their subnets require. However, with no knowledge of the source address, the routing switches have no way of knowing where to direct their joins. The DR for the source does not know where to direct the multicast traffic. Without these joins, the DR has no way of discovering which PIM-SM routing switches require the traffic.

To avoid this problem, the routing switches that are interested in a particular multicast group direct their joins to a router arbitrarily selected as the rendezvous point (RP) and establish an RP tree (RPT) rooted at that device. The source's DR also directs multicast traffic to the RP (you will see exactly how a bit later). Thus, the RPT connects the multicast traffic to the devices that need it. Depending on where the source resides, the RP might not provide the *best* root for tree, but what is most important is that it provides a well-known root. (The "Switching to the SPT" section explains how PIM-SM enables devices to establish more efficient paths.)

As you see in Figure 11-24, interested multicast receivers begin the process of establishing the RPT by sending IGMP reports for the group in question. The IGMP join triggers the DR to join the RPT, sending a (*, G) join toward the RP.

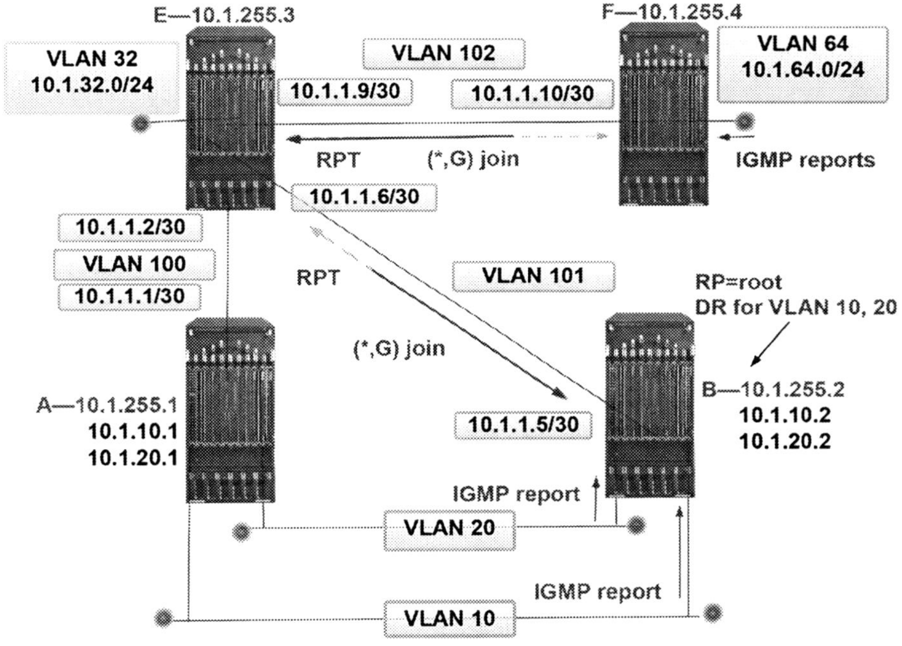

Figure 11-24: RPT

PIM-SM (*, G) entries

To establish RPTs, PIM-SM routing switches create a (*, G) entry for each multicast group for which one of their interfaces has a group membership, adding the interface with the membership to the downstream list (see Figure 11-25). However, unlike the PIM-DM (*, G) entry, this (*, G) entry has an upstream interface, which enables the PIM routing switch to explicitly join the tree. The switch discovers the upstream interface by looking up the forwarding interface in the unicast route to the RP.

The PIM-SM upstream interface transmits a join message, asking the upstream neighbor to add this interface as a downstream interface in the RPT tree. The upstream neighbor adds the interface in the (*, G) entry in question. It then discovers the upstream interface and sends a join message out that interface. The join messages travel up the tree in this manner until they reach the RP.

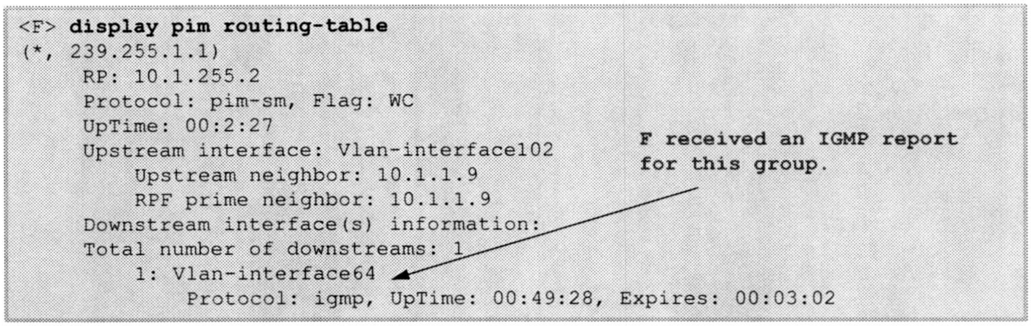

Figure 11-25: PIM-SM (*, G) entries

View information about the RPT on Comware switches

On Comware switches, you can use the **display pim routing-table** command to view information about the interfaces that have joined a multicast group. The output shown in Figure 11-25 shows that the VLAN 64 interface has received an IGMP report for a group:

display pim routing-table

```
(*, 239.255.1.1)
     RP: 10.1.255.2
     Protocol: pim-sm, Flag: WC
     UpTime: 00:2:28
     Upstream interface: Vlan-interface101
         Upstream neighbor: 10.1.1.5
         RPF prime neighbor: 10.1.1.5
     Downstream interface(s) information:
     Total number of downstreams: 1
         1: Vlan-interface102
              Protocol: pim-sm, UpTime: 00:49:28, Expires: 00:03:02
```

View information about the RPT on ProVision switches

Because ProVision software K.15.08 and above includes a CLI compatibility feature, which provides support for many Comware **display** commands, you can use the **display pim routing** command to view information about the RPT on these switches as well:

```
ProVision-1(vlan-1213-pim-sparse)# display pim routing
 VPN-Instance: public net
 Total 3 (*, G) entries; 1 (S, G) entry
 (10.1,2.100, 239.1.1.1)
     RP: NULL
     Protocol: pim-sm, Flag: SPT
     UpTime: 00:00:27
     Upstream interface: Vlan-interface214
         Upstream neighbor: 10.1.214.2
         RPF prime neighbor: 10.1.214.2
     Downstream interface(s) information:
     Total number of downstreams: 1
         1: Vlan-interface212
              Protocol: pim-sm, UpTime: 00:00:27, Expires: 00:00:02
```

RP registration

Figure 11-26 shows an RPT that is ready to deliver multicast traffic from the RP to all PIM routing switches connected to receivers for the group. The RP simply needs to receive the traffic.

When a source begins to stream traffic for this group, the DR in its subnet registers with the RP. The unicast register messages encapsulate the multicast traffic. The RP decapsulates the multicast traffic and forwards it on the multicast tree.

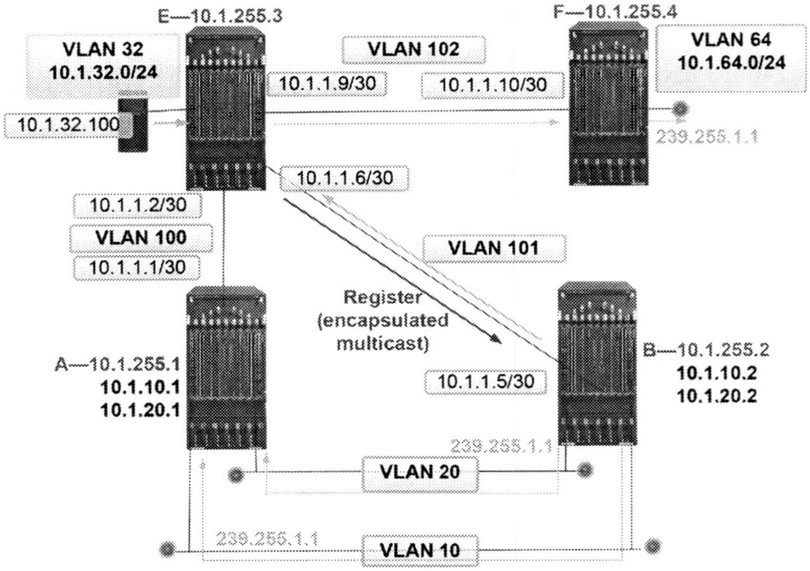

Figure 11-26: RP registration

PIM-SM RPT (S, G) entries

To forward this traffic, the DR, the RP, and any other routing switches in the path use (S, G) entries, which specify the active source and the multicast destination. Each router creates the (S, G) entry as soon as it receives multicast traffic on the RPT, copying the downstream interfaces from the (*, G) entry into the new entry. The DR also includes the tunnel on which it sends register messages as downstream interfaces.

The upstream interface depends on whether the routing switch uses the RPT. Figure 11-27 illustrates an RPT (S, G) entry as indicated by the RPT bit. However, you will almost never see this type of entry because the PIM-SIM routing switch quickly transitions to a more efficient tree.

```
<B> display pim routing-table
 (10.1.32.100, 239.255.1.1)
     RP: 10.1.2.2 (local)
     Protocol: pim-sm, Flag: RPT 2MSDP ACT
     UpTime: 00:03:11
     Upstream interface: Register
         Upstream neighbor: NULL
         RPF prime neighbor: NULL
     Downstream interface(s) information:
       1: Vlan-interface10
             Protocol: igmp, UpTime: 00:12:34, Expires: 00:03:02
       1: Vlan-interface101
             Protocol: pim-sm, UpTime: 00:4:44, Expires: 00:02:52
       1: Vlan-interface20
             Protocol: igmp, UpTime: 00:15:22, Expires: 00:03:14
```

The RPT is forwarding traffic (RPT and ACT flags)

RP receives multicasts in unicast register messages

Copied from the (*, G) entries

Figure 11-27: PIM-SM RPT (S, G) entries

Consider this point.

A more efficient traffic flow would have the DR as the root of the multicast forwarding tree. For example, in this network, multicasts flow from the source to routing switch E (the DR), which tunnels them in unicast messages to routing switch B (the RP). B forwards the multicasts back to E, which forwards them to F. If B could forward the traffic directly, the multicast stream would consume less bandwidth and impose less of a burden on the DR and RP.

Switching to the SPT

PIM-SM allows routing devices to transition from the RPT to the SPT with the DR at the root.

The arrival of the first multicast packet in a group triggers a switch from the RPT to the SPT on the DR, RP, and any PIM-SM switch that forwards traffic into a VLAN for IGMP hosts.

To switch to the SPT, a PIM-SM routing switch performs these tasks:

1. It determines the upstream interface for this tree, by using RPF to discover the forwarding interface for the shortest path to the source. It then sends an (S, G) join on this interface.

   At this point, to ensure that traffic flow continues uninterrupted, the switch continues to accept traffic on the upstream interface for the RPT.

2. It monitors the incoming interface for multicast traffic to determine when it can set the SPT bit.

   The SPT bit prevents the routing switch from accepting multicasts on any interface except the correct upstream for the SPT, preventing duplication of multicast streams. Routing switches can set this bit when they are certain that the SPT has been established above them—that is, they can receive traffic on this tree.

   Routing switches can determine that the upstream SPT has been established in a few different ways, which include:

- The switch receives a multicast on the correct upstream interface for the (S, G) entry.
- The switch is the DR so, as the root of the SPT, it sets the SPT bit immediately.
- The upstream interface for the RPT and the SPT are the same, so the switch cannot ever determine whether the upstream SPT has been established—but does not care because it is receiving traffic in the same way in either case. This switch can set the SPT bit immediately.

Note

In reality, the behavior is a bit more nuanced because, in a multiaccess network, the upstream interface might be the same but the neighbor different. However, this basic description suffices for our purposes.

3. A non-RP routing switch sends an RPT (S, G) prune on the upstream interface for the RPT (unless that is the same as for the SPT).

   Note that the PIM-SM routing switch does not prune itself from the RPT; it maintains its (*, G) entry and its place in the RPT so that it can continue to receive multicasts in this group from other sources. Rather, this prune removes the routing switch from the RPT *for the specific source.*

4. The RP sends register stop messages. These messages request the DR to stop sending the encapsulated multicast messages because it can receive the multicasts on the SPT.

Thus, PIM routing switches always ensure that the SPT has been established upstream of them before they prune themselves from the RPT. This process ensures that the multicast stream is never interrupted.

On Comware switches, you can disable the transition on DRs for multicast receivers with the **spt-switch-threshold infinity** command, entered from the pim view. However, it is generally recommended that you leave the default behavior.

Even if you enter the **spt-switch-threshold infinity** command on a Comware switch, the DR for multicast receivers will transition to the SPT. This can cause problems when the RP does not transition also because the RP will not receive register messages. Therefore, you must never disable the SPT transition on an RP. (In fact on some models, the device itself will automatically ignore the setting when it is RP. Nonetheless, you should avoid setting the spt-switch-threshold infinity command.)

Figure 11-28 summarizes how and when PIM-SM devices switch to the SPT.

<table>
<tr><th>PIM-SM device</th><th>When it switches</th><th>How it switches</th><th>When it sets the SPT bit (rejects non-SPT traffic)</th><th>How it completes the switch</th></tr>
<tr><td>DR for source</td><td>After receiving one multicast (cannot disable)</td><td>No further action necessary</td><td>Immediately</td><td>Sends an (S,G) prune on the RPF interface for the RP</td></tr>
<tr><td>RP</td><td>After receiving one multicast (can disable on Comware but should not)</td><td rowspan="2">Sends an (S, G) join to the source upstream neighbor</td><td>When it receives multicasts on the RPF interface for the source</td><td>Sends a register-stop message to the DR</td></tr>
<tr><td>DR for receivers</td><td>After receiving one multicast (can disable on Comware)</td><td>• When it receives multicasts on the RPF interface for the source
• Immediately if source and RP interface are the same</td><td>Sends a (S, G) prune on the RPF interface for the RP</td></tr>
</table>

Figure 11-28: Switching to the SPT

Select an RP or RPs

You now have a basic idea of how PIM-SM routers join an RPT in order to receive initial multicasts and then switch to an SPT for more efficient traffic flow. Besides enabling PIM-SM on the proper interfaces, the most important step for configuring the system is specifying the RP or RPs.

Every PIM-SM router or routing switch in the domain must always select the same RP for the same multicast addresses. Clearly, if some routing switches that need a particular type of multicast traffic established an RPT to one RP, and the DR then registers to another RP, the traffic will go astray.

It is recommended that you select backbone routers for RPs, which encourages devices to establish the RPT over backbone links. Traffic would need to flow over these links in any case, so placing the RP there does not redirect the traffic to an area that might not otherwise have to support it. For example, in the example network shown in Figure 11-29, routing switch E is the natural choice for the RP because all other PIM-SM devices connect directly to it.

In addition, because RPs might need to store many multicast route entries and handle a great deal of multicast traffic, you should also select high-performance routing switches capable of meeting the needs.

When the PIM-SM routing switches meet these criteria, you should also try to place them as near multicast sources as possible, which makes the transition to the SPT more efficient.

You can either specify a single RP for all multicast groups. Or you can configure the bootstrap mechanism, which enables the PIM-SM domain to elect an RP for each group dynamically.

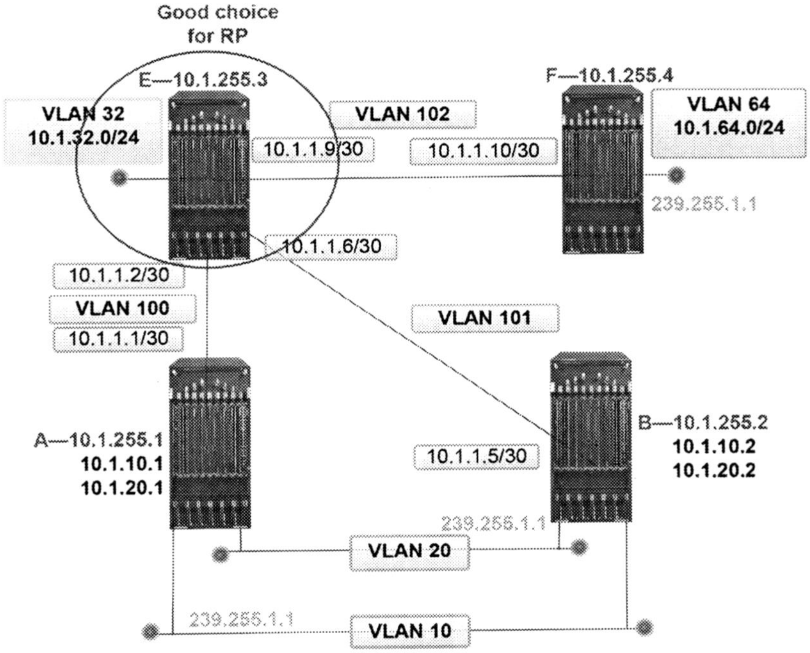

Figure 11-29: Select an RP or RPs

Specify a single static RP

You can specify a single static RP manually. In this case, you must set the same RP address on every PIM-SM device in the domain.

Because the domain will have only one RP, you should generally select the PIM-SM routing device that is the fewest hops from most other routers or has the highest bandwidth and throughput capabilities (see Figure 11-30). If you have a relatively small core and you know that multicast traffic will be originating from a specific location, you could alternatively set the DR for that location as the RP.

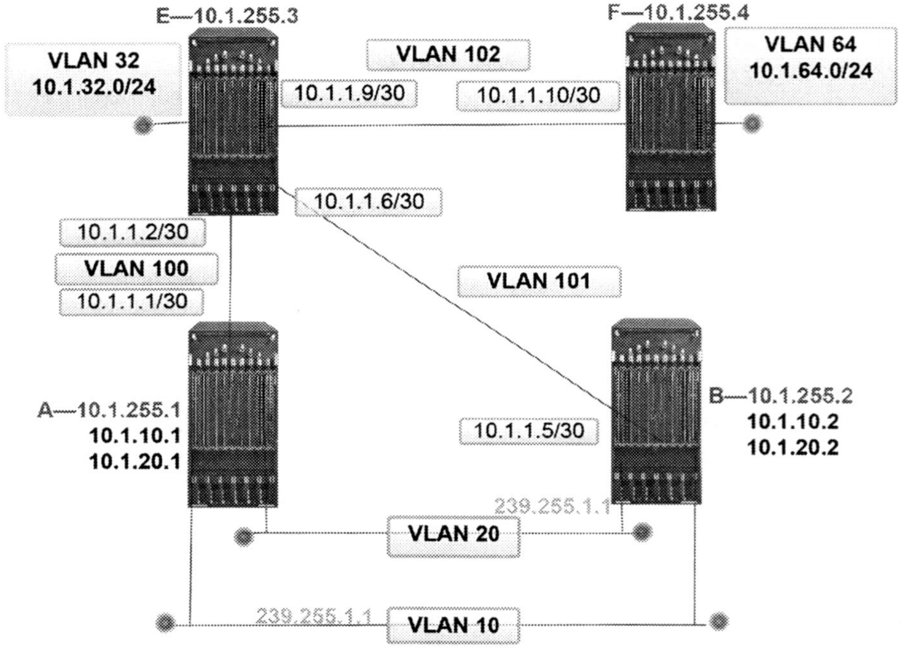

Figure 11-30: Specifying a single static RP

On Comware switches, enter this command to specify the static RP address:

```
[Switch-pim] static-rp <IP address>
```

On ProVision switches, enter this command to specify the static RP:

```
Switch(pim)# rp-address ip-addr [group-addr/group-mask]
```

You must enter the command on *every* PIM-SM router and routing switch in the domain; PIM neighbors do *not* advertise static RPs to each other.

Specify RPs dynamically with the bootstrap mechanism

Specifying each RP manually can create an administrative burden in a large domain. Even more critical, acting as RP for a network with a great amount of multicast traffic imposes a processing burden on the RP. Generally, you should distribute the RP role among several PIM-SM devices in the network backbone.

Instead of specifying an RP on all PIM-SM devices, you can configure each device with its role and make no configurations at all on devices that do not require a role:

- Candidate RPs (C-RPs) are PIM-SM devices that can act as RPs. In the pool of C-RPs, only one is elected the RP for a particular group. However, multiple RPs can be active at once, each supporting a different group.
- A Bootstrap Router (BSR) is a PIM-SM device that distributes information about RPs to other PIM-SM devices in the domain. You can configure a pool of Candidate BSRs (C-BSRs). Only one BSR is active, while the others function as backups.

In more detail, C-BSRs use their bootstrap messages, which they begin to send as soon as they are configured with the role, to elect the active BSR (see Figure 11-31). That is, when a C-BSR receives a bootstrap message with a higher priority than its own, it suppresses its messages. If the priority is the same, the C-BSR defers to the C-BSR with a higher IP address. Thus, after all C-BSRs have heard each other's messages, only the C-BSR with the highest priority, or tied highest priority and highest IP address, remains active.

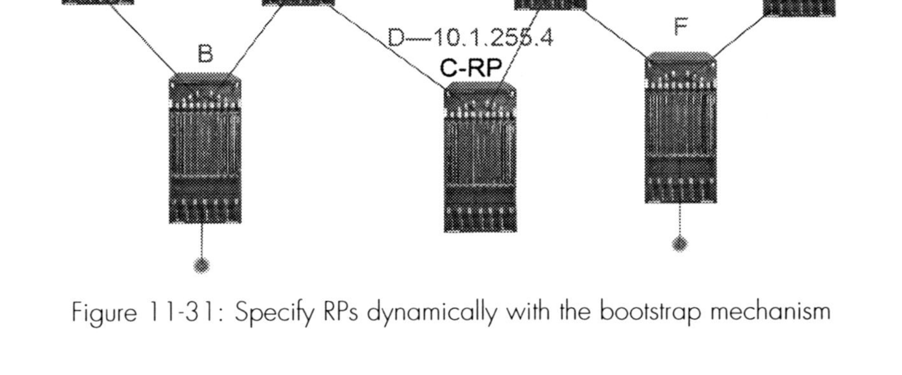

Figure 11-31: Specify RPs dynamically with the bootstrap mechanism

Note

The C-BSR uses the IP address of the interface that you specify with the C-BSR command. PIM-SM must be enabled on this interface or the BSR.

The C-BSRs continue to listen for the BSR's periodic messages. If the holdtime expires without a message, the C-BSRs become active again, electing a new BSR.

C-RPs send C-RP advertisements to the active BSR. The advertisements include the RP's IP address, the range of multicast addresses that it supports, and its priority for each range. The BSR compiles the information as an RP set, which it advertises to all PIM-SM routing devices in the domain.

Comware switches: Electing an RP

Comware switches that have identical RP sets always select the same RP for the same multicast address, following these rules:

1. Select the RP that advertises the most specific range of multicast addresses that includes the desired address.
2. If multiple RPs advertise ranges of equal specificity, select the RP with the best priority for the multicast group address.
3. The lower the priority value, the higher the priority.
4. If two RPs have the same best priority, use a hash algorithm to calculate a value for that RP and that multicast group address. The algorithm functions such that it assigns a new value to the RP for each range of multicast addresses. Thus, for each range of addresses, a different election occurs, and a different RP might be selected.

Note

The precise hash value for a specific RP and multicast group address = (1103515245 * (1103515245*(Group address AND mask)/12345) XOR RP address)/MOD2^{31}.

5. If multiple RPs have the same hash value, select the RP with the higher IP address.

Figure 11-32 outlines this election process.

The PIM-SM devices store the RP set, but only calculate an RP when they need to join an RPT for a group (when a connected IGMP host sends a report for the address or when a downstream neighbor sends a (*, G) join). This behavior preserves memory and processing power.

You can take a couple of approaches to selecting and setting up the RPs.

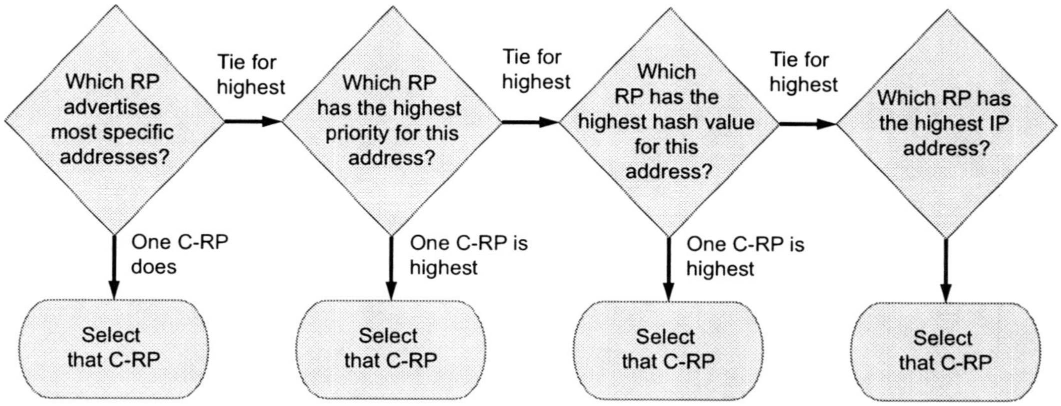

Figure 11-32: Electing an RP on Comware switches

ProVision switches: Electing an RP

ProVision switches use a similar process to elect the RP. ProVision switches use the C-RP's group-prefix configuration settings to identify the multicast groups a switch can support. If multiple C-RPs are configured with the same group prefixes, the following criteria are used to select the RP to support the group:

1. The C-RP configured with the longest group-prefix mask applicable to the multicast group is selected to support the group.

The remaining criteria mirror the criteria used by the Comware switches.

2. If more than one C-RP is using the longest group-prefix, the C-RP configured with the highest priority is selected.
3. If more than one C-RP is configured with the highest priority, a hash function (using the configured *bsr-candidate hash-mask-length value*) generates a series of mask length values that are individually assigned to the set of eligible C-RPs. If the hash function matches a single RP candidate to a longer mask length than the other candidates, that candidate is selected to support the group.
4. If the hash function matches the longest mask length to multiple RP candidates, the C-RP that has the highest IP address is selected to support the group.

Allow the hash function to automatically load-balance among RPs

For the simplest setup, you can have all C-RPs advertise support for all multicast addresses with the same priority.

On Comware switches, enter this command on each C-RP:

```
[Switch-pim] c-rp <interface type> <interface number> priority <0-255>
```

On ProVision switches, enter this command on each C-RP:

```
Switch (config)# rp-candidate priority <priority>
```

You should explicitly set the same priority on each C-RP because different models of switches use different default priorities. Use the same guidelines for selecting the interface, which determines the IP address advertised to the BSR, as you would use for selecting the IP address on a static RP. If you decide to specify the loopback interface, you *must* enable PIM-SM on that interface.

After you set up each C-RP in this way, the hash algorithm will automatically allocate multicast addresses among the various C-RPs in a more or less equal fashion.

Consider this statement in more detail.

As you just learned, the hash algorithm assigns each RP a value for a block of multicast addresses. The length of the range of addresses that produce the same value for the same RP depends on the length of the mask that is advertised by the BSR. When PIM-SM devices use the hash algorithm to select an RP for a particular multicast address, they apply an AND function on the multicast address and the mask (much as an IP address is combined with a mask to produce a network address). Therefore, a 32-bit mask would result in a different value for each multicast address. With a 31-bit mask, the hash would produce the same value for two addresses in a row (because the 31-bit mask reduces two addresses to the same value), and a 24-bit mask would produce the same result for 256 addresses.

You can set the hash length on the BSR to control the size of the blocks of addresses that are always assigned to the same RP. The smaller the size (the longer the hash length), the more likely it is that multicast addresses will be distributed equally among the C-RPs. If the BSR advertises a shorter hash that produces large blocks of addresses, the same RP might happen to be selected for a preponderance of the multicast addresses—particularly, if server administrators select multicast addresses in a row.

The default mask length is 30, which produces the same hash for blocks of four addresses. In other words, the PIM-SM domain will elect the same RP for multicast group 239.255.0.0, 239.255.0.1, 239.255.0.2, and 239.255.0.3. The domain might also elect that same RP for 239.255.0.4-239.255.0.7, or it might elect another RP, depending on the C-RP's IP addresses. However, the more multicast addresses that are used, the higher the likelihood that they will be distributed more or less equally.

As for determining how many PIM-SM devices to configure as C-RPs, you should consider the size of your network and the number of multicast receivers. In an environment with thousands of receivers, you should configure several RPs, which load-balance groups between them.

To configure the hash algorithm on Comware switches, use these commands:

```
[Switch-pim] c-bsr <interface type> <number> <0-32>
```

```
[Switch-pim] c-bsr <interface type> <number>

[Switch-pim] c-bsr hash-length <0-32>
```

To configure the hash algorithm on ProVision switches, use these commands:

```
Switch (config)# bsr-candidate hash-mask-length<1-32>

Switch (config)# router pim bsr-candidate hash-mask-length 1-32
```

Comware switches: Prefer specific RPs for specific multicast groups

On Comware switches, you can also assign a specific set of addresses to a C-RP.

Ideally, the RP for a specific multicast address is also the DR for the eventual source. Then the RPT matches the SPT, and the traffic flows efficiently from the onset of the stream to its conclusion. If possible, you should collect information about the active multicast groups in your environment. Do some of the groups have a consistent source or sources that connect to the same default router? If so, you should ensure that the PIM-SM domain selects that device as the PIM-SM RP for the groups in question.

> **Note**
> If the subnet has multiple routers, the sources' default router does not automatically become the DR. Rather, the device with the highest IP address (not counting a VRRP virtual address) becomes the DR. To ensure that the default router is elected DR, set its DR priority to be better than other routers' and routing switches'.

Identify the router that you want to act as the RP. If you want that router to be the preferred RP for all multicast addresses, you can simply set a higher priority (lower value) for the C-RP setting. Then the redundant C-RP is only used if the preferred RP is down.

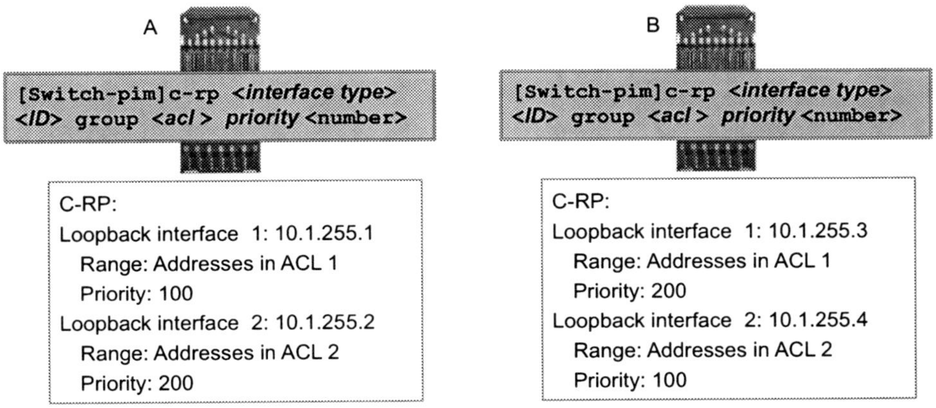

Figure 11-33: Setting multicast address limits and priorities on RPs

If you want different routers to become RPs for different multicast addresses, you can take this approach on Comware devices (see Figure 11-33):

- On each router, create two loopback interfaces. Run the unicast routing protocol and PIM-SM on these interfaces. (Loopback interfaces are not required but provide more stability.)
- Create a basic ACL that selects the multicast addresses for which you want this router to become the RP:

```
[Switch] acl number <2000-2999)
```

```
[Switch] rule permit source <multicast address> <0 | wildcardbits>
```

You can specify one address at a time or a range of addresses using wildcard bits:

- Create another ACL that selects the multicast addresses for which you want this router to act as RP if the usual RP is not available.
- Set one loopback interface as a C-RP, specifying the first ACL and a high priority (low priority value):

```
[Switch-pim] c-rp <interface type> <interface number> group <2000-
2999> priority <0-255>
```

- Set the second loopback interface as a C-RP, specifying the second ACL and a lower priority (high priority value).
- Follow the same steps on other routers.

These ACLs restrict the router from advertising itself as an RP for other multicast addresses. Often you want this behavior because you do not want the network to route any multicasts except the ones that you have identified. However, you can also take a more permissive approach. Instead of specifying an ACL for the loopback interface with the lower priority, simply advertise the loopback interface for all multicast addresses.

Also note that when a PIM-SM router needs to select an RP for a multicast group, it prefers the RP with the most specific range that includes the desired multicast group regardless of priority.

Summary

This chapter introduced you to the technologies that enable a network infrastructure to support the multicast traffic that is becoming more and more ubiquitous. You can implement IGMP and either PIM-DM or PIM-SM as appropriate for routing multicast traffic throughout your environment.

You can also add features such as IGMP snooping and multicast VLANs that streamline the traffic flow. Finally, you can control the multicast solution in advanced ways, limiting multicasts according to source, for example, or slicing a domain into administrative scope, each of which provides its own multicast services.

Learning check

Answer the following questions:

1. What advantages are provided by using the bootstrap protocol method of assigning RPs in a PIM-SM domain?

2. On which OSI layer does IGMP snooping operate and where do you enable this feature? On which OSI layer does IGMP operate and where do you enable this feature?

Answers

1. What advantages are provided by using the bootstrap protocol method of assigning RPs in a PIM-SM domain?

   Depending on how many ro uters and routing switches the domain contains, the bootstrap method might be easier (you only have to enter commands on the C-BSRs and C-RPs). The true advantage, however, is that the PIM-SM domain supports more than one RP, which provides load balancing and redundancy.

2. On which OSI layer does IGMP snooping operate and where do you enable this feature? On which OSI layer does IGMP operate and where do you enable this feature?

   IGMP snooping operates on Layer 2, and you enable it on the VLAN. IGMP operates on Layer 3, and you enable it on the VLAN interface.

12 Practice Test

INTRODUCTION

The *Deploying HP FlexNetwork Core Technologies* exam tests your knowledge, skills, and ability to configure, implement, and troubleshoot enterprise-level HP FlexNetwork solutions. You will be tested on networking technology topics such as Quality of Service (QoS), Virtual Router Redundancy Protocol (VRRP), Multicasting (Internet Group Management Protocol (IGMP), Protocol Independent Multicast-Dense Mode (PIM-DM) and Protocol Independent Multicast-Sparse Mode (PIM-SM)), dynamic routing (Open Shortest Path First (OSPF) and Border Gateway Protocol (BGP)), and security technologies such as Access Control Lists (ACL) and port authentication (802.1X, MAC-Auth, and Web-Auth). You will also be tested on network virtualization technologies such as HP Intelligent Resilient Framework (IRF), as well as your ability to enable and manage network resources, services, and users with HP Intelligent Management Center (IMC).

This certification exam is designed for candidates with "on-the-job" experience. The associated training course, which includes numerous design and hands-on lab activities, provides a foundation, but you are expected to have experience in the real world as well.

EXAM DETAILS

The following are details about the exam:

- **Exam ID**—HP0-Y47
- **Exam type**—Proctored
- **Number of items**—55
- **Item types**—Multiple choice (single response), Multiple choice (multiple responses), Matching, and Pull-down List
- **Exam time**—150 minutes
- **Passing score**—65%
- **Reference material**—*No online or hard copy reference material will be allowed at the testing site.*

HP2-Y31 testing objectives

31% Fundamental FlexNetwork architectures and technologies

- Describe unicast routing protocols, such as OSPF and BGP, and explain how these protocols are used to route traffic.
- Describe the purpose of multicast traffic, compare multicast routing protocols, such as PIM-DM and PIM-SM, and explain how these protocols route traffic.
- Describe network management and monitoring protocols and explain how technologies such as sFlow, NetFlow, NetStream, and Remote Network Monitoring (RMON) are used to monitor network traffic.
- Describe security technologies, such as ACLs, 802.1X, MAC-Auth, and Web-Auth, and explain how they are used.
- Describe QoS technologies, explain QoS mechanisms supported on HP switches, and explain how they are used.
- Describe and differentiate technologies that provide redundancy and availability and explain how they provide these capabilities. The technologies include:
    - PVST+
    - RPVST+
    - UDLD
    - DLDP
    - Distributed trunking (DT)

9% FlexNetwork products, solutions, and service offerings

- Describe two-tier and three-tier HP FlexCampus and HP FlexFabric solutions.

38% FlexNetwork solution implementation (install, configure, set up)

- Configure HP devices to implement technologies as per planned design and verify the configurations. The technologies include:
    - Mirroring
    - ACLs
    - 802.1X
    - MAC and Web authentication
    - QoS

    - OSPF
    - PIM-DM
    - PIM-SM
- Configure IMC to manage a network and configure HP switches so that they can be managed by IMC within an SNMPv2c, SNMPv3, or 802.1X solution.

22% FlexNetwork solution enhancement

- Optimize network availability using VRRP, IRF, IRF with Multi-Active Detection (MAD), and distributed trunking.

Test preparation questions and answers

The following questions will help you measure your understanding of the material presented in this study guide. Read all the choices carefully because there may be more than one correct answer.

Choose all correct answers for each question.

Questions

1. A network includes HP Comware switches as access layer switches that do not route traffic. What is the benefit of configuring IGMP snooping on these switches?

   a. The switches can act as backup IGMP queriers.

   b. The switches can filter multicasts on ports that do not need them.

   c. The switches can forward multicasts.

   d. The switches can enforce multicast boundaries, preventing multicasts from being forwarded across certain routed interfaces.

2. What is one difference between PIM-DM and PIM-SM?

   a. PIM-DM is better suited for environments with limited bandwidth than PIM-SM is.

   b. PIM-SM requires setup related to Rendezvous Points (RPs), but PIM-DM does not.

   c. PIM-DM uses IGMP to discover VLAN interfaces that require multicasts, but PIM-SM does not.

   d. PIM-DM can use any type of unicast routes to determine the correct upstream interface for multicasts, but PIM-SM must use static routes.

3. For which scenario can a network administrator create a full mesh of stacking or IRF links between switches?

   a. The backplane stack consists of four HP 2920 Series switches.

   b. The backplane stack consists of four HP 3800 Series switches.

   c. The backplane stack consists of six HP 3800 Series switches.

   d. The IRF virtual device consists of four HP 5500 Series switches.

4. Refer to the exhibit.

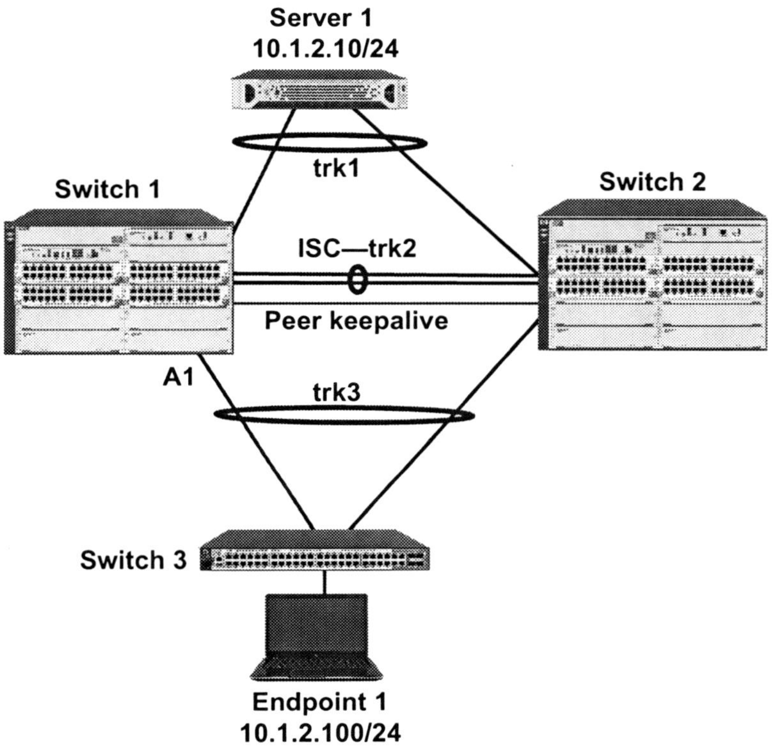

Figure 12-1: Exhibit for question 4

Switch 1 and Switch 2 have established an Interswitch Connection (ISC) and two distributed trunks. All of the switches have Multiple Spanning Tree Protocol (MSTP) enabled. An MSTP Bridge Protocol Data Unit (BPDU) arrives on A1 on Switch 1. What does Switch 1 do?

a. It drops the BPDU.

b. It processes the BPDU.

c. It does not process the BPDU, but it floods it out other ports in the VLAN.

d. It logs an alert and, depending on the settings, disables the interface.

5. Match the protocol to a use case for it.

a. Preventing issues from partially broken or cross-connected fiber cables	— Distributed trunking
b. Enabling two switches to establish a link aggregation to a downstream device	— Unidirectional Link Detection (UDLD)
c. Detecting Layer 3 connectivity between the device and a neighbor	— Bidirectional Forwarding Detection (BFD)

6. Refer to the exhibits.

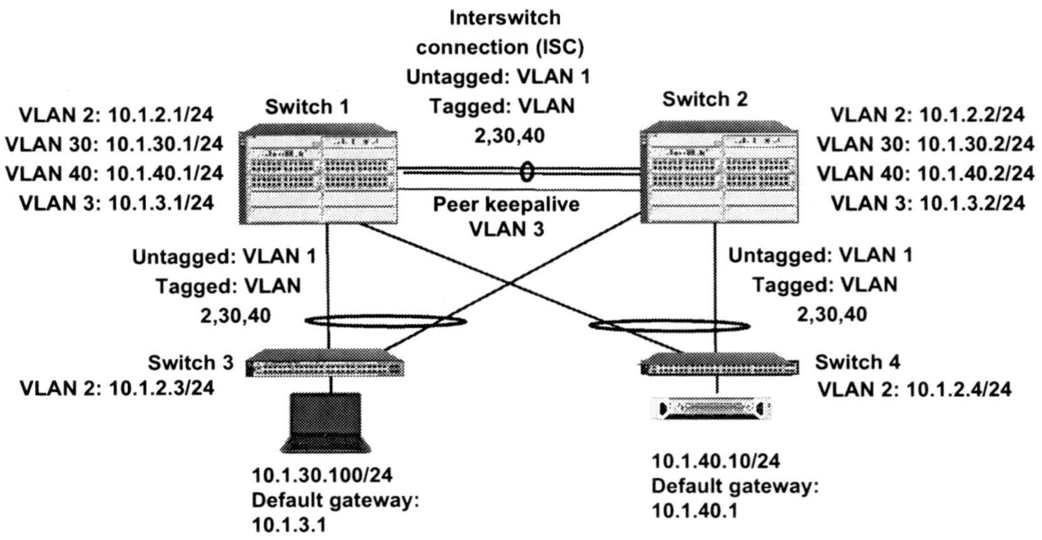

Figure 12-2: Exhibit 1 for question 6

Exhibit 2 (which is on the next page) shows the Virtual Router Redundancy Protocol (VRRP) configuration for VLAN 30 on Switch 1 during normal operation (when both Switch 1 and Switch 2 are up).

If Switch 1 fails and then comes back up, the network administrator wants Switch 1 to delay two minutes before taking over as the VRRP Master. The administrator must complete two steps to configure this behavior. What are they? (Select two.)

a. Enable VRRP preempt delays globally on Switch 1.

b. Enable the desired VRRP preempt delay on the VLAN 30 VRID 1 on Switch 1.

c. Enable the desired VRRP preempt delay on the VLAN 30 VRID 1 on Switch 2.

d. Configure a different actual IP address on Switch 1 and change it to a backup role.

e. Change the actual IP address on Switch 2 to match the IP address on Switch 2.

```
Switch1# show vrrp config
VRRP Global Configuration Information
 VRRP Enabled : Yes
 Traps Enabled : Yes
 Virtual Routers Respond To Ping Requests : Yes
 VRRP Nonstop Enabled : No


 VRRP Virtual Router Configuration Information
 VLAN ID 30
 Virtual Router ID : 1

Administrative Status [Disabled] : Enabled
 Mode [Uninitialized] : owner
 Priority [100] : 255
 Advertisement Interval [1] : 1
 Preempt Mode [True] : True
 Preempt Delay Time [0] : 0
 Respond To Virtual IP Ping Requests [Yes] : Yes
 Version [2] : 2
 Null authentication compatibility [False] : False
 Primary IP Address : Lowest

  Primary IP Address : Lowest

  IP Address      Subnet Mask
  --------------- ---------------
  10.1.30.1        255.255.255.0

 VRRP Virtual Router Statistics Information

  Vlan ID                   : 30
  Virtual Router ID         : 1
  State                     : Master
  Up Time                   : 50 min
  Virtual MAC Address       : 00005e-000101
  Master's IP Address       : 10.1.30.1
  Associated IP Addr Count : 1          Near Failovers             : 0
  Advertise Pkts Rx         : 0          Become Master              : 6
  Zero Priority Rx          : 0          Zero Priority Tx           : 4
  Bad Length Pkts           : 0          Bad Type Pkts              : 0
  Mismatched Interval Pkts : 0          Mismatched Addr List Pkts : 0
  Mismatched IP TTL Pkts   : 0          Mismatched Auth Type Pkts : 0
```

Figure 12-3: Exhibit 2 for question 6

7. What is a characteristic of a CLOS fabric?

   a. It features an arbiter that helps to load-balance traffic across internal switch paths.

   b. It features a three-tier switch fabric in which each tier includes multiple crossbar switches.

   c. It features multiple virtual switches that integrate the management and interface modules.

   d. It features an explicit congestion notification mechanism for ensuring lossless delivery.

8. Refer to the exhibit.

```
vlan 30
vlan 40
interface g1/0/1
  port link-mode bridge
  port link-type hybrid
  undo port hybrid vlan 1
  port hybrid vlan 40 tagged
  port hybrid vlan 30 untagged
  port hybrid pvid vlan 30
```

Figure 12-4: Exhibit for question 8

A Comware access switch running Comware version 5.20 connects to Voice over IP (VoIP) phones. The phones, in turn, connect to users' computers, so each switch port connects to a computer and phone.

The VLAN for data traffic is VLAN 30, and the VLAN for VoIP traffic is VLAN 40. The phones support Cisco Discovery Protocol (CDP). The administrator would like to use CDP to advertise the voice VLAN ID and priority settings to the phones.

The exhibit shows the VLAN and the switch interface configuration. The administrator needs to complete several more steps. What is one?

   a. Enable the voice function and priority settings on VLAN 40.

   b. Enable MAC-based VLANs on the physical interfaces that connect to phones.

   c. Enable voice VLAN 40 (voice vlan 40 enable) on the physical interfaces that connect to phones.

   d. Enable the switch interfaces that connect to phones to advertise voice VLAN 40 with LLDP (lldp voice-vlan 11).

9. Refer to the exhibit.

```
<Switch> display mac-authentication
MAC address authentication is enabled.
 User name format is MAC address in lowercase, like xxxxxxxxxxxx
 Fixed username:mac
 Fixed password:not configured
 Offline detect period is 300s
 Quiet period is 60s.
 Server response timeout value is 100s
 the max allowed user number is 1024 per slot
GigabitEthernet1/0/1 is link-up
 MAC address authentication is enabled
 Authenticate success: 0, failed: 0
 Max number of on-line users is 256
 Current online user number is 0
MAC Addr Authenticate state AuthIndex
#output for other interfaces not shown

[Switch-luser-aaaaaaaaaaaa] display this
 password simple aaaaaaaaaaaa
```

Figure 12-5: Exhibit 1 for question 9

A network administrator wants to configure a Comware switch to implement local MAC-Auth. The exhibit shows the relevant settings that the administrator has already configured. The administrator wants to permit an endpoint with MAC address aaaaaa-aaaaaa to connect to it. What must the administrator do?

a. Set the service type for the local-user account to lan-access.

b. Change the username for the local-user account to mac.

c. Change the username for the local-user account to aaaaaa-aaaaaa.

d. Remove the password from the local-user account.

10. Refer to the exhibit.

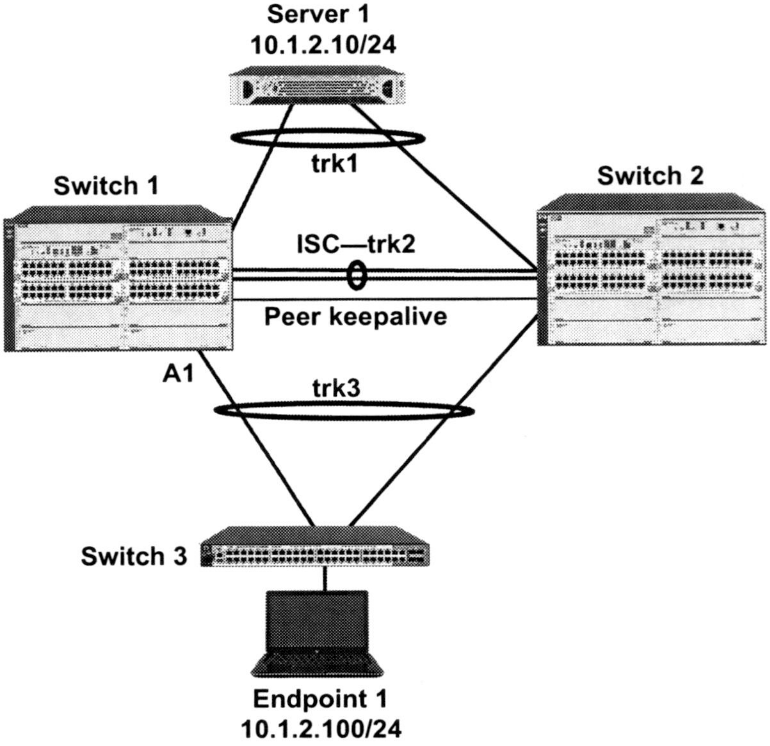

Figure 12-6: Exhibit for question 10

Switch 1 and Switch 2 have established an Interswitch Connection (ISC) and two distributed trunks. A frame from Endpoint 1 arrives on Switch 2, which forwards it to its destination, Server 1. Server 1 sends return traffic toward Endpoint 1, but it sends the traffic on the link to Switch 1. What does Switch 1 do?

a. It drops the frame.

b. It buffers the frame and sends a MAC Learn request to Switch 2.

c. It forwards the frame on A1 in trk3.

d. It forwards the frame on trk2 (the ISC).

11. Refer to the exhibit.

```
<Switch>display acl 3000
Advanced ACL  3000, named FromUsers, 1 rule,
ACL's step is 5
 rule 0 permit ip destination 10.1.3.0 0.0.0.255
 rule 5 permit tcp destination 10.2.1.10 destination-port 8080
 rule 50000 deny any
[Switch-Vlan-interface30]display this
interface Vlan-interface30
ip address 10.1.30.1 255.255.255.0
packet-filter 3000 in
```

Figure 12-7: Exhibit for question 11

Several packets arrive on this switch's interface g1/0/1, which is assigned to VLAN 30 (10.1.30.0/24). The switch is the default gateway for 10.1.30.0/24. What does the switch do?

Packet 1:

- Destination MAC address—aaaaaa-aaaaaa
- Source IP address—10.1.30.100
- Destination IP address—10.1.30.101
- IP protocol—TCP
- Destination TCP port—21

Packet 2:

- Destination MAC address—This switch's VLAN 30 address
- Source IP address—10.1.30.100
- Destination IP address—10.2.1.10
- IP protocol—TCP
- Destination TCP port—21

a. It drops both packets.

b. It forwards both packets.

c. It drops Packet 1 and forwards Packet 2.

d. It forwards Packet 1 and drops Packet 2.

12. Refer to the exhibit.

```
#partial running-config for Switch 1, Switch 2, and Switch 3

Switch 1

pim
 c-bsr loopback 0 priority 250

Switch 2

pim
 c-bsr loopback 0
 c-rp loopback 0 priority 10

Switch 3

acl number 2000
 rule 0 permit source 239.1.1.1 0
pim
 c-bsr loopback 0
 c-rp loopback 0 acl 2000
```

Figure 12-8: Exhibit for question 11

The exhibit shows the Protocol-Independent Multicasting (PIM) configuration for Switch 1, Switch 2, and Switch 3. Switch 4 is not configured with these commands. However, all four switches have PIM-SM enabled on the proper interfaces. They are PIM neighbors.

Switch 4 also has IGMP enabled on the proper interfaces. It receives an IGMP join from an endpoint for multicast group 239.1.1.1. What does Switch 4 do?

a. It sends a PIM join toward Switch 1, which it selected as RP.

b. It sends a PIM join toward Switch 2, which it selected as RP.

c. It sends a PIM join toward Switch 3, which it selected as RP.

d. It cannot send a PIM join because it cannot select an RP.

13. A company router connects to two Internet Service Providers (ISPs). The dual-home ISP connection provides redundancy for the company's public-facing services. Which routing solution meets the needs for the company router connection to each ISP?

a. Open Shortest Path First (OSPF), in which the routers have interfaces in a single area

b. Open Shortest Path First (OSPF), in which the routers have interfaces in different areas

c. Internal Border Gateway Protocol (iBGP)

d. External Border Gateway Protocol (eBGP)

14. A network administrator wants an HP router to use Border Gateway Protocol (BGP) to communicate with a connected router. What must the administrator do?

    a. Enable BGP on the interface that connects to the router, which enables the router to discover the peer.

    b. Enable BGP on the interface that connects to the router and also advertise a network in the BGP settings.

    c. Manually configure the peer IP address in the BGP settings.

    d. Configure a route map that includes the peer IP address. Then redistribute routes in the BGP settings with that route map applied.

15. Refer to the exhibits.

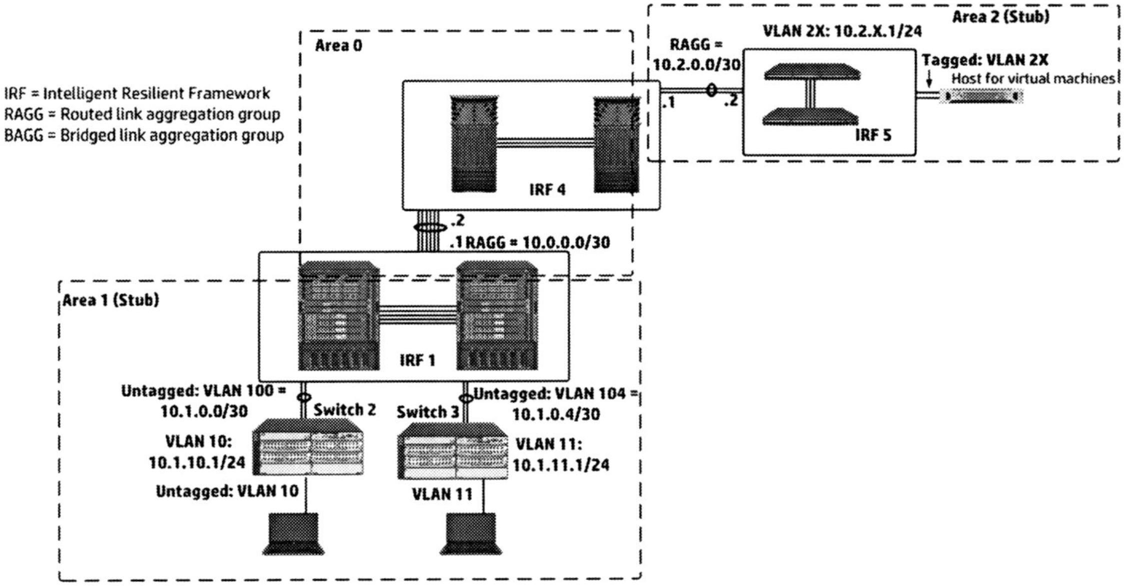

Figure 12-9: Exhibit 1 for question 15

```
[IRF4-ospf-1]display this

 ospf 1
  area 0.0.0.0
    abr-summary 10.0.0.0 255.255.0.0 cost 1
    network 10.0.0.0 0.0.255.255
  area 0.0.0.2
    abr-summary 10.2.0.0 255.255.0.0 cost 1
    network 10.2.0.0 0.0.255.255
    stub
```

Figure 12-10: Exhibit 2 for question 15

In Exhibit 1, all infrastructure devices are implementing OSPF on the interfaces. Exhibit 1 also shows settings for OSPF areas. Exhibit 2 shows some additional OSPF settings on IRF 4.

The network administrator enters this command on IRF 1:

```
[IRF1-ospf-1-area-0.0.0.1] abr-summary 10.1.0.0 16
```

The administrator verifies that the solution is functioning correctly. Indicate whether the Link State DataBase (LSDB) on a device should include an LSA for 10.1.0.0/16.

a. Includes the LSA	— Switch 2 area 1
b. Does not include the LSA	— IRF 5 area 2
	— IRF 4 area 0

16. A network using PIM has multiple potential paths for multicasts. How does Protocol-Independent Multicast (PIM) determine the shortest path tree (SPT)?

    a. It uses interface PIM costs to choose the preferred upstream neighbor.

    b. It uses the PIM neighbor priorities to determine the preferred upstream neighbor.

    c. It uses the unicast routing table to determine the correct upstream interface for joins or multicast traffic.

    d. It uses Internet Group Management Protocol (IGMP) to determine the correct upstream interface for joins or multicast traffic.

17. Refer to the exhibit.

```
acl number 3000 match-order auto
 rule 0 permit tcp source 10.1.30.0 0.0.0.255 destination 10.2.10.0 destination-port 8080
 rule 5 deny any
```

Figure 12-11: Exhibit for question 17

A network administrator accesses the ACL 3000 configuration view. The administrator enters these two commands in this order:

```
rule permit ip source 10.1.30.0 0.0.0.255 destination 10.1.10.0
0.0.0.255

rule deny tcp source 10.1.30.0 0.0.0.255 destination-port eq telnet
```

Which correctly explains the effect on how the switch handles traffic matched to this ACL?

a. The switch does not change how it handles any traffic based on these new rules.

b. The switch now permits all traffic between endpoints in 10.1.30.0/24 and 10.1.10.0/24.

c. The switch now permits only Telnet traffic between endpoints in 10.1.30.0/24 and 10.1.10.0/24.

d. The switch now permits all traffic between endpoints in 10.1.30.0/24 and 10.1.10.0/24 except Telnet traffic.

18. Match each characteristic to the correct authentication mechanism. If both mechanisms have that characteristic, you must select both.

a. Web-Auth or portal authentication (applied to the physical interface)	— Allows the application of dynamic VLAN assignment
b. 802.1X (PEAP)	— Requires a supplicant on the client endpoint
c. Both	— Often works with an external RADIUS server

19. Refer to the exhibit.

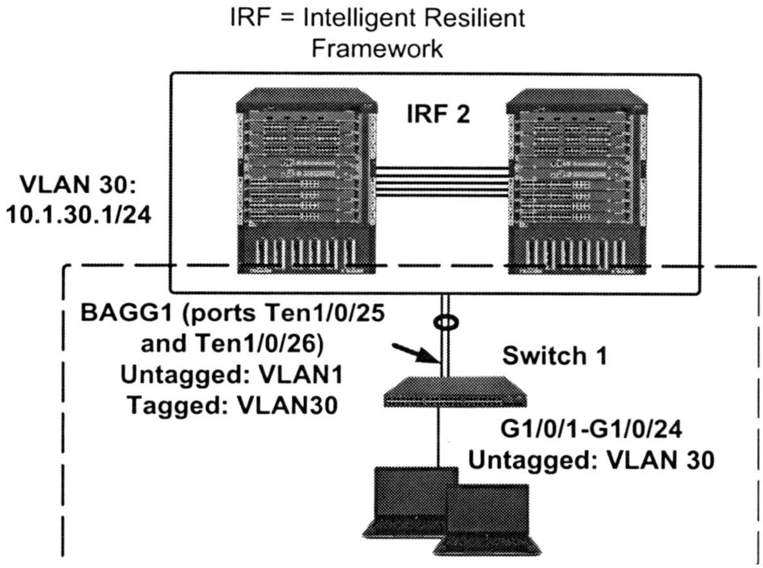

Figure 12-12: Exhibit for question 19

A network administrator is configuring a Comware switch.

The switch supports endpoints in VLAN 30. Most of the endpoint traffic should be best effort, but the switch should select some traffic and give that traffic a priority value.

The administrator has configured a QoS policy for this purpose. Where should the administrator apply the policy?

a. to Switch 1 interfaces g1/0/1 to g1/0/24 as an inbound policy

b. to Switch 1 interfaces g1/0/1 to g1/0/24 as an outbound policy

c. to Switch 1 interfaces ten1/0/25 and ten1/0/26 as an inbound policy

d. to Switch 1 VLAN 30 (Layer 2 interface) as an outbound policy

20. An interface on a Comware switch is configured to use Open Shortest Path First (OSPF) Bidirectional Forwarding Detection (BFD) in control mode. The administrator sets a minimum transmit interval. What is the effect?

   a. The interface always sends BFD control packets at this rate.

   b. The interface sends BFD control packets at this rate *except when* the peer has a lower minimum transmit interval.

   c. The interface will send BFD control packets at this rate *except when* the peer has a higher minimum transmit interval.

   d. The interface sends BFD control packets at this rate *except when* the peer has a higher minimum receive interval.

21. Refer to the exhibit.

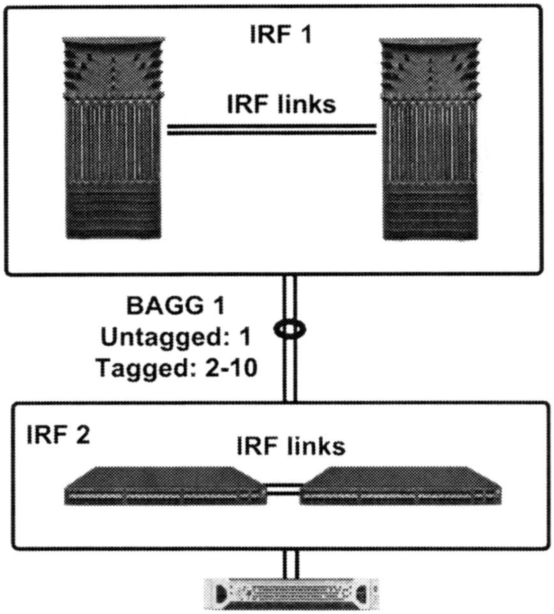

Figure 12-13: Exhibit for question 21

The network administrator wants to set up Bidirectional Forwarding Detection (BFD) Multi-Active Detection (MAD) on IRF 1. What is the recommended plan?

   a. Create a new link between the IRF 1 members; place that link in a new VLAN dedicated to BFD MAD.

   b. Create a new link between the IRF 1 members. Enable the link to carry VLANs 1–10 and make sure BFD MAD is enabled on all these VLANs.

c. Enable BFD MAD on a new VLAN. Add that VLAN to a link aggregation to a Comware switch.

d. Enable BFD MAD on VLANs 1–10, which are carried on the current link aggregations to other switches.

22. A network administrator is planning to apply a Generic Traffic Shaping (GTS) rate limit to priority queue 2 on an interface. What makes a GTS rate limit different from a simple rate limit?

a. The GTS rate limit can only be applied to inbound traffic.

b. The GTS rate limit only affects traffic that has a drop precedence of 1 or 2.

c. The interface buffers excess traffic and sends it at the rate limit.

d. The interface applies the rate limit to traffic on a session-by-session basis (each session receives that rate).

23. Refer to the exhibits.

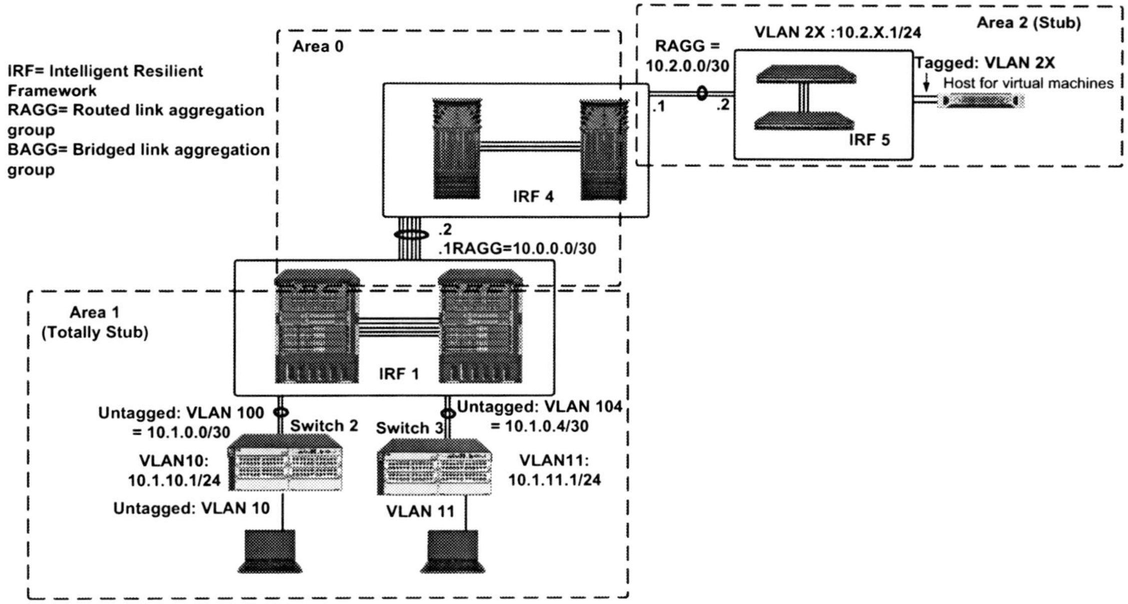

Figure 12-14: Exhibit 1 for question 23

```
[IRF1-ospf-1]display this

 ospf 1
  area 0.0.0.0
    abr-summary 10.0.0.0 255.255.0.0 cost 1
    network 10.0.0.0 0.0.255.255
  area 0.0.0.1
    abr-summary 10.1.0.0 255.255.0.0 cost 1
    network 10.1.0.0 0.0.255.255
    stub no-summary


 [IRF4-ospf-1]display this

 ospf 1
  area 0.0.0.0
    abr-summary 10.0.0.0 255.255.0.0 cost 1
    network 10.0.0.0 0.0.255.255
  area 0.0.0.2
    abr-summary 10.2.0.0 255.255.0.0 cost 1
    network 10.2.0.0 0.0.255.255
    stub
```

Figure 12-15 Exhibit 2 for question 23

The network administrator connects IRF 4 to an external router, configures static routes, and redistributes static routes into OSPF. Which devices receive the external OSPF route?

a. IRF 1 only

b. IRF 5 only

c. IRF 1 and IRF 5 only

d. IRF 1, IRF 5, Switch 2, and Switch 3

24. Refer to the exhibit.

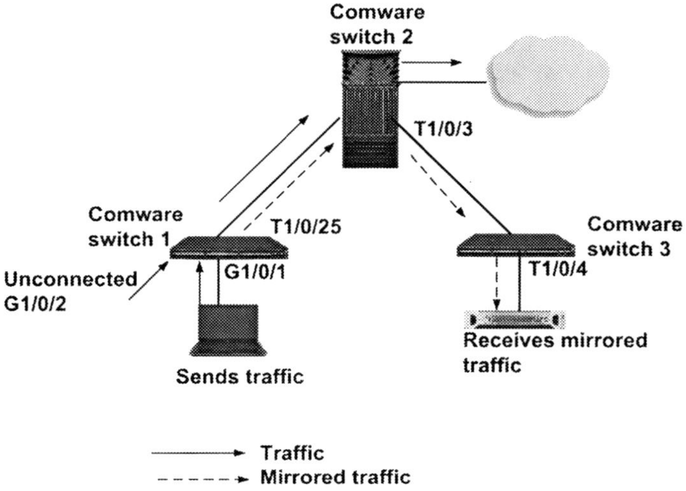

Figure 12-16 Exhibit for question 24

A network administrator is setting up remote mirroring on the Comware switches shown in the exhibit. The administrator must define several port roles while entering mirroring-group commands on these switches. Which port should be the monitor port?

a. Switch 3, T1/0/4

b. Switch 1, G1/0/1

c. Switch 1, G1/0/2

d. Switch 1, T1/0/25

25. What is a valid use case for an extended or advanced ACL?

a. Selecting traffic based on its source MAC address

b. Selecting traffic based on its destination IP address and TCP port

c. Selecting traffic based on its throughput

d. Selecting traffic based on the payload content

26. A network administrator is configuring Weighted Round Robin (WRR) queuing on a Comware interface. The administrator has the choice to set WRR to weight-based or to byte-based mode. What is the difference between these modes?

a. Weight-based mode always uses predefined weights for each queue; byte-based mode lets the administrator configure the weight for each queue.

b. Weight-based mode allocates bandwidth according to queue weights; byte-based mode allocates bandwidth according to queue guaranteed minimum bandwidths.

c. Weight-based mode lets higher weight queues transmit relatively more packets; byte-based mode lets higher weight queues transmit relatively more bytes.

d. Weight-based mode reserves the bandwidth allocated to a queue whether or not the queue is using it; byte-based mode lets other queues use more bandwidth if a queue does not need its share.

27. A medium campus LAN core consists of two HP ProVision switches. These switches implement Distributed Trunking (DT) on links to downstream switches. They also run Virtual Router Redundancy Protocol (VRRP) on the VLANs carried on those distribute trunks. How do these features interact?

  a. Traffic that needs routing might arrive on a DT link on the Backup VRRP router. The Backup router then routes the traffic.

  b. Traffic that needs routing might arrive on a DT link on the Backup VRRP router. The Backup router then sends the traffic over the ISC to the Master.

  c. Traffic that needs routing always arrives on a DT link on the Master VRRP router due to how downstream routers learn the virtual router MAC address.

  d. Traffic that needs routing can be directed to a DT link on the master VRRP router; the administrator must simply enable MSTP on the distributed trunks.

28. Both Switch 1 and Switch 2 run IGMP and PIM on VLAN 30 and on a connection to the data center core router. Both switches could forward multicasts from a server in the data center to endpoints in VLAN 30. What determines which switch does forward the multicasts?

  a. The switch that is VRRP master for the VLAN forwards them.

  b. The switch that is the active IGMP querier for the VLAN forwards them.

  c. The switch that has the lowest path cost to the source forwards them.

  d. The switch that has the lower PIM BSR priority forwards them.

29. An administrator wants to upgrade an IRF virtual device using the In-Service Software Upgrade (ISSU) process. During this upgrade, the administrator does not want to reboot any switches.

  The members of the IRF virtual device each have two MPUs, and all switches are running software that is compatible with the new software version.

  What is the correct ISSU process?

  a. Upgrade the standby MPU on the master, perform an active to standby switchover on the master, and then upgrade all other MPUs one at a time.

b. Upgrade the active MPU on the master, perform an active to standby switchover on the master, and then upgrade all other MPUs one at a time.

c. Upgrade the active MPU on the master and then upgrade all other MPUs one at a time.

d. Upgrade a slave, perform an active to standby switchover to make the upgraded member the new master, and then upgrade the former master.

30. Refer to the exhibit.

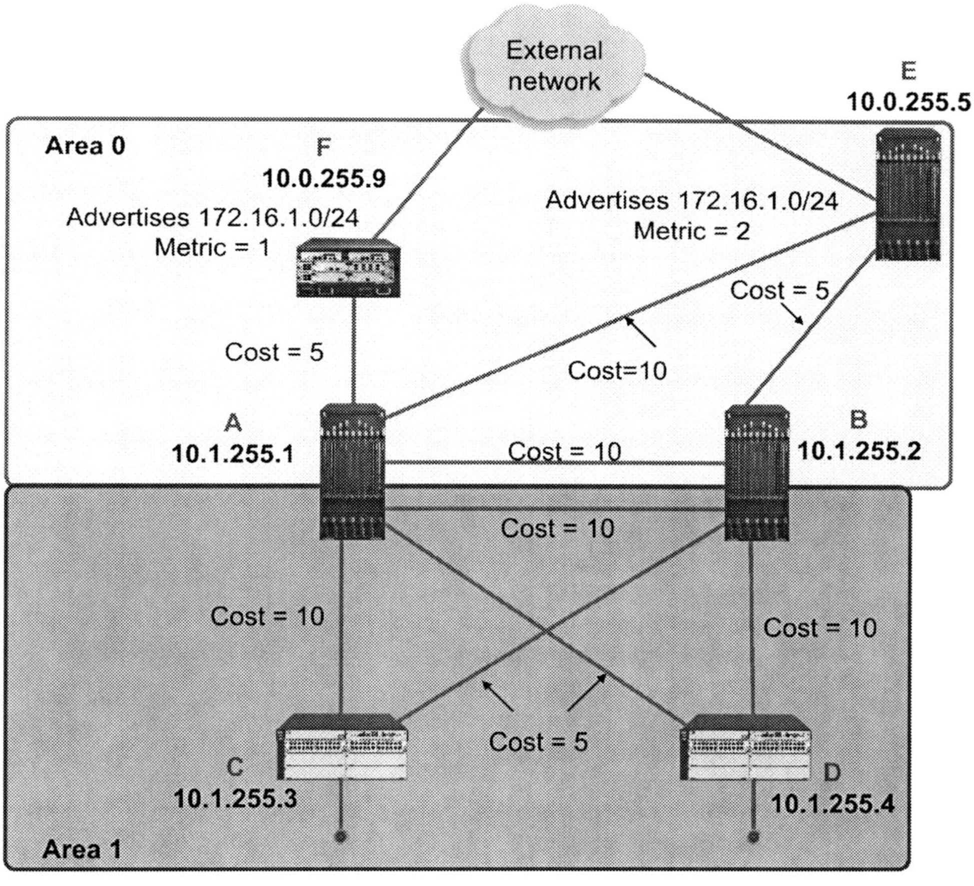

Figure 12-17: Exhibit for question 30

Device C receives two Type 5 LSAs for 172.16.1.0/24, one advertised by Device E and one by Device F. Both E and F advertise a type 2 metric. What does Device C do?

a. It selects the LSA from Device E and learns a route to 172.16.1.0/24 through Device A.

b. It selects the LSA from Device E and learns a route to 172.16.1.0/24 through Device B.

c. It selects the LSA from Device F and learns a route to 172.16.1.0/24 through Device A.

d. It selects the LSA from Device F and learns a route to 172.16.1.0/24 through Device B.

31. Multiple physical links are bound to a virtual IRF port. What is a possible factor that the switch uses to load share traffic across the links?

a. Manual assignment of sessions through traffic engineering policies

b. Source and destination MAC addresses

c. Current traffic load

d. 802.1p value

32. Refer to the exhibit.

```
#Partial config
interface GigabitEthernet1/0/1
 port link-type trunk
 undo port trunk permit vlan 1
 port trunk permit vlan 10 20 30
 qos apply policy MyPolicy inbound
 qos priority 3
 qos trust dot1p
acl number 3000
 rule 0 permit udp destination-port 8000
traffic classifier MyClass
 if-match 3000
traffic behavior MyBehavior
 remark dot1p 6
qos policy MyPolicy
 classifier MyClass behavior MyBehavior


#command output

<Switch> display qos map-table dot1p-lp
MAP-TABLE NAME: dot1p-lp TYPE: pre-define
IMPORT : EXPORT
0 : 2
1 : 0
2 : 1
3 : 3
4 : 4
5 : 5
6 : 6
7 : 7
```

Figure 12-18: Exhibit for question 32

This traffic arrives on interface g1/0/1:

- Includes 802.1Q tag
- 802.1p = 5
- DSCP = 0
- Destination UDP port = 8000

When the switch forwards the traffic, in which queue does it place the traffic?

a. 2

b. 3

c. 5

d. 6

33. Refer to the exhibit.

```
#Partial config
interface GigabitEthernet1/0/1
 port link-type trunk
 undo port trunk permit vlan 1
 port trunk permit vlan 10 20 30
 qos apply policy MyPolicy inbound
 qos priority 3
 qos trust dot1p
acl number 3000
 rule 0 permit udp destination-port 8000
traffic classifier MyClass
 if-match 3000
traffic behavior MyBehavior
 remark dot1p 6
qos policy MyPolicy
 classifier MyClass behavior MyBehavior


#command output

<Switch> display qos map-table dot1p-lp
MAP-TABLE NAME: dot1p-lp TYPE: pre-define
IMPORT : EXPORT
0 : 2
1 : 0
2 : 1
3 : 3
4 : 4
5 : 5
6 : 6
7 : 7
```

Figure 12-19 Exhibit for question 33

This traffic arrives on interface g1/0/1:

- Includes 802.1Q tag
- 802.1p = 5
- DSCP = 0
- Destination UDP port = 8000

When the switch forwards the traffic, what DSCP is in the IP header?

a. 0

b. 24

c. 40

d. 48

34. A network administrator wants to create an ACL rule that selects this IP traffic:

- From devices with IP addresses in the 10.1.24.0–10.1.27.255 range
- To devices with IP addresses in the 10.1.10.16–10.1.10.31 range

The rule syntax is:

rule 5 permit ip source _____1____ destination ___2_____

What are the correct values for the blanks?

a. 1 = 10.1.24.0 255.255.252.0
   2 = 10.1.10.16 255.255.255.248

b. 1 = 10.1.24.0 0.0.3.255
   2 = 10.1.10.16 0.0.0.15

c. 1 = 10.1.24.0 255.255.255.192
   2 = 10.1.10.16 255.255.255.240

d. 1 = 10.1.24.0 0.0.27.255
   2 = 10.1.10.16 0.0.0.16

35. Refer to the exhibit.

```
acl number 3000
 rule 0 permit source 10.1.30.0 0.0.0.255
traffic classifier 2000
 if-match 3000
traffic behavior police2000
 car 1 cir 1000000
qos policy police
 classifier 2000 behavior police2000
```

Figure 12-20: Exhibit for question 35

Currently, this policy drops traffic from 10.1.30.0/24 if that traffic exceeds 1 Gbps. The administrator wants to change this policy so that:

- When this traffic exceeds 1 Gbps, the 802.1p value is set to less than best effort.
- When this traffic exceeds 2 Gbps, it is dropped.

The administrator needs to reconfigure the CAR action. What are two changes for the new action? (select two).

a. It includes a PIR of 2 Gbps (the CIR remains the same).

b. It includes a CIR of 2 Gbps and a PIR of 1 Gbps.

c. It includes a CIR of 1 Gbps and a PIR of 3 Gbps.

d. It specifies that yellow traffic is remarked with 802.1p value to 1.

e. It specifies that yellow traffic is remarked with 802.1p value to 0.

36. Refer to the exhibit.

```
vlan 999
  untagged 1-20
aaa port-access authenticator 1-20
```

Figure 12-21: Exhibit for question 35

Employees who authenticate on these interfaces will receive dynamic VLAN assignments. The network administrator wants the interfaces to permit guest access in VLAN 999. The guests should not have to log in, but automatically receive access.

Either a guest or an employee might connect to any interface, but only one device will connect at a time.

What should the administrator do to make the configuration work?

a. Set the 802.1X client limit on the interface to 1.

b. Enable web-based authentication on the port.

c. Set the Unauthenticated VLAN to 999.

d. Add an Authenticated VLAN ID, which makes the untagged ID apply to unauthenticated users.

37. A network has HP switches. The company wants to implement 802.1X authentication with Protected Extensible Authentication Protocol (PEAP). HP User Access Manager (UAM) will act as the RADIUS server. The root certificate of the Certificate Authority (CA) has been installed on UAM and the clients.

What are the server and client certificate requirements for this solution?

a. PEAP does not require certificates on any of the devices.

b. PEAP requires a server certificate on UAM only.

c. PEAP requires a server certificate on UAM and client certificates on the supplicants.

d. PEAP requires server certificates on UAM and the HP switches.

38. Refer to the exhibit.

```
acl number 3000
 rule 0 permit tcp source 10.1.30.0 0.0.0.255 destination 10.2.10.0 destination-port 8080
 rule 5 deny any
```

Figure 12-22: Exhibit for question 18

A network administrator accesses the ACL 3000 configuration view. The administrator enters these two commands in this order:

```
rule permit ip source 10.1.30.0 0.0.0.255 destination 10.1.10.0
0.0.0.255

rule deny tcp source 10.1.30.0 0.0.0.255 destination-port eq telnet
```

Which correctly explains the effect on how the switch handles traffic matched to this ACL?

a. The switch does not change how it handles any traffic based on these new rules.

b. The switch now permits all traffic between endpoints in 10.1.30.0/24 and 10.1.10.0/24.

c. The switch now permits only Telnet traffic between endpoints in 10.1.30.0/24 and 10.1.10.0/24.

d. The switch now permits all traffic between endpoints in 10.1.30.0/24 and 10.1.10.0/24 except Telnet traffic.

39. Two Comware switches connect to each other. The network administrator wants to configure the switches so that they detect if the link is unidirectional but do not shut down the port if they detect the issue. How should the network administrator configure the switch ports?

a. Enable DLDP and configure it to use manual mode on the ports in both switches.

b. Enable DLDP on the port in one switch and configure it to use manual mode.

c. Enable DLDP on the ports in both switches; configure UDLD to use manual mode on one switch and auto on the other switch.

d. Enable DLDP and configure it to use auto mode on the ports in both switches.

40. View the SNMP settings in the exhibit. The Authentication and Encryption passwords are the same: "password"

    How should SNMP be configured on the Comware switches to support these security settings on HP IMC?

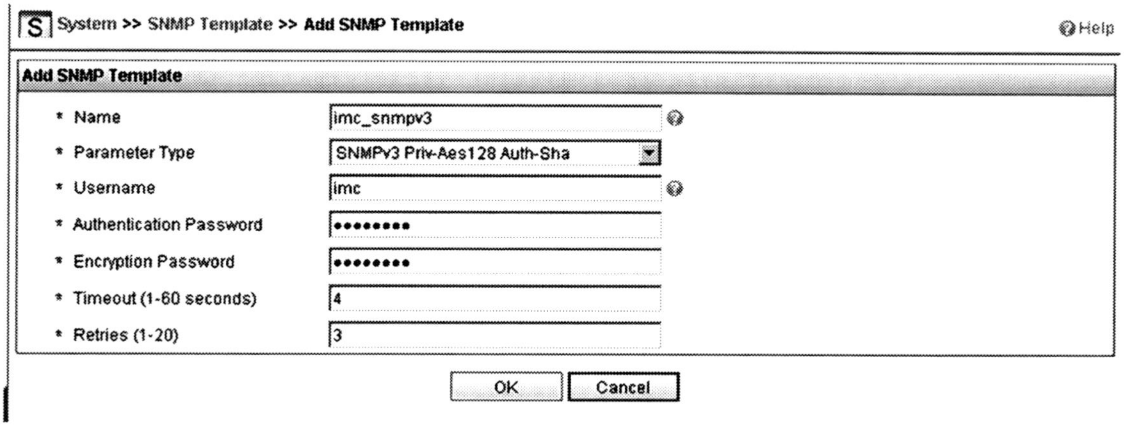

Figure 12-23: Exhibit for question 40

   a. snmpv3 user imc auth sha password priv aes password
   b. snmpv3 user imc auth sha password priv des password
   c. snmpv3 user imc_snmpv3 auth sha password priv aes password
   d. snmpv3 user imc_snmpv3 auth sha password priv aes password

41. An agent samples a percentage of traffic and sends their headers to a collector. The collector can use a statistically accurate algorithm to create a picture of network utilization based on the samples. Which technology has been described?

   a. NetConf
   b. NetFlow
   c. RMON
   d. sFlow

42. A switch is configured to use simple authentication for Open Shortest Path First (OSPF) on an interface. How does this setting affect OSPF packets sent by the switch interface?

   a. The switch uses MD5 to encrypt the key; it includes the encrypted hash in the OSPF header of the first packet it sends to a neighboring switch.
   b. The switch includes the key as plaintext in every OSPF packet's OSPF header.

c. The switch includes the encrypted key in every OSPF packet it sends to neighboring switches.

d. The switch includes an authentication field in the header of each OSPF packet, but that field is empty because the switch does not enforce authentication.

43. Two switches are configured to use Virtual Router Redundancy Protocol (VRRP). One router is configured as the master and one as the backup. Match the responsibilities to the VRRP role. You can use the same answer for than once.

a. Backup router

b. Master router

c. Both Backup and Master routers

________Listen for virtual router advertisements.

________Process and route frames sent to the virtual MAC address.

________Respond to ARP requests sent to the virtual MAC address.

________Send virtual router advertisement.

44. Administrators are implementing an 802.1X authentication. Endpoints connect to Comware switches. Supplicants on the endpoints will authenticate with Protected Extensible Authentication Protocol (PEAP) to UAM. Besides the supplicants, where must administrators select PEAP for the authentication method?

a. On the switch interfaces that connect to endpoints

b. In the RADIUS schemes configured on the switches

c. In an access policy for a service assigned to users in UAM

d. In the global UAM settings

45. Which role does the management plane in a switch play?

a. It queues packets and schedules them for delivery.

b. It makes changes to the switch configuration.

c. It programs MAC forwarding tables.

d. It builds routing tables and determines the correct next hop for routed traffic.

46. A routing switch supports Protocol-Independent Multicasting Sparse Mode (PIM-SM). It is configured with a static RP address, 10.1.1.1. A Bootstrap Router (BSR) also sends this RP set:

- 10.1.1.2 supports 224.0.0.0/4 with priority 50.
- 10.1.1.3 supports 224.0.0.0/4 with priority 100.

What happens when the switch needs to select an RP for multicast group 239.1.1.1?

a. It selects 10.1.1.1.

b. It selects 10.1.1.2.

c. It selects 10.1.1.3.

d. It cannot select an RP.

47. A network administrator is troubleshooting an 802.1X configuration. The administrator wants to capture packets related to the clients' Extensible Access Protocol (EAP) authentication process that are sent from the switch to the RADIUS server. What type of packets should the administrator capture?

a. EAP messages

b. RADIUS messages

c. Transport Layer Security (TLS) messages

d. Secure Socket Layer (SSL) messages

48. What is a possible disadvantage of Strict Prioritization (SP) as a queue scheduling mechanism?

a. Higher-priority queues receive the same weighting as lower-priority queues.

b. Switch ports do not honor existing QoS settings in the packets.

c. Most traffic is assigned to lower-priority queues, so the low-priority traffic can starve out high-priority traffic.

d. Lower-priority queues might be starved out.

49. Which is requirement for Link Aggregation Control Protocol (LACP) MAD?

a. The IRF switches must be connected a switch that supports extended LACP Data Units (LACPDUs).

b. The IRF switches must be connected to a switch that supports LACP in dynamic mode.

c. The IRF switches must be connected with a link that is assigned to a VLAN dedicated to MAD.

d. The IRF switches must be connected a switch that supports LACP in static mode.

50. Which statement describes PVST+?

a. It creates one spanning tree for each VLAN that runs PVST+.

b. It creates one spanning tree for each instance.

c. It is not completely VLAN aware, requiring you to make all switch-to-switch links members of VLANs that support PVST+.

d. It requires fewer BPDUs than MSTP because it is VLAN aware.

Answers

1. **B** is correct. IGMP snooping allows a switch without Layer 3 functions on a VLAN to listen for IGMP packets in that VLAN. The switch creates a table of ports that require specific multicasts based on the IGMP reports that it hears on those ports. It then filters the multicasts on other ports (which have not sent reports).

   A, C, and **D** are incorrect. A is incorrect. On Comware switches, IGMP snooping is specifically a Layer 2 function. It does not allow the switch to act as an IGMP querier.

   C is incorrect. Without IGMP snooping enabled, the Layer 2 switch can still forward multicasts. In fact, it floods the multicasts on all ports in the VLAN. In other words, without IGMP snooping, when one endpoint on a Layer 2 switch joins a multicast group, all endpoints in that VLAN and on that switch receive the multicast.

   D is incorrect. IGMP snooping does not enforce multicast boundaries on routed interfaces. This feature has nothing to do with forwarding (or filtering) multicasts across routed boundaries, which would require a multicast routing solution.

2. **B** is correct. PIM-SM only forwards multicasts toward routers that have explicitly asked for them, so an RP is required to receive the joins. PIM-DM forwards multicasts toward all routers unless they prune themselves, so it does not use an RP.

   A, **C,** and **D** are incorrect. A is incorrect; PIM-SM is usually better suited for environments with limited bandwidth because it floods fewer multicasts.

   C is incorrect because both PIM-SM and PIM-DM use IGMP to discover which interfaces require multicasts.

   D is incorrect because both PIM-SM and PIM-DM can use any type of unicast routes to determine the correct upstream interfaces.

3. **B** is correct. You can create a full mesh topology for a backplane of HP 3800 Series switches if the stack consists of five or fewer members.

   A, **C**, and **D** are incorrect. A is incorrect because the HP 2920 Series switches only support a ring (or chain) topology for the stacking links.

   C is incorrect because the HP 3800 Series switches only support a ring (or chain) topology if the stack consists of six to ten members.

   D is incorrect because IRF virtual switches only support ring (or chain) topologies for the IRF links.

4. **A** is correct. Spanning tree BPDUs is always filtered on distributed trunks.

   B, **C**, and **D** are incorrect. B is incorrect; because the BPDUs are filtered, the switch does not process them.

   C is incorrect because the switch does not forward the BPDUs either.

   D is incorrect. Distributed trunks automatically filter BPUDs but that is all.

5. Shown below are the correct use cases for the three protocols:

a. Preventing issues from partially broken or cross-connected fiber cables	b Distributed trunking
b. Enabling two switches to establish a link aggregation to a downstream device	a Unidirectional Link Detection (UDLD)
c. Detecting Layer 3 connectivity between the device and a neighbor	c Bidirectional Forwarding Detection

6. **B** and **D** are correct. **B** is correct because you must configure a preempt delay on Switch 1, which is the switch that will preempt the role. **D** is correct because a preempt delay configured on Switch *does not take effect when Switch 1 is the owner*. You must change the actual IP address on Switch 1 so that it is the Master, but not the owner.

   A, **C**, and **E** are incorrect. A is incorrect because you do not need to enable VRRP preempt delays globally.

   C is incorrect because you do not *have* to enable the preempt delay on Switch 2 for this particular scenario. As long as the preempt delay is configured on Switch 1, Switch 1 will delay taking over as master after failing. (You might want to enable the preempt delay for consistency and as a best practice).

   E is incorrect; it is not valid to configure the same actual IP address on both switches.

7. **B** is correct. A CLOS fabric features a three-tier fabric, and each tier includes multiple crossbar switches.

   A, **C**, and **D** are incorrect. A is incorrect because a CLOS fabric does not have an arbiter.

   C is incorrect. The CLOS fabric is a hardware architecture, and virtual switches are not involved.

   D is incorrect. A CLOS fabric uses credit-based mechanisms for ensuring lossless delivery.

8. **C** is correct. You must enable the voice VLAN with the correct ID to enable the interface to advertise the voice VLAN with CDP.

   A, **B**, and **D** are incorrect. A is incorrect because you do not set the voice settings on the VLAN on Comware switches; you set them on the physical interface. (This is another step that should be completed.)

   B is incorrect. This scenario does not require MAC-based VLANs because the physical interface only supports one untagged VLAN.

   D is incorrect because you should not use the LLDP voice VLAN command when the switch is using CDP to advertise the VLAN ID and priorities.

9. **A** is correct. If you want to use a local-user account to authenticate a device with MAC-Auth, the account must have the lan-access service type.

   B, **C**, and **D** are incorrect. B is incorrect because the output for the **display mac-authentication** command indicates that the switch is not using the fixed username setting. It uses the MAC address for the username.

   C is incorrect. The output for the display mac-authentication command indicates that the switch is using all lowercase and no delimiters for the MAC-address format. The username is correctly configured with this format already.

   D is incorrect. Accounts used for MAC-Auth do require passwords. When the switch is not using the fixed password setting, which it is not in this scenario, the MAC address is the password.

10. **C** is correct. When one DT switch receives a frame on a distributed trunk, it sends a MAC Learn to the other switch, which then learns the distributed trunk as the forwarding interface for the frame's source MAC address. Therefore, when Switch 1 receives the frame for Endpoint 1, it knows to forward the frame on its own link in the distributed trunk.

    A, **B**, and **D** are incorrect. A and B are incorrect because the switch can forward the frame and has no need to drop or buffer it. D is incorrect because the frame is forwarded on the distributed trunk, not the ISC.

11. **D** is correct. The switch forwards Packet 1 because this packet is not routed. Therefore, the ACL applied as an inbound routed ACL on the VLAN 30 interface does not apply to it. The switch drops Packet 2. This packet is received on VLAN 30 and routed to another VLAN, so the ACL applies to it. The packet does not match rule 0 because the destination IP address does not match. It does not match rule 5 because the TCP port does not match. It therefore matches the **deny any** rule.

    A, **B**, and **C** are incorrect. A is incorrect because the switch forwards Packet 1. (The ACL does not apply to the packet because the packet is not routed.)

    B is incorrect because the switch drops Packet 2. (The packet does not match the permit rule 5 because the TCP port does not match.)

    C is incorrect because the switch forwards Packet 1 and drops Packet 2. (See the explanations above.)

12. **C** is correct. Switch 4 receives the RP set from Switch 1, which is elected as the bootstrap router (BSR). The RP set includes Switch 2 and Switch 3. Switch 2 has the better priority for being RP in general. However, Switch 3 advertises itself specifically for 239.1.1.1. The more specific match always takes precedence, so Switch 4 selects Switch 3 as the RP.

    A, **B**, and **D** are incorrect. A is incorrect because Switch 1 does not advertise itself as an RP.

    B is incorrect because Switch 3, not Switch 2, is selected as RP for 239.1.1.1. Although Switch 2 has a better priority for being RP in general, Switch 3 matches this multicast group more specifically.

D is incorrect. As long as Switch 4 has PIM enabled on the correct interfaces, it does not require other configuration to learn the RP. It receives the RP set from Switch 1, which is the BSR.

13. **D** is correct. The company router is part of one Autonomous System (AS), and the ISP router is part of another. An External Gateway Protocol (EGP) such as BGP is typically used for a situation like this. eBGP is used because the routers are in different ASes.

    A, **B**, and **C** are incorrect. A is incorrect because you do not typically run OSPF between a company router and an ISP.

    B is incorrect for the same reason as A. In addition, you cannot run OSPF in different areas on two sides of the same link.

    C is incorrect. Although BGP is the correct protocol for this situation, the company router and ISP router are in different ASes. Therefore, eBGP, rather than iBGP, is used.

14. **C** is correct. BGP is different from protocols such as OSPF. You must manually configure a peer in order for your router to establish the TCP session with that peer.

    A, **B**, and **D** are incorrect. A and B are incorrect because you do not enable BGP on interfaces. You must manually configure the peer.

    D is incorrect. It does not correctly describe how to use route maps. In any case, you must manually configure the peer.

15. The correct answers are shown below:

a. Includes the LSA	b	Switch 2 area 1
b. Does not include the LSA	a	IRF 5 area 2
	a	IRF 4 area 0

    The area 1 LSDB on Switch 2 does not include the LSA. IRF 1 advertises the area 1 summary LSA in *other* areas (such as area 0), not in area 1.

    The IRF 4 area 0 LSDB and IRF 5 area 2 LSDB both include the LSA. IRF 4 received the LSA in area 0 from IRF1. IRF 4 then advertised the LSA in area 2, and IRF 5 received the LSA.

16. **C** is correct. PIM relies on the unicast routing table for path selection.

    A, **B**, and **D** are incorrect. A and B are incorrect because PIM does not define interface costs and neighbor priorities.

    D is incorrect. IGMP helps the router to choose the downstream interfaces for multicasts.

17. **D** is correct. The administrator did not enter a rule ID for these rules, so they are added with higher IDs than the rule with the highest current ID. The exhibit shows that this ACL is using the *auto* match order. The switch processes the second new rule first because it is most specific (matches a specific protocol and a specific port). This rule denies Telnet traffic.

The switch processes the second new rule second because it matches a more specific source address than the existing rules. This rule permits all other traffic between endpoints in 10.1.30.0/24 and 10.1.10.0/24.

A, **B**, and **C** are incorrect. A is incorrect. A would be correct if the ACL used the config match order. In that case, these rules would be processed last after the deny any rule and would not change the behavior of the ACL. But the match order is auto.

B is incorrect because the second rule entered is processed first. Therefore, Telnet traffic is denied.

C is incorrect. The second rule is a deny rule.

18. Shown below are the correct mechanisms that have the characteristics:

a.	Web-Auth or portal authentication (applied to the physical interface)	c	Allows the application of dynamic VLAN assignment
b.	802.1X	b	Requires a supplicant on the client endpoint
c.	Both	c	Often works with an external RADIUS server

19. **A** is correct. The QoS policy needs to mark incoming traffic on the interfaces that connect to endpoints.

    B, **C**, and **D** are incorrect. B is incorrect because an outbound policy on the endpoint interfaces would not affect traffic received by the endpoints.

    C is incorrect because an inbound policy on the uplinks would not affect traffic forwarded on the uplinks.

    D is incorrect. You can apply QoS policies to VLANs on ProVision switches, but you cannot apply QoS policies to VLANs (Layer 2) on Comware switches.

20. **D** is correct. BFD peers do not have to have the same *configured* minimum transmit and receive intervals, but they use the configured intervals to negotiate the actual intervals that they will use. A BFD interface transmits at the *higher* interval out of these two configured intervals:

    - Its own minimum transmit interval
    - The peer's minimum receive interval

    Therefore, **D** is correct. The interface transmits at the configured minimum transmit interval *except when* the peer has a higher minimum receive interface.

    A, **B**, and **C** are incorrect. A is incorrect because the configured transmit interval does not always match the actual interval due to the negotiation with the peer.

    B and C are incorrect because the peer transmit interval does not affect the interface's transmit interval. The peers can transmit at different intervals because each peer knows when to expect the other peer's transmissions. What matters is that this peer transmits at a rate that the other peer can receive.

21. **A** is correct. A dedicated link is recommended for BFD MAD. In addition, the VLAN on which BFD MAD is enabled must be dedicated to this function. Otherwise, issues may occur. (For example, spanning tree BPDUs are filtered on links that carry the BFD MAD VLAN, which could cause loops.)

    B, **C**, and **D** are incorrect. The BFD MAD link should not carry production VLANs, and those VLANs should not have BFD MAD enabled on them. This setup could cause a loop.

    C is incorrect. It is correct to create a dedicated BFD MAD VLAN. However, a dedicated physical link is typically recommended as well, and the item asked for the recommended setup.

    D is incorrect. Again, BFD MAD should use a dedicated VLAN.

22. **C** is correct. Instead of dropping traffic that exceeds the set rate, GTS buffers the traffic and then sends it at the rate limit.

    A, **B**, and **D** are incorrect. A is incorrect because GTS is always applied to outbound traffic on an interface.

    B is incorrect because GTS affects all traffic regardless of the drop precedence.

    D is incorrect because the configured rate limit applies to all of the traffic not to each session separately.

23. **A** is correct. OSPF routers in area 0 and normal areas receive the external route. IRF 1 is in area 0.

    B, **C**, and **D** are incorrect:

    - IRF 5 is in a stub area, so it does not receive external routes.
    - Switch 2 and Switch 3 are in a totally stub area, so they do not receive external routes (or inter-area routes).

24. **A** correct. The monitor port is defined in the destination switch's mirroring group. The mirroring traffic is transmitted out of this port. The exhibit indicates that the desired destination for the mirrored traffic is connected to T1/0/4 on Switch 3, so this is the correct monitor port.

    B, **C**, and **D** are incorrect:

    - G1/0/1 on Switch 1 is a mirror port, which receives the traffic that should be mirrored.
    - G1/0/2 on Switch 1 is a valid reflector port (or monitor egress port).
    - T1/0/25 on Switch 1 is not given a formal definition, but it carries the remote probe VLAN.

25. **B** is correct. An extended or advanced ACL selects IP traffic based on a number of criteria including:

    - Source and/or destination IP addresses
    - IP protocol type (such as GRE, ICMP, ESP, and so on)

- Source and/or destination TCP/UDP port
- TCP flag
- ICMP or IGMP type
- Fragmented packet
- VPN instance
- QoS mark (IP precedence, Differentiated Services Code Point [DSCP], or Type of Service [ToS] value)

A, **C**, and **D** are incorrect. An extended or advanced ACL cannot select traffic based on any of these criteria.

26. **C** is correct. In both weight-based mode and byte-based mode, each queue is assigned a weight. Each queue's weight relative to the total weight of all queues determines how much traffic the queue can transmit. In weight-based mode, scheduling is based on packets. For example, a queue with weight 10 transmits twice as many packets as a queue with weight 5. Byte-based mode is similar, but scheduling is based on bytes. A queue with weight 10 transmits about twice as many bytes as a queue with weight 5.

    A, **B**, and **D** are incorrect. A is incorrect because administrators can configure queue weights in either mode.

    B is incorrect because both weight-based mode and byte-based mode use weights. WRR does not let you configure guaranteed minimum bandwidths in either mode. (You can figure out the bandwidth by using byte-based mode and using this formula: bandwidth = (queue weight/total weights) * interface bandwidth. But you cannot set the bandwidth in kbps.)

    D is incorrect. In both modes, if a queue does not need its share of the bandwidth, other queues can use that bandwidth.

27. **B** is correct. From the downstream switches' point of view, the distributed trunks are normal aggregated links, and they use their normal methods for assigning traffic to the links. Therefore, traffic that needs routing might arrive on the Backup VRRP router. The Backup VRRP router is not allowed to route the traffic. It switches the traffic to the Master, which is available across the ISC.

    A, **C**, and **D** are incorrect. A is incorrect because the Backup VRRP router is not allowed to route the traffic. DT does not change that.

    C is incorrect because, as explained above, the downstream switch sees the distributed trunk as a normal link aggregation. It learns the virtual router MAC address on the link aggregation, so it might traffic destined to this MAC address over any link in the aggregation.

    D is incorrect. MSTP cannot be enabled on distributed trunks. If MSTP is enabled on a DT switch, the distribute trunks filter MSTP BPDUs.

28. **C** is correct. When more than one PIM router exists on an interface (VLAN 30, in this question), the routers use PIM asserts to determine which router forwards the multicasts for the VLAN. They use the unicast path cost to the source to determine the forwarder. (If the switches have equal path costs, the switch with the higher IP address is elected.)

    A, **B**, and **D** are incorrect. A is incorrect because the VRRP role has no relationship with the IGMP and PIM functions.

    B is incorrect. As long as both switches run IGMP on VLAN 30, they both learn VLAN 30 as downstream interface in the shortest path tree (SPT) for these multicasts. Which router is the active querier does not affect which switch is elected PIM forwarder.

    D is incorrect. The PIM BSR priority is specific to PIM sparse mode (SM). This setting determines which switch is elected BSR and distributes the RP set, and it has no relationship to the multicast path in this case.

29. **A** is correct. The MPU-based ISSU process updates and reboots one MPU at a time rather than reboots an entire switch. You begin the process by upgrading the standby MPU on the master. You then switchover the master role to the standby MPU and upgrade other MPUs.

    B, **C**, and **D** are incorrect. B and C are incorrect because you should upgrade the standby MPU before the active one.

    D is incorrect because it describes a chassis-based ISSU process. The scenario in the question points you toward an MPU-based process.

30. **C** is correct. When a Type 5 LSA has a type 2 metric, the metric remains the metric originally advertised by the ASBR. Device C selects the LSA from Device F because F advertises the lower metric. The next hop for the external route is Device A because A is in the lowest cost path between C and F.

    A, **B**, and **D** are incorrect. A and B are incorrect because E and F advertise type *2* metrics. The metric does not increment as it crosses the AS, so it does not matter that Device C has a lower cost path to Device E than to F (10 vs. 15). Device F's LSA still has a lower metric (1 vs. 2 *not* 16 vs. 12).

    D is incorrect because Device B is not in the lowest cost path between C and F.

31. **B** is correct. A switch can take a hash of traffic's source and destination MAC address and use the hash to assign traffic to the link. (You can also configure switches to use other factors such as source and destination IP address.)

    A, **C**, and **D** are incorrect. A is incorrect because you cannot manually assign specific sessions to an IRF link.

    C is incorrect because the current load on a link does not affect whether the link is selected. (When a network carries traffic many sessions, it becomes likely that the load will be distributed fairly equally without the switch specifically taking the load into account.)

    D is incorrect. The 802.1p value is not used to load balance traffic over IRF physical links.

32. **D** is correct. When processed against the inbound QoS policy on the interface, this traffic is selected by the classifier. The behavior remarks the traffic's 802.1p value. This QoS policy takes precedence over the 802.1p value, which, when the qos trust setting is dot1p, takes precedence over the port priority.

    A, B, and **C** are incorrect. A would be correct if the QoS policy were not applied to the interface and if the qos trust setting were dscp. In that case, the DSCP 0 setting would be mapped to 802.1p value 0, which corresponds to queue 2. However, neither of these circumstances are true.

    B would be correct if the QoS policy were not applied to this interface and the undo qos trust setting were configured. In that case, the port priority (3) would apply. However, neither of these circumstances are true.

    C would be correct if the QoS policy were not applied to this interface. In that case, the incoming 802.1p value (5) would be applied.

33. **A** is correct. The incoming traffic has a DSCP of 0. The QoS policy places the traffic in queue 6. However, it does not change the DSCP. If you want to change the DSCP, you must include another remark action in the behavior.

    B, **C**, and **D** are incorrect. None of the settings shown in the exhibit change the incoming DSCP.

34. **B** is correct. ACL rules on Comware devices use wildcard bits, which use a reverse logic from subnet masks. The ranges in the rules correspond with valid CIDR subnets, so you can use this rule of thumb to calculate the wild card bits:

    Subtract each octet in the first address from the corresponding octet in the last octet.

    10.1.27.255 -

    10.1.24.0

    =

    0.0.3.255

    10.1.10.31 -

    10.1.10.16

    =

    0.0.0.15

    A, **C**, and **D** are incorrect. A is incorrect because it uses the subnet mask for these ranges rather than wildcard bits. C uses a subnet mask, which is incorrect, and D uses incorrect wildcard bits.

35. **A** and **D** are correct. Originally, the CAR action had only a CIR. In that case, traffic above the CIR is set to red (and dropped). But when you add a PIR to a CAR action, traffic over the CIR is set to yellow and traffic over the PIR is set to red (and dropped). To meet the requirements in the question, the CIR should stay the same and the PIR should be 2 Gbps.

The yellow action must also be configured to meet the requirements. 802.1p value 1 is the priority less than 0.

B, **C**, and **E** are incorrect. B is incorrect because it is not valid to set the PIR lower than the CIR.

C is incorrect because the PIR is wrong (it should be 2 Gbps, the rate over which excess traffic is dropped).

E is incorrect because it sets the yellow traffic (traffic between 1 Gbps and 2 Gbps) to 802.1p value 0, which is the best effort priority, not a lower than best effort priority.

36. **C** is correct. In the original configuration, clients that did not pass 802.1X authentication were blocked access entirely. (They were not placed in the untagged VLAN, VLAN 999.) To place the clients that do not pass authentication in VLAN 999, you must set the Unauthenticated VLAN to 999.

    A, **B**, and **D** are incorrect. A is incorrect. The client limit of 1 sets the port to user-based mode and limits the port to one user. This setting works for this scenario but does nothing to place the client in the guest VLAN.

    B is incorrect. Enabling web-based authentication would force the clients that do not pass 802.1X to log into a web page. The question indicated that clients should receive guest access without authenticating in any way.

    D is incorrect. The Authenticated VLAN ID applies to authenticated clients only and does nothing to place unauthenticated clients in a VLAN.

37. B is correct. Although PEAP does not require certificates on the supplicant, it does require the RADIUS server (UAM) to have a certificate.

    A, **C**, and **D** are incorrect. A is incorrect because a certificate is required on UAM.

    C is incorrect because the client supplicants do not require certificates (they would require certificates with EAP-TLS but not with PEAP).

    D is incorrect. The HP switches primarily encapsulate and decapsulate EAP messages sent by the supplicant and server, and they do not need certificates to play their role.

38. **A** is correct. The administrator did not enter a rule ID for these rules, so they are added with higher IDs than the rule with the highest current ID. The exhibit shows that this ACL is using the default match order, which is config. In an ACL that uses config match order, the rules are processed in order of ID. Therefore, the new rules are processed last, after the **deny any** rule, and do not change the behavior of the ACL.

    B, **C**, and **D** are incorrect. All of these answers indicate some change in the behavior, but the rules, as entered, do not change the ACL behavior.

39. **A** is correct. For the feature to work, both sides of the link must support DLDP. Manual mode configures the switch to detect the unidirectional link and log an error but not to shut down the port.

B, **C**, and **D** are incorrect. B is incorrect because both ports in the link must implement DLDP.

C and D are incorrect because auto mode, on one or both ports, causes the port to automatically shut down if it detects an error.

40. **A** is correct. The username, authentication algorithm, and privacy algorithm all match on the switch and on IMC.

    B, **C**, and **D** are incorrect. B is incorrect because the privacy algorithm on the switch does not match the algorithm in the template.

    C and D are incorrect. The username on the switch should match the username in the SNMP template settings not the name (which is the name of the template).

41. **D** is correct. The question describes sFlow, which is a traffic sampling algorithm.

    A, B, and **C** are incorrect:

    - NetConf is a protocol that is intended to help management services configure switches.
    - NetFlow is also a traffic sampling technology, but it has NetFlow agents that compile statistics about all traffic in a flow.
    - RMON is a technology that ties SNMP traps to port statistics.

42. **B** is correct. Simple authentication does not encrypt the key. It simply includes the plaintext key in the header.

    A, C, and **D** are incorrect. A and C are incorrect because they indicate that the key is encrypted.

    D is incorrect because the authentication field is not empty but instead includes the key.

43. Shown below are the correct roles for each VRRP function:

    a. Backup router

    b. Master router

    c. Both Backup router and Master router

    ____a___Listen for virtual router advertisements.

    ____b____Process and route frames sent to the virtual MAC address.

    ____b____Respond to ARP requests sent to the virtual MAC address.

    ____b____Send virtual router advertisement.

44. **C** is correct. You must match the EAP method on the supplicant and RADIUS server, which is UAM. An access policy in a service is the correct place to configure this setting on UAM.

    A, **B**, and **D** are incorrect. A and B are incorrect because you do not configure this setting on the switches, which are authenticators in this solution. They simply pass EAP messages between the supplicant and server.

    D is incorrect because you do not configure this setting globally on UAM.

45. **B** is correct. The management plane provides the interface between administrators and the switch configuration.

    A, **C**, and **D** are incorrect:

    - Queuing and scheduling traffic is a data plane function.
    - Building the MAC forwarding table and the routing table are control plane functions.

46. **A** is correct. The static RP setting takes precedence.

    B, **C**, and **D** are incorrect because the RP set sent by the BSR does not take effect when the switch has a static RP setting.

47. **B** is correct. The switch encapsulates the EAP messages in the RADIUS protocol before sending them to the RADIUS server.

    A, C, and **D** are incorrect because the 802.1X process uses EAP, but the EAP messages exchanged between the switch and the RADIUS server are encapsulated in RADIUS protocol.

48. **D** is correct because if SP is used as the queue scheduling mechanism, the switch will always send traffic from high-priority queues first. If these high-priority queues have a lot of traffic, the lower-priority queues can be starved.

    A, B, and **C** are incorrect. Higher-priority queues get priority handling, SP does not affect existing QoS settings, and lower-priority queues will not starve higher-priority queues.

49. **A** is correct because the IRF switches must be connected to a switch that supports extended LACPDUs, which will contain information about the IRF devices.

    B and **D** are incorrect because neither dynamic mode nor static mode is required for LACP MAD.

    C is incorrect because a dedicated VLAN for MAD is not required with LACP MAD.

50. **A** is correct. With PVST+, the switch creates one spanning tree for each VLAN that is configured to support PVST+.

    B is incorrect because it describes the functionality of MSTP. **C** is incorrect because you do not have to manually make all switch-to-switch links members of all the VLANs configured on the switch. One of the advantages of PVST+ is that the switch-to-switch links are automatically made members of all VLANs that run PVST+

    D is incorrect because switches send BPDUs in every VLAN that is enabled for PVST+.

    For MSTP, switches send BPDUs for every instance, which can contain multiple VLANs.

Index

A

C

F

L

M

T

U

V

W